THE LIBRARY OF RACIAL THINKING

RACIOLOGY

THE SCIENCE OF THE HEREDITARY TRAITS OF PEOPLES

BY

VLADIMIR AVDEYEV

SECOND EDITION

MOSCOW 2007

"BELIYE ALVY"

As translated into the English language, from the Russian,
by Patrick Cloutier

Cover art by MDesign

"Raciology: the Science of the Hereditary Traits of Peoples", by Vladimir Avdeyev,
As translated into the English language, from the Russian, by Patrick Cloutier
375 pages, 127 photos and illustrations.
Copyright by Patrick Cloutier, 2015.

ISBN-13 978-1507787830

To the Great Racial Theorists,

From a Grateful Pupil

Vladimir Avdeyev

Vladimir Borisovich Avdeyev graduated from the Moscow Power Energy Institute in 1985 as a specialist in Military Space Technology. He was a technical officer in the Soviet Air Force and was stationed near the Chinese border. His rank was equivalent to Second Lieutenant. After military service, he started work as an engineer and manager for the military industry. Since 1993 he has been a member of the Russian Writers Union. He studied anthropology and raciology under the main experts of Russia. He lives in Moscow.

BBK 87.7
A-18

Vladimir Avdeyev
Classical Racial Theory: The Science of the Hereditary Traits of Peoples. Published as a second edition. The Series *Library of Racial Thinking*. M.: Beliye Al'vy, 2007.

-672 pages, illustrations.
ISBN 978-5-91464-001-6
ISBN 978-5-91464-002-03

The second presentation in the Russian language of the fundamental basis of classical racial theory, with the new author's explanation.

The term "raciology" was first suggested for use by V.B. Avdeyev, for it did not exist in the Russian language. Raciology cannot in any way be equated with customary anthropology. If anthropology examines the quantitative difference in the parameters of the body in members of the human race, then Raciology strives to evaluate those qualities, interpreting the inherited differences in the context of an historical process.

The study of the influence of race on human culture and history was formally prohibited during the time of the communist dictatorship; the classic social approach was feared like the plague, in any form of biological determinism in philosophy. However, the fall of the Soviet Union, with the resulting growth of ethno-racial separatism and banditry throughout all the post-Soviet territories, showed everyone the poverty and utter emptiness of the methods of sociology, culturology, and political science, which were unable to explain the occurrence of the given processes, and give corresponding recommendations for their containment and prevention. The fall of the USA as the result of a bloody civil war is no longer beyond the horizon.

And only such a science as raciology is able to provide the answer to the two endless and cursed Russian questions: "who is guilty" and "what shall be done?"

The book abounds with an enourmous number of anthropological facts, not known in the academic circle of the contemporary epoch, silenced by the considerations of false political correctness. There also happens to be dozens of statements from classical scientists about the nature of the human races, which are considered "unacceptable" in our time. The work of V.B. Avdeyev is written in the vivid Russian language, in an easily understood and clear form; in an extravagant, aphoristic manner, that makes it accessible to the majority of readers, who may not have specialized training in anthropology and psychology.

Table of Contents

Photo, Map, and Illustration Index

Foreword to *Raciology*
By Kevin MacDonald
Professor of Psychology
California State University–Long Beach
Long Beach, CA 90840
kmacd@csulb.edu

Psychologist J. Philippe Rushton once mentioned that science moves forward, continuing to gather data and refine its theories, but with one important exception: a century ago, there was a robust Darwinian science of race differences in a variety of traits, from differences in head shape and cranial capacity, to differences in intelligence and behavioral restraint. However, this young science was nipped in the bud, but not because it was displaced by a new and powerful, empirically-based theory — the demise of racial science came about because of intellectual movements, which were dominated by ethnic Jews and tightly linked to the political left — the topic of my book, *The Culture of Critique: An Evolutionary Analysis of Jewish Involvement in 20th-Century Intellectual and Political Movements.*[1]

This was a case of science being replaced by ideology — an ideology designed to oppose the idea that Europeans were in any way unique or superior to other human groups; an ideology designed to advance the interests of ethnic outsiders, who have their own strong sense of biological uniqueness and superiority. Ultimately, it was an ideology that rationalized the decline of Europeans and their culture — something that we see all around us today.

The new ideology decreed that humans are infinitely malleable creatures of their culture, and eventually became defined by the view that race does not exist at all. Franz Boas, the high priest of the new cult, was a strongly identified Jew and committed leftist. His famous study purporting to show that skull shape changed as a result of immigration from Europe to America was a very effective propaganda weapon in the cause of eradicating racial science. Indeed, it was intended as propaganda. Based on their reanalysis of Boas's data, physical anthropologists Corey Sparks and Richard Jantz do not accuse Boas of scientific fraud, but they do find that his data does not show any significant environmental effects on cranial form as a result of immigration.[2] They also claim that Boas may well have been motivated by a desire to end racialist views in anthropology:

> While Boas never stated explicitly that he had based any conclusions on anything but the data itself, it is obvious that he had a personal agenda in the displacement of the eugenics movement in the United States. In order to do this, any differences observed between European- and U.S.-born individuals will be used to their fullest extent to prove his point.[3]

As a result of the massive success of this onslaught, the science of race differences languished. Whatever truths it had uncovered were forgotten. In *Raciology*, Vladimir Avdeyev resurrects the vast tradition of research on the physical anthropology and psychology of race differences. His book is an exhaustive summary of research in the field, dating back to the 18th century to the present. It includes a great many summaries of the research of individual scientists, many of whom have been virtually forgotten. But it is far more than a compendium of research. It also vigorously defends the idea that "the problem of race is the nerve center of world history." It is intended, therefore, to influence how people think about race in the context of history and current events.

---

[1] Kevin MacDonald, *The Culture of Critique: : An Evolutionary Analysis of Jewish Involvement in 20th-Century Intellectual and Political Movements* (Blooomington, IN: Authorhouse, 2002; originally published by Praeger [Westport, CT, 1998]).

[2] C. S. Sparks & R. L. Jantz, "A reassessment of plasticity in cranial capacity: Boas revisited." *Proceedings of the National Academy of Science 99(23)*, 14637–14639 (November, 2002).

[3] C. S. Sparks & R. L. Jantz, "Changing Times, Changing Faces: Franz Boas Immigrant Study in Modern Perspective." *American Anthropologist 105(2), 333–337 (June, 2003).

Several themes recur throughout *Raciology*. Race is overwhelmingly the result of biological inheritance, not cultural programming. Beginning with Count Joseph Arthur de Gobineau, this body of theory and research proposed that the biologically based racial characteristics of Whites have led them to be originators of superior cultures. The White race evolved in the north of Europe and spread south and east, to become the main force behind the ancient cultures of Greece, Rome, Egypt, India, Persia, and the Hittites.

The ancestral type of the White race, originally called the *Nordic race* by Joseph Egorovich Deniker, is characterized by blond hair, blue eyes, light skin, tall stature, and dolichocephalic (long-headed) skull with a well-developed prefrontal area (the area of the brain associated with intelligence and decision making).   Houston Stewart Chamberlain may be considered paradigmatic of a theorist who proposed that northern Europeans are a superior people:

> All outstanding peoples that appeared starting in the 6th century, in the role of true deciders of the fate of humanity as founders of nations and creators of new thinking and original art, were namely of German origin. The creations of the Arabs stand out for their short duration; the Mongols only destroyed but they created nothing; the ingenious Italians of the Middle Ages were all émigrés, or of the north which was saturated with Lombard, Gothic, or Frankish blood, or they were Germano-Hellenes of the south; in Spain, the creative element was the Visigoths. The awakening of the Germans forms the foundation of European history, for their worldwide historical significance as founders of a completely new civilization and a completely new culture.

Nevertheless, Avdeyev notes that despite his views on the centrality of the Germanic peoples, Chamberlain advocated a union of Celtic, Germanic, and Slavic peoples in defense of the White race. Indeed, a theme of *Raciology* is that "the scientists of Germany well understood that the differences between the Germans and the Russians were extremely insignificant." Indeed, Avdeyev notes that Russians have a higher percentage of light hair and eyes, than the European population in general.

The idea that Whites had superior traits naturally went along with eugenic ideas of racial betterment. In the words of German racial theorist Hans F. K. Gunther, the question is "whether we have enough courage to prepare a world for future generations, [by creating a race] that has purged itself in racial and eugenic terms." Geneticist Fritz Lenz, writing in 1934, viewed creating and maintaining a superior race as the ultimate struggle: "Undoubtedly, one may lead our race to such an ascent and flowering like it has never achieved before. But if we lose heart, our Nordic race will utterly die. ... Before us stands the greatest task of history." That is, active efforts must be made to preserve the best elements and to rid the race of detrimental elements by discouraging reproduction of White people who are prone to criminality, low intelligence, or psychiatric disorders. Avdeyev expresses the fundamental goal of eugenics as follows: "Our main goal is crystal-clear: the creation of a new, super-perfected White Race, the moral and physical degradation of which has reached its limit." Compare American writer Lothrop Stoddard, writing in 1920:

> The eugenic ideal is ... an *ever-perfecting super race*. Not the "superman" of Nietzsche — that brilliant yet baleful vision of a master *caste*, blooming like a gorgeous but parasitic orchid on a rotting trunk of servile degradation, but a super *race*, cleansing itself *throughout* by the elimination of its defects, and raising itself *throughout* by the cultivation of its qualities. [emphasis in original][4]

However, despite the great flowering of culture emanating from Europe, and despite the knowledge that Europeans and their culture dominated the planet, there is also a pessimism that pervades this literature—the idea that White racial elites tend to become eroded over historical time, because of admixture with lesser types. It was common among these thinkers to assert that

---

[4] Lothrop Stoddard, *Revolt against Civilization: The Menace of the Under-man* (New York: Scribner's, 1920), 262.

the depletion of the Nordic racial stratum accounted for the decline of Greece, Rome, the Hindus, the Persians, and other Nordic civilizations. For example, Ludwig Woltman: "The blonde element of the people defines its cultural worthiness, and the fall of great cultures is explained by the dying out of this element." Eugen Fischer: In Greece, "the death of the families of fully-vested citizens and the admission of the descendants of slaves and the aboriginal population as citizens, led ... to collapse. Rome died of race mixing and the products of degeneracy. And finally, Otto Reche, writing in 1936:

> That which we call 'world history' is in essence nothing more than the history of the Indo-Germans and their achievements; the powerfully rousing and simultaneously tragic song about the Nordic race and its idealism; a song which tells about how the strength of the race did what seemed impossible and reached for the stars, and how the strength quickly dried up when the 'law of race' was forgotten, when the Nordic man ceased to preserve the purity of his blood and strongly mixed with races [that are] less gifted in cultural terms.

The psychological traits attributed to Nordics are principled moral behavior and idealism, high intellect, inventiveness, and, in the words of Gustav Friedrich Klemm, a proclivity to "constant progress" and science. "Members of that race most often strive for the unknown, for the sake of a pure idea, driven by the thirst of knowledge, and not self-seeking interest."

My view is that there is a strong empirical basis for this suite of traits, and that ultimately, these traits, particularly moral idealism and science, are the psychological manifestation of individualism as a response to selection pressures in the far north. Avdeyev notes that the "the home of the Nordic race may be located in the zone of a cool and moist climate, abundant with clouds of fog, in which water vapor is retained in the air [absorbing ultra-violet rays.] In this climate there should be strong and frequent fluctuations of temperature."

I first became aware of the idea that natural selection in the north was responsible for the unique traits of Europeans by reading Fritz Lenz, whose work is reviewed in *Raciology*. As do several modern theorists,[5] Lenz gives major weight to the selective pressures of the Ice Age on northern peoples.[6] He proposed that the intellectual abilities of these peoples are due to a great need to master the natural environment, resulting in selection for traits related to mechanical ability, structural design, and inventiveness in problem solving (what psychologists term 'performance IQ'), whereas he argued that Jewish intelligence was the result of intensive social living (what psychologists term 'verbal IQ'). There is in fact good evidence that intelligence in general is linked to mastering the natural environment,[7] and this is particularly the case among Northern peoples.

Lenz argued that over the course of their recent evolution, Europeans were less subjected to between-group natural selection than Jews and other Middle Eastern populations. Because of the harsh environment of the Ice Age, the Nordic peoples evolved in small groups and have a tendency toward social isolation rather than cohesive groups. This perspective does not imply that Northern Europeans lack collectivist mechanisms for group competition, but only that these mechanisms are relatively less elaborated and/or require a higher level of group conflict to trigger their expression.

Under ecologically adverse circumstances like the Ice Age, adaptations are directed more at coping with the adverse physical environment, than at competing with other groups, and in such an environment, there would be less pressure for selection for extended kinship networks and

---

[5] Richard Lynn, "Intelligence: Ethnicity and culture." In *Cultural Diversity and the Schools*, ed. J. Lynch, C. Modgil, & S. Modgil. London and Washington, D.C.: Falmer Press; J. P. Rushton, (1988). Race differences in intelligence: A review and evolutionary analysis. *Personality and Individual Differences* 9:1009–1024.
[6] Fritz Lenz (1931). The inheritance of intellectual gifts. In *Human Heredity*, trans. E. Paul & C. Paul, ed. E. Baur, E. Fischer, & F. Lenz. New York: Macmillan.
[7] Dan Chiappe & Kevin MacDonald (2005). The evolution of domain-general mechanisms in intelligence and learning. *Journal of General Psychology 132(1)*, 5–40.

highly collectivist groups. Evolutionary conceptualizations of ethnocentrism emphasize the utility of ethnocentrism in group competition. Ethnocentrism would thus be of no importance at all in combating the physical environment, and such an environment would not support large competing groups.

Europeans are therefore less ethnocentric than other groups, which makes them susceptible to being subverted by groups with a strong sense of ingroup solidarity. Individualist cultures show relatively little emotional attachment to ingroups. Personal goals are paramount, and socialization emphasizes the importance of self-reliance, independence, individual responsibility, and "finding yourself." Individualists have more positive attitudes toward strangers and outgroup members, and are more likely to behave in a pro-social, altruistic manner toward strangers. People in individualist cultures are less aware of ingroup/outgroup boundaries and thus do not have highly negative attitudes toward outgroup members. They often disagree with ingroup policy, show little emotional commitment or loyalty to ingroups, and do not have a sense of common fate with other ingroup members. Opposition to outgroups occurs in individualist societies, but the opposition is more 'rational' in the sense that there is less of a tendency to suppose that all of the outgroup members are culpable. Individualists form mild attachments to many groups, while collectivists have an intense attachment and identification to a few ingroups.[8] Individualists are therefore relatively ill-prepared for the between-group competition so characteristic of the history of Judaism.

Cultural anthopologists have located European groups as part of what is termed the North Eurasian and Circumpolar culture area.[9] This culture area derives from hunter-gatherers adapted to cold, ecologically adverse climates. In such climates there is pressure for male provisioning of the family and a tendency toward monogamy, because the ecology did not support either polygyny or large groups for an evolutionarily significant period.

The historical evidence shows that Europeans, and especially Northwest Europeans, were relatively quick to abandon extended kinship networks and collectivist social structures, when their interests were protected with the rise of strong centralized governments.[10] There is indeed a general tendency throughout the world for a decline in extended kinship networks with the rise of central authority. But in the case of Northwest Europe, this tendency quickly gave rise to the unique European "simple household" type, long before the industrial revolution. The simple household type is based on a single married couple and their children. It contrasts with the joint family structure typical of the rest of Eurasia in which the household consists of two or more related couples, typically brothers and their wives and other members of the extended family.

These cultures are characterized by bilateral kinship relationships, which recognize both the male and female lines, suggesting a more equal contribution for each sex as would be expected under conditions of monogamy. There is also less emphasis on extended kinship relationships and marriage tends to be exogamous (i.e., outside the kinship group). This tendency toward exogamy, combined with relative lack of ethnocentrism, could account for the tendency for genetic barriers between Nordics and others to break down over time and a general decline in the population, a point noted by several of the writers mentioned by Avdeyev.

In some of my recent writing, I have attempted to account for the Nordic tendencies toward idealism and principled morality, as also a result of selection pressures for individualism. In collectivist cultures, the standard of morality is "what is good for the group", as seen for example, in the common phrase, "Is it good for the Jews?" Judaism is a highly collectivist culture, in which the needs of individuals are subordinated to the needs of the group. In individualist cultures, on

---

[8] Harry C. Triandis. "Cross-cultural studies of individualism and collectivism." *Nebraska Symposium on Motivation 1989: Cross Cultural Perspectives* (Lincoln: University of Nebraska Press, 1990), 61.

[9] Burton, M. L., Moore, C. C., Whiting, J. W. M., & Romney, A. K. (1996). Regions based on social structure. *Current Anthropology, 37* (1996, 87-123).

[10] Kevin MacDonald, "What Makes Western Culture Unique?" *The Occidental Quarterly 2(2)*, 9–38, 2002.

the other hand, there is a tendency toward moral universalism, where morality is defined not as what is good for the individual or the group, but as an abstract moral ideal — e.g., Kant's moral imperative: "Act only according to that maxim whereby you can, at the same time, will that it should become a universal law." Individualism implies an equality of interest—that everyone has interests but no one has a privileged moral position. Arguments on morality therefore must necessarily seek an abstract sense of morality, independent of the interests of an individual or the group.

Moral idealism is a powerful tendency in European culture, apparent, for example, in the German idealist philosophers and the American transcendentalists.[11] Universalist moral ideals are erected and then steps are taken to achieve the moral vision by changing the world, often accompanied by a great deal of moral fervor.[12] This pursuit of moral ideals accounts for some of the dynamism of Western history.

The moral universalism characteristic of individualism is a liability in a struggle with other groups. Individualists are prone to acting against their own people on behalf of a moral principle, as in the American Civil War, where a great many Yankees were motivated to go to war against the South in order to eradicate slavery as a moral evil.[13] Such people place their moral ideals above ties of racial kinship. Here, US Supreme Court Justice John Paul Stevens expresses a typical sense of moral idealism common among Europeans:

> "The ideas of liberty and equality have been an irresistible force in motivating leaders like Patrick Henry, Susan B. Anthony, and Abraham Lincoln, schoolteachers like Nathan Hale and Booker T. Washington, the Philippine Scouts who fought at Bataan, and the soldiers who scaled the bluff at Omaha Beach," he wrote in an unusually lyrical dissent [in a 1989 flag burning case]. "If those ideas are worth fighting for—and our history demonstrates that they are—it cannot be true that the flag that uniquely symbolizes their power is not itself worthy of protection.[14]

Ideas are worth fighting for, but Stevens has no interest in advancing the cause of White people as a racial kinship group. Here he idealizes non-White Filipinos fighting alongside Whites to secure a set of principles. He is not concerned about his race, presumably because he thinks that what's important is that certain ideas will continue to guide the country, even if (as seems likely) people like him are fated to become a small minority of the country. These ideas are more important than the racial composition of the country.

There is an obvious sense in which such moral idealism can be fatally maladaptive. In the contemporary world of political correctness defined by the multicultural left, moral ideals incompatible with the interests of European-derived peoples are constantly trumpeted by elites in the media and in the academic world. Such messages fall on fertile ground among European peoples, even as other races and ethnic groups continue to seek to shape public policy according to their perceptions of self-interest.

The European proneness to moral idealism thus becomes part of the ideology of Western suicide. Similarly, science is an outgrowth of individualism, because it implies that scientists are independent researchers not influenced by allegiance to an ingroup or commitment to

---

[11] Kevin MacDonald, "American Transcendentalism: An indigenous culture of critique." **The Occidental Quarterly 8** (91-106, 2008).

[12] Kevin MacDonald, "Evolution and a Dual Processing Theory of Culture: Applications to Moral Idealism and Political Philosophy." *Politics and Culture* (2010[Issue 1], April). http://www.politicsandculture.org/2010/04/29/evolution-and-a-dual-processing-theory-of-culture-applications-to-moral-idealism-and-political-philosophy/

[13] MacDonald, "American Transcendentalism: An indigenous culture of critique."

[14] Jeffrey Toobin, "After Stevens: What Will the Supreme Court Be Like without Its Liberal Leader?" *The New Yorker* (March 23, 2010). http://www.newyorker.com/reporting/2010/03/22/100322fa_fact_toobin?currentPage=all#ixzz0tJXKtDE6

religious dogma. Scientists, like individualist moral actors, adopt a disinterested intellectual stance in which they independently evaluate evidence and are not influenced by an ingroup affiliation, such as their race or ethnic group. Real science assumes that groups of scientists which form around particular ideas (e.g., the theory of evolution in biology) are maximally permeable and highly subject to defection, when the empirical data do not support previously held views.

On the other hand, in the movements reviewed in *The Culture of Critique,* Jewish intellectual endeavor had strong overtones of ethnic group solidarity, as individual participants could always count on others to hold similar views and to present a united front against any unwelcome data.[15] As in the case of Boasian anthropology, "truth" could be manufactured to meet the goals of the group, and without any connection to the real world. This "truth" could then be disseminated from the most prestigious academic and media organizations, giving it an air of scientific respectability and a huge influence on the public.

Avdeyev makes brief reference to how Jewish identity influences the views of Jewish scientists, when they discuss race. Regarding the view of A. I. Yarkho that racial instinct has been lost among humans, he notes "It is particularly amusing to hear through the mouth of 'God's chosen' people, that incontrovertible racial and species solidarity is considered anti-semitism .... The very principle of Zionism is built on the racial solidarity of the Jews." He also mentions a need in recent times to defend Russian racial anthropology against a view, which is common in the West, that there are no races. In doing so, he makes it clear that his main opponents are Jews: "With authentic Russian patience and quick good sense, a convincing answer was given to the grandchildren and great-grandchildren of those same tailors and tavern-keepers" [i.e., typical Jewish occupations in the Russian Pale of Settlement of the 19th century].

Although the current state of anthropology in the West is far from monolithically Jewish, the strong influence of Franz Boas and his tightly compacted group of Jewish race deniers continues to have a strong influence. Anthropology, as well as other fields in the social sciences and humanities, are best described as "tribal moral communities" — communities based not on science, but on a shared moral vision, which is unified by the view that research on race and race differences must be suppressed at all costs.

Raciology is a most welcome development. It is clear that the anti-racial theorizing of Boas and his followers continues to bear fruit in the current era. Such views are, in their essence, political movements against European peoples masquerading as science, designed to disarm Europeans — to make them defenseless against the onslaught of other peoples and cultures. The reality is that the racial science that thrived in America until the 1920s, and in Germany until the end of WWII, coincided with an era of racial and cultural confidence among Europeans. It occurred at a time when Europe dominated the planet and was spreading its people and culture to all corners of the world.

On the other hand, the assault on this body of research has coincided with an unprecedented retreat of Europeans, not only from outposts like South Africa and Rhodesia (now Zimbabwe), but even in Europe itself, which is now being overrun by non-Whites. Countries like the United States and Australia, which were at least 90% European in 1950, are undergoing demographic transitions which predict that Europeans will be a minority with a generation or two. During this ongoing disaster of European retreat, racial science has remained undeveloped and largely forgotten.

It is to be hoped that a resurgence of racial science, as outlined in *Raciology*, will be part of a general resurgence of the European peoples. It is certainly a step in the right direction.

---

[15] Kevin MacDonald, *The Culture of Critique*, *Ibid.*

Foreword to the Second Edition

For thus sayeth Justice: "People are not Equal."
And they should not be equal.
Friedrich Nietzsche

When I first saw Vladimir Borisovich Avdeyev's book, [titled] *Raciology*, my first reaction to the very title was one of irony: 'now here is the next para-scientific opus about the mysticism of "blood" and "soil",' [I thought]. Instinctively, the unthinking character of this reaction points to the significant judgemental value and cultural burden of the term "race," which evokes negative connotations *a priori*. More simply said, *any* discussion about racial questions is taken from a position of assumed distrust, lack of scientificness, and even of reaction. This negative reaction is expressed particularly strongly among professional, intellectual-humanitarians (to whom this author addresses these lines), than among the public at large.

However, my original skepticism not only quickly evaporated as I carefully read Avdeyev's book, but it was replaced with sharp interest. This interest was all the more stronger, because the materials discovered in Avdeyev's work allowed me to conclusively formulate and partly to verify my own hypothesis about the nature of ethnic sociobiology. In turn, this hypothesis formed a methodological basis for my own book: *Russian History: a New Reading* (Moscow, 2005), which earned a reputation for eccentricity.

But the work of Avdeyev not only furnished me with valuable information, it also provoked disagreement on a number of points, and forced [me] to think over several important questions, to which I did not have an answer. The desire to challenge and discuss is the sign of a good book. A book which neither provokes disagreement, nor gives rise to thought, is not worth anything.

Somehow or other, this desire was so much greater, that I sought an opportunity to become acquainted with the author of *Raciology*, in order to discuss these interesting topics in person. The first meeting set down the start of a series of busy discussions, which brought to light our nearness on a number of scientific questions, and serious, principled differences on others. Although each party remained with his original opinion, it turns out that in this long dispute, the divergent opinions nevertheless changed slightly and became more perfected and nuanced. About the disagreement in concept, Avdeyev spoke further, but now I took note of two very important circumstances, which in my eyes were more important than the intellectual differences.

First. In the person of Vladimir Borisevich Avdeyev I discovered a person with an intellect [that was] consistently rational and devoid of any fakirism—such as "blood" mysticism, or religious mysticism, or the occult. In his opinions and hypotheses he stood on a solid ground of established facts and he stuck firmly to logic. And even a possible mistake in the author's conclusions would not change the consistently scientific character of his work.

Second. Avdeyev is a genuine enthusiast, driven overall in the field of research, by his above-average temperament, intellect, and will. I especially emphasize this circumstance, because similar evaluations—self-returning and selfless—in the name of truth [are] now in greater deficit, namely in corporate science [collectives]. One may differ with regard to the scientific views of Avdeyev, but one cannot fail to be impressed by the energy and persistence of this person, [who] single-handedly, and without assistant scientists, financial assistance, or grants—organized the book series and *Library of Racial Thinking*. For each one of its published tomes (and these amount to no less than a dozen) stands as a colossal work of investigation of the classical anthropological thinkers, edited translations, and quests for the illustrated materials described in the introductions. Again I emphasize: all of this was done by one single individual, who did not receive one penny for his work. "As a *psychological* and *cultural* type, Avdeyev is very close to that admired constellation of naturalists of the 19th and 20th centuries, who whole-heartedly devoted themselves to the service of science. (Considering this spirit of theirs, they very likely served as his spiritual and scientific standards.)

In this area, the skeptical reader will not fail to notice: enthusiasm is a good thing, but how [does one combine] scientific dispassion and professional knowledge? On the whole, one may now frequently observe the enthusiasm and investigative ardor of the dilettantes that are turning out works of intellectual idiocy, of para-scientific obscurantism, and not rarely, of intellectual charlatanism and open fakery.

In terms of scientific professionalism and knowledge, it never occurs to doubt Avdeyev—a guarantee of this is the dozen definitive reviews of his work by biologists, doctors, and geneticists. What may also serve as a measure of the quality of his professionalism, is the fact that in a number of medical institutions of higher learning in Russia, *Raciology* is included on the list of faculty literature.

And if several members of academic anthropology thumb their noses and call Avdeyev a "dilettante," they only demonstrate their own jealousness and hatred toward the work done by him, rather than an impartial, professional evaluation. Not through hearsay being familiar with the customs that reign among professional humanitarians, I can emphasize without equivocation: the hatred toward the professional achievements of others is an inherent trait of this corporation. [That] "Mankind is too human" more often than not determines the conduct of those, who, being summoned to serve the truth, would eliminate this "human" at every opportunity.

Moreover, it is unlikely that one would call a person who has dedicated himself to the study of physical anthropology for more than 25 years, a "dilettante." And the education received by Avdeyev in technical systems engineering has, in the case of anthropology, turned out to be an initial important advantage, as will be shown later. In general, the formal non-involvement of Avdeyev toward the professional anthropological guild is more of a plus than a drawback. The *Homo novus* is free of the inner-corporate limitations, the cultural prejudices, and oppressive pressure of experts, who permeate any branch of scientific knowledge. When a person from the outside is able to look at the customary intellectual landscape with a fresh, unencumbered view, the new vantage point opens a new picture. And I am not saying that the researcher, who is liberated from the necessity of being concerned about dissertation degrees and a career in science, is more free in his intellectual undertakings.

Finally, science, particularly socio-humanitarian science, absolutely should not come unwound within its inter-disciplinary frameworks. The same interesting and penetrating ideas are often born when and where the boundaries—disciplinary and corporate—conflict.

Raciology has two important aspects in equal degree—strictly speaking, scientific and cultural. I will start with science. I think that Avdeyev *already* wrote his name into the history of Russian anthropology, with the creation of a genuine compendium of racial theory and racial differences. This "dilettante" did what should have been done long ago, but as such was never done by official Russian science.

It needs to be made clear here, that contemporary physical anthropology in Russia resembles a great communal dwelling, the residents of which only turn to each other when forced to, preferring instead to forget about the existence of one another. Specialists in various areas of racial anthropology occupy themselves almost exclusively with their thematics, avoiding any venture into related scientific territory; they do not even attempt to connect any results obtained. There, we will say, is A.A. Zubov, the famous ontologist, who has worked on teeth his entire life, and knows about racial differences in this area, and probably in all of them. Or there is G.L. Khit'—a gifted expert on human dermatology. One can easily add similar personalities to this list. But not one of these excellent specialists works with Man as a whole. And after all, Man is not simply teeth, skull, body, and so on, or even the sum of their total. In the context of biology, the person, as with any living thing, represents a biological **system**. Therefore, studying the separate parts or the organs, no matter how deeply, would no more lead away from understanding the human being, than it would lead to understanding of him. It works out exactly according to Koz'me Prutkov: "The specialist is like a gum boil—his completeness is one-sided."

And so is the lop-sidedness of domestic anthropology, which Avdeyev attempts to overcome. Of course, he handles teeth worse than Zubov, and dermatology worse than Khit'. But in the matter of the systematization, classification, and synthesizing of a colossal [amount] of information, taken from the works of narrowly specialized fields, like anthropology, genetics, medicine, etc., his contribution is enormous and unsurpassed. Avdeyev is the first in Russia to portray Man as a biological system (this is where his system-technical engineering background becomes useful) looking through the specific prism of racial variances. The systematic look at the human being—the first (but not the only!) and principle merit of Avdeyev's work.

The second merit, possibly, which is not always noticed by the inexperienced reader, but well understood by specialists, may be the wide historiographic essay on racial theory. The first with such completeness and—love (I do not fear use of this word)—it reconstructs the formation and development of racial knowledge. In this regard, Avdeyev has made a work akin to archaeology: he collected and edited out biased and openly untruthful arguments and put together a complete picture from the separate fragments of the history of the science. This observation regards not only the historiographic branch of "Raciology," but the less important publications by Avdeyev, the two-tome *Russian Racial Theory until 1917*,[16] and the works of German racial thinkers.

And why is primary attention given to the Germans? The answer is simple. German science was in the vanguard of research on questions [of race] and laid the fundamental basis of knowledge in this area. Russian science for the extent of the 19th Century, and the beginning of the 20th Century, experienced the overriding influence of the German "think tanks."

It is very important to point out, that although the German racial research of the 20s and 30s of the last century was not free from the influence of political, ideological, and cultural contexts (and where in general can one find a science that is not dependent on external influences?), it did not the least bit serve the goals of the para-scientific justifications of Nazism. Many German scientists in this area of knowledge regarded Nazism negatively, and came under the suspicion of the regime. Thus, that seemingly natural conclusion from the historical-cultural context, of associating and even equating German racial scientific information with the Nazi regime, is most likely mistaken, or at a minimum, is in need of a different approach.

Avdeyev's *Raciology*[17] wholly and completely places itself on a course formed by the classics of racial thought. In the content of his work, he is in no way a revolutionary or overthrower of fundamentals, but a conservator, who cleanses original thoughts, reestablishing preeminence and pointing to the currency of the scientific classic, which remains a living source of thought.

In this area, I do not hold back, in order to share my personal impressions. Time brought me to become a scientific authority, during which period there have been several personages on the periphery of my interests, of whom Avdeyev has written, and the works of whom he has published. Thus, comparing their genuine texts and biographies with that which is known about them, [and] which was gleaned by me from the official Soviet compendium, I may state without equivocation: the Soviet share of knowledge was blatant lies and falsifications.

However complete the historical achievement and however impressive the systematization of Avdeyev's work, it incited a sharp reaction. His third important achievement lies in the fact that the author made far-reaching conclusions from the scientific information which was at his disposal. Official science closely approached those conclusions, but chose not to cross the cultural Rubicon. I have in mind the consequent development of the polygenic [theory of human evolution], and criticism of the monogenic theory. Stated in other words, Avdeyev proves that humanity is not represented by a single biological type; that the various races arose from different primates.

Not being a specialist in this area, I will not side unequivocally with Avdeyev, but neither will I unequivocally refute him. But I will not fail to remark that indirectly, his point of view is born out by those conceptual diagrams, which are now in use by official anthropology. The "mesh" theory of human evolution, taking the point of divergence of the human races far into the deep past (one million years ago, and earlier, still) creates the impression that anthropological science resists with all its strength, the necessity of making a long-overdue, definitive conclusion from the Mont Blanc of information accumulated by it: humanity is actually not a single type, its oneness is no more than intellectual fiction, a carefully cultivated and guarded cultural myth. This is exactly that classical situation, when a cultural axiom encircles science with a lie, forcing it to perform intellectual acrobatics and employ doubtful ruses.

And how! After all, the acceptance of the polygenic [theory of human evolution] and its introduction into the system of knowledge and culture automatically necessitates a second look at the past, present, and future. This would not be as much a scientific, as a cultural revolution, comparable in its scale and consequences with the appearance of humanity in the Modern Epoch.

Being independent of the connected class of organized, professional corporations, Avdeyev is candid in his shocking conclusions. On the other side of this liberty is his openness to criticism, the critics specifically turning their attentions toward him, and motivated by the ideological, rather

---

[16] *Russkaya rasovaya teoriya do 1917 goda.* Avdeyev.
[17] *Rasologiya.*

than by the scientific side of the issue. If different races represent within themselves different biological types, then that naturally suggests qualitative differences among them. Of itself, this throughsting conclusion serves as the start point of Avdeyev's claims on the creation of a new discipline—raciology. But this cannot summon accusations of racism in his address.

Although the qualitative differences of the races are discussed logically, overall no formal legal recognition of their inequality, nor any idea of the eternally secured superiority of one race over another, follows from it. Moreover, the increasing frequency of complaints right-wing Western intellectuals and politicians on the crisis of the white race, and about the dramatic threats to it, emanating from the side of the other racial types—and this opinion, as much as I understand, Avdeyev shares—points out that the white race in the modern world context no longer possesses that qualitative superiority, which it had in the Age of Discovery, in the Industrial Revolution, and during the conquest of the non-Western world. In other words, that quality which in other circumstances was the strength of the white world, in other circumstances becomes its weakness. The same can be said about the non-white world. But where is the racism here?

From my point of view, Avdeyev's worldview is correctly classified not as racist, but as a "racial way of thinking"—that is, the consideration of the world and its reality, through the prism of ontological racial differences. A similar worldview was widely held among the European and Russian educated public in the 19th and early 20th centuries. Avdeyev restores and carefully cultivates this worldview.

For me personally, reading Avdeyev's work once again confirmed the correctness of the maxim about failings being a continuation of merit. Biological determinism is natural, is understood, and is justified in the context of biology, but the interpretation appears excessive, oversimplified, and distorted outside of its scope. The author of these lines has been criticized quite a bit for the [discussion of] biological determinism in the book, *Russian History: a New Reading*, but compared to Avdeyev, I look like a devotee of the refined social theories.

For both of us, (for my own self, I can vouch—for Avdeyev, I can suggest) the initial methodological position served the thesis about the socio-biological (or bio-sociological) nature of the human being. Despite the trivial character of this assertion, in my view, it incorrectly treats the majority of the sciences, which essentially confess a single nature of Man, but in their understanding completely ignore the biological side or set the social [nature] against the biological, arriving at a dual essence by the very thing. A correct understanding of the socio-biological nature of Man should proceed from recognition of his *unified*, but *bipolar* essences. More simply stated, not only do biological impulses project to the sphere of sociality and culture, and form them, but precisely also, culture and sociality, drifting in the course of human evolution from its initial biological foundation, turned into independent historical factors that influence biology. Humanity built a grandiose structure of culture and sociality over biology; any biological impulses going through a multitude of cultured agencies, and not too often, manifesting themselves directly and immediately.

For Avdeyev, the connection of biology and culture is linear (from biology to culture), and shows rigid deterministic dependency. The projection of this view onto history creates a satisfactorily strange picture. Even if it is suggested that the origins of the different religions and ideological systems go back to—as Avdeyev postulates—different racial types, it is impossible to explain the spread and historical and protracted residence of these systems among the different racial types, as emanating only from biology. The assumption that their longtime influence in foreign and racial surroundings is the result of large-scale undermining operations, or the actions of racial "agents of influence" would lead us too far away from the boundaries of any science or sober intellectualism, in general.

Avdeyev combines biological determinism with methodological monism. In the intellectual perspective proposed by him, racial differences turn out to be not simply very important, but factually the *sole* explanatory principle. And in fact, in *Raciology,* the author declares himself the decisive enemy of religious and ideological monism—incomprehensibly so—because he does not extrapolate his pluralism to the area of history. Yes, and the definition by Avdeyev of raciology as a science, studying the "biological factors of world history...the biological premises of any activity in general" means that in history and in human activity there exist factors and premises, that are other than biological.

The dissatisfaction of biological monism is clearly manifested in his attempt to apply it as an explanatory meta principle, albeit to the last two centuries of history. The large-scale conflicts (for example, the First and Second World Wars) proceeded from the bosom of one racial type—the Europoid. And what? Did "racial agents of influence" make a timely arrival there? I think, for an infinitely more fruitful understanding of history, ontological and paramount ethnic differences are consistently covered by me in the aforementioned book, *Russian History, a New Reading*. It is not race, but peoples that comprise the main subjects of world history, where racial differences manifest themselves in a removed form.

And although the significance of racial differences may sharply grow in the 21st Century, which according to Avdeyev's opinion and our general thinking, will open a new era of large-scale racial collisions, ethnicity will not only be preserved, it will strengthen and be of paramount significance. The centuries-old history of Russia with the West proves that the racial singularity of the European and Russian peoples (about which Avdeyev persuasively writes, showing that in the narrow anthropological sense, Russians are more European than many European peoples), stood for nothing in the face of cultural, historical, political, and geo-political factors. And there are not serious grounds to think that the situation is capable of radical change. Even in the face of an advance by other races, Europeans do not see a racial counterpart in Russians. Several times in this past year, I heard how people close to the Kremlin called Russia "the last reservoir, the last hope of the white race." Yet, for Europeans we remain "wild Asians" from a country substantially foreign and hostile to the West.

Probably pushing the limit, the biological determinism and methodological monism of Avdeyev's book is not an intellectual error, but the result of natural enthusiasm for the defined theme, and a purposeful writer strategy (I do not know to which degree of consciousness). His book is not only a scientific work, but it is also an important cultural manifest. Its shocking impact is intended to be forceful, in order to penetrate the breaches of the cultural bastions.

The thought that "culture" conveys, and which Avdeyev conveys, is the following: humanity has arrived at a new epoch; the world created by the Enlightenment and the Modern Age, to the sounds of pleasing notions like "democracy," "equality," "progress," and "human rights" has irrevocably gone into the past. And with it goes the scientific concepts and the intellectual baggage that belongs to it. In its place comes a world founded on blood and soil, strength and hierarchy, which demands a new explanation, and new concepts.

Only in the context of this dramatic, truly world-wide historical shift is the resonance and readership success of Avdeyev's *Raciology* and publishing activity understood. *Knizhniy Biznes* declared the first publication of *Raciology* an "absolute bestseller in the category of philosophy, political science, and esoterics." And after all, its pages do not brandish the card board sword of "the USSR that we lost;" they do not expose another set of "Third Reich occult secrets;" and they do not propose the futuristic utopia of the death of the United States under the weight of its own crimes. The question is one of the topics of natural science, which is not easily understood, and therefore, boring to the general public. Ease of style is not lacking in the literature of this gifted author, and the numerous illustrations in part compensate for the difficult scientific terminology, the caustic argumentation, and the refined presentation of logic.

And if *Raciology* has already been read by 10,000 people, and still more are reading, it means that our society is not only mentally and culturally prepared for the refined perceptions in its ideas, it is in need of just such an explanation of today's realities—realities which literally cry out about the fundamental significance of racial and ethnic differences. In this way, Avdeyev's book is not only a manifest, it is simultaneously a symptom, a sign of a developing socio-cultural and intellectual revolution in Russia. Its deep thinking lies behind a return (incidentally, the word "revolution" means "return") to familiarity with the timeless truth: "For the life of every creature is its blood" (Leviticus, 17:14).

Valeriy Dmitriyevich Solovey
Doctor of Historical Sciences, Expert of the Gorbachev Fund

RACIOLOGY AND ANTHROPOLOGY:
IN ADDRESS OF THE DIFFERENCES

Every living thing knows without being told, to which species it belongs, for its inborn instinct signals this every instant. It is namely that simple and incontestable fact of natural science that lies at the foundation of a science such as raciology. The desire to study the breeds of living beings, with the senses and reasoning abilities of that breed, to which the researcher himself belongs, is the chief motivation for the creation of the given science. One might say that its premises are rooted in the instinctive essence of every organism. Evaluating by analogy with its flesh, spirit, desires, and passions, everyone and everything, the living being creates a hierarchy of values, taking itself as the standard. The inner biological origin is a vector, passed on by generations of ancestors, who perpetuated their kind through an incessant struggle for existence, creating a perpetual form of **race**. Race, or species—this is that bridgehead from which every living being begins its invasion of the world, filling it, at his appearance into the light, with the cry of the newborn, like a battle cry in the attack. "I exist, and You, and all others, are now obligated to reckon with this fact"—the infant announces to the world, barely gaining freedom of will and fate.

The idea of its place in nature, is, for every organism, its life-priority mission. And it is namely for this reason—to determine this place in the natural hierarchy of existence, that the cognitive methods of raciology are implemented. In consideration of races, it strives first of all, to define the quality of the human material, for it defines the biological value and viability of each individual. If, in the sphere of competence of physical anthropology, the study of this or any quantity of characteristics of the human body in its racial variances is included, then in raciology their qualitative evaluations are in the forefront. Differences in cephalic indicators among the members of different races, or concentrations of them or other hormones and proteins does not of itself interest raciology, but only in the context of the social, cultural, and political interpretations of this data. In this way, the establishment of regular connections between the quality of human material and its historic manifestations is the main task of raciology.

It is precisely this pivotal nature-philosophical accentuation that advances it in a number of very honored and privileged natural science disciplines, as raciology teaches every individual that the effective struggle for the possession of natural resources proceeds from the knowledge of his own inherited instincts, and the inherited instincts of biological competitors. It is substantiated by the strategy of the struggle for existence, through maximum use of the racial traits of one's group, in confrontation with the racial traits of biological competitors, for "a place in the sun." There is "us" and "them;" so it was, and so it will always be.

Analogously with the functions of an organism, in the structure of the natural sciences raciology fulfills the function of an immune system,. The sole and direct purpose is maximum successful resistance to foreign attack and infection. The objective use of the scientific method leads to the recognition and neutralization of all foreign influences, which are penetrating into the geographic range of one's race, at a level they would not have manifested themselves.

A person who does not possess racial instinct is similar to a dog, which does not know how to bark or bite; a cat which doesn't catch mice; or a tomato not having the taste or scent of a tomato. The human being who does not possess an expressed racial instinct, at a minimum finds himself at the first stage of degeneration, since racial instinct is at the heart of evolution. In full measure, human society submits to the laws of thermodynamics, and any form of race-mixing unavoidably leads to entropy. Natural inequality—this is the cause for any cultural, social, and political movement in general. The hierarchically built biological forms of existence, gathering to themselves any living impulse, inevitably sort them by qualitative significance. Death inescapably makes equal, what life and time separates into the higher and lower. There is no other way.

The word "race" [in Russian, *rasa*] comes from Sanskrit—an ancient language of the Aryans, at the time of the beginning of the formation of [their] community, when there were still no peoples in the contemporary social understanding, and when there was still no multitude of national languages, that now represent that community. For a biological unity of blood and a [unified] geographical range of habitation, was the basis of connection between peoples then. Social, political, and religious differences arose much later, and destroyed this community of Aryans.

Thus, the creation of the word "race" relates to the epoch of proto-history. In the authoritative, modern-day *Oxford Sanskrit-English Dictionary*, edited by Sir Monier-Williams, the word "rasa" is interpreted as "the better part of something; the essence, the nectar from fruits, syrup, potion, an elixir"—in a word, a concentrated expression of taste. Several dozen derivative words are derived from the root, still bearing in themselves the main principle thought: *Rasa*—this is a basic evaluative category, which [applies] not only in the physical medium, but also in the area of transcendental ideas and understandings. Rasa, race—this is the atomic, unchanging unity of taste and the wider, evaluative worldview in general. Figuratively speaking, Rasa is the comparative standard in all its numerous manifestations: from sight, from scent, taste, and touch, right through to the religious condition and the pure intellect free of emotions.

Rasa is a universal criterion, on the basis of which more difficult and concrete evaluative categories are formed. Thus, the term *rasa* denotes one of many concentrated understandings of an ancient language of the world, and goes back to the very beginnings of the language as such. When an Indo-European set about to compare different things, a certain basic evaluative category, a desired standard of perfection, was needed. Evidently, the first word *rasa* was so formed in the depths of its archtype.

Ancient Egyptian Racial Classification. 14th Century, B.C.

It is completely obvious, that in discussing the quality of the surrounding world, Man began to use the word [rasa] with reference, first of all, to his own counterparts, and from there [he] evaluated the people from neighboring and distant tribes. This is precisely how the term [rasa] was applied.

V.P. Alekseyev (1928-1992), a luminary of Soviet anthropology, laid down the following train of thought in the book, *Geography of the Human Races*[18]: "Any science begins from subconscious empirical observation. Paleolithic Man already included racial traits in the complex of knowledge about the human face. The vital critical period in opinions on racial differences between peoples, particularly and namely racial, and not incidental traits, evidently occurs in the epoch of the creation of the ancient states. The very appearance of the latter, the integration of great masses of peoples within the states, and their contrasting barbarian periphery, immeasurably widened the notion of peoples about their neighbors, see that all ancient states without exception (Egypt, Sumer, Akkad, Assyria, Babylonia, and the ancient states of the Indus Valley) arose in areas of mixed racial composition, and attracted members of different racial types into a sphere of mutual contact. Egyptian frescoes serve as a graphic example of this: depicted in them are peoples that brought tribute to the pharaoh. Despite all of their conventionality, the distinctive physical types of each people—the form and color of the hair, the color of the skin, and the structure of the face—are quite distinctly recorded."

Professor Boris Fedorovich Porshnev (1905-1972), a noted Soviet paleo-psychologist, discussed the same key [issue]. In his fundamental monograph, *On the Beginning of Human History*[19], [he writes]: "Much in the ancient history of Man receives additional light, if it is remembered that people developed by contrasting themselves to the anti-people living

---

[18] *Geografiya chelovecheskikh ras.* Moscow, 1974.

[19] *O nachale chelovecheskoy istorii.* Moscow, 2006.

somewhere near or far from the periphery—the "non-humans", the "non-living". This contrasting became more realized. It was the other side of the self-awareness of racial groups." Thus it is namely **biological determinism** that lies at the basis of the formation of the worldviews of the first human communities, when people measured other peoples within the framework of the dichotomy, "Us and Them," on the basis of real, substantial racial traits.

Worldview—this is the quintessence of life's experiences, and also a strategy for the conquest of the surrounding world, defined by a group of people having a commonality of origins and common, vital interests. The worldview tries to evaluate, and race can and should be evaluated. For in the very structure of myth that arises together with this or that race, the core of the worldview forms a system of values; and the reflective sum total of biological requirements for the development of the given race are encoded within it. Race and worldview are always interconnected.

Firmly ingrained in classical Indian culture, the word "rasa" first appears in Europe, only with the start of the Age of Exploration at the end of the 14th Century, in Italy and Spain, and then at the end of the 16th Century in France. In the 17th Century it appears in England, and in Germany in the 18th Century. But its significance from the very beginning was still not firm, and was highly vague.

Ancient Egyptian Racial Classification. 14th Century B.C.

In the modern, particularly biological sense, the word "Razza" is first encountered in the Italian language in 1552, when Federico Grizon uses it as a term in horse-breeding; and in 1600 Olivier de Serra uses the term "La race" in the same sense. In 1672, the German, George Simon Winter, uses it as a French loan word in the German language, for the needs of cattle-raising. The original fashionable foreign term only applied for the purpose of classifying the breeds of noble animals; however, the important French ethnographer and traveler, Francois Bernier, first adapted its use in 1684 for defining the differences among human tribes. The ingenious Swedish naturalist, Carolus Linnaeus (1707-1778), created the first classification of human races in 1746. For his part, the prominent German scientist, Johann Blumenbach (1752-1840), first used cranial measurement research of the skulls of different ethnic groups for the purpose of classifying races, in 1776; and the French naturalist George Cuvier (1789-1832) suggested using the color of skin as a basis for defining differences. Besides that in 1775 Immanuel Kant (1724-1804) and in 1784 Johann Gottfried Herder (1744-1803) substantiated the term "rasa" as a philosophical category. Later, many anthropologists created their own variants of racial classifications: Augustin Thierry (1817), Ettiene Geoffroy de Saint-Hilaire (1818), Boris de Saint Vincent (1827), Amedey Thierry (1828), James Pritchard (1836), Anders Retzius (1842), Robert Knox (1850), Charles Darwin (1859), Paul Pierre Broca (1860), Isidor Geoffroy de Saint-Hilaire (1870), Thomas Henry Huxley, and Paul Topinard (1878).

The last, in his notable essay, *Anthropology* (1878), wrote: "In anthropology, the term "race" has a completely real significance, and serves for the designation of the natural sub-divisions of the human species, independent of the time when they were formed."

However, for the extent of this long period, there did not exist a distinctly and unambiguously defining term, for scientists mixed the specially biological parameters with linguistic and ethnographic factors, because of which a jumble arose, and peoples having similar outward appearances and psychological characteristics, were recorded as different races on the basis of a given etymology, or the conclusions of comparative linguistics. Not rarely, peoples not having anything in common between themselves in the dimension of physical structure, became placed with a different race, solely on the basis of linguistic commonality. These contradictions and inaccuracies in systematization dearly cost the followers of racial theory, for they complicated the entire science as a whole. As a result, the identification of "peoples" and "races" grew into an absolutely absurd combination of words, such as the "Teutonic Race," the "German Race," and the "Slavic Race;" and the very number of races in different categories oscillated from a few to several hundred.

The situation was corrected by a Russian scientist of French origin: Joseph Egorovich Deniker (1852-1918). In 1900 he published the book, *The Human Races*, in the French and Russian languages. It is namely in this monograph, which to this time is considered a standard for the systematization of natural scientific information, that the principles for evaluating the differences between human races were first formulated. **Racial typology** arose in anthropology, thanks to which the classification of human races acquired the modern distinct form. Different readings disappeared, and the use of specific, anthropological terminology acquired a strict, scientific character. In his book, J.E. Deniker was the first in the world in practice to lay out a new synthetic principle of classification: "What touches on the classification of races is the fact that only physical traits are taken into account for it. By means of anthropological analysis of each of the ethnic groups, we attempt to determine the races that go into its composition. After that comparing the races with each other, we unite the races which possess the greatest number of related characteristics, and divide by race, those that demonstrate the greatest number of differences."

By the term "race" Deniker distinctly meant "somatological unity," and in this way, an end was put to idealism in anthropology. In essence, the entire book is committed to the division of ethnography and anthropology, which the author defines as phenomenon of a different order: the first is *sociological*, and the second is *biological*. He writes: "..I suggested the classification of races, based solely on physical traits (color of the skin, the quality of the hair, height, shape of the head, the nose, and so on)."

In essence, J.E. Deniker was the first to stand by the position of strict and consistent **biological determinism** in racial philosophy. In his opinion, the surrounding environment is powerless in the face of racial characteristics. He emphasized: "Racial traits are preserved with remarkable persistence, in spite of the mixing of races and changes in the conditions of civilization, the loss of a previous language, and so on. The only thing that changes is the ratio, in which this or that race goes into the composition of a given ethnic group."

Defining race as a "somatological standard," Deniker laid the basis for racial topology, which, without substantial changes, endures to this day. He was fortunate, because his contribution was immediately recognized by almost everyone. From the moment of recognition, Deniker's authority in anthropological literature, the notion of the **anthropological type**—constant and unchanging, once and forever given, and not subject to environmental influences—was firmly planted in anthropological literature. Historically, the fully developed combination of anthropological types was itself a product of social development—an ethnos, and the type that dominates subsequently forms the physical and spiritual character of every national community. This rule was adopted and became a foundation stone in raciology. Thus, concepts such as tribe, ethnos, people, and nation relate to the sphere of competence of the sociological sciences, while concepts of race relate to the area of the biological sciences.

The famous Polish biologist, Ludwik Krzywicki (1859-1941), distinctly formulated differences in psyche-types in research, belonging to these different areas of knowledge: "In ethnography conservative and religious mindsets have found a refuge of intellect; in anthropology, intellect is freethinking. In the first, they defended in their time, the unity of the human species; in the second they suggested that the human species arose in several centers. Some founded their opinion on historic fact, others considered the latest minor discoveries in comparison with anthropological indications."

V.P. Alekseyev, in his monography, *Man, Evolution, and Taxonomy*,[20] considered it necessary to emphasize: "The consistent substantive struggle with social Darwinism and racism, the awareness of qualitative, specific features and natures of Man as a social being, has aroused a rejection of biological laws among Soviet philosophers and scientists, even if it is only a limited role. Despite the fact that in anthropology such an extreme, nihilistic view does not find support, and in philosophical and socio-historical works it is not once expressed, it still finds its way into the popular conscious and popular literature.

It is absurd to impoverish history, to completely close the eyes to the biology of Man and its role in social development, to consider Man as only some sort of incorporeal substance of a society. Peoples differ not only in social-cultural particulars, but also in their biological traits. As to dreams of peaceful coexistence, if anthropologists ever had them, it is not to be. It turns out that the most peaceful epoch in the history of anthropology, not promising any explosions, was fraught with revolution in our views on the **races**, and that revolution turned out to be connected with the most deep ideological problems, not only in anthropology, but in general in biology, anthropo-genetics, the origins of Man, and the understanding of his place in the universe."

We see that the very **problem of race** unavoidably pushes itself to the surface of the social consciousness, revolutionizing and rearranging the entire structure of the worldview. But it is always located at its focal point. **The problem of race is the nerve center of world history**.

Still one more recognized authority of Russian science, A.A. Zubov, also emphasizes in his last monograph, *The Paleo-Anthropological Genealogy of Man*,[21] that "The races of Man appear, in essence, by their biological categories; they cannot be recognized in the dynamic of their genesis and development, without bringing in historical scientific data. And on the other hand, the materials of anthropological investigations are invaluable sources of information for the historical sciences that [in turn] stimulated the development of physical anthropology." Thus, for the scientist, the uninhibited "humanitarian dogmas" of the Age of Enlightenment, and the abstract word "culture" do not have such a magical, bewitching significance, for he understands that any culture is the result of the biological life activities of an historically established community, the quality of which is entirely defined by the quality of the human material that created the culture. The scientist, standing at the positions of the philosophy of biological determinism, is free from the idealistic notions of human nature in general. Therefore, the distinguished German racial philosopher, Ernst Krieck (1882-1947), correctly wrote in his book, *Science, Worldview, and the Reform of Higher Education* (1934), that: "It is necessary to dispute, first of all, the claim of so-called culture to the role of higher values."

In the first half of the 20th Century, the term "race" acquired stable scientific meaning. The great German racial theoretician, Hans F. K. Guenther (1891-1968), gave such a definition: "Race is a single group of people, standing out from other groups by a distinctness present in the combination of physical traits, and psychological attributes, and always reproducing with like kind." Another luminary of raciology, the distinguished German scientist Otto Reche (1879-1966), building on biochemical research findings and comparative morphology, gave such a formula:

---

[20] *Chelovek: evolyutsia i taksonomiya.* Moscow, 1985.

[21] *Paleoantropologicheskaya rodoslovnaya cheloveka.* Moscow, 2004.

"Race is the understanding of natural science systematics. Race is a group of living beings, which developed in isolation, thanks to natural selection, from one root, and without mixing of foreign elements—this thanks to a majority of inherited physical and spiritual traits, forming in their combination, a certain unity—and also thanks to the form of its outward appearance, it significantly stands out from the other groups of this family, and always reproduces only with its own kind. Race means the same thing as "harmony," "life style," and "character." Race is a subgroup of a species."

Eugen Fischer (1874-1967), Director of the Kaiser Wilhelm Institute of Anthropology, and Chairman of the German Anthropological Society, gave his definition of the term [race]: "In anthropology, race is understood as a significant portion of people, connected to each other by common bodily and psychological characteristics, passed down by inheritance, that stand apart from the characteristics of other groups."

Alongside the psychological traits of races, in his definition of Man, Arthur Guett included questions linked to behavior and values: "In [the word] 'race,' I understand a group of people, for which affiliation is defined by their physical and psychological distinctions, and their character; in every life situation they conduct themselves in a similar way, and they perceive it identically."

There are also more colorful and poetic definitions of race, although they are no less scientific. Leonardo Conti (1900-1945), thought: "Race is a wide, broad familial kinship." Achem Gerke pointed out: "Race is a stream of blood, carrying the genetic pool."

Summarizing all of the above from an evolutionary point of view, it seems possible to emphasize that a **race is a biological subject of an historical process.**

As is evident, [the definitions] are all highly similar, and we today have full right to use any formula. As to the notorious term "racism", which today receives widespread use in daily life, thanks to the power of liberal social scientists and democratic culturologists, it bears absolutely no relationship to [natural] science. That word first appeared in 1932 in Larousse's French Dictionary, and only represented journalistic slang, created by politically motivated people, and does not have any clear or understandable scientific meaning. Therefore, its use among educated people is not recommended.

For its part, the science of anthropology, which afterward took the term "race" into service, also has a rich and ancient history. The word "anthropology" in the modern contemporary sense is first encountered in literature, as the title of a book published in 1501, by Magnus Hundt (1449-1519), who was a doctor and a theologian. But the study of Man may be traced from times of antiquity. Anaximander, Hecate, Herodotus, Hippocrates, Aristotle, and many other husbands of science devoted attention to the nature of Man in their essays. The Greeks, the Romans, the Byzantines, the Arabs, Hindus, and Chinese left us a gigantic layer of material with interesting observations. The Greek philosophers more than any others came close to the exact modern use of this term, inasmuch as Aristotle already called the sages who studied man, "anthropologists." In modern Europe that term is often found in the following thinking: thus for example, Chamber's Cyclopedia of 1740 treats the term "anthropology" as "The study of the body and spirit, and the laws governing their combination." In 1772 Denis Diderot (1713-1784) and Jeanne D Alambert (1717-1783), used it for a designation in the *Tractate on Man*. In 1788 Immanuel Kant published a book under the title, *Anthropological Experiments*.

However, only starting with Johann Blumenbach, did the term "anthropology" receive that meaning which is attached to it today. Several medical doctors have used it for a designation in entire encyclopedias, which simultaneously cover anatomy, psychology, and hygiene. Then again, in our time, the sense of this word is completely distorted, as for example, in the title of the science, "**cultural anthropology**." Laying claim to an all-encompassing approach in the study of Man, it at the same time engages in the mythologizing of a certain abstract homunculus, not having [any] clear racial traits.

The stormy blossoming of physical anthropology in the 19th Century formed the boundaries of the use of the term. The recognized classics o fhis science gave it this definition:

"Anthropology is the monographic, natural history of Man, in the sense of any zoological monograph,"—Armand de Quatrefages de Breau.

"Anthropology is the clear and concrete science aimed at the complete cognition of the human species, considering first of all, from the viewpoint of its division into four representative groups (variety, race, appearance, and type, if such exist), comparison between them and the ties with their corresponding traits; and secondly, in its entirety and in regard to the rest of the animals"—Alphonse Bertillon.

"Anthropology is a science having the study of the human species as a whole as a subject, and in particular, its relationship to the rest of Nature"—Paul Broca.

"Physical anthropology is the definition and explanation of the presence of various types of peoples in various countries,"—Franz Boas.

"A science that specially engages the somatic particulars of the human family as a whole, compared to other animals, [and Man] in his diversity, is called anthropology."—J.E. Deniker.

"Anthropology is a branch of natural history that studies Man and the human races. The anthropologist studies the human breed in its unity, and in its relationships to other zoological groups, and then crosses over to the subdivisions, the customarily so-called races. Its means of observation are the same as those of the zoologist, but wider in scope. Several traits, not important on an animal, receive first degree significance on Man; brain function, for example. Thus, the anthropologist, in view of these tasks, studies the following: 1) physical traits; 2) physiological appearances which relate also to psychological life; 3) the particulars of social structure, and finally 4) historical phenomena. There cannot be any kind of doubt that the method of research for [studying] Man should be identical, as that for the other animals: judgment, *a priori*, and any manifestation of feeling should be definitively excluded from it. No matter how brilliant the role of Man was on our planet, and no matter what his situation was on the summit of organized life, he will constitute a separate area, whether it is a Human Kingdom, or whether he will only be the highest of primates, in whatever case he may be [classified], we should study [him] with the help of such methods."—Paul Topinard.

The outstanding Russian anthropologist and founder of the [Russian] Fatherland School Academy, Anatoliy Petrovich Bogdanov (1834-1896), in one of his main essays, *Anthropological Physiognomy*,[22] while still at the dawn of the formation of the science as a whole, nevertheless delineated everything particular to the mentality which should separate the anthropologist from the members of other natural science disciplines: "For the modern anthropologist-naturalist, the study of Man in general is not an easy task: it is a matter of anatomy, physiology, psychology, and philosophy. Important to it are those variations, which in their form and structure represent a tribe; important to the extent that they offer opportunity to distinguish and group these tribes; to find differences and similarities in them for their natural classification, for reconstruction of that family tree, by which they developed [differently] from one another, under the influence of different causes."

Thus, the Russian scientist, in defining the priority tasks of anthropology, reconciled the methodology with the hierarchy of the objectives of research, rather than unavoidably insert a subjective evaluation of the racial problem into the foundation of the whole. From the first, and in full measure, only talented dilettantes with humanitarian educations, saw the necessity of [this reconciliation]. Such were the Frenchman, Joseph Arthur de Gobineau (1816-1882), the German, Gustav Friedrich Klemm, (1802-1867), and the Russian scientist, Stepan Vasiliyevich Yeshevskiy (1829-1865)—the founders of an entire ideological alignment that was obtained from the title of **racial theory**.

The term "Rassenlehre"—racial studies or racial theory—was brought into use by the German philosopher Kristof Meiners (1747-1810), in 1786. But only in the second half of the 19th Century did widespread use of the term in literature begin. New designations arose on a wave of trendiness: "Rassenforschung"—racial research; "Rassenkunde"—racial behavior; "Rassengedanke"—racial thought. As is evident, it was namely in Germany, at the boundary of the 19th and 20th centuries, that the new socio-political tendency achieved its highest development, for all the concepts reflect a moral-ethical and qualitative accent in the resolving of the racial problem. Modification and completion of physical and psychological methods of research developed in parallel with the ethical interpretation of their results, and projection onto the evolutionary-historical process. A principally new branch of philosophy formed, which studied **biological thinking**, and was based on values other than the humanitarian rhetoric of the Enlightenment. The new ideology formed through the efforts of anthropologists, anatomists, medical doctors, psychologists, psychiatrists, historians, religious leaders, culturologists,

---

[22] *Antropologicheskaya Fiziognomika.* Moscow, 1878.

27

philosophers, and finally, political publicists. The interpretation of existence through the prism of physical realities rapidly acquired the contours of a new worldview. One of the pillars of racial theory, Houston Stewart Chamberlain (1855-1927), emphasized: "The normal course of development is directed not from race to the [disappearance] of race, but from the politics arising from the absence of race to an overall sharper manifestation of race, the highest product of which is the genius, the hero. Under "race" I understand that agitation of all essence, which is achieved under certain conditions of selection, mixing, and reproduction with a race—only under these certain conditions, but already without exception, that is consequently, with accuracy, a law of nature. I do not engage in the investigation of tombs, in order to discover there is some kind of "race;" I follow the great English naturalist to the stable and the bird sanctuary, and to the gardener I say: "It is indisputably clear, that things here give content to the word "race."

Ludwik Krzywicki formulated a novel approach [to race] with the following thought: "Racial historiosophia seeks to find the causative relationship between race, on the one hand, and the manifestations of social life on the other; in the opinion of racial historiosophia, if there would have been no given race, then no corresponding civilization would have existed. In this case, race is not only a patroness of social process, but [also] its source, strength, and creative foundation. Customary history writes only that which a given race accomplished, but does not examine the causative relationship." Ignaz Zollschan (1877-1948) wrote: "In the racial problem, we understand the question of the significance of the racial factor in historical and cultural development." Still one more major racial theorist, Walter Gross (1904-1945), emphasized in this regard: "Racial theory changes the picture of history as radically as Copernicus' theory changed the picture of the solar system."

Thus race firmly occupied a place of abstract general-human values in a new system of philosophical coordinates as a real value, as a physically and psychologically measurable principle. The founder of **racial hygiene**—Alfred Ploetz (1860-1940), pointed out: "Anywhere ethics claims space outside of the individual…where the politician consciously or unconsciously looks after basic life interests as a final object, there is always the organic whole of life, the notion of race." Fritz Lenz (1887-1976), a prominent specialist in the area of the biology of heredity,

defined the main idea of the given type of nature-philosophy with the following words: "Race is the carrier of everything. The individual personality, the state, the people—everything essential originates from it, and it is essence itself. The constellations of our fate are inside us. The meaning of our higher ideals is in our very own nature. Thus, the health of race serves the happiness of separate peoples, and thus common, basic, and constant happiness. A degenerative people is inescapably unhappy, even [when] possessing all the treasures of the world. We do not need race for the sake of happiness, but happiness for the sake of race."

Left: Ancient Greek vase shaped into the head of a Negro. (4th Century, B.C.)

Correspondingly, the racial quality of the researcher himself has primary importance for raciology as a science, for by the laws of the given worldview system, like only recognizes like.

The famous German racial psychologist, Eric Rudolf Ensch (1883-1940), justifiably noted: "Blood and race determine the pureness of ideas. Race and blood lie at the basis of everything. A single straight thread stretches from the structure of the capillary net to the worldview." Lothar Gottlieb Tirala developed this thought in a more radical and witty form: "Racially clean peoples have a worldview to which they belong; half-breeds are drawn here and there."

Now conforming with all of the above, we come to a simple and clear conclusion: that raciology is principally different from classic anthropology. Raciology examines the qualitative differences between large human groups. It begins where physical anthropology ends, confining itself simply with the establishment of the very fact of [racial] differences. The distinguished German scientist, Walter Scheidt (1895-1976), noted in the book, *Basics of Racial Theory* (1927) that raciology is a science "describing race and its living manifestations." Still another luminary of raciology, Baron Egon von Eichstadt (1892-1965), emphasized in the fundamental monograph *Raciology and the Racial History of Man* (1937-1943), that "Without bodily form there is no race and no constitutional psychology; without a tangible picture of manifestation, there are no inherited traits; without connections to the form of the body and the spirit, there is no historical influence, no spatial and cultural ties; there are no historic strengths and changes. Only a living form is the logical and constant center of raciology. The raciological task is the study of Man himself, concluding in the research of his biological groups; and the goal is the knowledge of the various forms of expression of his natural completeness and the presentation of the causes, essence, and influence, of the diversity of his manifestations. At this time, anthropology only studies existing forms of these biological groups. Therefore, raciology should occupy a particular place, and with the clean logic of scientific work, one should strive to avoid mixing it with anthropology."

Concerning the principle of the study the qualitative differences between the races, through their living manifestations, Hans F.K. Guenther points out in the book, *Raciology of the German People* (1922): "For the social consciousness, raciology is something completely other than simply another science: it is concerned with what closely touches every person, and to what every person is sensitive to—[it is concerned] with the unchanging, inherited physical and psychological traits of Man, that legacy which is passed down."

Anthropologists usually declare that the subject of their studies is the human races; however, at the same time, they permit themselves to discuss "Man in general" or "Man as such;" and instead of bringing this jumble themselves to scientific research, they make an impossibly synonymous evaluation of the phenomenon of organic nature. In contrast to this vague position, raciology poses the question otherwise. The prominent racial philosopher, Ernst Krieck, emphasized in his fundamental three-tome composition, *National-Political Anthropology* (1936-1938), that: "A racial representation of Man does not [recognize] the equality of everyone who has a human appearance, but does recognize the differences in dependence of racial character, and the life orientation based on that and in particular it [recognizes] aptitudes toward political creativity. The place of the equality of peoples is replaced by their differences in racial aptitude and historic achievement: thus the races are placed according to rank."

Eugen Fischer, in the book, *Anthropology* (1923), gave the following definition: "When talk turns to the influence of racial traits on the fate of a nation, they talk about **historical anthropology**. Very often, the attempts to explain the fate of a people by its anthropological composition is briefly termed **racial theory or raciology.** In his later essay, *Race and the Origin of Races* (1927), he developed and deepened the idea: "The fate of peoples, tribes, and states depends to a decisive degree on the racial nature of their carriers. World history—is part of racial history. A race or a combination of the race bearers of a people and state [set] the conditions of their fate. Thus, racial biology and the question of the origin of races leads to the most difficult questions in the life of humanity in general. The natural history and cultural history of Man are not two lines, but one line."

In completely the same spirit, Soviet anthropologist V.P. Alekseyev, stated in his book, *Man, Evolution, and Taxonomy;*[23] that: "One of the fundamental problems in the area of racial science is the problem regarding the results of the study of Man as a physical type with the results of the development of the whole complex of socio-historical disciplines; briefly said, the problem of the relationship of anthropology and sociology, of anthropology and history."

---

[23] *Chelovek, Evolyutsia, i Taksonomiya.* Moscow, 1985.

In the peak of his creative activities, the famous German anthropologist-evolutionist and paleontologist, Gerhard Heberer (1901-1973), headed the Instititute of General Biology and Anthropology, in Jena. In his particularly natural-scientic works, significant space is set aside for philosophical comprehension of key questions in biological development. In the work, *The Theory of the Origin of Species and Modern Biology* (1942), he emphasized: "In the general theory of the origin of species, or philo-genetics, all fields are synthetically united in a causal explanation of the history of living things, all the way to Man and the rise of his racial diversity. Thus, the common history of organisms crosses to the history of the human races. The theory of the origin of species is simultaneously that soil, in which **racial theory** has its roots."

Still another luminary of German science, Professor Walter Scheidt (1895-1976), [who was] the Director of the Institute of Anthropology in Hamburg, gave the following definition in his monograph, *General Raciology*[24]: "Raciology is the science of the formation of human racial characteristics, and the reasons and results of such concentrations."

In Soviet philosophical dictionaries by various editors, professors of Marxism-Leninism repeatedly undertook attempts to present racial theory as a sum total of pseudo-scientific, namely social and political views; but that does not correspond to reality, since even according to the indications of Soviet and foreign anthropologists and biologists, it is a higher synthesis in explanation of complex patterns of biological evolution.

In this question, all communist ideology has traditionally come down to ceaseless insinuations of social metaphysics, aimed at physical anthropologists. And moreover, this entire regime constantly identified itself as materialistic, and animated by the spirit of Darwinism. The given substitution of scientific values and methodological bases, in order to satisfy imaginary humanitarian sympathies, distorts and disfigures the whole enormous system of knowledge about the world. Conditionally speaking, in their opinions, sociologists and biologists see and evaluate the angel and the demon in the single human nature in different ways. So what the physical anthropologist, who studies races, considers the evolutionary basis of their development and resultantly interprets as an adaptive and positive factor, the social philosopher prefers to interpret as a subconscious base for the development of "unhealthy" judgments about the supremacy of one race over another. And vice versa, when the scientist, who is able to think in terms of biological categories and through investigation, bravely announces that a certain phenomenon in the nature of Man is hereditary degeneracy, culturologists, who are animated by the ideals of humanism, for some reason always try to teach that this [degeneracy] is the higher freedom of the spirit. Where the former is able to see the decay of natural instincts, the latter tries to see a miraculous, supernatural revelation. For one culture, there is an unconditional core of the biological consolidation of a species—for the other, this is a justification of an unnatural wave of egotistical instincts. This inconvenience, founded on the disparity of the pretensions of both worldviews, deforms the picture of the world.

Classical anthropology tortures itself all the time with the politicized chimera of the unity of mankind, at times trying to substantiate this worldview, even when it is contrary to the biological facts and elementary logic. For raciology in general there exists no such task, for if the focus of research is directed toward the study of the variations [in Man] and their manifestations, then the discussion of equality and unity loses any meaning or value. Raciology teaches attentiveness and sensitivity toward nuances and small things, but does not work by abstract categories. Facts are valued more than formulas in its system of coordinates. These deep differences, included in the basic worldview and methodological specifics of both sciences, do not in any way speak to any qualitative differences in the researchers, who belong to the fields of anthropology and raciology.

The anthropologist is similar to the incorporeal, Biblical archangel, who on a whim, intrudes into the lives of people—a little-studied foreign substantiation. The raciologist is the manual laborer of race, recognizing his own with its indissoluble genetic relationship in fate; he is answerable for his deeds and hypotheses in full measure. The anthropologist is the indifferent wanderer of academic natural studies, while the raciologist is the valuable guard dog of race, warning about strangers, even when the master does not throw him a bone. The anthropologist aloofly sorts out racial traits, as if a collector of antiques; the raciologist is involved, like an investigator or a judge. For the former, [racial affiliation] is completely absent in the process of

---

[24] Published 1925

30

research; the latter begins his analysis precisely from a position of [racial affiliation]. The anthropologist establishes the very fact of racial differences, and upon this, assumes his task is done. The raciologist, on the other hand, evaluates [the racial differences] and creates a hierarchy from them, based on degrees of importance, in order to teach his fellow tribesmen to use them in the struggle for existence, only for the advantage of the members of his race. Raciology as a science is not afraid to give a direct answer to the two most traditionally treacherous questions: "who is guilty?" and "what is to be done?"

Proceeding from the above, one may apply such a definition to the given science.

It is necessary to understand under raciology, a single philosophical system, located at the boundary of the natural, precise, and humanitarian sciences, by means of which all social, cultural, economic, and political manifestations of human history are explained by the influence of the hereditary racial differences of peoples on the given history of creativity. The abundance of facts, accumulated by anthropologists, biologists, geneticists, psychologists, and related disciplines, about the inborn racial differences of peoples, projects itself into the sphere of their spiritual lives. At the basis of every historical phenomenon, raciology tries to single out the biological source of its cause; that is, the hereditary, specific traits of the members of the human races. For their part, differences in biological structure lead to differences in behavior, and also— to differences in the evaluation of phenomena. Thus, raciology—the science—studies the biological factors of world history. More broadly, it researches and evaluates the biological premises of any activity in general.

It would not be superfluous to point out that the very term, **raciology**, is proposed for use by the author of these lines, for the first time in the Russian language and that it has heretofore not been used in Russian scientific literature. Its meaning is drawn according to the general wording of the translation of German terms known to us: "Rassenlehre", "Rassenforschung", "Rassengedanke", and "Rassenkunde".

Not only in Western, but also in Russian scientific literature, we may easily discover an analogous complex of ideas. Thus, our important scientists, V.V. Ginsburg and T.A. Trofimova, in the report, *Paleo-anthropology in Middle Asia*,[25] wrote that: "The races of Man, as with the subspecies of animals, appear by categories; that is, as biological essences. It is not abstract peoples that appear in the arena of history, but concrete, social-ethnic unions of varying scale— tribes, peoples, and nations, occupying defined territories and characterized by one or another racial type. Studying the racial composition of the populations of various regions in various epochs, one may see its dynamic: the long-time existence of [racial] types on a territory, mixing with other races, or replacing them. This is established by anthropological methods, sometimes very accurately, and it allows light to be shed onto the ancient roots of peoples, layers of immigrants, and so on. The stability, or on the other hand, the changing of the racial composition of a population, reflects the history of separate ethnic groups and the social unions cultivated by them (tribes and peoples). The study of the dynamic of racial genesis may provide the key to the understanding of ethno-genesis. The role of anthropology in the study of ethnogenesis concludes in such an analysis. By the same, anthropology, archaeology, ethnography, and linguistics mutually complement each other with their material, sources submitted for history."

One of the world-renowned classics of Soviet science, V.V. Bunak, already emphasized in 1938 that "the denial of categories of race in Man is inwardly unscientific and a contradiction." Another standard of Soviet anthropology, the academic V.P. Alekseyev, wrote in the monography, *Historical Anthropology and Ethnogenesis*,[26] that: "In a final analysis of the data about the biology of Man, no detailed and in-depth analysis of ethnic inter-relations is possible without an account of the racial situation; no research of ethnogenesis and ethnic history can be genuinely comprehensive and all-encompassing without drawing on anthropological data. The morphological and physiological traits that are studied by anthropologists are genetically conditioned, and therefore, the biological attributes of populations are tightly interwoven with many aspects of their lives; this essentially enriches the picture of human history."

---

[25] *Paleoantropologiya Credney Azii.* Moscow, 1972.

[26] *Istoricheskaya antropologiya i etnogenez.* Moscow, 1989.

The given positions of our anthropological science, always considered as progressive, can be recognized as an axiom, and on the part raciological analysis, as necessary starting conditions.

One of the key concepts in raciology is the concept of **racial traits**. Eugen Fischer, in the book, *Race and the Origins of Races*, gave the following definition: "What characterizes races and determines the affiliation of a person with a concrete race, are the physical, anatomical, physiological, and psychological traits passed on by heredity. Racial traits are inalienable. Traits on the basis of which we mark boundaries between races, are called racial traits. By definition, these should be hereditary traits. Besides that they should be present in a large group of people. Individual distinctive features do not vary chaotically, but with defined regularity, in accordance with Gauss' probability curve. They group around a defined average quantity. All racial traits of Man coincide with those by which breeds of domestic animals differ."

Below: *Racial Diagnostic.*

V.P. Alekseyev, in the book, *Man, Evolution, and Taxonomy,* gave a more complete definition: "The complex of racial characteristics represents a certain aggregate of population and individual variations. Race is the aggregate of populations. The notion, according to which every individual is a carrier of the traits of a defined race, received the name "**typological concept of race**." The task of racial analysis leads to the definition of the race type of an individual. Further, it calculates the percentage correlation of various represented racial types; the percentage correlation is the basic racial characteristic of any group, and any people; and with the help of this characteristic, various peoples are distinguished, one from another. The typological concept of race leads genetics back to the notion of archtypes in morphology, and that in turn—if one gets to their genetic roots—leads back to the ideas of Plato. The transition from phenotypical variability to genotype, basically constitutes the essential task and basic course of racial-genetic research."

The head of the Polish School of Anthropology, Jan Czekanowski (1882-1965), is considered one of the founders of the typological concept. According to the concept, the individual is the carrier of racial characteristics, and race is the arithmetical sum of individuals; consequently, in order to break it down into its component elements, it is necessary to descend from the group level to the individual level. Jan Czekanowski and his numerous followers maintained that racial traits are inherited as an entire complex. There are no independent, hereditary racial traits that exist in isolation from others.

V.V. Bunak emphasized in his fundamental work, *The Genus Homo, his Origin, and Subsequent Evolution*, that "The works of the Polish school occupy an important place in anthropological literature. Its founder, Jan Czekanowski, proposed that the characteristics that comprise a racial complex are found to be indissolubly connected to each other; that these connections are genetically conditioned, and once created, remain stable, changing only as a result of mixing with different types. The determination of the racial composition of a population is figured by calculating the frequencies of combinations of traits, corresponding to the earlier established types. Therefore, the method itself received the name "**typological**." In every population, several combinations of traits always stand out, which characterize the group's racial composition."

The result of an anthropological determination of a racial type is the **racial diagnosis**.

Egon von Eichstadt gave an excellent definition in his fundamental research: "Race is not merely chance, not merely the makings of something that should still appear, but is itself an immediate reality. To dispose of race as [just] a hypothetical inclination, means to delete it from life and reality. The racial type and the racial core cannot exist without individual variation, variation of traits, the appearance of transitional forms, and mixing. The [racial] type as such, is the highest integral, always somewhat more; it is really simply the sum of separate traits that constitute the living reality of race.

Thus, if we determine a type and take the diagnostic of a racial category as a separate individual, we should begin from separate traits, in order to determine the given individual as a whole, and then from the individual whole, cross over to a higher whole, that is, race. Defined talents are necessary for this "**systematic view**." There is the normal type, described mathematically, graphically, and biologically; the [race type] originates from it. The racial diagnostic and anthropology are, in general, such sciences which presume the presence of defined abilities and long-term training."

In accordance with this, it appears possible to us to give an able definition of the given concept. **The racial diagnostic is a multi-dimensional set of measured (that is, quantitative) and descriptive (qualitative, that is) traits employed for the revelation and classification of hereditarily distinct racial types.**

The given sum total is formed as a result of a systematic synthesizing of the results of specific aspects of the racial analysis, which include:

1) anthropometic measurements
2) descriptive traits
3) morphology of the body
4) x-ray graphing(principally of the hands)
5) oximeter (a technique of the revelation of the degree of intensity of the metabolic processes in various race types)
6) genetic markers
7) anthropological photographing
8) odontological cast
9) dermatological imprints of the hands and feet

However, it is necessary to understand first of all that the above-enumerated aspects of the analysis have unequal significance and may not form into a single picture of the racial diagnostic on general grounds, as the question is one of genetic traits having different "weight" in the genome of an individual. According to Egon von Eichstadt's accurate observation, it is necessary to possess scientific ability and a certain amount of natural intuition, in order to reflect all the nuanced secrets of the work of nature in the enduring, stable forms, which may be impartially recorded by biological verdict, under the name of racial diagnostic.

However, one has to note with regret that in the world of anthropology today, a spirit of Middle Ages obscurantism and intolerance reigns, since all raciological research is discredited under the false slogan of "political correctness." Moreover, it denies the very fact of the existence of race, and also places the legitimacy of the use of the term "race" under doubt. To the credit of the modern Russian school of science, it must be recognized that its consolidated position toward this key question is fairly objective and consistent. It is namely this [principled] obstinacy in the face of the conditions of today's minimum-possible financing of science in Russia, which raises open indignation on the part of George Soros-clients and other grant-dependents. But even the collapse of the USSR, and the neglect of the needs of fundamental science that followed, in sum total with foreign political pressure, did not sway the convictions of Russian scientists; and that automatically brought them to the leading position in the world, in the matter of racial research.

One may consider the conduct of the 1st International Conference "Race: Myth or Reality?"[27] held in Moscow from 7 to 9 October 1998, as graphic proof, which supports the correctness of the given thesis. This important representative forum, conducted under the aegis of the Russian Department of the European Anthropological Association, and the Russian Academy of Science, undoubtedly will go into the annals of the science about the races. According to the materials of the work of the conference, the program document, *Problems of Race in Russian Physical Anthropology*,[28] was published.

In the introduction, the team of authors, headed by the scholar T.I. Alekseyev, it was considered necessary to immediately emphasize that "the races of Man actively exist; [that] raciology is one of the important fields of anthropological and other humanitarian sciences, and the data of anthropology is irreplaceable in historical reconstruction." In development of this thesis, N.A. Dubova announced in her presentation that: "[It is absurd] to deny the objectivity of race on the basis of large numbers of transitional variants—one may...then assert that the colors red, blue, and yellow do not exist in the [light] spectrum, insofar as they are all combinations of parts of the general spectrum, within which distinct boundaries do not exist between the specific

[27] "Rasa: mif ili realnost'?"

[28] *Problema rasi v rossiyskoy fizicheskoy antropologii.* Moscow, 2002.

[colors] comprising the spectrum." L.T. Yablonskiy went even further in his report, emphasizing: "The attempt to avoid the question of race and inter-race relations through simple disregard of the very existence of racial diversity within the species *Homo Sapiens*, is an "ostrich" tactic; moreover, this is how it appears to a majority of Russian anthropologists; it is completely unjustified from the point of view of factual scientific objectivity." Naturally, such adherence to principle in a major question may not rouse respect, [but] our monograph—*Raciology*—is dedicated to the maintenance of the position of authority of Russian science in a difficult moment of its existence.

For classical physical anthropology, all companion reports and many chapter titles may appear unusual, but once more we remind, that this is the specific evocative nature of raciology as a science. In customary anthropological research of the general type, there is a regular plan of account for evolutionary-historic information, and also a sequence of chapters dedicated to the description of morphological differences. By virtue of that in raciology, all this abundance of intrinsic data has no self-value, but is subject to logic; the explanations of socio-cultural processes through the prism of hereditary biological differences, and the presentation of material in reports of such a genre, are carried out according to the disclosure of the theme, and in accordance with the ideological importance of the subjects being discussed. In raciology, anthropological facts fulfill the role of illustrative material, in the explanation of this or that process in society. By virtue of this, the whole structural design of the text is built according to other laws, [rather] than [according] to classical anthropology. Not to record, but to evaluate and explain—this is the mission of raciology. In Russian scientific literature, there is not an analogous type of report, for until now, one could not be rid of the old illnesses of the Marxist worldview, and honestly and openly rise to the defense of the interests of that original source, from which the Europoid race came.

Not only foreign affairs and the general cultural situation, but also the very magic of language, nudges modern Russian science today toward fulfillment of the great Providential mission: the strengthening of the self-awareness of the White Race. By force of circumstances, the English language became popular and international; but in that language, the term "race" means both "race", as in a consistent physical type, and "race", as in a form of sports competition. This linguistic ambiguity is not the reason for the lively discussions between supporters and opponents of the existence of race and racial differences. Of all the modern languages of Indo-European geographical distribution, it is only in the Russian language that this term—which is key to our self-awareness—retains its earlist sound and archtypical, original Aryan meaning. Realization of this fact places upon Russians—the guardians of this magical word—the sacred burden of a missionary task: safeguarding the purity of that timeless substance, which is **race**.

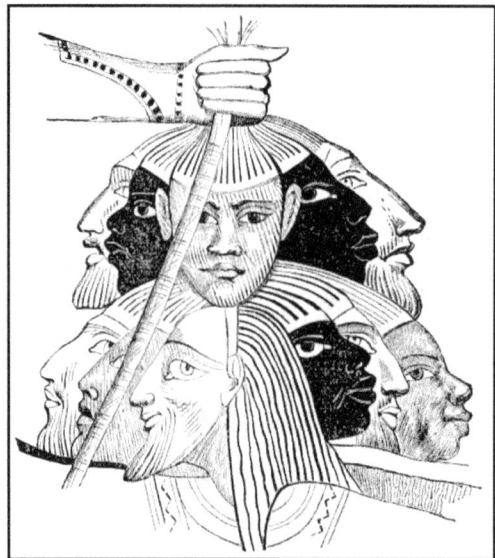

Racial Chaos in the eyes of the Egyptians.
Bas-relief. 14th Century B.C.

Classification of Races according to Deniker

Main Trunk	Their Characteristics	Affiliated Local Native Races	Their Characteristics
Group A	wool-like hair; wide nose	Bushmen	yellowish skin; steatopygia; low height; long-headed
		Negrito	Reddish-Brown Skin; Very Low Height; Relaitively Average Width of Head
		Negro	black skin; average height; long-headed
		Melanesian	blackish-reddish skin; average height; long-headed
Group B	curly or wavy hair	Ethiopian	reddish-brown skin; narrow nose; tall height; long-headed
		Australian	cocoa-brown skin; wide nose; average height; long-headed
		Dravidian	black-brown skin; wide or narrow nose; short height; long-headed
		Assyroid	light, swarthy skin; black hair; narrow, arched nose with thick tip, wide head
Group C	wavy, dark or black hair; and dark eyes	Indo-African	light-brown skin; black hair; narrow, straight or arched nose; tall height, long-headed
		Arab or Semitic	swarthy skin; black hair; eagle nose; jutting rear of head; oblong face in the shape of an ellipse; long skull; tall height
		Berber	swarthy skin; black hair; straight, coarse contoured nose; long skull; oblong face, quadratic; tall Height
		Southern European	swarthy skin; black hair; straight, narrow nose; oblong face, oval form; tall height; relatively average width of head
		Ibero-Islander	swarthy skin; black hair; short stature; long-headed

Group C	Wavy, Dark or Black Hair, and Dark Eyes	Western European	dull-white skin; dark, light-brown hair; short height; round head; round face
		Adriatic	dull-white skin; dark, light-brown hair; round head; tall height, oblong face.
Group D	wavy or straight hair; blond with light eyes	Northern European	pinkish-white skin; often wavy, chestnut hair; tall height; oblong face
		East European	pinkish-white skin; straight, flaxen hair; short height; round-headed
Group E	Straight or wavy black hair; dark eyes	Aynossian	Light-brown skin; hairy bodies; wide, caved-in noses, long headed
		Polynesian	Yellow skin, not very hairy bodies, often arched noses, tall height, ellipsoid faces, often round headed
		Indonesian	Yellow skin, not very hairy bodies, short height flat, sometimes caved-in noses, jutting cheekbones, rhomboid faces, long-headed.
		South American	Yellow skin, not very hairy bodies, straight or caved-in projecting nose, often long-headed.
Group F	Straight hair	North American	Bright yellow skin, straight or eagle nose, tall height, relatively average width of the head
		Meso-American	Bright yellow skin, straight or eagle nose, short height, round headed
		Patagonian	Bright yellow skin, straight nose, tall height, round-headed, quadratic-shaped face
		Eskimo	Yellow-brown skin, short height, round, flattened face, long-headed

Group F	Straight hair	Ugric	Yellow-white skin; straight or concave nose, cheekbone jut, short height, often long-headed
		Turko-Tatar	Yellow-white skin, straight nose, average height, round-headed
		Mongolian	Pale-yellow skin, jutting cheekbones, epicanthus, round-headed

The Chief Morphological Traits of Differentiation of the Races
(according to V.V. Bunak)

Descriptive		Measured	
On the Living	On Skeletons	On the Living	On Skeletons
Form of the Hair: Pigmentation of: a) skin b) hair c) eyes Soft parts of the face: a) epicantus b) eyelid folds c) width of eye slits d) thickness of the nose tip e) height of the alae f) position of the base of nose g) protrusion of upper lip h) profile form of upper lip i) thickness of lips Development of Tertiary Hair Cover: a) on the face b) on the body Form of nose bone: a) height of nose bridge b) profile of the slope Shape of Forehead: a) slope b) development of superancillary arches (brow ridges)	Height of the Nose Bridge: a) slope b) development of eyebrow ridges Horizontal Profile Prognatism	Height of the Face... Width of Cheekbones Height of the Nose.. Width of Mouth Slit.. Height of the Head.. The least frontal width of the Head.. Width of Lower Jaw.. Height Standing... Proportions of the Head, Face, and Body	Depth of the Canine Fossae Lower Edge of the Eye Orbit Apertures Height of the Face Width of the Cheekbones Height of the Eye Orbit Apertures Length of the Skull Width of the Skull Height of the Skull The least frontal width of the Skull Width of Lower Jaw.. Proportion of the Skull, Face, Skeleton. Length of the base of the Skull. Angle of the Slope of the Forehead Angle of Projection of the Face

Horizontal Shape Prognatism Form of breast glands on Women.			Angle of Projection of the Nose. Height of Eye Orbits Width of Orbits Size of Lower Jaw Width of Nose Bones Length of Crown Width of Crown

THE BLOND AND FAIR RACE:
HISTORIOGRAPHY AND ANTHROPOLOGY

"Almost every known black parent makes black offspring."
Richard Lewontin

"Among the man-like apes there are no blue eyes."
Ludwik Kryzwicki

Culture and its material bearer: what is their relationship? Since antiquity, sages have attempted to answer this question. It was the age-old desire of the creators of both primitive myths and world religions, to produce spiritual traits in their works, in accordance with the physical conditions of the man-creator. Tales and legends all over the world abound with the footprints of the common powers of observation, revealing the corporeal basics of the spiritual gifts of each creator of cultural values. From century to century, popular lore strives to depict every figure of creative significance as the concentrated embodiment of the physical and psychological features of its ethnos, emphasizing by the very same that the strength and uniqueness of his genius springs from the depth of the centuries, and is rooted in a hereditary mass of ancestors.

However, in the 19th Century, the stormy golden age of natural studies substantially corrected that archaic, leveling view of world history; on account of the development of positive statistical methods, it became clear that all human bio-types were far from possessing the ability to create culture in equal degree. From the standpoint of contributions to the world treasure room of culture, it was objectively revealed that biotypes exist, which are more, or less, "worthy."

On the level of the philosophy of history, the outstanding English scholar, Edward Gibbon, author of the multi-tome work, *The History of the Decline and Fall of the Roman Empire*, was probably the first in modern time who formulated the thought that the death of great civilizations in general, and ancient Rome in particular, singularly arose through the fault of washing out the more worthy blood of the social organism, and replacing it with less worthy [blood.] The history of the decline of the ancient world, according to Gibbon, comes down to a numerical decrease of the members of the culture-creating white race, in comparison with colored races, which are not good at creating high culture. The racial traits of the emperors and high aristocracy of the Empire, etched in the numerous statues coming down to us, are a graphic testimony to this; they clearly show the physiognomic symptoms of the racial substitution that appears in the 2nd and 3rd centuries, A.D. The European—primarily northern—race created the unprecedented, gigantic organism of world empire; and less-worthy—in culture-biological terms—races from Near Asia, Africa, and the Mediterranean gradually occupied leading positions, not even knowing how to preserve this priceless creation—leading it to its degradation in wild, perverted orgies of the East.

The term "race," (brought into European use by the famous French ethnographer and explorer, Francois Bernier), did not at first have wide use in humanitarian sciences. Immanuel Kant and Johann Gottfried Herder first gave it a philosophical basis in the beginning of the 19th Century, and Christopher Meiners (1747-1810) made an attempt to apply it in the context of historical process. And so namely they established the foundational beginning of **racial theory**.

Under racial theory, or raciology, it is taken today to mean a single philosophical system, located on the boundary of humanitarian and natural sciences, by means of which all social, cultural, economic, and political phenomen of human history are explained by the influence of hereditary racial differences in peoples—in the creators—on the given history. The abundance of facts, accumulated by anthropologists, biologists, geneticists, psychologists, and related disciplines, about the inborn racial differences of peoples, projects into the sphere of their spiritual and social lives. Racial theory strives to single out the biological source, that is, the hereditary differences of the members of the various races, at the basis of every historical phenomenon. For its part, differences in biological structure lead to differences in behavior, and also to differences in the evaluation of phenomena. Thus, racial theory—this science—studies the biological factors in world history.

However, until the start of the 19th Century, in European scientific practice there was no system of views enabling one to draw up a hierarchy of cultural worthiness from the specific physical features of human material. Culture as such, according to the opinions of the ideologists of the Enlightenment, arose literally out of 'nowhere,' under the influence of the caprices of God, or the urges of some Absolute. In this process, real Man fulfilled the role of a passive guide, and his race, ethnicity, and hereditarily conditioned characteristics were, in general, not taken into account. With the help of "dictates of heart," it was considered possible to explain any fine point of the creative process, in which connection the academic science of the time gladly supported many of the intellectual speculations of theological scholastics.

1. The Origin of the Science about the White Man

But the insight was unavoidable. The dictatorship of the heavens was sooner or later to end in a revolt of the flesh. The Englishman, Sir William Jones (1771-1845), laid down the basis for **comparative linguistics** in 1796, and the German Sanskrit scholar, Franz Bopp (1791-1867), published his monumental work, *Comparative Grammar*, in 1835. It is namely these two scholars who were the first to point out the genealogical unity of the Indo-European languages.

Max Mueller (1823-1900), a luminary of philology and religious studies, published the famous *Lectures on the Science of Language* in 1861, where he first mentioned not only the ancient Aryan language, but also the "Aryan race" and the "Aryan family of languages." He wrote of the time "when the first ancestors of the Indus, the Persians, the Greeks, the Romans, the Slavs, Celts, and Germans lived not only in one settlement, but under one roof."

However, linguists of that time based [theories] on their scientific intuition, and did not connect them with the data of anthropology and archaeology; as a result of that they placed the home of the Indo-Europeans in Asia. It was only the first representatives of the field of anthropology (just scarcely forming into an independent, academic discipline in the middle of the 19th Century) who cleared up the situation of the matter. Paul Broca (1824-1900), the founder of the French school of anthropology, wrote in 1861: "The ethnological value of comparative linguistics is little, overall. In point of fact, it can most likely lead to error, than to anything else. But philological facts and deductions catch the eye, more than painstaking measurements of skulls, and that is why the conclusions of philologists attract exaggerated attention." His famous countryman, Paul Topinard (1830-1911), supported him: "Craniology teaches us that the races which live in present-day Europe, have lived there continuously since the start of the Neolithic Period, when the northern deer and wild horse roamed Europe."

Paul Broca

Paul Topinard

In 1851, Robert G. Latham (1812-1888), was the first of the philologists to rebut the everyday opinion that the Indo-Europeans originated in Asia, for which he was ridiculed by his colleagues. In 1874, the German philologist Gen sarcastically wrote in this regard: "It happens that in England, a nation of eccentrics, the thought came to one original mind, to place the cradle of the white race in Europe." And only in 1868 was the voice of Robert Latham heard; in that year, August Frick (1833-1916) published the monograph, *The Comparative Dictionary of Indo-European Languages*, and with this laid the foundation for **linguistic paleontology**. Thus, the concept of the Asian origin of the Aryans was subjected to fundamental revision.

August Friedrich Pott (1802-1887), the founder of the comparative-historical method in linguistics, established an entire line of strict sound correspondences between the Indo-European languages. From 1859 to 1876, he published a fundamental six-tome report, *Etymological Research in the Area of the Indo-Germanic Languages*, laying in it a firm foundation for a subsequent comparative dictionary reserve of related languages, decisively parting from the old tradition, dating back to antiquity, of etymologizing by external similarities. His conclusions were based on a comparative analysis of Sanskrit, Zendic, Avestanic, Greek, Latin, German, Celtic, and Slavic languages.

Theodor Benfey (1809-1881), a prominent German philologist, wrote in the comprehensive work, *History of Linguistics and Eastern Philology in Germany* (1869): "[Since] geological research has established that since prehistoric times, Europe was the habitation of Man, all evidence advanced in favor of an Asiatic origin of the Aryans is refuted." With analogous emphasis, Wilhelm Geiger (1829-1870), a researcher of ancient Iranian languages, also wrote in his book, *History of the Origins of Man*: "From the two opposing theories (the European origin or Asian origin [of the Aryans]), one can only rely on evidence; not one argument has led to proof of migration from the east to the west. The latter is unlikely, and is, one can say, in and of itself impossible, if one should have to suggest that it was accomplished in gradual waves." The arguments of Benfey and Geiger were based on the common names for plants and animals in the Indo-European languages.

The prominent scholar, Rudolf Poech (1870-1921), emphasized that anthropology and archaeology should add to, and correct, the conclusions of philology; and the given point of view in science, starting with this period, at last came to be dominant. The synthesis of humanitarian and natural sciences bestows the right to possession of the truth.

Adepts of the concept of a European ancestor of the Aryans are highly indebted to the many studies of a scholar such as Karl Penka (1847-1912). In his book, *Textbook on Germanic Ancientness* (1880), he wrote that the common original vocabulary of the Aryans did not incur any kind of eastern influences on itself. He based his conclusions about the northern origin of the blonde race, in part on the fact that in the languages of peoples belonging to it, they only encounter names for northern animals and plants, such as the bear and the wolf, the pine, and the beech and oak; but those languages have no mention of lions, tigers, or camels, wheat, or palm trees. These languages are united by common terms for snow, ice, and winter. And in 1883, Penka published the book, *Origins of the Aryans*, which subsequently became a classic of the genre. Based on comprehensive historical, archaeological, and anthropological materials, it demonstrates that the ancestors of the white race were located in Scandinavia: "Pure-blooded Aryans are represented only by northern Germans and Scandinavians—the most fruitful race, endowed with large physique, large muscles, and strength, energy, and bravery. The shining natural talents of this race were used to conquer the weak races of the east, south, and west, and to take their language to these peoples." In the English-speaking scientific world, John Reese (1883-1901) immediately echoed this new trend toward the "Nordicization" of the Aryan ancestor; in 1836, he suggested that the Aryans may have originated somewhere in the limits of the Arctic Circle - in part, in the north of Finland. And in 1889 Gerald Randall applied the following definition to the Aryans: "The dolichocephalic [long-headed] race of blondes originated around the Baltic Coast. The Aryan represents the type of intelligent man—the basic product of race, in which the particular qualities of dark and light, North and South, emotionality and practicality, mixed and united into a higher, transitional stage of mind and body."

Thus, linguistics gave a jolt to the development of new studies in the area of archaeology, and that in turn stimulated the stormy development of anthropology. Determining their position [with regard to] the ancestor of the Aryans, scholars put forward the question about their original race type. The famous German linguist and historian, Otto Schraeder (1855-1919), put forward this thesis in 1884: "The Aryan race originally corresponded to the blonde, northern races, among which the Aryan language and culture developed. It was carried to other, non-Aryan peoples through migration and cross-breeding." Thomas Henry Huxley (1825-1895), the classical English naturalist, divided the Europoid race into a lighter, xantochroide race, and a darker, melanochroide race. Finally, they recalled the completely forgotten French ethnographer, Abel Remieux, who in 1820 was already one of the first to bring to the attention of the academic community, those Chinese chronicles about the ancient and powerful *Dinlini, Khakasi,* and *Usunyi* [peoples], who lived to the northwest of the Chinese. The Chinese chronicles tell of peoples with blonde hair and light-blue eyes. Alexander Ekker (1818-1887), a prominent German anthropologist, discovered skulls of a "northern type" in graves in southern Germany in the 1860s, and established their identicalness with the skulls of modern Germans. The skulls of a pure "northern type" were discovered throughout Scandinavia and northern Germany by the prominent Swedish anthropologist, Anders Retzius (1790-1860). Precisely on the basis of these numerous

Alexander Ekker

Jean Louis Armand de Quatrefages de Breau

craniological series, the proposal was expressed that the modern "northern type," by its structure, traced back to the Cro-Magnon type of Paleolithic Europe. Jean Louis Armand de Quatrefages de Breau (1810-1892), of the classical French school of anthropology, even designated the ancient Cro-Magnon as blonde, in the modern sense of the word. The elongation of the skull (dolichocephalic), the elongation of the limbs, the general gracefulness and proportion of the physique, enabled another classical German anthropologist, Herman Klaatsch (1863-1916), to introduce his synonym for the definition of the northern type: the"**gracile type.**" It carried an especially morphological meaning, and emphasized the special racial features of the physique. French anthropologist Paul Topinard pointed out in his monography, *Anthropology,* that "light blue eyes serve as an indicator of the presence of the blonde type in the blood." In regard to this, one of the most recognized authorities of the Italian school of anthropology, Giuseppe Sergei (1841-1936), wrote in *Species and Variety of the Human Genus*: "The skull is most important for classification. With one skull, one can differentiate the ethnic elements, which go into the composition of a mixed group. An initial classification is possible, even with one stable feature. The most stable are the brain and facial sections of the skull. From the most ancient times, up to our time, no new skull forms have appeared. An important trait for classification is the inner capacity of the cranium; it is directly tied to the form of the skull. The capacity of the human brain does not increase in proportion to the evolution of society. The average capacity and form of skull types remains as before. The increase in the capacity of the human skull is a myth."

Left: Rudolf Ludwig Karl Virchow

The noted Polish scientist, Ludwik Krzywicki, emphasized in the monograph, *Anthropology* (1900): "I believe that the person having black hair and eyes, never gave rise to light-haired descendents, and I am certain that the same may be said regarding the skull; [it may also be justifiably said] that the long-headed blonde laid the foundation of Aryan speech."

Johannes Ranke   **Julius Kohlmann**

Being a subject of the Russian Empire before the Bolshevik Revolution, he had the opportunity to participate in numerous archaeological and paleo-anthropological expeditions, carried out with great intensity in the gigantic expanses of the Russian state. Thus, in part, while excavating graves of the Neolithic epoch on the shores of Lake Ladoga, he discovered that their cranial index was 72.1, which meant: the tribes which arrived much later, composed of western and eastern Slavs, were hyper-dolichocephalic—carriers of the pure "northern type." L. Krzywicki's conclusion was simple, and at the same time, persuasive: "In the Kurgan Age, from Olonyetsk to Kiev, from Kursk to Moscow to Poland, there lived a markedly and racially pure people, in all likelihood, the blonde Rusi."

Rudolf Virchow (1821-1902), the founder of the German Anthropological Society, systematized the data from excavations he carried out throughout Europe. He came to a completely straightforward conclusion: "The Germans and Slavs were originally blonde; but from mixing with the Celts, they absorbed a greater or lesser number of elements of the dark [haired] type." Another luminary of the German school of anthropology, Johannes Ranke (1836-1910), developed [Virchow's] idea. In the large, two-tome report, *Man* (1901), he emphasized: "In the ancient typical German form, as with the Slavic, the skull was long-headed, dolichocephalic. Therefore, desiring to explain the short-headed [brachiocephalic] [trait] in a certain area of Germany, we should not think exclusively of the Slavs, who, in all likelihood, originally had the long form of skull, like the Germans, and changed as a result of mixing with other short-headed peoples. Just as we encounter the main region with a proliferation of blondes in north-central Europe, we see in the north of the Slavic and German worlds, a sufficiently compact core of long-headed skulls. This population group with a predominance of long-headedness, is drastically surrounded by short-skulled tribes on all sides. Thus, the distribution of both chief physical characteristics mutually combine: blond, long heads, with brunette, short heads. From this, we should conclude that [it was] in northern Europe that the general causes of the formation of local differences in complexion, and the form of the skull, predominated. Light-complexioned skin and hair, together with light-blue eyes by no means comprise the distinctive feature of German people, but are spread out over a wide area, which takes in completely different—and moreover, anthropologically different—layers of population. All of modern Finland is settled, in large measure, by blondes, and even extremely blonde peoples. Only in Lapland does dark complexion begin [to be encountered]. The Slavs to the north and east have remained blonde until this day, and perhaps, they were always such. Next come the Germans, who were blondes, and the so-called white Celts, and finally, the Caledonians in Scotland."

The noted German anthropologist, Professor Ferdinand Birkner (1868-1944) of the University of Munich, stated in his huge, encyclopedic work, *Race and the Nationalities of Humanity*, that among Neolithic skulls discovered in the territories of Germany and Hungary, the percentage of brachicephalic forms did not exceed 6%. Skulls discovered in Sweden have basic and clearly expressed dolichocephalic cranial indexes, within the limits of 68-70. During the time of the Roman Empire's rule over German territory, the percentage of brachicephalic [skulls] there did not exceed 13%. During research of skulls from ancient Slavic burial sites from the first centuries after Christ, Birkner discovered that the percentage of short-skulled brachicephalic forms did not exceed 8.5%. On the basis of data from excavations, Alexander Ekker and Julius Kollman (1834-1918), came to the conclusion that during the time of Roman sovereignty, the long skulls comprised the largest quantity. On this basis, Anders Retzius came to the generalization, which afterwards became classic in science: "The purer the race, the fewer the mixed forms."

It would be appropriate here to emphasize that the well-known and widespread accusation of German [political] authors on the 'racial inferiority of the Slavs', by reason of their mixing with Mongols, Turks, and Finno-Ugric [peoples], in point of fact has nothing in common with classical German science. Not one propagandist of similar kinds of falsehoods was in any way connected to anthropology. "Culturologists," "philosophers," "historians," as well as people without fundamental training, gladly spread the myth, mixing and equating German-ness with Aryan-ness; this categorically contradicted the true and recognized works of anthropological science. Racial theory in and of itself does not pose a danger to a humanitarian system of worthiness, but it is necessary to remind those who are accustomed to defaming any utterance connected with that 'terrible' word, *race*.

Ferdinand Birkner

Anders Retzius

2. The Russian Classical School of Anthropology

And now, in honor of Russian science, it should be noted that it also developed in the direction of the Nordic idea.

The founder of Russian academic school of anthropology, Anatoly Petrovich Bogdanov (1834-1896), like his European colleagues in the 1860s, proceeded to compile a systematized series of craniums, on the basis of archaeological excavations, using materials from the Kurgans in the European portion of Russia. And if one takes into account that a great part of the history of the white race unfolded namely on this territory, then the conclusions of the Russian scientist acquire weighty significance, which supports the ideological completeness of the entire concept as a whole. Bogdanov wrote: "The long-skull type was not accidentally or arbitrarily scattered around Russia; the more the skulls of the Kurgans from various times and places are discovered, the more clear the particular significance of this type in the settlement of Russia in the most ancient age, becomes. All excavations indicate, that the older the gravesite, the greater the percent of long-skulls, and the newer [the site], the greater the traces of short skulls. From several excavations, one may even say that there are localities where the population was homogenous—long-headed. Another prominent Russian scholar, Aleksander Vasilyevich Yeliseyev (1858-1895), also emphasized in *Anthropological Observations about the Finns*[29]: "It is proven that the original people of Europe and Scandinavia, in the north of Europe, [was] a long-skulled [people], which was replaced by a brachicephalic [people]. The initial long-skulled population served them for awhile, in which [capacity] they were circulated, and on that account, new peoples of secondary generations developed."

Left: Anatoliy Petrovich Bogdanov

Today it sounds strange, but a fact remains a fact: until the 19th Century, archaeologists who excavated the Kurgan graves and settlements, were hunting primarily for traces of material culture, completely ignoring—and sometimes destroying—the physical traces of its creators. Fragments of spears, pieces of pottery, and bone scrapers were examined separately by the scientific husbands of the time, outside of, be that as it may, any connection with the racial traits of the excavated individual, who created these hand-crafted masterpieces. No sort of cause-effect relationship between the specific cultural object and the creator exists in the interpretations of the scientists of the time. Culture was common to everyone, and was assumed, in the direct sense of the word, to have come out of nowhere. Many years would pass before common sense triumphed.

---

[29] *Antropologicheskie zametki o Finnakh.* Moscow, 1880.

Below: Aleksandr V. Yeliseyev

Determining the racial type of the original population of Europe, Russian specialists re-established the racial dynamic of historical processes on the whole Eurasian continent. Aleksandr Ivanovich Wilkins pointed out in *Anthropological Themes in Middle Asia*[30]: "It is known to us that the chief mass of population in Middle Asia developed from the mixing of the branches of two great tribes: the Aryans and the Mongols. This population is the ethnic result of the age-old struggle of the noble Iranian with the barbaric Turanian." Conscientious Russian scientists of the time saw the fundamental biological context of world history, namely in the confrontation between long-headed Europoids and the short-headed Mongoloids with the half-breeds.

The racial purity of the long-standing Russian population has been written of in many scientific works. The first that should be mentioned is A.P. Bogdanov's *Anthropological Physiognomy*,[31] in which the author set down the theoretical foundation of the notion of "characteristic Russian facial traits." Uniting and synthesizing anthropological and psychological data, and applying them to the description of the ideal national type from Russian folklore, in sum total with age-old Russian morals, A.P. Bogdanov directly concluded that the Russian people have a northern origin.

"Very often [one hears] such phrases as: she is a pure **Russian beauty**; this is **the very image of a Russian**; a **typical Russian face**. Perhaps, as they apply to private situations, the expressions meet with disagreement between observers, but denote a number of similarities in the definition of Russian physiognomy. One may be persuaded that there is nothing fantastical [about this], and that reality lies in this common expression of **Russian physiognomy**, **Russian beauty**. This is all the more clearly expressed in the negative sense, in encounters with the physiognomy of those from other tribes, which formed differently historically; for example, in comparison of foreigners with Russians. In such cases, '**no, this is not Russian physiognomy**' is decisively said and very convincingly heard - and with great conviction. In each of us, in the area of our "unconscious," there is a sufficiently defined notion about the Russian type, about Russian physiognomy." His conclusions were extremely candid and shocking for the science of the time. A.P. Bogdanov considered it appropriate to illustrate [the matter] with the forceful words of the Russian ethnographer and historian, N.I. Nadezhdin, which were already spoken in 1837: "The physiognomy of the Russian people, [who are] Slavic in foundation, is etched with the natural traces of the northern environment. [The color of] Russian hair is where the very name "Rusi" came from, in ancient times."

In another of his works, *Materials for Anthropology of the Kurgan Period in Moscow Gubernaya*,[32] A.P. Bogdanov wrote: "The native, local Kurgan population is sharply and typically characterized as long-headed; it extended entirely along the middle zone. Since the dolichocephalic [type] is encountered in different European countries, to the west and to the north in Austria, Germany, Sweden, and probably in Denmark, I propose a name for the original dolichocephalic Europeans [which] corresponds most to their history. There were no differences in the peoples of central and eastern Europe; they had more similarities in the skull, in facial traits, and in other anthropological indicators, than with the peoples of the south, which originated from a completely different prehistoric race. This similarity becomes understood in the sense of the unity of the original long-headed population, which ranged from Switzerland to Sweden, from the Baltic to eastern Russia, and the Urals. If we reconsider from this point of view, the question of *ancient Finns* and *ancient Germans*, *ancient Slavs* and *ancient Danes*, *ancient Swiss* and so on, we go back to the most remote times, [when] these peoples converge into *ancient Europoids*—the Kurgan long-headed, dolichocephalic [type]."

---

[30] *Antropologicheskie temy v Sredney Azii*. Moscow, 1884.
[31] *Antropologicheskaya Fiziognomika*. Moscow, 1878.
[32] *Materialy dlya antropologii kurgannogo perioda v Moskovskoy gubernii*. Moscow, 1892.

In the article, *Studies of the Skulls and Bones of Stone Age Man from the Shores of Lake Ladoga*,[33] the scientist pointed out with all clarity, "For the white European race, we must recognize a single original ancestry, and for this hypothetical race there is even a name—the Aryans."

In 1915, archaeologist D.K. Tret'yakov discovered skulls in Kurgan [sites] near Odessa, belonging to burial sites of the 3rd and 2nd millennia, B.C. The overwhelming majority of them were dolichocephalic, and carried in themselves the clearly expressed traits of the Nordic Race. With this, Tret'yakov emphasized that they sharply contrasted with the skulls of the Mediterranean Race, and all brachiocephalic skulls found were considered by him to be part of the Alpine Race. The Russian scientist denied the presence of even a minimal percent of Mongoloid admixture in the cranial series.

Later, in 1931, prominent Soviet anthropologist G.F. Debets (1904-1969), studied these same skulls and completely confirmed D.K. Tret'yakov's conclusions about the total absence of Mongoloid traces in the skulls from the Kurgan burial sites.

Thus, the names "Rusi" and "Russian people" have an ancient, and particularly racial-anthropological origin, that traces back to the main racial trait of the northern race—light brown hair.

With the given conclusions, we do not in any way want to offend any of the other European races. It is only a statement of an evident fact.

Still another luminary of Russian anthropology, Nikolay Mikhailovich Maliyev, emphasized in the brochure, *Anthropological Research*,[34] that "the ancient skulls are undoubtedly of Slavic origin, as are, for example, the Kurgan [skulls] of the Smolensk Gubernaya, and the skulls of ancient Kievans; and the Scythian skulls from our southern provinces also represent the long-headed structure. And in antiquity, in eastern [European] Russia, along the Kama and Volga [rivers], there lived a long-headed tribe, [which] by its anatomical structure, was similar and perhaps genetically linked with the tribes that populated the central zone of Russia." In the book, *On the Question of the Ancient Population of the Ryazan Gubernaya*,[35] A.G. Rozhdestvenskiy pointed out that the majority of Russian skulls from graves dating to the start of the Mongol invasion were dolichocephalic, and some of the skulls in the excavation retained fragments of blonde hair.

Therefore, it becomes completely evident that the primordial creators and carriers of culture on the entire territory of Europe, and the European portion of Russia, were always one and the same racial type—the long-headed, light blue-eyed blonde.

On the basis of this original thesis, which subsequently became the basis of classical racial theory, the following monographs by Russian scientists were published: *Anthropometric Research of the Male Great Russian Population of the Vladimirskaya, Yaroslavskaya, and Kostromskaya Gubernayas*,[36] by N. U. Zograf; *On the Anthropological Composition of the Population of Russia*,[37] by A. A. Ivanovskiy; *Anthropological Data on the Great Rusi of the Staritskiy Uezd of Tverskaya Gubernaya*,[38] by Y. D. Galay; and *Geographical Distribution of Cranial Forms and the Color Index of the Peasant Population, primarily of Greater Russia, in Connection with its Colonization by the Slavs*,[39] by E. M. Chepurkovskiy. All of the said works represent a systematic code of data on the racial anthropometrics of the Russian people, and to this day, they are not outdated.

---

[33] *Izuchenie cherepov i kostey cheloveka kamennogo veka poberezh'ya Ladozhskogo ozera.* Sankt-Peterburg, 1882.

[34] *Antropologicheskie izyskaniya.* Kazan, 1881.

[35] *K Voprosu o drevnem naselenii Ryazanskoy gubernii.* Ryazan, 1893.

[36] *Antropometricheskie issledovaniya muzhskogo Velikorusskogo naseleniya Vladimirskoy, Yaroslavskoy, and Kostromskoy guberniy.* Moscow, 1892.

[37] *Ob antropoligicheskom sostave naselelniya Rossii.* Moscow, 1904.

[38] *Antropologicheskie dannie o Velikorusakh Staritskogo uezda, Tverskoy gubernii.* Moscow, 1905.

[39] *Geograficheskoe raspredelenie formi golov i tsvetnocti krestyanskogo naseleniya preimushchestvenno Velikorossii v svyazi s kolonizatsiey ee slavyanami.* Moscow, 1913.

Left: Aleksey A. Ivanovskiy

At various times, "nomads from culture-philosophy [departments]" have intruded on the space of the historical world vision of the Russian people, attempting to prove [that] the Russians are racially non-homogeneous. Pointing to an imaginary deuterogenesis of the Russians and their mixing with the Finns and Turks since time immemorial, it originates from enemies of the white race. Numerous historical works by such standards of Russian science as Dmitriy Ivanovich Ilovayskiy (1832-1920), Vladimir Ivanovich Lamanskiy (1833-1914), and many others, are dedicated to the rebuke of all these "Western" and "Eastern" biases. To this day, the actual or indicative [evidence] of this aspect is meager in volume; but the extraordinarily clear and persuasive article, *On the Greater Russian Tribe*,[40] by the distinctive historian, Ivan Dmitriy Belyaev, exposes the same faulty algorithm of the dishonest shuffling of Russian history on a racial-biological level.

The famous Russian geographer and cartographer, Aleksandr Fedorovich Rittich, wrote several serious studies on the question of the geographic range of the Slavs. In the book, *The Slavic World*,[41] he cites a long list of populated points, and the natural boundaries on the territory of western and central Europe, which previously had Slavic names, that show that a large portion of the continent owes its history to Slavic, and in particular, to Russian influence, which is etched in a great number of geographical place-names. But if an original racial type from the Russian people is anthropologically established, then there would subsequently be no doubt about the racial origin of the peoples of the entire European continent.

In the racial regard, the Slavic influence, which was conscientiously indicated by classical [anthropologists], including Rudolf Virchow and Johannes Ranke, did not damage anyone, as we were earlier convinced. Thus, the various national schools of anthropology, not giving in to the provocations of political prostitutes, drew the same picture of the origin of the population of Europe, and proved the legitimacy of the anthropological approach to history.

It is necessary to take note – separately - that Russian anthropologists actively participated in the formation of the scientific concept of not only the history of the Russian people, but of all the numerous and diverse tribes, which went into the composition of the Russian Empire, [as well as] the tribes bordering it. As a result of the titanic work of dozens of specialists on ethnographic and archaeological expeditions, [there is] a vast and highly reliable canvas of the racial and ethnic history of the Eurasian continent, right down to detailed descriptions of the evolutionary characteristics of the relic tribes populating these boundless expanses.

The works of ethnic anthropology left by Anatoliy Petrovich Bogdanov, Dmitriy Nikolayevich Anuchin, Nikolay Yur'evich Zograf, Aleksey Nikolayevich Kharuzin, Mikhail Andreyevich Tikhomirov, Vasiliy Nikolayevich Benzengr, Nikolay Dmitriyevich Nikitin, Aleksandr Ivanovich Tarentskiy, Lazar Konstantinovich Popov, Nikolay Mikhaylovich Maliyev, and Aleksandr Ivanovich Wilkins, has not lost its significance to this day, in view of the thoroughness of the treatment of the factual material, including in questions of the origin of the blonde-haired race.

---

[40] *O velikorusskom plemeni.* 1869.
[41] *Slavyanskiy Mir.* Sankt-Peterburg, 1885.

Lazar Konstantinovich Popov   Vasiliy Nikolayevich Benzengr   Nikolay Dmitriyevich Nikitin

Aleksey Nikolayevich Kharuzin   Nikolay Mikhailovich Maliyev   Nikolay Yurevich Zograf   Mikhail Andreyevich Tikhomorov

The ethnographer and explorer, Grigoriy Efimovich Grumm-Grzhimaylo (1860-1936) is unique in his contribution to Russian and world science. Studying the Pamir country, the Zabaikal, Mongolia, and the coastal area and northern China, the Russian researcher came to a straightforward conclusion: the original biological type, which created the cultures on these gigantic expanses, was the long-headed blonde.

Left: Grigoriy Efimovich Grumm-Grzhimaylo

Numerous mummies from the northern provinces of China graphically testify to this. Finally, Confucius himself—one of the pillars of Chinese culture—cannot possibly be considered as pure Mongoloid, for as is known, in all contemporary canonical portrayals of him, he is drawn with a luxuriant beard—while Mongoloids are characterized by slight hair growth on the face. This may minimally testify to a high percentage of European blood in Confucius. Being an authentic encyclopedist, like an absolute majority of Russian scholars of the time, G.E. Grumm-Grzhimaylo analyzed ancient Chinese chronicles, and came to the conclusion that the original racial type that created the culture of northern China was inarguably Europoid. This was wonderfully substantiated in his monograph, which was characteristically titled, *Why Do the Chinese Draw Demons With Red Hair?*[42] In it, he wrote, "One of the pre-Chinese peoples populating the Yellow River Basin, was the red-headed *Di* people." In their records, the Chinese conscientiously admitted that they were not indigenous to these localities, which by custom are linked to traditional Chinese culture. On numerous scenic portrayals of the epoch of earlier dynasties, one can find detailed portraits of demons from hell—"gui"—which are portrayed with red hair, light blue eyes, and other anthropological traits characteristic of the Europoid race. Characteristically, much later, in the Age of Discovery, when the Chinese first saw European explorers, the influence of the racial stereotype was so strong in their consciousness that they immediately called them "Yan-gui-tszi"—"devils from across the sea."

In the 25th Century B.C., the Chinese occupied an insignificant portion of the territory of present-day China, and called themselves the "people of 100 families," or "the black hairs;" all neighboring peoples differed not by place of habitation, but by racial characteristics, of which several written testimonies remain. It is namely from their foreign-race neighbors that the Chinese learned agriculture, irrigation of fields, construction of dikes, the very complicated irrigation arts, and other engineering skills, as well as the construction of towers, palaces, and other architectural structures. Of all the tribes, the Chinese particularly distinguished the numerous, red-head *Di* tribe. Finally, the red-heads were members of one of the early Chinese dynasties, the

---

[42] *Pochemu kitaytsi risuyut demonov ryzhevolosymi?* From *K voprosu o narodakh belokuroy rasy v Sredney Azii.* Sankt-Peterburg, 1899.

Chzou (1122-225 B.C.) In the opinion of Grumm-Grzhimaylo, that indicates Chinese interbreeding with members of the mysterious *Di* tribe, since the hieroglyph "di" actually consists of two hieroglyphs, literally meaning "fiery dog." He points out, "...the *Di* belonged to the white (and probably, the blonde) race, this being confirmed by the circumstance that among them were giants. A similar suggestion does not contain anything impossible. In prehistoric times, the white race had a completely different geographical range, than today. Its remnants in various gradations of miscegenation are now preserved on the islands of Polynesia and Sunda, in Indo-China, in southern China, in Manchuria, in Japan, on the extreme northeast of Siberia, and in North America; finally, in northern China and in present times, the long-headed type has been preserved. Traces of the blood of the white race are evident among several parts of the population of Bhutan, Nepal, and Kashmir; this incidentally explains their long-headedness, their straight-set eyes, and narrow, straight noses."

The Aryan origin of the *Di* tribe is also indicated by the fact that their leaders and rulers were called *As*.

In the 7th Century, B.C., the *Di* tribe split into two branches—the "white" and the "red." The "whites" received the name *Dinliny*. And a number of modern peoples, who now populate the gigantic expanses of Middle Asia and the Far East, arose from mixing with the "reds". To this day, many are encountered that have completely European facial traits.

Reconstructing the racial-psychological appearance of the ancient *Dinliny*, Grumm-Grzhimaylo came to the following conclusions, which are highly characteristic in the plan of our presentation. The male *Dinliny* stood out by their tall height, athletic build, endurance, and their stern and war-like customs; their women, on the other hand, were refined, gracious, and maintained their chastity. Their soldiers were fierce and they never parted from their weapons, not hesitating to use them for any reason. But they fought with intelligence, not numbers. They forged armor, helmets, and daggers, and swords they used were of extremely high quality. Their craftsmen built wood cabins, knew how to craft expensive clothing, and how to create delicate jewelry decorations from gold, silver, and semi-precious stones. Their favorite pastimes were boisterous meals with wine and dances. The men always went clean-shaven, while the women plaited their luxurious, blonde curls with ears of grain, beads, and sea shells. All forms of *Dinliny* authority were elected. The dog was considered a sacred animal. A penalty, equal to the penalty of manslaughter, was prescribed for killing one. A spirit of freedom and independence permeated their whole tenor of life, and unrestraint of temper went into many Chinese sayings. Common talk of "The Heavenly Empire" sings their praises, not otherwise, as "shaggy buffaloes." No one could compel a girl to marry against her will. Besides that only among the *Dinliny*—of all the peoples of this gigantic Asian territory—was monogamy the primary and basic form of marriage; lewdness and idolatry were absent; on the contrary, the ancestor cult and chivalry were very widespread.

This begs the valid question: what race was this? Grumm-Grzhimaylo—a researcher of the given region, himself came to the straightforward conclusion, that by all outward descriptive traits, their combined psychological portrait, and also by details of their everyday living arrangements, and specific modes of conduct, this could only be a Europoid race, with an evident, dominant substrata of northern origin in its biomass. Answering the question about the origin of the indigenous population of Central and East Asia, he reconstructed a dynamic of historical processes in this region, which advances his theory for the decrease in the numerical composition of the blonde and fair race.

The long-headed blondes lost the fight in the struggle for living resources in Asia to the short-headed brunettes, not because of insufficient physical and psychological organization, but on the contrary, due to excessive concentration in a region not conducive to their [drive for] discovery and exploration. They were moved by a feeling of individual freedom, and were unable to submit; endowed by nature with initiative and a rich, creative imagination, they unavoidably had to lose to the droves of shiftless beings, capable only of Asiatic slavery and blind obedience. Devoid of lustfulness, and endowed with chivalrous, noble sentiments towards women, the fair blondes emphasized a single form of marriage for themselves—monogamy, which predetermined that the numerical weight [would shift in favor of] the Mongoloids and mestizos. The specifics of the [Mongoloid] sexual-psychological constitution of polygamy, guest marriages, and other similar forms of sexual activity, were completely incomprehensible to the understanding of the white man.

Grumm-Grzhimaylo confidently developed these brave, revolutionary views of history in his next thematic work, *The White Race in Middle Asia*.[43] First of all, he quotes the words of that French authority in the area of anthropology, Professor Paul Topinard, pointing out: "One may consider the existence of a race with green eyes and red hair, in central and northern Asia in bygone days, to be proven."

"The excavations of graves within the limits of the Altai-Sayan Highlands indicate to us, that the mountainous region was a continuous encampment of the long-headed ones. Here, it is necessary to think that they should have, for the most part, been moving around, if not natives of the Trans-Baikal region; that the subsequent long-skulled population of this region belonged to a similarly long-headed, Altaic, taller race, most probably even European; that is proven by the form of their skulls, and by their gypsum masks, on which many notable facial beauty traits—that are completely European—stand out."

Excavations of Kurgan graves in the Selenga River Valley revealed the existence of two racial antipods in this region in prehistoric times: a short-headed type with a cephalic index of 93.6, and a long-headed type with a cephalic index of 68.4.

Besides that classical Chinese treatises speak about tribes populating Middle Asia, outside the Great Wall of China: the *Usuni*, the *Khagyasi*, the *Dinliny*, and the *Boma*, with emphasis that they had light-blue eyes and red hair. It is noteworthy, that among the Chinese, all known non-black-haired tribes, beginning with blonde and ending with dark light-brown, went by name of "red-headed." The treatise, *Bei-Shi*, directly confirms that the southern edge of the Gobi Desert was the land of the *Dinliny*. To designate the many tribes inhabiting this vast expanse, the Chinese used hieroglyphs, which in exact translation mean "white" and "freckled."

It was only toward the end of the 4th Century A.D., that the Altai-Sayan Highlands were flooded with Turks, who mixed with the *Dinliny* people, forming the Uigur people, who were referred to as the "golden-headed" in Chinese records. Touching on the Kirghiz, who lived nearby, information is preserved relating to their canons of racial beauty, which still existed among them at the beginning of the 9th Century; [they valued] tallness, white-colored skin, a rosy face, red hair color, and light-blue eyes. And that basic type prevailed so much that black hair was considered a bad trait, and people with brown eyes were simply considered descendents of the Chinese. By the 17th Century, when Russians undertook to settle Siberia, they collided with a completely different people, represented basically by black-haired, swarthy types. Chinese chroniclers also noticed that by the 18th Century, the majority of subjects encountered among the Manchurians had "light-blue eyes and aquiline noses."

Summarizing all this rich information, Grumm-Grzhimaylo came to the conclusion, that "the *Dinliny* comprised an isolated branch of the blond and fair race."

Aleksandr Ivanovich Wilkins, mentioned above, published a monograph about the results of his research expeditions, titled, *Anthropological Themes in Middle Asia*.[44] In it, he states, "I examined a great number of mountain Kirghizes, who populate the interior part of the Tyan-Shanya. During the journey along the Kashgar border, I could not help but notice several specimens that turned attention to themselves, because of their unusual (among Middle-Asian Kirghizes, that is) traits. These were light-brown haired individuals, even almost blonde. Besides that they had grayish-blue eyes. It even appeared to me that the faces of these particular [individuals] were more regular, particularly the build of the eye orbits, than those of their countrymen, with typical black hair and brown iridescence. That such characteristics may be inherited from the ancient inhabitants of the Issyk-Kulsk coast, should not be subject to doubt; we recall that in this place, still before our own era, there lived a light blue-eyed and blonde tribe—the *Ussuny*. The *Ussuny* were dolichocephalic. Unintentionally, tales about blonde, blue-eyed peoples that we heard among the Pamir tribes came back to me; and comparing the observations cited above, it occurred to me, that in essence there are no grounds for rejecting the possibility of reconstructing the branches of the now-extinct mountain Aryans, with the help of a thorough analysis of the common traits that set them apart—light-brown hair, light-blue eyes, and elongated skulls."

[43] *Belokuraya Rasa v Sredney Azii*. Sankt-Peterburg, 1909.
[44] *Antropologicheskie temy v Sredney Azii*. Moscow, 1884.

During extensive field work in Central Asia (from 1895 to 1899) Russian scholar S.D. Maslovskiy also distinguished a northern, light-haired type. Another Russian, N.A. Aristov, wrote about the mountain dwellers of the Pamir region in 1900: "In the eastern part of Middle Asia, there existed a particular race, as recognized by anthropologists, that had light-colored skin and hair, with green or light-blue eyes (*Dinliny, Ayny*)."

A recognized standard of Russian anthropology, D.N. Anuchin, wrote in the article, *A Preliminary Account of the Journey to the Ostyaks of the Yenisey in 1905*,[45] that he "saw several Ostyaks, who by outer appearance, could in many ways be considered among the Aryans. Their children were often blonde, but with years their hair darkens; and curly hair is encountered. The cut of their eyes, not rarely gray or blue, is open and horizontal. Their noses are straight and narrow, but one comes across subjects with aquiline noses, and even pug noses."

Finally, still one more recognized authority of the Russian anthropological school, Aleksey Ivanovich Kharuzin, studying on expeditions to Persian territory, definitively confirmed the basic assertion of racial theory, which later became classical, that **always and everywhere in world history, the starting racial type—the creator of culture—was the blonde and fair race of men. Therefore, it is namely the most worthy, biologically.**

Ludwik Krzywicki, a famous Polish anthropologist and Russian subject, emphasized in his monograph, *Anthropology*:[46] "The blondes constantly appear in history in the capacity of a restless element. The colonization of North America, the Crusades, the great migrations of peoples, the invasion of Hispania and North Africa, right down to Egypt in the 15th Century B.C, and perhaps even the 25th Century B.C.—all is the affair of the blonde. Even in the invasions of the Irano-Indus and the Hellenes he played a role, in the opinion of scholars, in the capacity of a main force, particularly in the aristocratic social strata. The color of the hair of a majority of Greek heroes, the family names fo the Roman patriarchs, the portraits of members of the old German and French nobility, all point to the blonde as a type in which masculine energy and heroism was embodied. From there the conclusion is made, that the blondes, as a motive element, brought their language to the comparatively more passive, short-headed brunettes."

Russian scholar A.N. Krasnov's article, *Materials for the Anthropology of the Russian People*,[47] has a completely inestimable role in the context of the theme discussed by us.

The uniqueness of the given study lies in the statistical anthropometric measurements carried out by the author at selected points, located in the principle territory of central Russia. The author wrote in this regard, "Summing up the measurements from ten different guberniyas (states), and twenty-one uezdi, we cannot help but be amazed by the homogeneity of the composition, which they characterize. Everywhere it hits the eyes: the preponderance of the blonde, light-eyed type. Blondes comprise 20% to 50% of all those measured; therefore, assuming all possible occurrences in the composition of each separate party, it is impossible not to recognize that in the ten designated guberniyas, the basic element of the Greater Russian population should be a blonde-haired, light color-eyed race, which despite mixing with black-haired [types], yields numerically few hybrids of a transitional color in the eyes and hair, preserving in its pure facial appearance, so great a number of absolute blondes.

Its influence is strong in cross-breeds, such that the number of grey eyes is still greater, and grey eyes predominate among those hybrids, whose hair takes a darker coloration, under the influence of the impure blood of a more pigmented race. In addition to that blonde members are more homogeneous. In them we find more usual, so to say, more typical physiognomies for Great Russians, which are constantly repeated on the vast extent covered by the ten designated guberniyas, so that switching pictures, one would be at a loss to say from which guberniya it was taken. It is nothing impossible to say that these Russian dolichocephalics are only a variant of the Scandinavian race.

[45] *Predvaritel'niy otchet o poezdke k eniseyskim ostyakam v 1905 godu.*
[46] *Antropologiya.* 1900.
[47] *Materialy dlya antropologii russkogo naroda.* Russkiy antropologicheskiy zhurnal, 1902, N3.

Left: Dmitriy Nikolayevich Anuchin

Somehow or other, from all that has been said, the result clearly ensues, that the dark-haired race cannot be called Russian. This extraneous element is derived in large measure from the Finns and Turks, and perhaps from the southern and western nationalities, who came into contact with the main, blonde elements of the Russian people."

Of the more than 300 tomes of the publication, *News from the Imperial Society of Lovers of the Natural Sciences, Anthropology, and Ethnography*,[48] which was put out until 1917, one may glean information about the revolutionary innovations of the creators of Russian anthropological science, which today is stubbornly kept quiet. In 1931, the world-famous aircraft builder, Igor Ivanovich Sikorskiy, by now living as an emigrant, published a consolidated tome of the works of his father, the philosopher and psychologist, Ivan Alekseyevich Sikorskiy (1842-1919), who was engaged in the comprehensive problems of race. In this comprehensive tome there is a notable work under the title, *Anthropology*, in which the Russian scholar lays out the quintessence of his view of world history: "The Aryans belong to the most gifted branch of humanity, standing out by strength and depth of talents, breadth and versatility of ability, and an inborn idealism and ideal direction of life. In this sense, not one other branch of the human species is comparable to the Aryans. The talentedness of the Aryans strengthened them for their first role in the possession of the world. With the keenness of their intellect, the Aryans deeply fathomed the essence of things; gifted in the sciences and arts, correctly foreseeing the distant future, and preparing for the long-term with corresponding measures and actions. Their characteristic idealism gives understanding and strength to progressive organizations for the future advancement of humanity. The Aryans created exemplary literature, museums, book depositories, art galleries, schools, all manner of governing institutions, academics, and societies for the improvement of life in all respects. In accordance with these progressive programs, they implemented the right to trial, and good legislation. The Aryans create and constantly refine the whole material condition of everyday human life, in conformity with the demands of science, art, and life experience. Their entire existence, in all its steps and through the arts, makes art of life, comprehensive guidance of science, hygiene and technology, with constant care about the distant future. Almost all Aryan peoples conduct life according to a national type; such a life has chances to hold out in the future, in the course of many centuries. Inasmuch as the Aryan peoples have their place of residence in Europe, then Europe and all things European are a synonym of anything Aryan, or the highest [of anything].

Another important Russian philosopher-evolutionist, V.A. Moshkov, in his fundamental monograph, *A New Theory on the Origin of Man and his Degeneration*,[49] researched the problem of the emergence of the white race, on the basis of a synthesized generalization of the data from zoology, geology, archaeology, anthropology, ethnography, history, and statistics; as a result of this, an entire chapter in his essay was given the highly characteristic title, *Traces of the White Race are everywhere in the World.*[50] With the abundance of facts laid out by him in support of his theory, he ingeniously concluded, that not only on Easter Island, but on a multitude of different

---

[48] *Izvestiya Imperatorskogo obshchestva lyubiteley estestvosnaniya, antropologiya i etnografii.*

[49] *Novaya teoriya proiskhozhdeniya cheloveka i ego vyrozhdeniya.* 1907.

[50] *Sledy beloy racy est' vo vsem mire.*

islands in the Pacific Ocean, and also in regions of Equatorial Africa, the [stone] idols have the same, consistent European facial traits. This can only speak to one thing: the cultural-creative abilities of the white race were recognized in all parts of the world.

3. The Creation of Racial Theory

And so, relying on the historiography of the question, we gradually become convinced of the approximately similar conclusions that linguists, archaeologists, and anthropologists of the second half of the 19th Century came to on the racial origins and the racial dynamic of the history of Mankind. But there was still one group of scholars, which the reading public recognized as a cabinet of odd fellows, with an exotic theory that was read in a variety of light entertainment literature. They fulfilled the role of a prime mover in the development of a new social-political tendency; namely, they welded the research of specialists from different fields of knowledge in a single academic field, the name of which is **racial theory**.

By common recognition, the Frenchman, Count Joseph Arthur de Gobineau—a talented philosopher, poet, and diplomat—is considered the founding father of racial theory. Leaving behind himself a multitude of reports in very different genres, he entered history with the help of his fundamental monograph, *Experiments in the Inequality of the Human Races*,[51] in which he was the first in European science to undertake the bold attempt of interpreting world history on

the basis of the struggle of races. In the foreword of the book, he wrote that his "particular approach" lay in showing the logic of historical processes as a sort of "historical chemistry." He distilled the whole, gigantic living organism of the history of humankind, in essence, into a simple flask, in which the process of chemical reaction united biological races. Bubbling, mixing, and separating, they carried on an implacable struggle in the body of peoples and nations, creating history [in the process].

Left: Count Joseph Arthur de Gobineau

Flowing into the blood of a people, one race, as if a chemical reactant, adds creative strengths to it, vital activity; and another leads to premature degradation and social laziness. One racial admixture adds idealism and nobility to the citizens of a given state, and another oppresses with vices and an addiction to vulgar, carnal excesses. One rules over peoples with a feeling of civic duty, and unites them into a community, transforming [it] into gigantic empires and entire civilizations, which give birth to thinkers and leaders. Another, on the other hand, eats away all spiritual bonds from within, like rust, pushing peoples to embrace strife, bloody fratricidal wars. It brings forth generations of parasites and infantile dreamers into the world. One blood creates world religions, another— endless troubled epochs. Thus is created, in the opinion of de Gobineau, the history of the human genus.

Such a view, so large-scale and well-argued, was unfamiliar, and therefore, not accepted by contemporaries. However, the time of open borders in the area of the natural sciences in the 19th and 20th Centuries, all the same compelled the enlightened public to change its attitude toward de Gobineau's fundamental work. His well-earned recognition had come.

The essence of his social-racial concept is best illustrated by the following maxim: "The purer the race, the less vulnerable its social base, inasmuch as the racial logic does not decrease. The organized character of any civilization is defined by the very obvious trait of its dominant race. A civilization changes, or transforms, by measure of how this race itself endures change. In the framework of a civilization, in the course of a more or less continuous period, an impulse, which at some point the vanishing race gave to the civilization, continues to act. Consequently, the system forming a society is a demonstrated fact, which most clearly testifies about the concrete

---

[51] *Opyt o neravenstve chelovecheskikh rac.* (1853-1855.)

abilities and the level of the people—this is a better mirror, in which to reflect the individuality of a people."

Touching on the hierarchical worthiness of races, the French aristocrat was categorical, for he emphasized: "History shows that any civilization takes its start from the white race, and none can last long without its participation; that a society may be great and flower, only in that measure that it preserves its creative, noble group; and that group itself belongs to the most developed branch of our species. In order to definitively make these truths clear, it is sufficient to enumerate the civilizations which existed on the earth; I assure you, the list will not be very long. The white race is skilled in the mechanical arts, strives to turn the affairs of war into genuine art, in contrast to the wild scuffles of savages; in domestication of numerous kinds of animals, it stands on such a high level, in comparison with the rest of the human families, that it should be understood once and for all: any comparison is senseless, for one single reason only: even in its childhood years, there is no trace of barbarity in this race."

Proceeding from these principles and correlating them with the rich historical material gleaned from numerous legends and documented testimonies, de Gobineau quite obviously stated his opinion about the physical characteristics of that racial substrata, which created culture on the gigantic expanses of Eurasia: "The Aryans had white-pink skin: such were the very ancient Greeks and Persians, and so the ancient Hindus were likewise recorded. Hair, as a rule, was light—we recall that the Hellenes gave preference namely to this color of hair, and considered [those with this hair color] among the favorites of the Gods [of Olympus]. Among the Hindus, the ideal of beauty to this day is associated with light skin and light hair, to which the description of selected children in Buddhist legends so often testifies. The legends describe, for example, divine creation in an infant with gold-colored skin. He has long arms, a wide forehead, closely placed brows, and a protruding nose."

The French racial theorist also positively points out, with regard to the gigantic territories of the Far East, that "not at all long ago, numerous peoples with blonde hair or red hair, and light blue eyes, lived along the western borders of China." Characteristically, the Chinese called these Aryan tribes "people with long, horse-like faces," for the dolichocephalic shape of their skulls, which is namely a characteristic of the blonde, northern type. Finally, in his opinion, members of the white race actively took part in creating the great civilization of Egypt; bas-reliefs record pharaohs and pagan priests with lighter skin, than the general mass of the population—and with light-blue eyes.

Gobineau based his findings on studies of ancient, written sources, and the data of comparative linguistics. Academic anthropology then was still in a state of formation; however, that did not hinder him from making correct conclusions, and he did not fall into the error of "pure" linguists. In this regard, he wrote: "The essence of language is tightly connected with the form of thinking of a people, and from the very beginning it retains in itself, assumes in its initial stages, all the necessary means of conveying the most varied characteristic traits of a form of thinking. As a result of this, what exists is a sufficiently tight connection between the language and its carriers—a race. The race is preserved far longer than the corresponding peoples preserve their statehood. All this is used to make the irrefutable conclusion that not one single people can have a language, which stands at a level higher than itself. The hierarchy of languages is located in strict accordance with the hierarchy of race."

Despite the fact that modern, liberal-minded culturologists include de Gobineau in the halls of the founders of racism, and [call] him the founder of the "hegemonistic pretensions of German fascism," we recommend attentive reading of his fundamental work. In regard to the ancient German-Slav dispute, he quite clearly writes: "Touching on Europe, the pragmatic element, which was introduced to the most active German tribes, constantly became stronger in the north, on account of an influx of Celts and Slavs." Commentary, as the saying goes, is unnecessary.

Count Joseph Arthur de Gobineau expressed the main body of his political ideology with the following words: "The superiority of the Aryans is tied with an exclusive development of moral qualities—laid down in principles, from which these qualities flow."

If the name of Count Joseph Arthur de Gobineau—the French philosopher, historian, poet, and diplomat, is known to a wide circle of people interested in the history of various social-political doctrines, then the name of another official founder of classical racial theory is known only to a narrow group of specialists. Gustav Friedrich Klemm (1802-1867), a modest German librarian, historian, and collector of antiquities, went down in history as the author of the multi-tome work, *The General Cultural History of Mankind* (1842-1852).

In his work, he set before himself the task of clarifying the laws of interaction between Man and Nature. Analogously, just as people are divided into sexes, all of humanity, in his opinion, is divided into two parts: active and passive. Under the active part, he had in mind a rather concrete racial substrata—people of the northern type: the long-headed, blue-eyed blondes. The first tomes of the given fundamental work represent a detailed ethnographic directory on a racial basis, which was absolutely unfamiliar to the science of the time. In his book, the author wrote: "The blonde race, outstanding for its light skin and a more developed forward part of the skull, separated into two or three elements, and it turns out, that namely the northern peoples are members of the most active workers in the sphere of politics and culture. The original home of this race was in Europe, and they had already spread out to the south and east in ancient times; in short, the blonde and fair race constitutes the first and oldest cultured race."

Left: Gustav Friedrich Klemm

In the opinion of Klemm, the spiritual aspect of this half of humanity stands out by a developed will; a striving for authority and power; independence and freedom; it is characterized by unceasing activity, an incessant striving toward development and progress; and also an inclination toward investigation and a keenness for anything new. Its pride and confidence in itself stands out. These psychological traits manifest themselves quite distinctly in the history of those nations, which form the given, active part of humanity, namely: the Hellenes, the Romans, the Germans, the Slavs, and the Persians. "These peoples constantly migrate, they destroy the old, well-organized states, and build new ones; they stand out as skilled sea-farers; they have a particularly developed freedom of direction, the basic element of which is constant progress; among them, knowledge, research, and thought take the place of blind faith; science and art blossom in the midst of these people, and in this area, these nations have done more than all the rest. Their home is situated in the middle latitudes, from where they spread out to different areas and conquer them." Such was Gustav Klemm's basic outline of the concept of race.

Left: Stepan Vasiliyevich Yeshevskiy

The name of a third founder of classical racial theory has generally been consigned to oblivion. So that historical justice may triumph, the time has come to correct this regrettable oversight, and restore rights to the Russian genius, Stepan Vasil'yevich Yeshevskiy (1829-1867). His unique contribution to the foundation of racial theory lies in the fact that in 1862, being a professor in the history department of Moscow State University, he began to teach the first general history course in the world, on a racial basis. Part of the lectures were formulated by him in the form of a separate work, *On the Significance of Race in History*.[52] Only sudden death prevented the author from making a reality of all his brave and innovative thoughts. But even for this one work, one may assert that this is the first Russian canonical essay on racial theory. Moreover, in contrast to this, the works of the French and German researchers recall artistic works, unburdened by reliable scientific information. Those researchers were rather geniuses of intuition and insight. [But] Yeshevskiy was the first to create a full-fledged scientific labor that met strict academic standards. All of Yeshevskiy's

---

[52] *O znachenii ras v istorii.*

57

ideological passages are founded on archaological and anthropological data. His racial doctrine of history may be reduced to the following basic theses:

1) The basic moving force of history is the struggle of peoples, who have hereditary racial differences.

2) As a result of this struggle, natural inequalities of peoples arise, which are recorded in their cultural and social-political history.

3) Racial types are stable, and are not subject to the influence of the outer environment.

4) The basic human races developed at different times, and in various places, independently of one another.

5) The members of the northern blonde and fair race have higher [intrinsic] racial-biological worthiness, from the point of view of their contribution to world civilization.

6) Racially mixed cross-breeds, on the contrary, by the sum total of their biological and consequently, socio-cultural characteristics, possess less [intrinsic] worthiness.

7) In the name of humanitarian interests, justice, and social stability, the members of the "higher" races should control the members of the "lower" races.

There was not a similar rigid categoricalness in a summary with scientific argumentation, in either the works of de Gobineau or Klemm. The high bar, set by Stepan Vasilyevich Yeshevskiy, was only achieved again by the better minds of Europe, toward the start of the 20th Century. It was namely on the boundary of the 19th and 20th Centuries, that racial theory formed as an independent, nature-philosophical and socio-cultural field.

In 1895, the talented philosopher and historian, Ludwig Schemann (1852-1938), created the Society of Gobineau, with the goal of mass propagation of the ideas of the forgotten French academician in Germany. In 1900 there was a second rediscovery of Gregory Mendel's Laws of Inheritance, in connection with which the idea of "racial purity" was demonstrated on a genetic level. In the same year, the biological basis of the existence of different blood types was first advanced. Around this time, the titanic work of the German Anthropological Society was finished, under the direction of Rudolf Virchow (1821-1902); it compared the skulls of modern Europeans with excavated samples collected by paleontologists.

Also in 1900, a great cumulative work on Russian anthropology, *The Human Races*, produced in France by Joseph Egorovich Deniker (1852-1918), appeared in the Russian and French languages; in it was the first application in scientific practice, of new statistical principles of racial classification.

The concept of race lies at the basis of racial theory. It was introduced to European science in 1684, by the French ethnographer and explorer, Francois Bernier (1625-1688). For the extent of two centuries, there was no clear and unambiguous definition of the term, for scholars mixed biological parameters with linguistics and ethnographics, as a result of which a jumble constantly arose, and peoples, having identical outward looks and psychological characteristics, were included in various races on the basis of etymological data, or the conclusions of comparative linguistics. Not rarely, peoples having nothing in common between themselves in the context of physical structure, were tied to another race solely on the basis of shared linguistic commonalities. These contradictions and inaccuracies in systematization dearly cost the adepts of racial theory, for they compromised the entire science as a whole. As a result of identifying "peoples" with "races," completely absurd combinations of words arose, such as "Teutonic Race," "Germanic Race," and "Slavic Race."

It was namely the Russian scholar Deniker who corrected the situation, when he published his monograph, *The Human Races*,[53] which to this day is considered a model of the systematization

---

[53] *Chelovecheskie rasy.*

of natural scientific information, and the first in which the basic anthropological principles of evaluating the differences between peoples was formulated. **Racial typology** arose in anthropology, thanks to which the classification of human races acquired its modern, distinct look. Alternative readings disappeared, and use of the particularly anthropological terminology acquired a more strict, scientific character.

Under race, Deniker understood "somatic unity," which put an end to idealism in anthropology. The whole book in essence was devoted to the separation of ethnography and anthropology, which the author defined as phenomena of different orders. He defined ethnography as sociological, and anthropology as biological. He wrote: "Several years ago, I proposed the classification of races, based solely on physical traits (color of skin, quality of hair, height, head shape, nose, and so on)."

I.E. Deniker was the first to stand by the position of strict and consistent **biological determinism** in racial philosophy. In his opinion, the surrounding environment was powerless in the face of racial characteristics. He emphasized: "Racial traits are preserved with remarkable stubbornness, regardless of mixing of races and changes in the conditions of civilization, the loss of a former language, and so on. Only the ratio changes, in which this or that race enters the composition of a given ethnic group."

Left: Joseph Egorovich Deniker

The work of the Russian scientist was recognized by the whole scientific world community. Thanks to that work, the concept of the **racial type,** constant and unchanging, once and always a given, and not subject to the influence of the environment, firmly took root in anthropological literature. The historical combination of racial types that has developed, is itself the product of social development—an ethnos, and the type which dominates it, afterwards forms the physical and spiritual character of every national community. This rule was firmly developed, and became the basis for racial theory.

Turning to the basic theme of our narrative, we particularly distinguish I.E. Deniker's contribution to the strengthening of the position of anthropologists, in their struggle with ethnographers and culturologists, in the dispute about the ancestral homeland of the white race. He closed the book on the dispute about the Aryans, which at that time had reached its culmination; he introduced a new term which had nothing in common with the romantic conceptions of linguists: "The long-headed, very tall, light-haired race may be called **Nordic**, inasmuch as its members are grouped primarily in the north of Europe. Its main characteristics are: very tall height, 1.73 meters on the average; blonde, wavy hair; light eyes, usually blue; and oblong head (cranial index, 76-79); rosy-whitish skin; the face—oblong; and the nose prominent and straight." Terminological incorrectness in racial theory ceased, and the term "Aryan" gracefully departed the sphere of culturology, sociology, and religious studies: "There can be no talk of an Aryan race; rather it is permissible to speak only of a family of Aryan languages, and probably, of a primeval Aryan civilization."

Simultaneously with this, the evolutionary studies of Charles Darwin began to have the most influence on public life. Thus, one of his followers, the outstanding German philosopher and

Left: Ludwig Scheman

naturalist, Ernst Haecker (1834-1919) was the first to carry the basic biological laws of the struggle for existence into the area of sociology, and by that occasioned the beginning of a new philosophical-political doctrine, which received the name, **Social Darwinism**. In his fundamental works (*The Natural History of the Universe*, (1868); *Anthropogenics*, (1874); and *Naturalist in the Tropics* (1876)), which earned him mass popularity, he in essence substantiated the idea advanced by racial theoreticians, proving the cultural inequality of the human races in the process of their evolutionary development. The white race received priority in his nature-philosophical concepts.

The prominent Russian biologist, Ilya Ilyich Mechnikov, supported Ernst Haeckel's concept. His fundamental work, *The Struggle for Existence in the Wider Sense*,[54] was a lengthy masterpiece of Russian Social Darwinism. In it he emphasized: "The natural inequality between individuals, tribes, and races is a common principle in the organized world." On the cultural inter-relations between "higher" and "lower" races, the Russian scholar answered in the following way: "The artificial safeguarding of the lower savages cannot accomplish anything, except at the expense of present and future Europeans."

4. Anthropo-sociology

Thus, the synthesis of data from the natural sciences, with the philosophical concepts of racial theoreticians, gradually led to the creation of still another scientific field at the end of the 19th Century, which received the name of **anthropo-sociology**.

A constellation of talented scholars—Georges Vacher de Lapouge (1854-1936); Otto George Ammon (1842-1916); Ludwig Woltmann (1871-1907); and Ludwig Wilser (1850-1923), formed it with their essays, and gave it a fashionable luster (for the times) of a philosophical idea.

The basic idea of the French founder of anthroposociology—Georges Vacher de Lapouge—is best reflected in two of his basic essays: *Social Selection* (1896) and *The Aryan and his Social Role* (1899).

In them he wrote that in the basic racial differentiation of the European continent, there lay three subdivisions: *Homo europaeus*—the long-headed (dolichocephalic), blonde and fair tall race, outstanding in psychological terms for vanity, energy, intellect, and idealism. Next, a dark, a short-headed (brachicephalic) and short-statured race, encountered throughout Europe, that

Lapouge called *Homo alpinus*; he ascribed conservative, cautious, and low intellectual traits to it. The third main race of Europe—the Mediterranean—is dolichocephalic, but darker and morally standing lower still than the dark, brachicephalic [type].

Left: Georges Vacher de Lapouge

"The long-headed blondes fill the role of the brain and nerves in the societal organism, while the short-headed [types] and their half-breeds play the role of the bones and muscle. A thousand brachicephalic types are not worth a thousand dolichocephalics."—and that is the heart of his anthroposociological concept. Also new was that Lapouge lowered Man from the pedestal of Beloved Creation of God, and subjected him to biological laws, common to all animals. Therefore, his paper about *The Aryan* begins with the words: "This book is a monograph about *Homo Europeaus*, that is, about that form to which they gave various names for

---

[54] *Bor'ba za sushchestvovanie v obshirnom smysle.* 1878.

the light-haired dolichocephalic race, [such as] the German or the Aryan. Everywhere I will mean its scientific term, given by Linneus. This is the most reliable means, in order to constantly remind the reader that in essence…Man is not a completely special animal, but goes into the common system of Nature, and is subject to common, biological laws. Mercy in human affairs exists only in the imagination of mystics, while Darwinist political science or anthroposociology aspires precisely to replace metaphysics and mystic notions of philosophical sociology with concrete concepts."

Precisely on the basis of these universal biological laws, Lapouge re-interpreted world history. The long-headed blondes, in his opinion, were not good at low, systematic work, and therefore, they everywhere and always try to form a ruling caste, leaving the primitive work to the short-headed brunettes. Being enterprising, they made up their minds on everything, and appeared for battle out of love for it, not for a calculated profit. Their range of vision and interests was wide, their desires and intentions brave, and their deeds matched [their intentions]. Among them, progress was an inborn passion. They demanded respect for individual freedom, and they sooner tired of lifting themselves, than in putting others down. Generally speaking, in a mixed community, they were the unfailing active beginning.

Lapouge ascribed the education of the upper classes in Egypt, Chaldea, Assyria, Persia, and India, to the blonde, dolichocephalic race, just as he ascribed to them great influence on all Greco-Roman civilization.

"Indeed, on the monuments of Egypt, Chaldea, and Assyria, all high-ranking persons are represented as blonde, blue-eyed, and of tall stature. Greeks in Egyptian portrayals are also represented as tall, blonde, and long-headed. The Greek heroic type, without a doubt, was also such. The Gods and heroes of Homer are always blondes of tall stature, and with light eyes. In the first song of the Iliad, Minerva seizes Achilles—the hero champion—for his blonde hair. And this is expressed another time in the 12th Song, when Achilles brings his hair as a sacrifice to the mortal remains of Patroklos. King Menelaus is also blonde. In *The Odyssey*, Meleager, Mentor, and Rhadamanthus are blondes. Virgil portrays Dido as a blonde, although she is a Phoenician, and would therefore, probably have had black hair; Minerva, Apollo, Mercury, Komert, Camille, and Lavinia are also recorded by him as being blonde. All of Ovid's main heroes are blonde. In the Roman aristocracy, the blonde type was predominant, as the names *Flavius* and *Fulvius* attest: they come from the word *flavus*—Latin for "yellow." Dante and Petrarch praise blonde heroines: Beatrix, Matilda, Laura. In general, it is enough to look over a gallery of paintings from the Renaissance, in order to be convinced of how blondes predominated then, particularly among women. Protestantism—an evolution of Catholocism—spread primarily among the blonde peoples of Europe, but not among the black-haired [peoples]."

Lapouge goes further to state, that the civility of peoples is in direct proportion to the number of blonde dolichocephalics that make up the ruling classes. In general, "in the evolution of Mankind, the dark-haired brachicephalics and the product of their cross-breeding play the role of ordinary soldiers in the main headquarters, which consists of blonde dolichocephalics."

Before the conquest of Gaul by the Romans, Lapouge indicates that its estimated population was 5-6 million people, [both] short-headed and blondes. Around one million died in the wars with Caesar, and just as many were sold into slavery. The dead who fell in this struggle were primarily the energetic blondes; therefore, after the defeat of Gaul, it became the most industrious, but most servile province. A spark of revolt broke out only in the north, where the light-haired elements were more numerous. And the state of affairs remained that way for centuries: wealth grew, but not glory. Gradually, however, first as allies, then as conquerors, the long-headed types penetrated into the country in the 5th Century A.D., and in succeeding centuries, the country came back to life with them. Just several hundred thousand of the new [blonde] arrivals were quite enough to get the servile population of several millions in the mood for war. Thus arose, according to Lapouge, the France of modern times.

Afterwards, through the course of several centuries, the light-haired giants scattered to neighboring countries (via the Crusades, expeditions, and feudal wars). In a much later period, these same racial elements took part in the Reformation Movement, and created a gigantic net of Frankish colonies across the Earth. But all those campaigns, the struggle for ideas, the movement of crusaders, and the Inquisition, destroyed the most energetic members of the long-headed types, and when they died, the short-headed brunettes saved their strength and conquered with the help of passive waiting.

The great French Revolution, at the head of which stood primarily blondes, only legitimizes anthropological fact: the fruits of initiative of the long-headed blondes passed into the hands of the short-headed brunettes. The anthropological result of the Napoleonic Wars was the decrease in the average height of the male population of France by about 10cm, and its drastic darkening. The political insignificance of France at the end of the 19th Century, according to Lapouge, was also the result of the dominance of the short-headed brunettes [in that country].

The scale of biological worthiness of the races was created in the following manner: the cultural achievements of each of them is hereafter a universal measuring stick, easily enabling a prognosis of the social activity of this or that communal organism, according to the percentage ratio of initial biological components.

Taking into account the new treatise of history, Georges Vacher de Lapouge formulated his famous **eleven anthropological laws:**

1. **The Law of Distribution of Wealth:** in countries with a mixed population of *Homo Europaeus* and *Homo Alpinus*, wealth increases in inverse proportion to the cranial index number.
2. **The Law of Height:** in areas where *Homo Europaeus* and *Homo Alpinus* exist together, the first type is localized in lower places.
3. **The Law of Distribution of Cities:** important cities are almost exclusively localized in dolichocephalic areas, or in the least brachicephalic parts of brachicephalic regions.
4. **The Law of the Cranial Index of a City:** the cranial index of a city population is lower than in an agricultural population.
5. **The Law of Emmigration:** among populations that are beginning to disassociate (to come apart, to separate), the least brachicephalic element emigrates, before displacement.
6. **Law of Marriage in New Places:** the less brachicephalic element is particularly inclined to marriages outside their country.
7. **The Law of Concentration of Dolichocephalics:** the migrating element is attracted to centers of dolichocephalics, which are more enriched by dolichocephalics. This law may be formulated this way: in areas, where the brachycephalic type exists, it strives to localize in forests, while the dolichocephalic type strives to localize in cities.
8. **The Law of Urban Elimination (Exclusion):** city life produces choice in favor of the dolichocephalic elements and destroys, or eliminates (excludes) the most brachycephalic elements.
9. **The Law of Stratification:** the cranial index in one or another locality extends, decreasing from the lower classes to the upper classes. The average height and number of tall people increases from the lower classes to the upper.
10. **The Law of the Intelligent:** the skulls of the intelligent are more developed in all directions, particularly in width.
11. **The Law of Index Increase:** beginning with prehistoric times, the cranial index everywhere continually increases.

The list of Lapouge's laws concludes with a number of observations, summarized in the following way: "However surprising these laws, we are only in the beginning of discovery. There still remains much more, about which we do not suspect. The necessity for a complete anthropological inventory of each city is clear, and if it will be conducted in a proper way, it will give political results, the importance of which may be scarcely foreseen."

His German colleague, Otto Ammon, came to practically the same conclusions in his works; he bolstered his theoretical calculations with abundant data from measurements of new [army] recruits throughout Germany. Already in his first significant work, *Darwinism against Social Democracy* (1891), he divided European society into four anthropological classes, in an analogy with the ancient caste system of India.

"In the first class are the innovators, inventors, and pioneers, opening new paths to humanity. They have a level of intellect higher than average; this is a people with character; they are untiring and daring creators; on well-beaten paths they do not feel like themselves...Mankind is indebted to them for all progress.

The second class—intelligent and skilled people, who do not possess a creative spirit, but know how to grasp, work out, and improve the ideas of others....The first two classes mutually complement each other.

Otto Ammon

In the third class are people with an average or lower than average intellect. For them the characteristic condition is called "the herd mentality." They give in to training, and not having their own ideas, they adopt the ideas of others. They cannot themselves develop any mastered ideas, and oppose any innovations. They think that they possess the universal truth, and preserve adherence to it with the inertia of mass.

The fourth class is inferior people, incapable of producing, discovering, combining, or adapting to an unfamiliar culture."

In accordance with the given classifications, Otto Ammon correctly concluded that the significance of peoples, their worthiness to world culture, and the superiority of one people over another is the greater, when the first and second classes of people are stronger within a people.

"A person of the first class is worth 1,000 simple, manual laborers, and he improves the well-being of thousands. People of the third and fourth class only follow the paths laid down [for them] by the first class of people. People of the first class are locomotive engines, people of the third and fourth class are train cars."

His next book, *Natural Selection among People*, was published in 1893. In it, he added the *Theory of Embryonic Plasma*, (developed by August Weisman and Thomas Morgan), to anthropological measurements. "The determinants of embryonic plasma, which define the skull and the rest of the skeleton, present themselves as a stable form, equal to those which determine the color of the eyes, the hair, and the skin." Besides that in this report, Ammon was the first to derive the ratio between height and the cranial index. He also emphasized that among the ruling classes, the percentage of people of tall height with light hair and blue eyes was substantially higher, than among members of the lower classes.

And in his book, *Social Structure and its Natural Foundation*, published in 1895, Ammon developed that general thought with absolute certainty. In his opinion, the death of the ancient world occurred namely because of the erosion of the racial elites: "In accordance with anthropological concept, [the ruling elite] belonged to the higher, Aryan race—the people of the North—which in prehistoric times arrived in Greece and Italy, and ruled over the dark-skinned natives, who possessed weaker character, then gradually mixed with them."

Afterwards, young Ludwig Woltman's star rose in 1903, when he published his fundamental essay, *Political Anthropology*, and instantly became popular with the enlightened public. In it, he brought the racial problem to a qualitatively new level of discussion. The books of theoreticians of the given field began to have greater circulation, and public resonance increased. Woltman's book was subtitled, *Research on the Influence of Evolutionary Theory on the Science of the Political Development of Peoples*; this answered to the author's intents to emphasize the globalness of his scientific range of vision. Besides that the given essay was a response to the first social request in the area of racial theory, for the world-famous German steel firm—Krupp—announced a contest to carry out global research. Thus, racial theory began to take shape as an independent scientific field, and acquired the status of an academic discipline. The period of non-professional discussion, conjecture, and intuition began to gracefully recede into the past, and the methodical formation of the canons of racial theory—with which the good students of the world were compelled to reckon—began. And the important politicians of the day clearly recognized, that on the proving ground of public ideas, weapons of unprecedented destructive power came and went, and that victory went to the one who could control their mass application.

Left: Ludwig Woltman

In his book, Woltman threw out all of the indecisive rhetoric of previous amateurs in racial theory: "Among governments, only those remain longer than others, on the summit of the prosperity achieved by them, who have not kept to one-sided natural selection, and did not yield to energetic cultural outbursts; but spared the portion of his more gifted population and protected them in as healthy living conditions as possible, as its natural source of talent. The Aryans are the last arrivals and conquerors, who as a result of their higher physical strength and intellect, conquered the lower people. Everywhere they form a government caste, and attach great significance to unequal marriages, which is why they despise half-breeds. The lightest race is at the same time, the most gifted and noble race."

For evaluating the social significance of historical events, Woltman traced a detailed concept of the world, based on new racial-biological criterion. He recognized the strength and novelty of his new method, which prompted science to establish the important periodical publication, *Political-Anthropological Review*, of which he became a major editor.

Besides that in his next two monographs, *Germans in the Renaissance in Italy* (1905), and *Germans in France* (1907), he analyzed the sources of the unprecedented cultural upsurge of the age, on the basis of the history of art, linguistics, patronymics, and anthropology, and laid down the source of a new science—**cultural biology**.

Woltman graphically and convincingly showed, that all the key figures of the Renaissance belonged to the northern racial type. The stylistics of the psychological experiences of namely this race are most seen in the masterpieces of the Renaissance, which in essence was a visible embodiment of the realization of the potential of the northern European spirit. In both books, the main anthropological theme also clearly dominates everything: "The blonde element of the people defines its cultural worthiness, and the fall of great cultures is explained by the dying out of this element."

Culture does not arise out of nowhere, in and of itself; it is not created by capricious Gods, and it is not possible to create as a result of a favorable distribution of annual norms of precipitation or successful speculation on the stock exchange. Culture is always the result of the tension of the higher physical and psychological strengths of an individual of a concrete racial type. Culture by itself is not the cause, but it is always only the result of the actions of the hereditary traits of Man, bumping heads with the realities of the surrounding world. At the base of any creative act lies the inborn system of values of its creator, which in essence fulfills the function of a drawing, which is later brought to life in the creation itself. But the system of values of the creator is inherently conditioned and entirely determined by his race.

"In the present time, there can be no doubt that the **husky, blonde, fair, and light-eyed race** everywhere created the foundation for that level of development, that we call an ideal level of culture. Undoubtedly, the Mediterranean and Mongol races possess some cultural ability, and denial of this fact would an outright exaggeration. What has most significance is the fact that the northern race, in all its branches, always achieves **a higher degree of civilization**, the same of which can never be said about the Mediterranean and Mongol peoples. The outstanding geniuses of humanity belong to this [Nordic] race, or arose from mixing of races, with a pre-eminent quantity of northern blood. The better people of the new, spiritual history were such in greater part, such as Duerer, Leonardo da Vinci, Galileo, Rembrandt, Rubens, Van Dyke, Voltaire, Kant, and Wagner. Others show admixture of a dark race, expressed in darker pigmentation, or more rarely, a wide skull, such as Dante, Rafael, Michael Angelo, Shakespeare, Luther, Goethe, and Beethoven."

The false, illusory ideals of "liberty, equality, and fraternity," and those founded on them, were first publicly disgraced from the time of the French Revolution.

The fast and untimely death of the still rather young Woltman in 1907 was not able to erase his accomplishments.

A similar system of views and argumentation was also adhered to by such classics of socio-anthropology as Ludwig Wilser (1850-1923). In his book, *The Germans* (1904), he detailed an examination of the cause-effect relationship between culture and its bearers, for which he took the problem to a qualitatively new level of the evolutionary morphology of the various human races. "On the basis of morphological and physiological considerations alone, one may come to the correct conclusion, that the tall person with a prominent skull, long-head, and light color, that is, of the northern European race, presents himself as a perfect member of the human race, and as a higher product of organized development, since the noble, outward bearing, that is, the higher development of the body according to a basic, psycho-physiological law, undoubtedly corresponds to a very fine organization of the brain. On the average, the northern race is the largest and the strongest. With the strong build of the body, it maintains a complete proportionality of limbs, the convenience of which reveals itself in a perfect cut, and uniformity of distribution of organized material. From studies on the racial beauty of women, it is evident that the more beautiful proportion is observed on blondes. The Mediterranean race, although it also stands out with the same physique of the limbs, does not reach the same height and strength of the northern type; it also does not have the light skin and blonde hair, while strong general pigmentation leads to high loss of energy, at the same time that light races preserve it for use in the nerves and muscular strength. Along with this expedient use of energy, the later appearance of sexual maturity acts beneficially on the growth of intellect."

Wilser developed the ideas of Karl Penka with great enthusiasm. In his book, *Ancestors of the Aryans*, Penka came to an original conclusion, on the basis of new synthesized methods, in which the conclusions of a group of scientists is obligatorily double-checked by the methods of another. With such an approach, many facts appear before us in a completely different light. Thus, he wrote: "Egypt was settled many times by the white race in prehistoric times, and afterwards in the invasions of the Libyans and the Sea Peoples. It is a fact that these peoples belonged to the northern race; that is, they were Aryans; this is proved by the fact that megaliths belonging to the time of the Stone Age were also found in Egypt. It is easy to show the common [northern] origin of the Dolmans, [that are] dispersed throughout Europe, Asia, and Africa; they came from the [same] northern native country, from which the Aryan race came. From the example of Egypt, it is clear that anthropology should be formed on the basis of political and cultural history. Extremely interesting questions arise, the answers to which, from the point of view of science, are not so difficult: did the idea of building pyramids not originate in the minds of the blonde and blue-eyed pagan priests? It is obvious, that the pyramids are only identified with further development of the megaliths."—Penka

Wilser was fully versed in the methods of anthropometry; therefore, in order to double-check the culturological assertions of Penka, he resorted to a simple and radical method: study the preserved skulls of the mummy of the Egyptian Pharoah, Ramses the Great. Such was his astonishment, when the dolichocephalic skull of the ancient ruler was discovered to have the remains of blonde curls; this enabled Wilser to count the Pharoah as "the last offspring of the northern race" in that region.

The main theoretical conclusion of Ludwig Wilser's fundamental work, was that it was necessary to search for the ancestors of the blonde race - who appeared always and everywhere in the capacity of "natural-born bringers of world civilization" - in northern Europe, rather than in the south or the east.

5. German Raciology

The activities of English aristocrat Houston Stewart Chamberlain (1855-1927), not only made deep impressions in the development of racial theory, but in the formation of social-political ideas at the boundary of the 19th and 20th Centuries. He was naturalized in Germany, and is famous today as a German racial philosopher. His two-tome essay, *The Foundations of the 19th Century*, was published in 1899, and became a milestone in the development of racial theory and provoked lively discussions.

In his work, the author sufficiently exposed the history of the world to scrupulous analysis, from the viewpoint of the contributions [made] to it, by defined racial types. "All outstanding peoples that appeared starting in the 6th Century, in the role of true deciders of the fate of humanity, as founders of nations, and creators of new thinking and original art, were namely of German origin. The creations of the Arabs stand out for their short duration; the Mongols destroyed, but they created nothing; the ingenious Italians of the Middle Ages were all émigrés, or of the north, which was saturated with Lombard, Gothic, or Frankish blood, or they were Germano-Hellenes of the south; in Spain, the creative element was the Visigoths. The awakening of the Germans forms the foundation of European history, for their worldwide historical significance as founders of a completely new civilization and a completely new culture."

He also characterized the massive Reformation Movement as an "outrage of the German spirit against non-German torture."

In the context of our narrative, it is necessary to point out the following fact: many political prostitute-scientists have, for 100 years, stigmatized the name of Houston Stewart Chamberlain, recording him as the forerunner of "German fascism," and they also represent him as one of the founding ideologues of militarism.

Left: Houston Stewart Chamberlain

First of all, we recommend to all his angry critics, that they should turn to primary sources for clarification of the question. By Chamberlain's self-admission, the term "Germans," for defining the given bio-type, was borrowed from the ancient Roman historian Tacitus, who wrote that they, most of all, guarded the purity of [their] blood, and they never mixed with other peoples. However, with this, Chamberlain openly recognized: "There have never existed peoples, which call themselves Germans."

Chamberlain quotes the words of Tacitus, which describe the Germans as a people having "blue eyes, light hair, and tall figures," but with this he conscientiously emphasizes that two more racial-ethnic groups could fall under this description: the *Celts* and the *Slavs*. In ancient times, at the start of the migrations, the Slavs were clearly marked as dolichocephalic and of tall stature. Besides that Chamberlain referred to Rudolf Virchow (a prominent author on physical anthropology), and to his broad investigation of hair and eye color. On the basis of this he came to the conclusion that the Slavs originated from the same center of the region, from which all members of the given racial type spread.

Chamberlain did not approve of Linnaeus' term, *Homo europeaus*, which he considered extremely abstract and already moving into the field of history. He did not use it in his classification of race, or the definitions *Indo-European* and *Aryans*, by virtue of their vagueness.

At the end of his book, giving a prognosis for the future, Houston Stewart Chamberlain prophetically wrote that a European world, which is good for the entire white race, could only be achieved on the basis of a Celtic-Germanic-Slavic union. Therefore, we simply consider all accusations leveled in the address of that German philosopher to be ridiculous and based completely on ignorance of the sources.

In whole, the boundary of the 19th and 20th centuries was the amazing time of the Titans of political philosophy. Books on racial theory that went into circulation, only numbered from one to 3,000 copies, but this turned out to be sufficient, in order to completely alert the reading public to new, explosively dangerous ideas, and to abruptly change public opinion. On the basis of the works of Otto Ammon, differential methods of selection were created for recruits to the German Army. The numerous publications of Ludwig Wilser and Ludwig Woltman subsequently warned of the degradation of all the West European social democracies, and made the existence of communist regimes in the developed European countries impossible. Kaiser Wilhelm II himself applauded him, and more than once afterwards sought the advice of the scholar on questions related to the drawing up of German foreign policy.

The Russo-Japanese war had scarcely begun, when Kaiser Wilhelm II sent a letter to Nikolay II, in which he called his cousin—the Russian Tsar—the "Ruler of the East," and himself "Ruler of the West," emphasizing that with all strength he would assist Russia in the suppression of the rising "yellow menace." The German monarch clearly saw the racial basis of this conflict and gave the Russian monarch to understand that he openly stood on the side of a racially close people. It is also a fact that shady liberal scientists emphasize with complete falsity and groundlessness, that Chamberlain and similar racial theoreticians sanctioned aggression against [Russia]. To the contrary, Chamberlain himself, and many of his colleagues, wrote numerous letters in which they accused England of unleashing the First World War, which in essence became the first fratricidal, civil war slaughterhouse of the peoples of the white race. After all, it was namely England, during the time of the Russo-Japanese War, who stood at the side of the Asian empire, openly helping it financially and technically. With this, it is necessary to remember that Chamberlain himself was a descendent of an English aristocrat, and his open and consistent Germanophilia should not be considered a betrayal of his country, but the uncompromising purity of position of an authentic racial philosopher.

Not all of the above-listed authors were professional anthropologists; however, motivated by their racial intuition, they were able to change the whole social-political situation in the civilized world, at the boundary of the 19th and 20th centuries. As a result, even scientists who considered themselves "pure academic anthropologists," were compelled to cite them and their open philosophical discussions, about the authentic order of things in the organized world in general, and about Mankind in particular, on the pages of their works.

The famous anthropologist, Karl Eugene Uyfalvi (1842-1904), a subject of the Austro-Hungarian monarchy, who was later naturalized in France, became famous for writing a series of books, the so-called "iconography of race." With the personal permission of the Tsar, he travelled a significant portion of the Russian Empire, with the aim of studying the racial aspects of Iranian influence on the racial culture of Middle Asia. With this same goal, he later travelled to Iran and the Near East, where he studied the images of gods and heroes, from a racial point of view, as a result of which he came to the following conclusion: "The Hindus, Persians, and Scythians originally stood out as a northern type. That the Romans, Greeks, Etruscans, and Gauls treated toward them as [such]...there can be no doubt. There is information, pointing to the connection of the northern race with the Egyptians." For his part, Franz Pruner-Bey (1808-1882), a Belgian anthropologist, studied numerous skulls of Egyptian mummies in the 1860s; he clearly set apart two persistent and highly contrasting racial types from them, and ascribed to the "noble" Europoid the honor of creating all the great culture of Ancient Egypt.

Still another famous German anthropologist, George Bushan (1863-1942), pointed out the same connection in his monography, *The Science of Man* (1911). He wrote: "On ancient Egyptian tomb monuments of the Age of the 18th Dynasty, we encounter images of warriors with light-color skin, blonde hair, and light-blue eyes, the so-called "Tamakhu," or, "men of the north," as they were called directly.

In his fundamental book, *Race and the Peoples of Mankind*, Ferndinand Birchner (1868-1944) compared and contrasted the data of comparative craniometry, ancient historical sources, and also preserved images. He supported the concept of a Nordic origin of the white race: "In the descriptions of ancient authors, the Celtic and Germanic tribes play a significant role. According to their testimonies, these were tall, strong people, with white-colored skin, with golden or reddish hair, colors which the Celts and Germans strived to enhance artificially. Their eyes, according to a description by Tacitus, were of a light-blue color. Strabo recorded them in a similar way. Procopius described the Wends [Venedi] in exactly the same way, as an extremely tall and strong tribe, and although the color of their hair is not very blonde, they are not found to have a tendency toward dark-colored hair."

While researching the Japanese contemporary to them, German raciologists Erwin Beltz (1849-1913) and Karl Stratz (1858-1924) discovered a racial type that more resembled Europoids in general grace and proportion of body, that belonged to the aristocratic layer of Japan. The given type is characteristically more light-skinned, has a longer face, and is called "kho-shiu." Beltz and Stratz, independently of the Russian scientist G.E. Grumm-Grzhimaylo, also came to the conclusion that this racial type originated from the ancient *Dinliny*. They observed an analogous picture in Northern China and Tibet, where a more refined, Europoid type was encountered far more often among the important officials, than among the common folk. Besides that they noticed one highly visible and characteristic nuance; it turned out that Chinese and Japanese artists obligatorily attributed an acquiline nose to each great person, without any regard for the historical truth. Originating from racial stereotypes, they strived to Europeanize the portrayal of greatness in the public consciousness - something that could only be caused by the bygone influence of the cultural creativity of the white race, on the aborigines of the given territories.

Karl Stratz' observation was developed into an entire series of works, dedicated to the questions of racial standards of beauty. Thus, in his popular essay, *Feminine Racial Beauty* (1904), he wrote: "The Chinese beauty tries, when given the opportunity, to put rouge and whitener on herself, in order to hide her yellow complexion. Not giving any further details on Mongol toiletry, we may already conclude from this example, that the Chinese woman strives for an ideal of beauty, enhancing her own charms by way of imitating those of higher races." A similar set of questions is observed with another important German anthropologist, Gustav Fritsch (1838-1891), who engaged in working out the racial canons of the human body. Quoting these racial canons, which he named the "**Fritsch Key**," Karl Stratz emphasized: "From the time that Montelouis proved that the Swedes were already inhabitants of the north 4,000 years ago, and already possessed a comparatively high culture then, I consider myself, as a German, as having a right to claim the northern race, and even all ancient tales about Gods and heroes not of Germanic, but of northern heritage. This light-haired and blue-eyed branch of the great white race, probably since ancient times, took root in Scandinavia, for if we compare statistical data and graphic portrayals of blue eyes and light hair, which fades progressively toward the south, east, and west, the center of expansion is located in Scandinavia. Alongside the Scandinavians, the northern Germans, Frisians, the Dutch, part of the English, and the northern Russians, are related to this tribe."

The matter went to such zeal, that even famous racial theoreticians of European origin, like Henrich Hertz (1832-1894) and Ignaz Zollschan (1877-1948) undertook to explain the presence of blonde Jews by an admixture in them of noble Amorite blood, which occurred in very old times, when the hordes of "Gog and Magog" invaded Judea. In his monography, *The Racial Problem* (1914), Zollschan wrote in the spirit of his time, "The type of the northern race is also encountered in the Inca, in the Indians of North America, the inhabitants of Java, the islands of the Pacific Ocean, and the Malaysian Archipelago." Still one Jewish racial theorist, Salomon Reinach (1858-1932), published a scientific ode in praise of the white race, titled, *The Aryans* (1892).

A student of Africa, George Schweinfurt (1836-1925), discovered the Akka people in the center of that continent; they have light hair. The Polish anthropologist, Ludwik Krzywicki, in his book cited by us,[55] emphasized: "The name *White Type* extends to a great number of sufficiently different peoples, having however, several common traits, namely: soft, long hair, sometimes curly, more or less rich hair on the face, a prominent and comparatively narrow nose, and finally, the color of the face, which goes to a dark-complexion, but always with redness in the cheeks. White types in the form of remnants are encountered in the northeast corner of Asia, and northwest America; they are scattered here and there in Japan and Indo-China, and finally, they are still encountered as Indonesians on the islands of Borneo and Sumatra, and in Polynesia. These remnants are important to us as testimony, that at one time, the distribution of anthropological types on the globe was completely different. The caste system of India is called "varna;" in Sanskrit this means "color." In point of fact, it is nothing less than a hierarchical organization of anthropological elements. The lower the caste that we observe in a known locality, the more clearly curly hair, fat lips, and dark coloration, particularly among the Dravidians, is observed. On the other hand, the higher the caste, the lighter the color of the body becomes, and the smoother the hair."

As a form of proof of the extent of light-pigmented racial types in antiquity, one may cite the names of several peoples; for example, in Senegal there lives the nomadic Fulakhs tribe, whose native self-designation means "light."

It is particularly worth emphasizing that anthropologists of the time, who called themselves "pure" scientists, did not refrain from forming definite laws between the biological characteristics

---

[55] *Anthropology.*

of racial groups, and their known cultural worthiness, in their findings. This enabled them to make highly philosophical generalizations. Thus, Ludwik Krzywick openly emphasized: "Chroniclers recount to us, that the feelings of the blondes toward the short-headed brunettes were not defined by any particular gentleness. Even after the time of the Merovingians, a marriage of a leader of the blonde German race, with a woman of the Celtic race, or some other woman with dark hair, was considered unequal. The legend of Bertwolf, cited by many historians and ethnographers, testifies to the existence of racial differences: Bertwolf's mother was outraged that her son disgraced his noble German race and forever dirtied the clean blood of his descendents, by entering into marriage with a girl, however beautiful, but with black hair and black eyebrows. After all, in her opinion as many such crows as his heart desired, could be found for his amusement on his own estates."

Felix von Luschan (1854-1924), was a prominent German anthropologist whose name is tied to the methods of racial measurements of skin colors, on the basis of extensive material dug up by him on expeditions. He significantly enriched the historical picture of the range of biological types, substantially correcting many positions of classical ethnography. In his famous monograph, *Peoples, Races, and Languages*,[56] he pointed out: "In the west of North Africa, it is firmly established that at least 10%, or maybe far more, are blonde people. That the nationality of the Fulbe, for the greater part, completely covers its face, as a result of which they are called by their Arab neighbors "le mulathemin" (the covered ones), this attests to their appearance from out of the far north. Also, from northern Arabia, I know many blondes with light skin, wavy or gentle curly hair, small height, with narrow lips and small noses; that is, they represent people who completely recall the good type of the Mediterranean race, someplace in Corsica or Sardinia."

Left: Felix von Luschan

Setting the culture-making Europoid racial substrata apart in the social hierarchy of the states of North Africa and the Arabian Peninsula, Felix von Luschan discovered the same pattern on the territory of Asia Minor. "The original Kurds were all blondes, blue-eyed and long-headed, and only under the influence of new geographic conditions and mixing with Turks, Armenians, and Persians, they became little by little, more and more brunette and short-headed. Thus the Kurds are descendents of migrants from northern Europe; they preserved the pureness of their language over the course of 3,300 years, and in several parts of the region, they maintained their proliferation, and also their somatic characteristics. Where, then, is the actual home of the blonde-haired, blue-eyed, long-headed Kurds located? Clearly, only where in general there is just [one] place on all the earth for the blonde-haired, blue-eyed, and long-headed peoples—in northern Europe. Undoubtedly, it is not at all my task here to examine the Aryan Question, and I feel myself completely free of Teutonic or Pan-German sentiments, like de Gobineau or Chamberlain; but all the same, I consider the independent existence of the defined, long-headed, blue-eyed, and blonde type of people indisputable, and I do not think that these particular characteristics could be accidentally found somewhere else, in a different place, without having a relationship to the northern European type."

Emphasizing his ideological impartiality, the German raciologist nevertheless, in a very unequivocal manner, supported all the basic postulates of classical racial theory, in regard to the ancestors of the Europoid race, and also of its priority influence on the culture of organized societies. Even in such a heterogeneous and racially diverse region as the Caucasus, he clearly discerned all the traces of that same racial dynamic: "Of all the nationalities of the Caucasus, where the most foreign elements are among the Ossetians, one encounters many long-headed [types] and approximately 20% blondes. Therefore, it is difficult to doubt in a strong, northern European admixture."

A correlation of the data of anthropologists and ethnologists, and also information gleaned from classical written sources, most strongly focuses the picture as a whole. Thus, von Luschan emphasized that in negotiations dating from 1370 B.C., the Mittani people and their king called

[56] *Narody, rasy, i yaziki.* Leningrad, 1925.

themselves *harri*, and 900 years later Xerxes and Darius also called themselves *har-ri-ya*—"Aryans of an Aryan family."

At approximately the same time, another luminary of German raciology, Wilhelm Muellman (1904- ?), published a series of works. From them it was obvious that the ancient settlement of the Indo-Aryans reached east to Polynesia. Traces of Vedic culture are easily found in Oceania, and on the very distant islands of the Pacific Ocean. The Europoid race, in his opinion, spread from India all over this wide region, forming an upper, aristocratic strata, which to this day preserves many of the characteristic traits of its original psyche-type and physique, in the midst of a surrounding black-skinned population.

In the 1910s-1920s, the Czech, B. Grozniy created a new science called **Hittitology**. Based on the results of racial ciphers of ancient Hittite cuneiform tablets, he produced a coup in the science of the time, and substantially corrected the presentation of the role of the Indo-European peoples in the history of the formation of culture in the Near and Middle East. In his work, *Proto-Indian Writings and their Ciphers* (1915), he wrote: "When we think of the migration of peoples, we usually have in mind the great migration of the Germanic and Hunnish peoples, in the 4th-6th centuries, A.D., which led to the collapse of the Roman Empire. But even before this, in the history of Mankind we meet with similar overthrows, and the resettlements of entire peoples. Examples are the invasion of the so-called northern peoples of Outer Asia, which took place around 1,200 B.C., and destroyed the Hittite state; the invasion of the Indo-European Phrygians, Armenians, Thracians, and Midians of Asia Minor, and the related invasion by the Phillistine people of Palestine; the invasion of the Shardana od Sardinia, and the Etruscans of Italy. Approximately 1,000 years earlier, an invasion by the Indo-European peoples occurred (the Hittites and Aryans of Outer Asia; the invasion by the Lycians and Hittites of Asia Minor and the invasion of the Indian sub-continent, and the invasion by the Mittani of northern Mesopotamia). Around the end of the 4th Millenium B.C., the first Indo-European conquerors—the Hittites—appeared in the Caucasus region, northern Syria, and in the eastern portion of Asia Minor. They arrived there from the region of the Caspian Sea, and probably carried the name "Kush." Our clarification of a proto-Indian letter indicates to us, that even in deep antiquity, in the first half of the 3rd Millenium, B.C., northern India was controlled by the ancient—the most ancient, one might say—Indo-European people, who left us brilliant monuments about their activities on Indian soil."

Later, in his work, *The Hittite Peoples and Languages*, B. Grozniy came to even more revolutionary conclusions: "The racial ciphers of the Hittite writings—cuneiform, and later hieroglyphics, caused a complete overthrow of our views on the history of ancient mankind. I think that we may speak of an overthrow in ancient history; clearly we see today, that not only the Sumerians, Babylonians, Assyrians, and Egyptians, but even the Indo-European peoples 4,000 years ago played an important role in the history of the Ancient East. Thanks to the race ciphers of the Hittite writings, we succeeded in identifying six peoples of the ancient East, which until now were unknown. Four are of Indo-European origin; that is, related to the peoples living in modern times, in Europe and India. Thus, a completely new science arose, called Hittitology, which today is considered to be on the same level as Assyriology and Egyptology, which are devoted to peoples of foreign origin. Hittitology arouses significantly more interest in Europe than the latter-named sciences, thanks to the Indo-European origin of the majority of the newly-discovered peoples. A common blood and common language connects us to these newly discovered peoples, who even 4,000 years ago strongly influenced the history of ancient Asia, and who used words like "*kuis*," which means "who," and is recalled in the Latin *quis*—"who"; then "*nebis*," which means "nebo," and resembles the Russian, *nebesa*; and "*dalugasti*", which means "long", and recalls the Russian, *dolgiy*—long, and *dolgota*—length."

In the meantime, at the start of the 20th Century, qualitative changes took place in European raciology. Not only the question of the ancestors of the white race, but questions about the use of adequate scientific terminology were finally brought to a close. The loud arguments between linguists, anthropologists, and ethnographers subsided. All the related sciences, once confusingly mixed in the framework of a single racial theory, now clearly gained legitimate limits of application in her bosom. And despite the stormy growth of chauvinism that was brought about by the First World War, the use of incorrect terms, such as "German race", "Teutonic race", and "Slavic race", which denoted improper equation of the categories of ethnology and biology—had ceased. The successful resolution of the problem, suggested by the Russian raciologist, I.E. Deniker, received universal recognition. The term *Aryan*, by virtue of the awkwardness of its application in racial classifications, based exclusively on a description of physical traits, gracefully departed for the areas of linguistics, culturology, and comparative religion. For the name of the anthropological type of the members of the northern European race, the term "Nordic" began to be used.

Felix von Luschan thus described the essence of the problem: "Any notion of an *Aryan* race should be rejected. One may, it is true, and not with complete certainty at that speak about an Aryan language, although this notion has several meanings, and by several is understood as embracing modern Persian and Armenian, with closely related languages, and with others it is spread significantly wider; but it is completely intolerable to speak about an Aryan race or still about an Aryan skull, or about an Aryan type of face; that as even Max Mueller observed, is also silly, just as wanting to set apart the language of long-headed types, or the grammar of brunettes would be. Particularly absurd is the fashionable use of the word "Aryan" as an antonym to the word "Jewish", since we quickly see, that modern Jews somatically are more likely to be affiliated with Outer Asians, than with Semites, and in this sense they are close relatives of the Armenians, that is, of a people, which speak a language that is Aryan, in the most narrow sense of this word."

One of the most popular racial theorists at the time of the Weimar Republic, and also the Third Reich, was Hans F.K. Guenther (1891-1968). In his book, *The Raciology of Europe*, he gave an explanation in completely the same spirit: "In philology, the word *Aryan* earlier referred to Indo-European languages; today, that term is usually only used to apply to the Indo-Iranian branch of that language family. In the beginning of racial studies, they sometimes called the white or Caucasian race *Aryan*; later, the Aryans began to be called *peoples* speaking in the Indo-European languages, and finally, the *Nordic* race. Today, the term *Aryan* has fallen from scientific use, and its use is not recommended, particularly now, as it has become currently popular among ignoramuses as a way of setting themselves against *Semites*. But anthropology also rejects the term *Semite*, since peoples of quite different racial origins speak in Semitic languages."

6. The Shaping of the Nordic Idea

In the beginning, the world scientific community, and then wider public-political circles, journalists, and writers took to using the term *Nordic*, meaning a type of tall, long-headed, blue-eyed blonde. In his book, *The Nordic Idea*, Hans F.K. Guenther honestly recognized that "the term *Nordic race* was first introduced by the Russian raciologist, I.E. Deniker." Still one authoritative German scholar, Walter Scheidt (1895-1976), focusing attention on the contribution of the Russian researcher, considered naming one of his essays: *The History of Anthropology, from Linnaeus to Deniker* (1928). The Austrian racial specialist, Erich Fegelin, in the book, *Race*

and Government, also emphasized that "the term *Nordic race* was first introduced by Deniker."

Left: Hans F.K. Gunther

Examples that denote the "maturing" of the entire science as whole, and its departure to a qualitatively new methodological level, can be brought up in multitudes. Apart from the previously used principles of craniology and somatology, that is, the determination of racial differences on the basis of measurements of the skull and the proportions of the body, dermatoglyphics (the study of variations based on fingerprint patterns on the hands and feet), and odontology (the determination of differences in the structure of the dental system) came into use, and also a great number of new biochemical and genetic methods of reseach. Besides that analysis of mental racial differences more and more often began to be conducted not on the basis of social interpretations, but namely [on the basis of] biological factors. From a narrow specialty field in anthropology, racial theory more persistently and actively began to turn into a powerful Nordic movement, called upon to renew the world perception and viability of, the northern race. This was the time of the greatest flowering of the concept. In this period, Hans F.K. Guenther created a whole series of works, directed at a comprehensive opening of the theme. In the books *Raciology and the German People* (1922) and *The Raciology of Europe* (1924), he made a detailed racial portrait of the inhabitants of the continent, tying together the constitutional traits of each race with its psychological characteristics. And in the works, *The Nordic Idea* (1925), *Race and Style* (1926), *Racial History of the Hellenic and Roman Peoples* (1927), and *The Nordic Race among the Indo-Germans of Asia* (1933), he connected the specific psychological experiences of the various member races, with their culture-making capacities, and on that basis came to the conclusion [that the] Nordic race had the greatest biological worthiness. Much later, in the book, *Racial Elements of European History*, published in many languages after the war, he developed and broadened his ideas, emphasizing: "The question is not in what measure, we, the people living today, are Nordic, but whether we have enough courage to prepare a world for future generations, which has purged itself in racial and eugenic terms. The de-Nordification of the Indo-European peoples will always take centuries; the will of people with Nordic thinking should throw a bridge across the centuries. When there is talk about selection, it is necessary to consider the multitude of generations; and contemporary people with Nordic thinking may wait for the extent of their lives, for only one reward for their efforts: consciousness of their own courage. Racial theory and research in the area of heredity gives strength to a new aristocracy of youth, striving for the highest goals, like Faust follows the appeals from the spheres, emerging beyond the limits of individual life. Inasmuch as this movement does not strive for profits, it will always be a movement of a minority. But the spirit of any age, among them the spirit of that Age of the Masses, in which we live, is always formed by just a minority."

One of the recognized standards of anthropology, Eugen Fischer (1874-1967), although he always tried to stick to the canons of "pure" academic science, nevertheless openly expressed the opinion...that "today every clear emphasis of the Nordic point of view brings profit." And in 1927, in co-authorship with Hans F.K. Guenther, he published the important study, *German Skulls are of the Nordic Race*, as a visible substantiation of his thesis.

Left: Eugen Fischer

In the work, *Race, and the Rise of the Races of Man* (1927), Eugen Fischer clearly singled out the essence of the problem: "One of the most substantiated hypotheses is such: the Nordic race—the builders of megaliths and Dolman burial sites in Scandinavia, Denmark, and so on—originated from the Cro-Magnon race. In accordance with the said hypothesis, the Nordic race arose as a result of a modification of the Late-Paleolithic race in the north, by measure of liberation from the ice [that was] in the now-inhabited areas. Here arose the Nordic race, at which time it also acquired its typical qualities. This is the best explanation of the Nordic race." And in his work, *Anthropology* (1923), he developed this idea in the following way: "The Indo-European peoples, which included among themselves a strong contingent of the Nordic race, imposed themselves on the pre-Hellenic population—and after some time, Greece started to shine. The death of the families of fully-vested citizens, and the admission of the descendents of slaves and the aboriginal population as citizens, led—sometime later, to collapse. Rome died of race-mixing and the products of degeneracy; the Nordic Germans arrived, and sometime later—[came] the Renaissance; Byzantium, preserving its Greek standards and traditions, did not experience the Renaissance. In Italy there was no re-birth, but a new birth, thanks to a new race. It is proven inductively, that the racial differences of separate groups render a huge influence on the achievements and fates of peoples and states. Woltman showed that the lion's share of great people from the Age of the Renaissance (artists, sculptors, statesmen, and scientists) belonged mainly to the Nordic racial type judging by appearance, and according to biographies and portraits. Out of 125 such people, 102, according to Woltman, had blue or grey eyes; 68 out of 108 had light hair, and 26 had chestnut hair. The northern ancestry of a single Indo-European people is today the most authoritative theory supported by the data of anthropology, linguistics, and archaeology.

Left: Ludwig F. Clauss

Fritz Lenz (1887-1976), a prominent specialist in the area of racial hygiene and biological inheritance, wrote *Race as a Value Principle* (1934). In it he created a theoretical foundation of ethics, on a racial biological basis: "Race is the carrier of everything: personal, government, and people. From it all existence derives, and it is essence itself. Thus, for us, everything springs from an ideal race—culture, development, personality, happiness, salvation—and everything returns to it. In it we find the unity of our essence, the unity of life, and unity in the highest sense of the word. In a higher value there is no room for compromise. The ethic ideal demands of us, that we lay down our whole lif in service of it. We are only small waves in a great torrent, but a multitude of waves forms a torrent. We say, after the manner of de Gobineau, the conveyor of racial theory: in defiance of everything, besides the certainty of death, there is still the most radiant and proud hope. Undoubtedly, one may lead our race to such an ascent and flowering, like it has never achieved before. But if we lose heart, our Nordic race will utterly die. Together with it, we will survive not for centuries, but for millennia. Before us stands the greatest task of world history. We stand on the eve of a turning point of all world history. The noted racial psychologist, Ludwig Ferdinand Clauss (1892-1974), very elaborately and exactly described the archetype spirit of the Nordic race of Man, in his monography, *The Nordic Spirit* (1936), by describing the stylistic nuances of his experiences. For the number of books published in Germany at the time, Ludwig Ferdinand Clauss securely occupies second place, after Hans F.K. Guenther. He achieved this mostly because his narrative style, like Guenther's, unites beauty of style, precision of observations, depth of conclusions, and clarity; it was easily understood by a very wide readership. Not burdening his books with an abundance of difficult terminology, Clauss, it turns out, turned directly to the original archetype of the reading public, and aroused its sympathy, mostly by penetrating the nuances of the style of psychological experiences. Delicately and masterfully he played those chords of the spirit, which until then, were considered inviolable. In the book, *The Nordic Spirit*, which became a bestseller, he wrote: "The Nordic spirit is fixed onto the south and in part, but to it the south is like a light to a moth. The more corrupting influence of the south manifests itself in the vanishing desire to strive toward something. When we speak of a "spirit" contrasting itself to the world, we are already talking about the Nordic style. Not every spirit is characteristically so contrasting. The ability to "objectify" the world is a Nordic talent. To the peoples of non-Nordic races, Nordic man often appears cold and impassive. But at the same time, this outer coldness often conceals strong passions. The "coldness" of the Nordic Man is explained by his effort to maintain distance between himself and the surrounding world.

Left: Fritz Lenz

It is not necessary to emphasize distance to the Nordic man, even in regard to inferiors. All Byzantine ways and forms of Eastern despotism are alien to him.

The Nordic spirit in all its fullness is self-sufficient, and is not in need of anything different. The Nordic man bravely looks his fate in the eye and greets it, whatever it may be."

For his part, Paul Schultz-Naumberg (1864-1949), a prominent raciologist, wrote a book characteristically titled, *Nordic Beauty* (1957). In it, he analyzed the anthropo-aesthetic canons of the body of the members of his race, and on their basis formulated common aesthetic precepts that move its members to be creative. "In any art, race is expressed more strongly than anything, for art is an expression of our racially conditioned, inner aspirations.

Take Classical Greek art, take art of the Middle Ages, and you will feel this interconnection. For us, for the extent of millennia, the Nordic race was the model and ideal form...In all small things, even in simple declarations and any advertisement, in order to produce the desired effect, an ideal, Nordic type of man should somehow be used...people with narrow faces, long skulls, a graceful body, light hair and blue eyes, that belong to the Nordic race...

Left: P. Schultz-Hamburg

In Greece, the Nordic tribes imposed themselves on the lower strata, which then was composed in part of people of the Mediterranean race, and in part perhaps, of the Near Asian race. The military aristocracy set itself apart from the dominant strata of farmers. [Some] attempt to reject the presence of Nordic racial elements in Greece, with...the argument that in the artistic trades, the influence of other races is clearly seen. But this is a false interpretation. The artistic trades, pottery, and others, were the business of the lower class, and naturally they expressed the traits of the races that were affiliated [with those trades]. Only much later, when peaceful times came, [did] the spiritual aspirations of the Nordic type find their reflection in sculpture and architecture. Even the famous Venus de Milo is an ideal image of the built, light-haired, young Nordic woman...

With reference to the dominant strata, we constantly encounter the word "*ksantos*", meaning "light-hair," "light," "golden," and "glowing." This sudden burst occurred in the 5th Century B.C. And only with Socrates, of whom we know through the Nordic Plato, does a different type begin to dominate Greece. In wisdom, Socrates is not Nordic, just as he is not in his outer appearance. Gradually, a different, non-Nordic worldview begins to predominate. The courier Cleon was the first democrat. The end of aristocratic government and the end of the leader principle had arrived...

The notion that peoples may grow old is biological nonsense. In the biological sense, there is no "old age." Biologically, each person is only a link in a chain; there is no inherent biological illness, degeneracy, extinction, or wearing out of distinguished peoples. Once and for all, put an end to this pseudo-science now.

When certain people from our nobility are called "outcasts", it is not a matter of the "aging" of their people, but the wear and tear of a bad hereditary mass. That is a situation particular to aristocracies that hinder natural selection. Only this makes the penetration of bad racial characteristics, which lead to "degeneracy" possible, and about which we can only speak of in different circumstances...

In the graphic arts, we once more experience the wonderful manifestation of the Nordic race in the Age of the Renaissance...

As I already said, ancient families died out. Rome died. But later, the Nordic Longobards, Vandals, and Goths arrived from the north. In the Middle Ages, the dominant class in Italy was Nordic.

Leonardi da Vinci had light hair and blue eyes; the images and problems that stirred him are testimony to the presence of Nordic blood. The grace of the Middle Ages also conveys the Nordic ideal of beauty...

Left: Richard Walther Darre

A racially gifted people defines its art. Racially foreign, degenerate art is an attack, with regard to the spiritual life of a people."

And the prominent public-political figure of Germany at the time, Richard Walther Darre (1885-1953), wrote the ingenious essay, *Hogs as a Criterion among Nordic Peoples and Semites* (1933), in which he subjected the gastronomical biases of a given race to comprehensive analysis, with particular consideration for such important components as animal proteins and cereals. From that he made the legitimate conclusion: "The homeland of the Nordic race was the forested zone of northern Europe, with a temperate climate." Besides a strict, scientific analysis of the biochemical processes of metabolic activities, he also supported his conclusions with numerous forays into the areas of history and folklore.

Prominent anthropologist and biologist, Otto Reche (1879-1966), supported this conclusion in the book, *Race and the Homeland of the Indo-Germans* (1936), on the basis of a synthesized generalization of the data of several related disciplines, with particular attention on new studies in the area of physiology and the science of the racial distribution of blood groups. The nature-philosophical views of Otto Reche led to a conclusion of the following character: "That which we call *world history*, is in essence, nothing other than the history of the Indo-Germans and their achievements; the powerfully rousing and simultaneously tragic song about the Nordic race and its idealism: a song which tells about how the strength of the race did what seemed impossible, and reached its hand for the stars, and how the strength quickly dried up, when the *law of race* was forgotten, when the Nordic man ceased to preserve the purity of his blood and strongly mixed with races [that are] less gifted in cultural terms. In ancient Europe there lived only long-headed races, the brachicephalics really could only reach Europe at the end of the glacial period, from Asia. What concretely concerns the Nordic race, is that in the last inter-glacial period, it was cut off in northern Europe from the ancient European long-headed type, where it acquired light pigmentation. I came to the conviction that the Indo-European languages were created by the Nordic race in northern Europe. This is supported by the hypotheses of de Gobineau, Wilser, Woltman, and others. It tears down the old theory about the imaginary origin of the Indo-Europeans, and in general [the theory] that [European] culture came from Asia. The Jewish-liberal slogan, "The Light of the East" is exposed as the "Mirage of the East." We are not Asians, we belong to a completely different human type; our Motherland is in our Old Europe, and we are indebted only to our millennial culture, and our own strengths, particularly to the gigantic creative gifts of the Nordic race."

Franz Schattenfroh, the author of *Will and Race* (1943), engaged in study of the issue from the point of view of reflexology and the history of legislation, and came to this conclusion in his essay: "In Doctor Guenther's exemplary work on the description of various human races, of which the most important are the Nordic, Dinaric, Western, and Eastern, it turns out that almost all the peoples of Europe are mixed, but the most worthy of them is the blood of the Nordic race. The greatest geniuses of all time, not only in Europe, but beyond its limits, were of that Nordic blood, or of a strong Nordic admixture. The Nordic blood was itself immortalized by their works in the area of religion, philosophy, and the mathematics of the ancient Hindus, who before our era mixed more and more with dark-skinned races. The collapse of the ancient Greeks was tied with the depletion of its Nordic strata. So also [the collapse of] the ancient Persians, the Midians, the ruling class of the Amorites and Phillistines, the Scythians, and of course, the Romans...the depletion of Nordic blood in these peoples (as a result of mixing, wars, and inner conflicts) brought complete collapse in its wake. The pre-Slavs were also of pure Nordic blood. Together with the Celts and Germans, they swept over Europe in the latest wave of Nordic blood. Europe is indebted to them for its high culture. The less admixture of Nordic blood in various peoples, the less significant a place they occupy in the world."

Respected reader, turn your attention to the fact that the above book was published, with the approval of the ideological high management of the Third Reich, even though Stalingrad and the Battle of the Kursk Salient had already concluded, and the situation at the fronts did not lend itself to good sentiments in the address of the Slavs. However, no Slavophobia is shown; supposedly it had a place in German political propaganda, but there was no [such] talk. That was a much later fabrication by the forgers of communist and liberal myth. The Third Reich did not struggle with Slavdom, but with the threat Bolshevism [posed] to the foundations of European Civilization. Incidentally, to this day, not one official German document from that time has been published, in which the Slavs are called a "race of sub-humans," something which devoted warriors of anti-fascism like to broadcast. By "sub-humans," in the anthropological sense of the word, Himmler's department was referring to Bolshevik commissars, like Lev Mekhlis, and open racists like Ilya Ehrenburg, an instigator who hid behind the backs of Russian soldiers, shouting "Kill the Germans!"

The contribution of Professor Phillip Leonard (1862-1947), a Nobel Prize Laureate, and world-renowned physicist, was highly important in the development of the philosophical fundamentals of racial history. In his book, *The Great Naturalists* (1929), he analyzed the racial origin of dozens of the great scholars from ancient times, to the first half of the 20th Century, whose discoveries changed the course of world history and founded the very type of European civilization in its modern meaning. The conclusion, based on the study of portrait characteristics, and also on psychological behavior, did not reveal anything new: the nature-philosophical foundations of the modern technical world were created by peoples with absolutely predominant Nordic blood.

The fundamental essence of F. Leonard's book was expressed with the following phrase: "In this science, as in all other respects, for the individual, everything is conditioned by his race and blood." Besides that he often cited Houston Steward Chamberlain, who for his part emphasized: "Perfection of intellect, the ability to analyze, as well as passion, attendant with his craving for self-study—all these characteristics are indicated in higher degree in our Nordic race."

Leonard characterized the Nordic scholar as a scientist, who was able to receive happiness just from the very process of research, and as well from experimental repetition, which serves to reveal truth. He also observed that only the Nordic scientist is able "to take pleasure in the struggle with the object of study, like the mystery of the hunt." Therefore, for the Nordic student of science, this is simultaneously a dialogue with nature, and a competition with reality. At the same

time, in the process of research, the non-Nordic scientist only satisfies himself with the posing of a problem and its results; but morally and ethically, a strong-willed look into the essence of existence falls around him.

Left: Phillippe Leonard

International science exploits the object of research; racial [science] aspires to give the subject a sense of usefulness, in the context of the existence of race. Soul-less pragmatism is equal to dogmatism; it is contradictory to the goals and tasks of racial science, whose principles are based on the vast genetic potential of the cognitive abilities of ancestors. This is what is known by the simple Russian definition of "inborn shrewdness;" this means that a person gifted with it is able, in the process of cognition, to repeatedly strengthen personal intuition with the genetic experience of [his] ancestors, and this synthesis gives the effect of a trampoline, in the shot for mastery of the unknown.

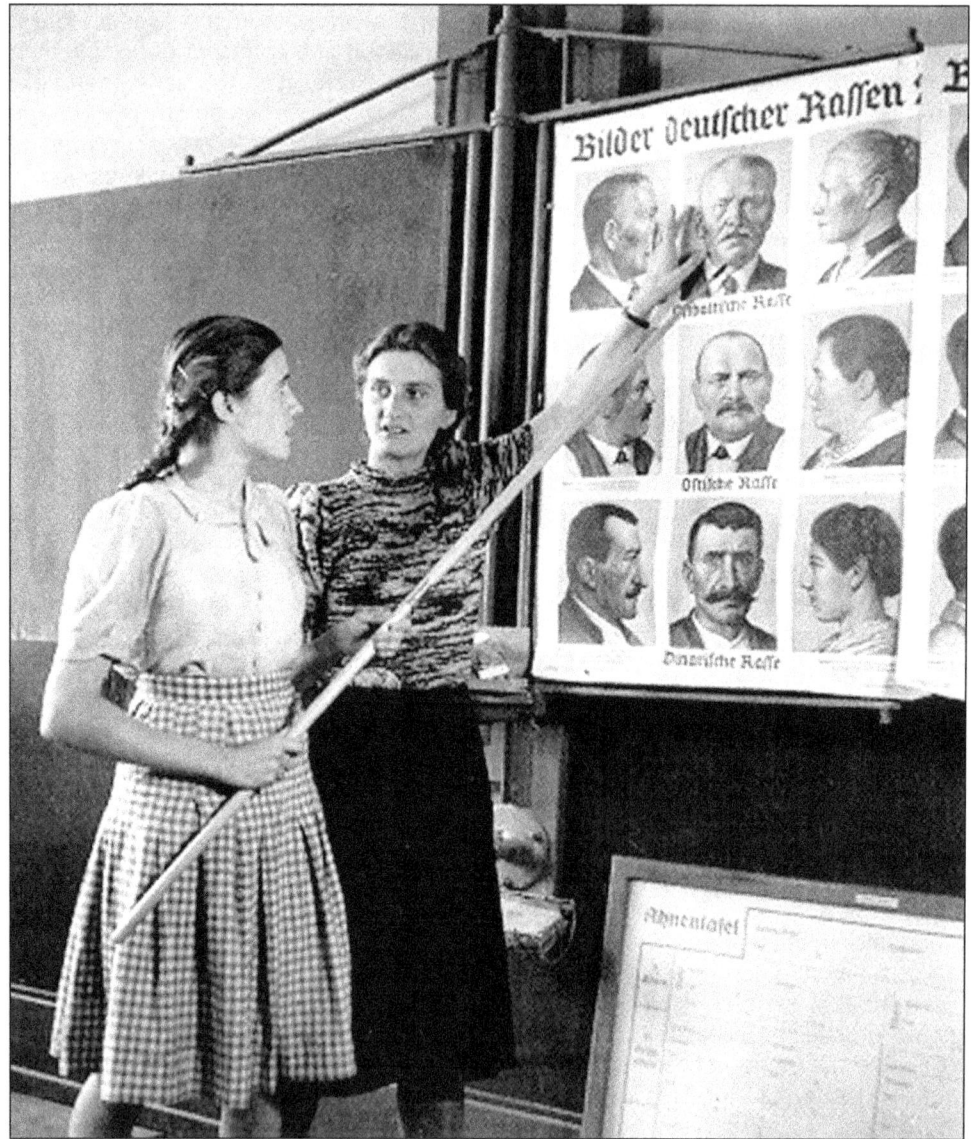

A lesson in Racial Theory in a School of the Third Reich

Phillip Leonard, and many other scientists of the first order in Germany, headed the *Aryan Physicists* movement in 1936; it was directed toward struggle with the tendencies of cultural pessimism, and with Einstein's *Theory of Relativity*, which is a clear expression of anti-racial science.

For good reason, the Swedish raciologist, Gaston Backman, emphasized in this regard, "If we measure a civilization not by the absolute number of creative personalities, but relatively, then the pure-blooded peoples of the north surpass all the other countries of Europe."

Professor Johannes Stark (1874-1951), a luminary in the area of physics, as well as a Nobel Prize Laureate, published numerous works about the theoretic fundamentals of science; among them is the fundamental monography, *National Socialism and Science* (1938). In it, he put forward the thesis that only for members of the Nordic race does the relationship toward scientific fact have independent psychological value, and is not a means for materialistic, self-interested speculation, as it is for members of the southern race, who experience genuine ecstasy from endless sitting at a noisy and dirty bazaar. The psychological type of the academic scholar, grasping the beauty and secrets of existence in the quiet of a study room, decorated with the busts of predecessors and numerous folios, is inherently an archetype member of namely, the Nordic race. "The pathos of distance," as the great Nietzsche said, identifies itself in its understanding, not only in contact with people, but with nature itself. And it is namely from this pathos of distance that an authentic respect for the world is born, [something] completely alien to the Asiatic trickery and devices for the surrounding environment. The Nordic man does not create for the sake of the environment, but in defiance of it. Racial instinct prompts him that the highest pleasure lies in the intellectual and strong-willed struggle with the puzzles of nature. Not for worthless metal, but for truth, is it necessary to give oneself without reservation. Not for tinsel, applause, and not for expensive, triumphal apparel does the Nordic man create, but for the realization of his inner rightness, strength, and primacy. It is not people, but God, who is his real rival. In his essay, Johannes Stark pointed out: "The ability to observe and respect facts with complete disregard of one's "self" is a very characteristic trait of the scientific activity of the Nordic type. They experience happiness and satisfaction from gaining scientific knowledge, because it is namely that which interests them. Only under pressure do they decide to make their discoveries public; to the Nordic man commercial and propagandistic use of [his discoveries], appears as a degradation of his scientific work."

Left: Johannes Stark

Dr. Bruno Kurt Schultz, a professor and SS Hauptsturmfuerher, was one of the prominent theoreticians and practical workers of his time, and his work to this day has not become outdated. Dr. Schultz worked out the criterion for the selection of personnel for elite military units. Curiously, he was the best and favorite anthropological luminary of Rudolf Martin. Besides that the morphological principles of evaluation of physical proportions that he proposed, are used to this very day for selection of the officers and soldiers of the Kremlin Regiment, the unit which carries out parade guard duty in the center of Moscow. This may be shown as an historical oddity, but a fact remains a fact: the "ideal Slavic type," which is assigned to carry out the duties of the honor guard around "the mausoleum of the leader of the proletarian world," were, for almost the entire extent of the Soviet period, selected according to the criterion put forward by the **Chief Racial Anthropometrist of the SS, Dr. Bruno K. Schultz.** And in this there is nothing conclusively unnatural, for the scientists of Germany well understood that the differences between Germans and Russians were extremely insignificant. Still another more interesting fact: the rulers of the Kremlin, not being members of the Nordic race, nevertheless selected namely this racial type as a canon for the sacral representation of their regime. What concerns Dr. Bruno K. Schultz is that on an equal footing with his colleagues, he was not noted in propaganda as a " rabid Russophobe," as political prostitutes would try to convince us. In his book, *The Science of Heredity, Raciology, and Racial Hygiene* (1933), he emphasized: "Europe, and the areas adjacent to it, the countries around the Mediterranean Sea and Near Asia, are the homeland and geographic range of distribution of the higher races, which stand out as progressive and most distant from animal forms, by their physical and psychological qualities. Among the higher races, the most highly developed is the Nordic race. The Nordic race is clearly an expressed race of lords; it has effected so much more influence on world history, than any other race. Along with its permanent worth, it is worth mentioning a strong inclination of the people of this race to act on their own, a frequent absence of understanding of common interests, and an unwillingness to submit, which was often a cause for lack of unity between Nordic peoples."

Left: Bruno Kurt Schultz

As we see, Dr. Bruno K. Schultz did not fall into primitive chauvinism; he pointed out the qualities of the Nordic race, which manifest themselves positively in one situation, and in others, lead to its weakening. Somehow or other, everything is truly relative. The modern development of China graphically testifies that the absence of egotistical aspirations, characteristic of members of the Nordic race, are a positive evolutionary factor in the conditions of a group strategy for survival. The views of the German scientist on the biological nature of the Slavs are also completely scientific, and do not differ strongly from the conclusions of Soviet science: "Among aristocrats, the Slavs were a significant percent of the Nordic race, the rest of the population belonging, in great part, to the Eastern race, or the Eastern-Baltic type. Therefore, we often equate the Eastern with the Slav, although the skeletons of Slavic princes from the Middle Ages are very similar to the skeletons of Germans from the Great Age of the Migration of Peoples."

Again, any discussion about a certain, almost inborn German chauvinism, does not find any kind of substantiation, including, it would seem, such that are on the level of political prostitutes. But a confirmation of the objectivity and impartiality of German science is found with ease.

The raciologist Richard Eichenauer (1893-?), in the book, *Race and Music* (1932), formulated basic postulates of racial musicology and determined that the psychophysiology of the Nordic man is right for harmonious music, and not for the jagged syncopation of negro jazz, with the beat of tom-toms; nor the irksome wailing of Turkish [music], which conveys the centuries-old sadness of nomads at the sight of the barren desert. Symphony, as a reflection of completely pagan colors of existence, maximally assists the appropriate racial cultivation of feelings in the truly white person. For his part, Siegfried Kadner (1887-?), in the book, *Race and Humor*, substantiated the racial attributes in such a subtle substance. And really, it doesn't take a special racial psychologist to know the difference between an anecdote from a small town outside Odessa and the salon humor of English gentlemen, by the characteristic traits of the narrators.

Left: Otto Hauser

The summary of observations of such a person enabled Otto Bangerdt (1900-?) to write a study with the fitting title, *Gold or Blood*, in which he concluded that for the people of the Nordic race, gold always was, and always will be, filthy lucre, because for those with real virtue, it is understood that it cannot be purchased with money, but can only be acquired by means of inheriting noble blood. Virtues cannot be taught or traded; people are born with them, and with them they die. On the basis of a metaphysical interpretation of worthy principles, another important raciologist, Wilhelm Erbt, (1876-1944), created the fundamental report, *World History on a Racial Basis* (1934), in which he advanced and substantiated a complete sacral-biological category—"Nordland." This territory, in his opinion, is the monastery for the bearers of high culture on Earth. Books by important scholars, like Otto Hauser's *The Blonde Man* (1930), and Willibald Henschel's, *Darwin: A Look at World History from the Position of the Aryan* (1918), are dedicated to working out a unified worldview, and the evaluation of principles of historical development, on a racial basis. Other prominent German scholars made weighty contributions to the substantiation of the Nordic idea in racial philosophy: the historian Gustaf Kossinna (1858-1931), who published the essay, *The Ancient Germans* (1921); the linguist Herman Alfred Hirt (1865-1936), with his book, *The*

Indo-Germans (1907); and also the archaeologist Karl Schuchardt (1859-1943), with the publication of the monograph, *Ancient Europe: Culture, Race, and Peoples* (1935).

Left: Friedrich Keiter

Finally, of the creators of an entire field in racial theory, which received the name **culturobiology**, Friedrich Keiter (1906-1967) wrote a three-tome report, *Race and Culture* (1938), in which he substantiated the criterion for the biological evaluation of the cultural-creative abilities of separate races, [with the help of] extensive historical and natural scientific materials.

The Light Hair of the Nordic Peoples in Ancient Times (1935), by Wilhelm Ziglin, is particularly worthy of consideration; it is small in size, but completely shocking in its detail of factual material. First of all, in the introduction to the main body of the book, the author emphasizes that he develops the ideas of his teacher, the world-famous Friedrich Ratzel—one of the creators of **ethnology**.[57] According to the Ratzel-Ziglin concept, the Nordic race arose from the so-called "Russian Island"—that portion of the territory of Russia that was cut off for 10,000 years from the south by the Black Sea, and from the north by glaciers. It is worth mentioning, that at that time in Germany—according to the assurances of Soviet propaganda—unbridled anti-Slavic hysteria reigned. However, one of the official [German] racial theorists wrote a fundamental theoretical work, which concluded that the appearance of the blonde racial substrata, which comprised part of the German people, is from Russian soil. If this is indeed an example of anti-Slavic propaganda, then it is not perfected to a very high degree.

Characterizing the purpose for writing his work, W. Ziglin remarked: "I started this work with the goal of clarifying for myself the very question: were all Indo-German peoples light-haired, in their original appearance? Was light hair an exclusive trait that set them apart from their neighbors? For this purpose, I collected evidence from ancient authors on the color of the hair of their peoples and that of their neighbors, but did not limit myself to data about separate peoples; I also collected information about persons who are known to us from literary sources or from works of art. I also took into account the gods, heroes, and literary personalities. This is eloquent evidence of the impressions of the speakers of their age, about their outward appearance." The Hellenes, the Italics, the Gauls, the Germans, the Scythians, the Armenians, the Persians, the Alemanni, the Batavians, the Franks, the Goths, the Longobards, the Rugi, the Saxons, the Suevi, the Teutons, the Vandals, the Celts, the Alans, the Albanians, the Ante, the Arimaspi, the Gedi, the Sarmatians, the Libyans, and still numerous other peoples, are described by ancient authors and portrayed in graphic productions of art, as peoples consisting of almost completely light-haired individuals. The ruling class of the Hindus, from the time of their conquest by the Aryans, also consisted of blondes.

Analyzing the outward appearance of the ancient Greek gods, Wilhelm Ziglin came to the conclusion that Aphrodite, Apollo, Ares, Arethusa, Asclepius, Athena, Dionysius, Eros, Europa, Harmonia, Helios, Hera, Hermes, Hymen, Nike, Pan, Persephone, Pluto, Satyr, Celine, Serapis,

---

[57] *Voelkerkunde*, in German

81

Typhon, Zeus, and so on—more than 60 were pure blondes, by a general count; and from mythological personalities, there were: Achilles, Adonis, Agamemnon, Amphion, Andromeda, Ariadna, Electra, Heracles, Helen, Jason, Cadmus, Medea, Medusa, Meleager, Menelaus, Nausica, Odysseus, Oedipus, Orestes, Pandora, Patroclus, Penelope, Perseus, Phaedria, Theseus, and many others with blonde hair, up to a general count of 140; and from the ranks of important historical figures from ancient Greece, the following had blonde hair: Alexander of Macedon, Anacreon, Apollonius of Tyan, Aristotle, Dionysius of Syracuse, Sappho, Pythagorus, and still dozens of philosophers, colonels, scientific and cultural figures, the general number of which reaches up to 200, from a number that is historically fixed.

In that same time, the number of dark-haired gods in existence numbered twenty-nine. Among mythical characters, a total of nineteen brunettes was revealed, and among the historical figures of the Iliad, only twenty had completely black hair.

Thus, it doesn't take any work to come to the simple conclusion that people of the Nordic race, having primarily blonde hair, were the creators of the unsurpassed ancient Greek culture.

In the ancient Roman pantheon, blonde gods numbered twenty-seven, and ten mythological personalities also had light hair. Apart from the dozens of important historical personalities, among the Roman emperors that definitely had blonde hair were: Augustus, Nero, Trajan, Titian, Hadrian, Domitian, Vitalus, Commodus, Caracalla, Gallinius, Gordian, Honorius, Valentinian, Julian, Theodosius I, and Theodosius II. Many other deciders of the fate of the ancient world also stood out for having light hair. The very archetypical symbol of ancient civilizations, from the viewpoint of the canons of beauty, is definitely tied to the biological type of the Nordic race. The appearance of the god-like, the heroic, and the supernatural is always shrouded in easily assimilated traits of light-skinned flesh, embodying in itself a higher concentration of sunny substance and blessings. Wilhelm Ziglin considered it necessary to sharpen the reader's focus on the fact that the ancient Vedic god Indra was a blonde. And a complete historical oddity that confronts us is the fact that Africa—the white-skinned goddess of the ancient Libyans—is now, with the passage of time, associated with black-skinned types.

And if the sources of higher ancient culture and civilization are rooted in the biologically hereditary essence of the blonde race, then the wave of anarchy, democracy, and sexual license in decaying societies is steadily accompanied by a drastic darkening of hair pigment. By the testimony of many historians and eyewitnesses, the final fall of ancient states usually arrived with the complete disappearance of light racial types from the ruling class. Regretfully, we are observing the sad experience of antiquity, in the conditions of decay in modern, so-called Western Civilization.

One may analyze the correctness of the German raciologist's harsh conclusions, with a visit to any historical museum that has a good collection. The Greek and Roman halls of Sankt-Peterburg's *Hermitage*, for example, have pure samples of the Nordic, recorded in the correct proportions of the bodies of athletes, the deep thinking facial expressions of the faces of philosophers, and the graceful, stately bearing of the emperors. Further in the course of the exposition, the statues of historical persons belonging to the 3rd Century, A.D., already have clearly legible traces of degeneracy, caused by the chaos of race-mixing; finally, they lead to caricatures of ugliness in the grimaces of the Christian saints in the Middle Ages. All Byzantine - and equally so - West European iconography is a straight-laced, powdering of the physiological decay of simple reflexes. To this day, the lifeless-eyed, powdered-nosed monsters, as masterpieces of spatial perspective, and the weak, rickets-stricken, dry-breasted figures, are presented to us as a personification of "high spiritual style." The decadent art, which has twisted around European culture since the end of the 19th Century, continues this depressing tendency of masking the harmful nature and elementary degeneracy of the artist, as "creative inspiration." The neurology of modern times showed the fact of degeneracy with all obviousness. Therefore, the leading racial theorists of Germany at that time appeared as an ideological security unit, in the hosting of the famous "Degenerate Art" exhibition in Munich, in 1936.

The book, *Nordic Man* (1939), by the leading Norwegian scholar, Halfdan Bryn (1864-1933), played a special role in the formation of racial theory. From the viewpoint of evolution, Bryn generally set apart the members of the light racial types in a special biological species: **homo caesius**—literally, "gray-blue eyed Man." Bryn pointed out that in Norway, they retained areas where 100% of the male population has white skin; 98.5% has blue eyes; and 99% has light or red hair; this enables one to make the legitimate conclusion: "In recent times, many spoke about the homeland of the Nordic race. The fact that in our days, it is represented in almost pure form on the Scandinavia peninsula, suggests the thought that [Scandinavia] was its homeland." The notion of a principle biological difference of *homo caesius*, from other types of peoples, was supported by an earlier observation of the Polish anthropologist, Ludwik Krzywicki, who validly observed: "Among the anthropoid apes, there are no blue eyes."

Left: Egon von Eichstadt

However, the highest development in the indicated period of the given field, was archived in a monumental report, under the title, *Raciology and the Racial History of Mankind*, by the famous world scholar, Baron Egon von Eichstadt (1892-1965). The two 900-page tomes were published from 1938 to 1943, and embraced the origin of the human races as a common problem of history, and numerous particular problems, connected with racial morphology, pathology, the theory of heredity, immunology, and so on.

In this circumstance, the report contains an abundance of information, dedicated to the evolutionary specifics of the Nordic race. To this day, the majority of it has not become obsolete. Besides that it is worth emphasizing that in the section dedicated to the history of the development of anthropology as a science in various countries, an entire block of information is dedicated to Russia. With emphatic respect, the idea of the contribution of Russian scholars is developed, and also the uniqueness of the Russian school of racial anthropology is highlighted. It is worth noting that all this took place during the culmination of the war on the Russo-German Front; however, in the academic circles of German academic science, there were no signs of indiscriminate Russophobia to be seen.

7. Soviet and post-Soviet Science in the Service of Racial Theory

From the side of ideological opposition, a negative, aggressive attitude was formed toward classical raciology, with the efforts of such persons like Arkady Isaakovich Yarkho of the Soviet Union, and many others. A Soviet classic Marxist variant of the science was developed, receiving the name of **ethnogeny**. In scientific academic literature, the lowering of style, and moreover, openly insulting attacks are not permitted; the freshly-baked Soviet race scientists immediately undertook to brand their German colleagues, calling them "anthropo-fascist," "raceniks," and "Nordomen," and naturally did not bother themselves to a make a conscientious analysis of their ideas.

However, the "researchers" of Marxist stylistics betray the biological origin of the critics, with more to spare, according to all laws of classical racial theory. We cite only several "eloquent" titles of these opuses, as an example of proletarian bad taste: A.A. Shiyk's *The Racial Problem and Marxism* (1930);[58] G.I. Petrov's *Racial Theory in the Employ of Fascism* (1941);[59] G.A. Shmidt's *The Truth about Races and Racism* (1941);[60] V.A. Vasilenko's *Racial Ravings of Fascist*

---

[58] *Rasovaya problema i marksizm.*

[59] *Rasovaya teoriya na sluzhbe u fashisma.*

[60] *Pravda o rasakh i rasizme.*

83

Bandits;[61] B.M. Zavadovskiy's *Racial Madness of German Fascism* (1942);[62] X.S. Koshtoyants' *Science against the Fascist Madness about the Races* (1942);[63] and M.A. Moskalayev's *Racial Pseudo-Science of the Fascist Robbers* (1942). We note that Goebbel's Ideological Propaganda Department did not permit itself to slide down to the level of vulgar profanity, for which many Soviet race scientists stood out, while wearing the mantel of academic regalia.

However, for the sake of justice, we emphasize that not all Soviet scholars included themselves in this primitive Comintern agitation. Despite the fact that a science career under the conditions of the Bolshevik regime was for many, linked even with the threat of physical extermination, the better scientists preserved their academic impartiality. From the beginning, ethnogeny was thought of by communist party functionaries as a class [struggle] answer to bourgeois raciology; therefore, the continual discrediting of the postulates of racial theory, among them the significance of the Nordic race in history and the formation of world culture, was part of the chore list of scientific tasks for Soviet race scientists.

Nevertheless, in the article, *Once More about the Blonde Race in Central Asia*,[64] the distinguished Soviet-era anthropologist, Georgiy Frantsevich Debets considered it necessary to emphasize: "At the end of the 1st Millennium, B.C., and at the start of the 1st Millennium, A.D., Chinese sources spoke of tall, blue-eyed, red-headed tribes inhabiting the territory embracing the Altai-Sayan hills. In that same age, and a little earlier, there lived a people on the territory of Minusinkogo Kraya, that were anthropologically, unconditionally, Europoid. A predominant portion of the skulls are cranially extremely close to the northern race." The said declaration completely supports the general postulates of race theory, just like the private research of G.E. Grumm-Grzymaylo, and the German raciologist, Hans F.K. Guenther.

Another luminary of domestic Russian science, Victor Valerianovich Bunak, described the the characteristic cranial traits of the Nordic race in the article, *On the Question of the Origin of the Northern Race*.[65] Analyzing contemporary scientific objections to the given topic, he came to this conclusion: "We should recognize the highly probable existence in Paleolithic Europe, of two types, namely: Cro-Magnon and Aurignacian, and see in them the main elements of the developed type of the Northern Race. Thus, culturally and schematically, the hereditary link of the races of the Paleolithic with the Neolithic, is established; in these we already find the cranial prototype of the northern race."

The given conclusion also supports the basic postulates of racial theory.

Now it is especially worth emphasizing that the very term "racism" first appeared in 1932 in LaRousse's French Dictionary, as a negative value judgement on research of the differences in the human races. Since 1945 this word began to be more actively used throughout the world, in connection with the fall of the Third Reich, in order to cast suspicion on and portray as evil, anyone who was able to differentiate the natures of peoples, just as, analogously speaking, we all differentiate the nature of dogs and cats.

But here we take into our hands the book, *History of the Ancient East*,[66] by the leading Soviet scholar, V.I. Adiyeva. It was published in 1948 by the State Publisher of Political Literature, and sold 100,000 copies. And it was accepted by the Ministry of Higher Education of the USSR, as a textbook for the historical departments of state universities and pedagogical institutes. In a section dedicated to Ancient India, we again observe the instructive, chopped style of Soviet propaganda, but now its aggressive attacks make a 180-degree turnaround. The Soviet historian proposed: "Defending the interests of the affluent strata of the population, lawmakers will strive to limit the personal freedom of the Aryans. Barbarians are not forbidden to sell or mortgage their descendents, but for Aryans there should be no slavery. The word "caste" is of Portuguese origin, and means "of pure tribal origin." In the Hindi language, "caste" appears as the word, *dzhati* (birth), or the word, *varna*, which means "color." People belonging to the first three castes called themselves "twice-born" or "twice-born Aryans."

[61] *Rasovie bredni fashistskikh banditov.*

[62] *Rasoviy bred germanskogo fashizma.*

[63] *Nauk protiv fashistkogo breda o rasakh.*

[64] *Esche raz o belokuroy rase v Tsentral'noy Azii.* Sovetskaya Aziya, 5-6, 1931.

[65] *K Voprosu o proiskhozhdenii severnoy rasi.* Antropologicheskiy zhurnal, N 1, 1934.

[66] *Istoriya drevnego vostoka.*

Later, in an original way, V.I. Avdiyev applies classic Marxist theory to the study of castes, noting that the first *Brahmins* came from the mouth of the first man, Purushi. Thus, according to ancient tradition, holiness and truth belong to them only, and so therefore, the study of sacred books became their main occupation, [as well as] the teaching of people, and completion of religious ceremonies. Members of the lowest caste, the *shudri*, were created from the feet of Purushi, and therefore were obligated to grovel in the dirt. By the laws of Manu, the son of a Brahmin woman and a *shudri* fell to a very low social group of *chandal*, and was called the "very lowest of people." "The living quarters of the *chandals* should be located outside of the village; they should have particular utensils, and their property should only be dogs and asses. Their clothes should be clothes of the dead; black iron [should be] their ornaments, and they should always wander from place to place. A man who fulfills religious obligations should not seek relations with them; their affairs should be between them and their spouses—with those like them. Their food should be given to them by others in broken dishes; at night they should not walk about the villages or cities." In the sacred text of the *Makhabkharate*, it says that mixing of castes is the result of lawlessness. In the laws of *Apastamba*, it maintains that every caste stands higher by birth, than the next, and honor should be shown to them, who belong to a higher caste. The main conclusion in this chapter, given by V.I. Avdiyev, led to the following: "The goal of the caste system was to strengthen the prevailing situation of the Aryan conquerors, over the conquered native *Dasyu* population." Openly savoring the racial and caste inequalities on the basis of the ancient Vedic texts, the Soviet scholar nowhere permitted himself even a hint of criticism of the given system, [something that was] always considered a common obligation in the style of presentation, in the communist period. The given fact speaks about the ideological support for the "Aryan theme," by the upper Stalinist entourage, after 1945.

In general, for the sake of fairness, one should note the powerful surge in the development of Russian anthropology, ethnology, ethnic history, paleontology, and paleo-linguistics, during the brief period of the late, post-war age of Stalinism; much of the emphasis of Soviet science almost completely agreed with the basic postulates of classic racial theory, which is more than a little ironic, considering the fact that Soviet science had been committed to opposing it. This is highly demonstrated in the context of Georgiy Frantsevich Debets' fundamental monograph, *Paleoanthropology of the USSR*,[67] in which he follows A.P. Bogdanoviy and many German raciologists. He demonstrated that the skull of the Nordic race is precisely **proto-European**. He wrote: "In the 3rd and 2nd Millennia, B.C., the range of the European race tree extended east to the Yenesei River. The skull of the proto-European type can be determined as Northern European. On the territory of the central regions of Russia, and also in the Ukraine, the cranial index remains practically the same for a long time (from 2,000 B.C. until the 7th-14th Centuries, A.D.) Similar types of skulls have been discovered in Egypt, dating from the 4th Millennium B.C. Available paleo-archaeological material shows that Turkish nomads of the Ukraine are related to a significant degree, to the Turanian type that arrived from Asia."

Besides that , G.F. Debets points out that from the 3rd to 7th centuries, A.D., the mountainous region of the Crimea was populated by Goths—a Europoid people with clearly expressed traits of the Nordic race: "In the Gothic series of skulls, there is inarguably no Mongoloid type to be found." Many authors—including Debets—consider the Goths to be the aboriginal inhabitants of the Crimean Peninsula, in which connection it is noted, that the skulls from common Germanic graves are very similar to [those of] ancient Slavs.

In a collective anthology of works, *The Origin of Man and the Ancient Settlement of Humanity*,[68] G.F. Debets writes in the article, *The Settling of Southern and Near Asia, According to Anthropological Data*,[69] that "Among the Indo-Aryan peoples of northwest India, to this day there is almost no mixture of types to be observed, of the Negroid-Australoid race. Hair is gently wavy or even straight, [with] strongly developed tertiary hair coverings. The skin is swarthy, approximately like that of southern Europeans. The eyes and hair are dark, almost completely without an admixture of mixed shades. The faces are narrow, and right-angled; the noses are also narrow, and now and then are hunched. The heads are dolichocephalic (72-74), the

---

[67] *Paleoanthropologiya CCCP*. Moscow, 1948.

[68] *Proiskhozhdenie cheloveka i drevnee rasselenie chelovechestva*. Moscow, 1951.

[69] *Zaselenie yuzhnoy i peredney Azii po dannym antropologii*.

forehead is slanted, and the brow ridges more strongly developed, than in other places of India. The body is large (169-175). The penetration of Europoid elements into India began, evidently, in deep antiquity. Paleo-anthropological materials allow one to speak of Europoids from the 4th Millennium, B.C., but it is highly probable, that their first appearance dates to a significantly earlier period.

In another article in the given anthology, *Anthropological Data on the Settlement of Africa*,[70] he points to an admixture of ancient Europoid elements in Asia, and northern and even central Africa, making a characteristic conclusion: "The admixture of Europoid elements traces to the south of Ethiopia, among the negro Bantus. The Europoid disseminated in North Africa, Asia Minor, and Egypt, more or less retaining its Cro-Magnon traits. The mixing of Europoid types with negroids occurred, evidently for the entire extent of the history of Africa, and continues up to our days.

Problems of the Settlement of Europe, According to Anthropological Data,[71] a large and thorough joint article by G.F. Debets, T.A. Trofimoviy, and N.N. Cheboksarov, is an authentic embellishment and true masterpiece of the publication. In the article, it emphasizes: "Somehow or other, the settlement of Europe is dated to the very earliest period of human history. The cranial type of the "northern race" already existed in Scandinavia, while in the East they still retained in full, the Cro-Magnon features of the proto-European type. Research of paleoanthropological materials from excavation sites in Mesopotamia, the Caucasus, Asia Minor, Iran, and the Pamir, leads to the firm conclusion of the general dissemination of the dolichocephalic skulls in the past in these areas. The kinship of all Europoid types cannot be subject to doubt. This kinship suggests the existence of a common, original form."

Thus, Soviet anthropologists again and again upheld the basic postulates of racial theory, according to which, namely the Nordic racial type was the original in the formation of all the white races, on the gigantic expanses of Eurasia. Not escaping their attention is the specific and highly significant problem of the megaliths, which were interpreted from the racial point of view, by the German anthropologist, Ludwig Wilser. It is noteworthy, that Soviet science unpretentiously and firmly labeled him as nothing but a racist, but for some reason came to the very same conclusions [that Wilser came to].

"Megaliths appeared very early in North Africa and on the Pyrenean Peninsula, and sometime later in France, where they are particularly numerous in Bretagne, and still later on the British Isles; and in northwest Germany, in Denmark, and Scandinavia. It is important to note, that sharply long-headed, comparatively gracile, narrow-faced and narrow-nosed skulls predominate in the Dolman burial sites of that time, while in mass graves beyond their range, more massive, moderately dolichocephalic skeletons of the Cro-Magnon type are encountered more often." On the basis of the conclusions of West European and Russian science, [it is believed that] the megaliths fulfilled the role of a cultic institution, summoned to collect the biological power of the Nordic, dolichocephalic race, and consequently, hinder miscegenation with other, wider-headed and dark-pigmented races. In support of this, although a daring, but obvious thesis, Soviet scientists destroyed the myth [of Russians as] heterogeneous, Asiatic "Scythians with horrible crossed-eyes." The authors in the article emphasized: "The skull, as settled, of the nomad Scythians is related to the long-headed European type." They further substantiated the stability of the picture of the racial types of not only the Russian people, but of all Europe: "The population of the Neolithic and Bronze Age cultures of Europe are genetically linked with the population of the preceding ages. The anthropological composition of the population of Europe as a whole did not undergo any substantial changes. There is no need to search without fail for the extra-European roots of the tribes of the Neolithic and Bronze ages." The authors of the article based their views on the works of Russian classics: "By the 1870s, it was established by Bogdanov, the prominent Russian scientist and founder of Russian anthropology, that in Slavic burials in the middle of Russia in the Kurgan period, an extremely long-headed anthropological type predominated. This long-headed and narrow-faced type was considered by A.P. Bogdanov as the single, original type, not only for the Slavs, but for the whole ancient population of Europe, independent of

[70] *Antropologicheskie dannye o zaselenii Afriki.*
[71] *Problemy zaseleniya Evropy po Antropologicheskim dannym.*

different ethnic groups. A.P. Bogdanov did not see any defining differences or traits in its physique, in separate localities or among separate, varied ethnic groups."

It is commonly known that the Kurgans of the central portion of Russia had an invented lineage, and also used them toward cultic ends. But in consideration of the given problem, from a racial point of view, everything once again ends up standing in its legitimate place. Analogous to West European dolmens and megaliths, the ancient Russian, hand-built Kurgan burial sites fulfilled the function of spatial generators of the power of the Nordic race, and were intended to support its ruling status. They represented a type of reservoir of the strength and wisdom of ancestors, from which the descendents could have drawn a vital burst of energy from the future. According to the information of many historians and archaeologists, the custom of preserving the skulls of ancestors as amulets and sacred symbols of authority, traces back to the Paleolithic Age, when, according to the assertions of V.P. Alekseyev, racial consciousness had securely formed and established itself in Man.

And finally, the last passage in this brilliant work again leads to a split with the class dogmas of Marxism-Leninism, for the great hysteria of the migration of peoples is explained, not by the whims of the economic moment, and not by the absolute power of abstract historic laws, but by biological consequences of the primeval struggle of races. "The great migration of peoples also caused the movement of the northern Europeans. Here and there, the preponderance of the light European types coincides with the dissemination of Germanic toponyms or Germanic languages, but in some places this tie has completely disappeared. Some Slavic peoples in the past and present are usually closer in physical type to their non-Slavic neighbors, than to territorially remote Slavs. Between various Slavic peoples, gradual transitions in the anthropological types of separate groups are observed. Anthropological materials do not give confirmation about the spread of Slavic languages that accompanied the massive migrations."

In a much later, but significant book in the history of the science, by the classic Soviet anthropologist, Viktor Valerianovich Bunak (1891-1978), *Of the Skulls of Man and the Stages of his Formation in Excavated Peoples and Modern Races*,[72] many given positions were confirmed and substantiated on a still qualitatively higher level: "The Europoid complex of characteristics is established on ancient skulls, far from the limits of Europe, on the wide expanses between the Volga and the Yenisey [rivers]. Very revealing is the increase in the cranial index in a series of skulls from the 14th Century, from the Central Zone of the Russian plain. In the previous age, on the territory between the Oka and Volga [rivers], no significant centers of brachicephalic types existed, by the extent of which, one may explain the higher index, in comparison with the Kurgan cranial type."

Left: Halfdan Bryn

From there another untraditional position flowed out, for it was discovered that in the anthropological sense, the notorious Mongol-Tartar yoke did not exist completely. One of the basic anthropological traits of the Mongoloid race, which is not subject to stimuli from the outside environment, is brachicephalia, or short-headedness; but pockets of its distribution appear once in central Russia in the 16th Century, when that very yoke itself, according to the generally accepted assertions of historians, had dwindled to nothing. In the author's conclusion, there is nothing that is maintained that is extravagant or contradictory to science.

Thus, the important modern Russian anthropologists, V.E. Deryabin and A.L. Purundzhan, clearly point out in the book, *Geographic Particulars of the Physique of the Body of the Population of the USSR*,[73] that: "As horrible as it was in its historical and social consequences, the Tartar invasion left practically no trace, anthropologically speaking."

---

[72] *Cherep cheloveka i stadii ego formirovaniya u iskopaemykh lyudey i sovremennykh ras.* Moscow, 1959.
[73] *Geograficheskie osobennocti stroeniya tela naseleniya SSSR.* Moscow, 1990.

V.P. Alekseyev also summarized in his book, *The Geography of the Human Races*,[74] that: "The cranial material, by its closeness to modern eastern Slavs, particularly the Russian population, paints a picture of craniological closeness between separate territorial groups of Russian people."

Jan Czekanowski

[74] *Geografiya chelovecheskikh ras*. Moscow, 1974.

In the context of our position, the magnificent fundamental report, *Paleoanthropology in Middle Asia*,[75] by the famous Russian scientists V.V. Ginsburg and T.A. Trofimova, will not be neglected. Relying on enormous archaeological and craniological materials, they lay emphasis first of all, on the assertion: "The races of Man, like the subspecies of animals, are categories; that is, [they are] essentially biological." Going over to a description of the processes of racial dynamics in the very heart of Eurasia, the authors reconciled them with the phenomena of social and cultural life, acting according to the methodology of classic racial theory. At the beginning of the 2nd Millennium B.C., the first states took shape in the south of Middle Asia: Hirkaniya, Parthia, Margiana, Bactria. "In Middle Asia, cultures of the Bronze Age arose in the 2nd Millennium B.C. Evidently this movement comprised the first significant wave of Iranian-language Indo-Europeans, penetrating Middle Asia from the northwest. Skulls discovered in graveyards of the 3rd and 2nd Millennia B.C. have two European types: Mediterranean and proto-Nordic. Skulls of the Bronze Age from Kazakhstan, taking into account individual differences, may also be placed with the two different types of the great Europoid race."

Left: Valeriy Pavlovich Alekseyev

Thus, Soviet researchers confirmed the basic postulates of racial theory, saying, that on the gigantic expanses of Eurasia, it was namely the Europoid type that fulfilled the function of culture-creator, and in which role the Nordic element was its biological base.

Considering the racial base of concrete ethnic communities, the authors pointed out that the *Saki* and *Sauromatians* of the Priural'ya belonged to the Andronovo culture, and on the basis of cranial materials, Mongoloid admixture was not seen. Shifting the zone of racial analysis deeper into Asia, which historians and ethnographers for some reason tie with the geographical range of the Mongoloid race to this day, Ginsburg and Trofimova refute [that position]: "The population of Pamir in the Bronze Age was also very homogeneous, and without Mongol admixture. The Europoid race, with insignificant Mongoloid admixture, comprises the basic anthropological type of the Usuney of Semirechýa and in Tyan-shanya. Mongoloid admixture as a whole is insignificant."

Thanks to the initiative of Russian ethnographer L.N. Gumilev, a popular picture of the Huns as a Turkish tribe with clearly Asiatic traits, has formed. But that view does not correspond to physical anthropological facts. The Huns were racially heterogeneous; among them a large community of **Ephtaliti**—White Huns—stood out, and for them dark hair was generally considered an abnormal phenomena. In the 4th-5th centuries, A.D., the **Tocharians** were subjected to Ephtalite influence. It turns out that "Tochar" literally means "white hair" or "white head." The northern racial type is easily seen on coins with portrayals of Kushan and Ephtalite kings. Russian scholars again turn our attention to ancient Chinese records, which report that the members of the Andronovo culture were light-pigmented *Dinliny*. Moreover, among the population of the mountainous Pamir, no Mongoloid admixture is to be seen in general, to this day. "The Europoid type of inter-river Middle Asia is well-marked in cranial materials, right down to modern times, and is now best represented in the mountain Tadzhiks, and the population of the western Pamir."

As a whole, a strong surge of Mongoloid blood in Middle Asia begins in the 13th Century; that is, from the times of the Mongol-Tartar invasion. "In the middle of the 1st Millennium A.D., in connection with the movement from the East of a new wave of Turkish nomads, there is a growth of Mongoloid admixture in the make-up of the different nomadic groups in Middle Asia, and in the settled population. In the 13th-14th centuries, the Mongoloid characteristics of the population of Kazakhstan, as in all the territory of the plains of Middle Asia, more strongly increased, as an immediate result of the Mongol invasion. The Turkification of the population of inter-river Middle Asia only began in the 1st Millennium A.D., inasmuch as the "Huns" and later Turkish tribes originated mainly from the areas of dissemination of the Mongoloid race type. The majority

---

[75] *Paleoantropologiya Sredney Azii*. Moscow, 1972.

themselves belonged to that type; and Mongolization of the local population went parallel with the Turkification of the language. Admixture of Mongoloid traits among the population of inter-river Middle Asia in the 1st Millennium and the start of the 2nd Millennium was very insignificant. A strong increase in the Mongoloid component in the racial type of the Uzbeks occurred, evidently only in connection with the Mongol conquest of the 13th Century."

It is noteworthy that the spread of Islam in Middle Asia is completely tied with the appearance of a higher percentage of Mongoloid admixture, and that the limits of [Mongolification] correspond exactly with the limits of the spread of Islam. **Thus, it becomes obvious that namely a change of the concentration of these or those racial traits facilitates an acceleration or slackening of the advancement of an ideology, [including] religious ideologies.** When the population of Middle Asia was more Europoid, it adhered to Zoroastrianism, or faiths like fire-worshipping cults that preached racial segregation and forbad race-mixing, in accordance with caste laws. The invasion of the Mongol tribes increased the percentage of racially-mixed people; in areas where racial segregation was absent, the road was opened to the advancement of Islam. Religious science, as we can see, is not in need of the fundaments of sociology, but of racial biology.

In part, confirmation of the social-political results of miscegenation is supported by another aggregation of facts from the given detailed book, for V.V. Ginsburg and T.A. Trofimova point to the custom of many peoples of the above-named territories, of deforming skulls. In the burial sites of that age, skulls that were subjected to lifetime artificial deformation predominate. Sometimes they were deformed circumferentially, sometimes with an occipital combination. Deformation is encountered more often in the skulls of women, than men. Characteristically, Mongoloid admixture in the female skulls appears to be stronger than in the male skulls. The point is, that the given type of skull deformation influences not only the form of the skull, but also several facial sectors of the skull, imparting a more European look to them. As a whole, one may say that the given style of deformation provided for leveling and smoothing out of the Mongoloid admixture in the population of those regions. Consequently, the given custom arose from a desire on the part of mestizos to look more like Europoids.

The social-political aspect of the given report is revealed by the authors without effort, in another scientific publication. In the anthology, *Problems of Anthropology of the Ancient and Modern Population of Soviet Asia,*[76] T.I. Yablonskiy wrote in the article, *Mongols in the Cities of the Golden Horde (according to Stately Muslim Tombs),*[77] that: "Up to the 15th Century, the greater part of city-dwellers of the Golden Horde was composed of people of the mixed type. The Europoid component predominated. Judging by everything, in the provinces, just as in the capitol of the Golden Horde State, the process of anthropological mixing was directed at the assimilation of the Mongol conquerors. In wealthy brick burial vaults, located on the territory of mosques of mausoleums, people with completely Europoid features are buried. By all appearances, the son of a Mongol and a Polovtsiy woman, for example, could occupy a high social position and consider himself a Mongol, and all the while have a Europoid appearance."

Again, we are convinced that in any talk of the distinctive character and uniqueness of the culture created by the Mongol race, there are several exaggerations, for in all degrees of its development, it continually relied on the insemination of the creative blood of the European race, in which, for its part, the Nordic racial type fulfilled the function of the most worthy culture-creating element.

A surprisingly keen observation in this sense [was made] by the famous Russian historian, Alexander Fomich Wel'tman (1800-1870), who wrote in the book, *The Magi and the Midian Kagans*: "Is it known that it was the name of the Mongols that conquered the Rusi? No. In the course of the centuries of the pre-dominance of the so-called Mongols, neither the Rus, nor the Grand Prince who went to the Horde, pronounced that name, and only in 1567 did that name appear in Russian chronicles, when Tsar Ivan Vasiliyevich commanded the Siberian atamans and Cossacks to reconnoiter the Mongol lands of the Chinese Imperium, located beyond Siberia." In his opinion, the ancient geographical reports are worth subjecting to more thorough scrutiny, in view of the natural cleverness of the "Mongols", or "Mongolmans", or any Eurasians. Thus, the Monk Ryuysbrek from Brabant, sent by King Ludovik to Tartary in 1253, reported: "The Tartars, in

---

[76] *Problemy antropologii drevnogo i sovremennogo naseleniya sovetskoy Azii.* Novosibirsk, 1986.

[77] *Mongoly v gorodakh Zolotoy Ordy (po materialam musul'manskikh nekropoley.*

order to give foreigners to understand about the power and vastness of the authority of their khans, have a custom of circling [the realm] with [them], instead of going from place to place, as the crow flies..."—and he still adds—"You say one thing to [the interpreter], but he translates what comes into his head." Namely with the help of mountain climbing such as this, Europeans formed opinions about the cultural and political achievements of the East.

The anthology of fundamental works, *The Bronze and Iron Age Siberia*,[78] is dedicated to the problem studied by the author of this book. The classical Soviet anthropologist, V.P. Alekseyev, quite clearly emphasized in his article, *New Data about the Europoid Race in Central Asia*,[79] that: "The material studied by us widens the circle of factual data, by which one may judge of a wide distribution of the Europoid race in Central Asia, right up to western Mongolia, in the early Iron Age; analogies to this material and its comparative study demonstrate that the period of penetration into Central Asia by Europoids may hypothetically be moved to the Chalcolithic [Period], and their geographical range may be extended to Inner Mongolia." V.P. Alekseyev, just as G.E. Grumm-Grzhimaylo, K. Stratz, G. Fritsch, F. Weidenreich, and Hans F.K. Guenther before him, considered it vital to reinforce his daring cultural conclusions with references to ancient Chinese written sources, in which the authors honestly recognized that they borrowed fundamental cultural, civilized, and technological innovations from the members of the Europoid race. **There can be no more talk of the original significance of the culture of the Mongoloid race, in view of the fact that it acquired independent historical significance comparatively recently. In general, what cultural impact by the Mongoloid race on the European mentality can one speak of, if no one in antiquity had ever heard of the very term "Mongol"?**

Another recognized Soviet scientific authority, A.L. Mongayt, in the monograph, *The Archaeology of Europe. The Stone Age*,[80] concretely confirmed the basic postulates of Georges Vacher de Lapouge and the school of anthroposociologists. Relying on modern material, he emphasized: "In the Neolithic [Period], Europe was populated by tribes, among which there were all the known anthropological types, and which are preserved among the modern Europeans. To the time of the developing Neolithic [Period], the number of dolicho- and brachicephalics became similar. At the end of the Neolithic, the number of brachicephalics once again decreased somewhat. Relating to the Neolithic brachicephalic type, *Homo Sapiens Alpinus* (in Central, and partly in Western Europe), gave rise to two groups of dolichocephalics: the Northern and the Mediterranean. The Neolithic dolichocephalics of Western Europe were divided into: 1) the Cro-Magnon types; 2) the Mediterranean type; and 3) the Nordic type. The last settled Scandinavia, and part of the territory of Switzerland and Germany. The people were tall and well-built. Modern Scandinavians are the direct descendents of the local Neolithic population."

As we understand, one of the main assertions of racial theory, advanced by Count Joseph Arthur de Gobineau, is that the ascent of any historically significant civilization occurs as the result of the flow of the fresh culture-creating blood of the Nordic race in the social organism of a society. Soviet scientists K.F. Smirnov and E.E. Kuzmina, again and again asserted in the book, *The Origin of the Indo-Iranians in Light of New Archaeological Discoveries*,[81] that: "The second quarter of the 2nd Millennium B.C. was a stormy period in the history of the Old World: in Egypt, this was the time of the Hyskos, which was connected to the development of horse-breeding in the Nile Valley, and the time of the consolidation of the 18th Dynasty, in which Egyptian art reached its fullest flower. And in Near Asia, this was the time of the first appearance of the Indo-Aryans, the spread of horse-breeding in the Mitani Kingdom, and of war chariots, which became an important innovation in the military affairs of Babylon during the Kassite Dynasty. In Asia Minor, this was the period of the rise of the Hittite Empire and its clearly distinctive culture, which was the first in the Ancient East to firmly establish the fire cult. And in Greece, this was the time of the creation of the Mycenaean Civilization by the Achaeans, in which the use of the horse-drawn war chariot was an important factor."

In the USSR in 1977, the International Symposium on Ethnic Questions of the History of Central Asia in Antiquity (2nd Millennium B.C.) took place, the works of which were separately

[78] *Bronzoviy i zhelezniy vek Sibiri.* Novosibirsk, 1974.

[79] *Novie dannie oevropeoidnoy rase v Tsentral'noy Azii.*

[80] *Arkheologiya Zapadnoy Evropy. Kamenniy Vek.* Moscow, 1973.

[81] *Proiskhozhdenie indoirantsev v svete noveyshikh arkheologicheskikh otkritiy.* Moscow, 1977.

published in 1981. Important scientists from eleven countries took part in the symposium, in order to discuss various aspects of the "Aryan Problem." Soviet delegate B.G. Gafurov pointed out in his speech, *Several Problems of the Ethnic History of the Peoples of Central Asia in the Ancient Period*, that: "Data of the Indian and Iranian languages testifies about their origin from a single common source; the systematic and deep traits of similarity in religion and culture, of social and political organization, in agri-economy and way of life of the Iranian and Indo-Iranian tribes at the dawn of their written history, [and] their common self-designation, testifies to a community of ancestors of the Indian and Iranian tribes in a common Aryan period. Indo-Iranian unity is consequently not only a linguistic artifact, it represents an actual historical whole, existing in a definite period on a single territory. As a result of economic and social development in this period, an expansion of Aryan tribes into different territories began. The Aryan problem is complex, but by its content, it is first of all, an historical problem."

Prominent Russian anthropologist V.P. Alekseyev, in his article, *The Anthropological Composition of the Population of Ancient India*,[82] from the anthology *India in Antiquity*,[83] made the following conclusion on the basis of a wealth of anthropological material: "Members of the Europoid race appeared here from the north, evidently at the end of the Upper Paleolithic, or in the Mesolithic Period, and burst into the geographic range of distribution of of the Negroid race."

Finally, racial-anthropological analysis magnificently clears up the picture in the question, which in history and ethnography received the vague, amorphous name, "The Great Migration of Peoples."

T.A. Todt and B.V. Firschstein emphasized in the work, *Anthropological Data on the Question of the Great Migration of Peoples. The Avars and Sarmatians*,[84] that: "The Sarmatians as a whole are related to the Great Europoid race. A very small portion of the skulls of Sarmatians from burials are characterized by the traits of the Mongoloid race (21%) or by mixed Mongoloid-Europoid [types] (10%). The basic mass of skulls from the Sarmatians that have non-deformed craniums (60%) are related to Europoid types. The least number of skulls (23%) are related to the northern type of the Europoid race."

But the Sarmatians were located at the eastern extent embracing the zone of the "Great Migration of Peoples;" therefore, in their composition a significant percentage of Mongoloid admixture was noted. Among these peoples, generally named in chronicles as "barbarians" and "vandals", and occupying the territories to the west of this extent, the absolute majority turned out to be light-haired, so much that there was not even talk of an insignificant admixture of Mongoloid blood among them.

That historical notion, which we know as the "Great Migration of Peoples," was only a trick by the fakirs of liberal ethnography. In point of fact, this was a routine wave of expansion of the great white race, directed at the forced redistribution of living space, in which the Nordic racial type was traditionally the initiative-taking, governing catalyst.

In the post-Soviet period, better, conscientious scholars continued the tradition of studying the biological fundamentals of civilization. The distinguished ethnologist and historian, Valentin Vasilyevich Sedov, pointed out in the monography, *The Ancient Russian People*,[85] that : "The assertion of linguists about the Iranian or Indo-Aryan origin of the ethnonym **Rus'**, acquires a reliable, true historical basis. It either goes back to the Iranian stem *rauka*, *ruk*, meaning "light or "white", or it arose from the regional Indo-Aryan stem, *ruksa*, *russa*, meaning "light" or "white".

But it is plainly obvious that by "white" it is namely the people that were populating the given enormous territories [that were meant], the word itself pointing to their racial affiliation. The middle zone of Russia does not at all resemble the snow-covered tundra, and so itself could not be called "white." *Rus'* is a racial name that testifies about the Nordic origin of its age-old inhabitants.

The fundamental anthology of articles, *The Eastern Slavs, Anthropological and Ethnic History*,[86] is dedicated to discussions in the same spirit. The given anthology, by virtue of its objectivity, wide scope, and depth of study, can be recognized beyond all doubt as the best

[82] *Antropologicheskiy sostav naseleniya drevney Indii.*

[83] *Indiya v drevnosti.* Moscow, 1964.

[84] *Antropologicheskie dannie k voprosu o velikom pereselenii narodov. Avary i Sarmaty.* Leningrad, 1970.

[85] *Drevnerusskaya narodnost'.* Moscow, 1999.

[86] *Vostochnie slavyane. Antropologiya i etnicheskaya istoriya.* Moscow, 1999.

Russian work on the physical anthropology of the Slavs, as typical members of the Europoid race. In the foreword a thought is set aside, which is highly important in the context of this narrative: "The anthropological particulars of the population, thanks to its conservativeness, allows one to trace the different stages of the formation of the physical characteristics of the people, when in some phases of its history, it is not represented by anthropological data." This says that the modern complex of methods of racial diagnostics enables one to very accurately re-establish the make-up of an ethnic community at any stages of its development.

In the named anthology's first chapter, titled *The History of the Study of the Anthropological Composition of the Eastern Slavs*, written by academic peer T.I. Alekseyev, it points out that the leading Polish anthropologist, Jan Czekanowski, and the distinguished German scholar, Ilse Schwidetzky, suggested that the original Slavic type was Nordic. It will not be superfluous to emphasize again that I. Schwidetzky argued this assertion and laid it out in the book, *Racial Studies of the Ancient Slavs* (1938), which was published in large numbers in the Third Reich, at a time when, as "professional anti-fascists" attempt to ensure us, a wave of literal anti-Slavic hysteria reigned. Another prominent Soviet anthropologist, V.V. Bunak, relied on data from a geno-geographic study of Eastern Europe, and came to the conclusion that the original "proto-Slavic type" was highly consistent, and by its roots went back to the Neolithic Age, and perhaps even to the Mesolithic. Academician V.P. Alekseyev highlighted the extreme degree of morphological similarity in all the cranial series of the modern Russian people. All local regional variants deviate quite insignificantly from a single race type, which covers an enormous territory from Archangelsk to Kursk, from Smolensk to Penza. The author of the article, T.I. Alekseyev, stated on the basis of this material that: "By hair and eye color, the general Russian type deviates from the Central-West European variant. In Russian groups, the portion of light and average shades are significantly increased, and the portion of dark shades, on the other hand, is decreased." Consequently, the concept of a Nordic racial basis of the Russian people is confirmed again and again. This also happens to be the opinion of the pre-[Bolshevik] Revolution

anthropologist, E.M. Chepurkovsky, who points to the large percentage of genetic material of the ancient population in modern Eastern Slavic groups. And the given point of view is maintained by the author of the article, on the basis of new seriological measurements, that also testify about the homogeneity and nativity of the original Russian racial type.

Left: Viktor Valerianovich Bunak

Another standard of Russian science, V.E. Deryabin, writes in the article, *Modern Eastern Slavic Peoples*:[87] "In a comparison of expectation values of anthropological traits for the peoples of Europe and for Russians, it is made clear, that by many racial characteristics, they occupy a central position among Europeans. This is observed in body length; in the dimensions of the head and its form; by the height and width dimensions of the face; and their correlations. In other words, by many traits, Russians are very typical Europeans. By pigmentation of the eyes and hair, Russians as a whole turn out to be lighter than the average European type." Thus, according to the calculations of V.E.

---

[87] *Sovremennie vostochnoslavyanskie narodi.*

Deryabin, among Russians light eyes (gray, blue-gray, light blue, and blue) are encountered at [a rate of] 45%, while the average level for the rest of Europe is only 35%. Dark eyes (dark or light brown) are encountered in 5% of Russians, while in the population of Europe it is an average 45%. Dark hair among Russians is encountered on average in 14% of cases, while in the population of trans-Russian Europe, it is 45%. The fashionable opinion about "snub-nosed" Russians is unsubstantiated. Thus, in 75% of cases, the straight-profile nose is encountered among them.

It is known that one of the characteristic traits of Mongolism on Eurasian territory is the presence of epicanthus. In groups of typical Mongoloids, epicanthus is encountered among adults quite often, in 70-95% of cases. Among more than 8,500 inspected Russian men, epicanthus was discovered in twelve cases in all, in which it was observed in a vestigal form. V.E. Deryabin came to the conclusion: "Thus, by their [racial] composition, Russians are typical Europoids; by a majority of anthropological traits, they occupy a central position among the peoples.
of Europe, and stand out somewhat more for the light pigmentation of [their] eyes and hair. It is also worth recognizing the significant singularity of the racial type of Russians, in all of European Russia."

In the article *Dermatoglyphics of the Eastern Slavs*,[88] N.A. Dolinova also came to an eloquent conclusion, on the basis of an analysis of skin patterns on the palms and soles of the feet. For evidence for a racial diagnostic, she uses such standards as the Northern Europoid Complex (NEC), which reflects the degree of expressed northern-Europoid characteristics in a group. Among Russians of the European portion of Russia, this index is never lower than 0.41; that enables the author of the article to confidently speak of the "morphological unity of Russians."

At the start of our research, we cited a statement of the famous ethnographer and historian, N.I. Nadezhin, who declared as early as 1837: "The physiognomy of the Russian people, [which is] fundamentally Slavic, is marked by the natural hues of northern nature: light-brown hair, from which in old times the very name, *Rusi*, originated."

With the passing of 150 years, **the fundamental thesis of racial theory, which says that the racial base of the Russian people, and in equal degree, that of other European peoples, is inarguably Nordic, is confirmed again and again. Namely all European civilization is indebted to the culture-creating abilities of the Nordic race, for its origin.**

Y.G. Rychkov, E.V. Balanovskaya, S.D. Nurbaev, and Y.V. Scheider definitively declared in the article, *Historical Genogeography of Eastern Europe*,[89] that on the basis of a comparison of archaeological and genogeographic maps of the given region, "the nucleus of the Russian genetic pool is located in the northwest Russian ethnic geographic range."

In the article, *The Odontological Aspect of Ethnogenesis and the Ethnic History of the Eastern Slavic Peoples*,[90] R.U. Gravere states on the basis of the morphology of the dental pattern system of the given region, that: "The northern branch of Slavdom evidently formed in Central Europe, possibly in the areas near or adjacent to the Upper Vistula River, brushing with the Balto-Slavic and Baltic-Germanic communities in their pre-history."

And in *Ecological Genogeography of Eastern Europe: the Genetic Pool, Health and Sickness of the Agricultural Population of European Russia*,[91] O.V. Zhukov, E.V. Ogryzko, T.P. Pankova, Y.V. Scheider, and Y.G. Rychkov explained in a simple and easily understood style, the reasons for the biological activeness and tribal ardor of members of the Nordic race, which manifested in full measure in the time of their victorious march along the endless spaces of Eurasia, beginning from the Late Paleolithic Age. "The main qualitative characteristic of the modern geography of illness in the agricultural population is the decrease in the number of cases towards the north; one may think that this was set down namely in the Late Paleolithic, through selection in favor of high vigor, with nearness to the boundary of the glacial ice sheet."

---

[88] *Dermatoglifika vostochnikh slavyan.*
[89] *Istoricheskaya genogeografiya Vostochnoy Evropy.*
[90] *Odontologicheskiy aspekt etnogeneza i etnicheskoy istorii vostochnoslavyanskikh narodov.*
[91] *Ekologicheskaya genogeografiya Vostochnoy Evropy: genofond, zdorovye i bolezni sel'skovo naseleniya Evropeyskoy Rossii.*

Finally, in the article *The Ancient Population of Eastern Europe,*[92] T.I. Alekseyev and S.I. Kurtz asserted in the same spirit that: "In the Mesolithic Age, the most numerous (judging by the existing distribution of our data) was a population connected in its genesis with the northwest territories of Europe. It is characteristically dolichocephalic, with a wide face that flattens in the upper section, and a sharp profile on average and a strongly projecting nose. The primary concentration of these facial traits in the north and northwest of Europe gives the basis to trace their bearers to a circle of northern Europoids."

In one of the most modern theoretical works, in the anthology, *Anthropological and Ethnographical Information about the Population of Middle Asia,*[93] the famous Russian anthropologist, L.T. Yablonskiy, writes that in the Bronze Age, the population of the southern Aral Sea coast was formed on the "proto-Europoid anthropological model," and in the Early Iron Age (8-7th Centuries, B.C.) the Aral Sea coast skulls also "stand out by very strong appearance of Europoid characteristics."

Thus, in the very heart of Asia, in the period of the formation of the first states, the creation of writing, religion, culture, technology, the fundamentals of civilization, and law-making, there was no mention of the Mongoloids, the Negroids, or their half-breeds. All epochal creations belong entirely to the white man of pure race, his will, his genes, and his sagacity. For the extent of all history, the later infusion of newly arriving foreign blood only weighed down or sent in the opposite direction, the very biological process of the creation of high culture. Racial chaos was always and everywhere identically manifested: through anarchy, civil strife, social parasitism, the destruction of monuments of culture, and derision of the sacred.

For a biologically-based reason, in the politics of survival under the Mongol-Tartar yoke, the Arab Caliphate, the Ottoman Empire, and the Asiatic states, there would always be the same policy: the murder of the white males and the stealing of white women into harems. With this constant, centuries-old practice of genetic parasitism by the colored of the white man's genes, all the talk by culturologists about the "originality" and "distinctive character" of different cultures is simply a cynical lie, and arouses the proper disgust of every sensible individual.

A comprehensive racial-biological analysis of the facts of the history of Nordic Man graphically shows us that any person, independent of political persuasion, religious affiliation, and even the color of his skin, who acts in detriment to the members of the white race, in the final analysis acts against himself and his descendents. Anyone today, who either out of stupidity, by virtue of old offenses, or some other reasons, acts in harm of the white race, is like a madman who squanders great wealth; [in this case] the culture-creating genes of northern type Man [are squandered]. Without them, no rational or postulated move forward through the thorny path of evolution is possible. Racial chaos was never a foundation for authentic greatness. No worthy retinue, nor knightly order ever came about from a pie-bald horde. Any ascent begins from within, when pure noble blood sublimates in a vessel of pure lives and majestic thoughts.

8. "The Crucible of Creation"

Such a poetic expression was applied namely by the famous Soviet scientist, N.I. Vavilov, in order to indicate an important anthropological category: **focal point of race formation.** Each one of the races existing today went through its crucible of creation, and acquired its inherent physical, psychological, and moral traits. Best of all in Russian science, V.P. Alekseyev worked out the given problem in part, in the book, *Geography of the Human Races.*[94] He wrote: "Under the crucible of race formation, I understand to mean portions of the Earth's surface, within the limits of which the race-forming process is characterized by a definite intensity and a specific direction."

One may say that the focal point of race formation is a region, where racial traits arise, are reproduced, and sharpened. The Soviet scientist emphasized that in the process of race formation, the significance of geographical barriers, particularly in the later stages of racial genesis, are traditionally exaggerated, while on the other hand, the role of natural selection is

---

[92] *Drevneyshee naselenie Vostochnoy Evropy.*
[93] *Antropologicheskie i etnograficheskie svedeniya o naselenii Sredney Azii.* Moscow, 2000.
[94] *Geografiya chelovecheskikh ras.* Moscow, 1974.

underestimated. One may speak of the focal point of race formation, as of the elementary singularity of the natural surroundings. The focal point of race formation has a structure [that is] spatially organized. Its structure is expressed, in that it embraces several populations, most often submitting to one another hierarchically. A defined race-forming modus corresponds to each focus of race formation. [Even] with local changeability, the focus preserves its stability for a prolonged time. It may also repeat along the direction of changeability, since the character of what forms in correspondence to the focus, naturally changes with time in such a complex of traits, but the focus itself—in its boundaries and structure—remains constant. With the supremacy of a typological changeability, the focus of race formation quickly loses shape and later transforms into new, subsidiary foci."

Thus, it comes about that in the conditions of the geographic isolation of the focus of race formation, several similar biotypes carry on a concrete struggle in the beginning, and then they form up into a hierarchical caste structure, constructed in accordance with the biological and cultural worth of these biotypes. Later, the given hierarchical structure implements the social sublimation of one united biotype, [which brings] a corresponding strengthening of all [its] racial traits. It is namely this result which subsequently becomes a standard model that is recognized in the process of conflict with other races, in the struggle for existence. It is namely this initial biotype which symbolizes this or that race, which is afterwards described in written historical sources, inscribed in numerous artistic portrayals, and endowed with all positive qualities in tribal epics, and made sacred and exalted in religious cults. The graceful head of the ancient hero, the slanting, moon-like face of the Chinese Mandarin, or the protruded jaw in the profile of the black-skinned negro—all are the result of the complex work of the evolutionary vessels of nature, realizing millennia-old work through the development of the races.

Yuriy Grigorevich Rychkov, a distinguished Soviet geneticist, also dedicated a special monograph titled *Anthropology and Genetics of Isolated Populations*,[95] to the question under consideration. Based on extensive material from field studies, he came to the following conclusion: "In the forming of an isolated population, there is a tendency toward a decrease in morphological changeability." This means that in the process of race-formation, a gradual consolidation of the dominant racial type—its morphological isolation—occurs, with a crystallization of the traits that are characteristic to it. The Soviet scientist emphasized: "The process of refinement in anthropology is indeed known, but we do not know, and we can hardly theoretically assume the possibility of the transformation of the northern blondes and southern brunettes, with a developed hair covering and other complexes of racially constitutional and physiological traits.

Gracialization is the process of the ongoing structural refinement of a racial type, with a resulting genetic reinforcement of its characteristics.

G.F. Debets emphasized that the ancient proto-Europoid form is more characteristic for the northern Europoids than the southern Europoids—i.e. in the structure of the skull, skeleton, and the fundamental chemistry of metabolic exchange. V.P. Alekseyev also pointed out that the Nordic biotype is the original for the white race. "What concerns the race-forming focal point for the northern race, is the fact that it abutted the Baltic Sea from both the north and the south. In connection with a specific geographical environment on the Scandinavian Peninsula, it was namely there that the fundamental traits of the northern race could conclusively form."

Again, we are convinced that regarding this key question, Soviet science has completely confirmed the basic postulates of classic racial theory. Advancing this hypothesis in the article, *Problems of the Settlement of Europe, according to Anthropological Data*,[96] G.F. Debets, T.A. Trofimova, and N.N. Cheboksarov asserted that: "The craniological type of the "northern race" already existed in Scandinavia, when in the East they still completely retained the Cro-Magnon characteristics of the Proto-European type."

---

[95] *Antropologiya i genetika izolirovannikh populyatsiy.* Moscow, 1969.

[96] *Problemy zaseleniya Evropy po antropologicheskim dannym.* From the anthology, *Proiskhozhdenie cheloveka i drevnee rasselenie chelovechestva* (The Origin of Man and the Ancient Settlement of Mankind). Moscow, 1951.

The Nordic racial type was the initial type, and gave a push to the evolutionary development of the white race. The principle logic is traced in the formation of other basic races, for separate biotypes afterward gave growth to the entire racial tree.

We recall once more, that the classics of the Russian school of anthropology—A.P. Bogdanov and G.F. Debets—called the skull of the Nordic race **proto-Europoid.** In his fundamental monography, *Paleoanthropology of the USSR,*[97] G.F. Debets stated: "The skull of the proto-European type may be characterized as northern-European."

In the essay mentioned by us, V.P. Alekseyev wrote that to this very day, the Danes and Swedes are carriers of the pure **gracial type** of the Nordic race: "On hand is the continuity of the physical traits of the ancient and modern population of the territory of Scandinavia, for a period of a minimum 4,000-5,000 years." It is namely the knowledge of the fundaments of the morphological evolution of biotypes that enables us to avoid mistakes: "The morphological similarity between starting and ending members of a series—between the Neolithic and modern population—enables one to say that the discovered differences between populations in the Neolithic Age repeat themselves into the modern age."

By the logic of G.F. Debets, the process of the branching of the racial tree created new variations: "In dependence of how far separate skulls and groups digressed from the original prototype, they may be diagnosed as "northern "or "Mediterranean." We recall that the great French anthropologist, Armand de Quatrefages de Breau, called Cro-Magnon "the forerunner of the white-skinned, light blue-eyed European."

On the general level of the Cro-Magnon type, the Nordic proto-Indo-European began to stand out for aggregate morphological traits, strengthened by a norm of psychological reactions. A.P. Bogdanov thus described the essence of the given process: "In savages, we see a receding lob; a rudely protruding occipital rear; the rude development of the places of the attachment of the muscles, and the development of heavy brow ridges. These traits wear away with the development of culture, with a more extended and provisioned life. The lobe increases in height and width, the occipital area of the head acquires a more complete build/structure, the cross-section diameter increases, the length of the head decreases, or more probably, the proportion of this diameter decreases in relation to the width of the skull." All these visible traits, by which painters and sculptors have been captivated for the extent of millennia, and stereotypically connect with the appearance of the white race, are in fact a morphological result of a complex evolutionary process in the development of its original biotype.

It has been shown many times, that the phenomenom of gracialization, or the humanization of appearance, is entirely connected to the ongoing development of the creative and productive abilities of Man, which essentially differ in all races. No kind of neutral index of the aptitudes toward culture exists in principle, therefore, the differences in the physiognomy of the separate races is so great.

In the book, *Peoples, Races, and Cultures,*[98] N.N. Cheboksarov and I.A. Cheboksarova point out that the phenomena of **gracilization** (Latin - *gracilis* - thin, gentle), as a result of which there is a decrease in the general mass of the skeleton, the angle of the forehead, the prominence of the brow ridges, the lengthwise diameter of the skull, the width—and to some extent—the height of the face—is tied to the development of reasoning and spiritual capabilities. Studies by G.F. Debets have shown, that in Europe, gracial changes already started in the Neolithic Period, at the time when people transitioned from hunting and gathering to farming. In the book, *Of the Skulls of Man and the Stages of his Formation in Excavated Peoples, and in the Modern Races,*[99] V.V. Bunak wrote: "By gracial complex, in craniological literature we mean a complete or partial combination of several features: a small size in the diameter of the brain or facial sections; a small thickness of the bones of the skull cap; a weakly developed brow ridge, temple, and occipital lines; and weakly expressed cranial sexual differences."

Jan Czekanowski, the head of the Polish Typological School, held to the point of view that the anthropological types of modern Europe already existed in Paleolithic times, and did not undergo any substantial changes for that time.

---

[97] *Paleoantropologiya SSSR*. Moscow, 1948.

[98] *Narody, rasy, kultury*. Moscow, 1985.

[99] *Cherep cheloveka i stadii ego formirovaniya u iskopaemykh lyudey i sovremennykh ras*. Moscow, 1959.

But from the point of view of objective raciology, it is completely obvious that the slender and refined form of ancient Nordic Man was accepted not only on a level of exterior physical forms, but also on physiological and psychological levels of organization, for they are all interconnected.

In addition to studies of craniology and somatology, German anthropologist Otto Reche left many interesting works on the analysis of racial differences in blood, and also in questions of evolutionary differentiation of physiological traits, that enabled him to substantially judge about the zone of the origin of the Nordic race from the point of view of a systematized approach.

In his work, *Racial-Physiological Indicators to the Homeland of the Human Races* (1937), he wrote: "Successes in the study of heredity have helped to overcome the theory of environment, and to make a more clear idea of race. We now know: inherited racial qualities, physical and spiritual—and that means the very races themselves—are not accidental phenomena, but the results of evolution. Every important racial quality has its particular sense and its goal, and it developed because it was useful for the given race in its living space, and had its value for [natural] selection. Every race thus possesses a complex of inherited qualities favorable to it, and represents itself as an attempt at the most expedient adaptation to the conditions of life in the area of its origin, in the home of its race."

Reche begins his analysis of physiological traits with the color of skin and the pigment located in it—melanin, which absorbs a definite part of the spectrum of the sun's light, particularly ultraviolet [light]. "Melanin protects the skin from ultra violet rays, particularly from rays with a wavelength of 280-313 microns. The home of the Nordic race should be a climatic zone, where because of a low amount of sunlight or its strong filtration, particularly ultraviolet rays, dark skin would be superfluous, or even a disadvantage."

The color of the eyes is a result of formation under the action of similar factors: "The eyes of the Nordic race are not dark, like in the polar beasts, but on the contrary are very light; that is, their irises contain very little pigment. They defend poorly against reflected [light] from the surface of the snow and ice. A race with such eyes could not originate in southern Europe, or in sunny tropical regions, where men of the Nordic race are compelled to carry dark glasses. Negroes constantly carry such "dark glasses" in the form of the dark iris of the eyes."

The light hair of the Nordic race is also adaptive to life with a small amount of light. Besides that a German scientist justly brought attention to the fact that the mucous membrane of people of the Nordic race is extremely sensitive to dust. Consequently, and on the basis of this trait, one can make the conclusion that the home of the Nordic race was located in a wet zone, and not a dusty zone, since the mucous membrane of its members bears humidity well, particularly ocean air and sea fogs containing salt.

Finally, this race is resistant to all illnesses from colds, but is very susceptible to illnesses from hot zones. In contrast to tropical races, the Nordic race is **eurythermal**; that is, it has a wide temperature spectrum, and bears temperature drops well and even their frequent and quick changes. It even needs such change and the definite "climatic energy" included in low temperature. Without it, in Southern Europe or in an unaccustomed climate in the tropics, this race quickly loses its spiritual and physical strength. The Nordic Man is energetic in winter, even though winter is cold and the weather strongly changes.

The Negro and Arab categorically cannot bear such Nordic exoticness that is tied with a sharp change in temperature regimes. Even without the contrasting cold bath, they cannot withstand the heat of a Russian steamroom.

"The Nordic race should appear in a climate, in which "changeable weather" is so to say, a normal situation; that is, in a cold sea climate, located under the strong influence of wind currents from the sea. The development and function of skin glands in the people of the Nordic race corresponds to this; they cannot work equally as the skin glands of negroes; perspiration for us is an extremely unpleasant condition, and our fatty glands may eliminate only some fat, therefore, we use cream for defense from the sun.

We will tally the results. The home of the Nordic race may be located in the zone of a cool and moist climate, abundant with clouds of fog, in which water vapor is retained in the air; almost without remains, it absorbs ultra-violet rays. In this climate there should be strong and frequent fluctuations of temperature; that is, we are speaking of a climate, the opposite of the steppe, such that the home of the Nordic race could not be any kind of steppe, tundra, nor steppe of moderate latitudes, nor hot and dry tropical regions. We are speaking of a typical climate, which is definitely close to the ocean, with frequent changes of weather; approximately such a climate that we have today in Northwest Europe, particularly in Ireland, England, on the coasts of the North and Baltic Seas, and partly in Scandinavia.

Above: Reconstruction of the appearance of the ancient Indo-Europeans.

Above: Reconstruction of the appearance of an ancient Indo-European.

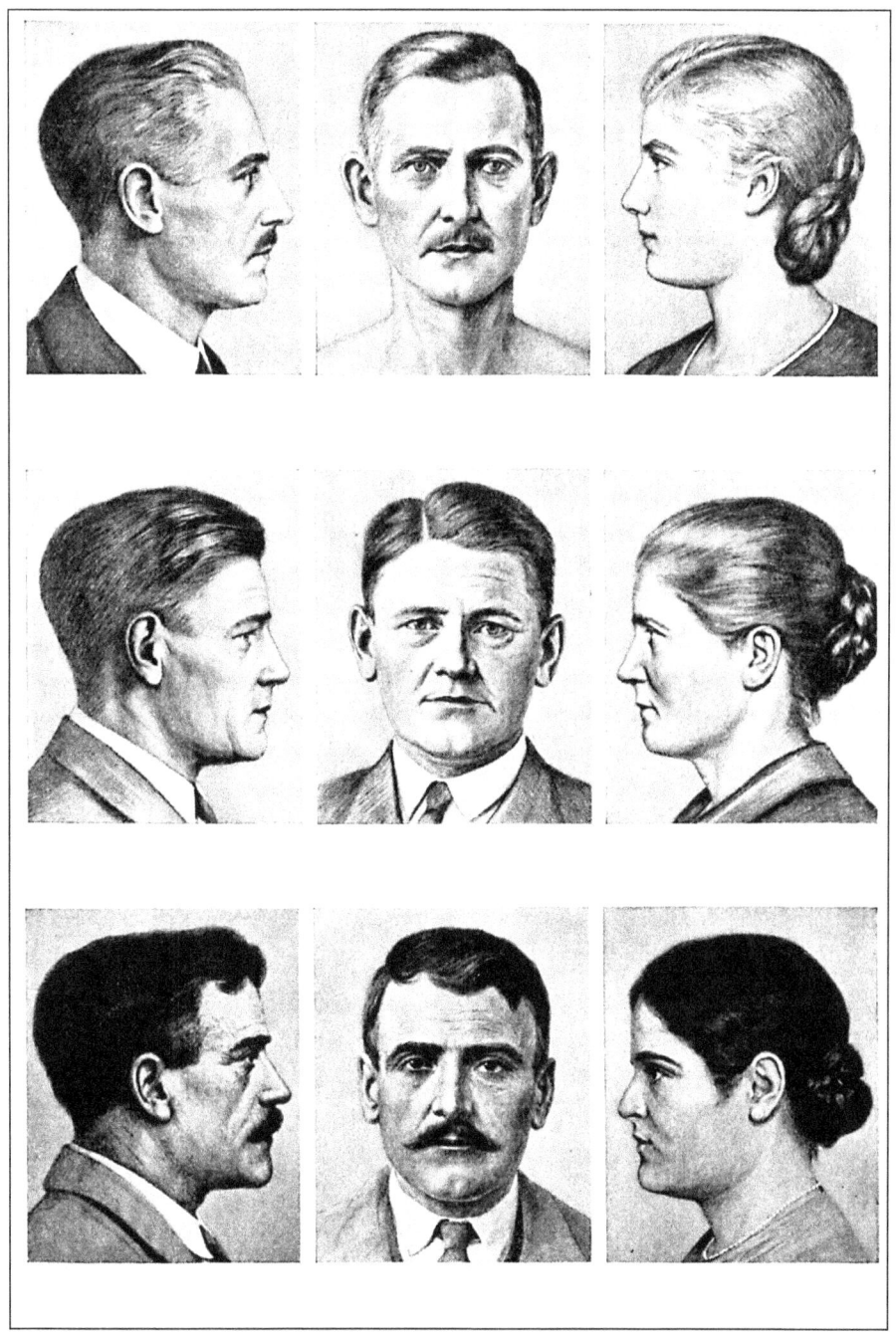

Above: The Basic Racial Types of Europe, according to Otto Reche: Top-Nordic; Center-Pfalzish; Bottom--Western

If this was a young race, arising in the post-glacial period, it could appear in modern Northwest Europe. But inasmuch as it could arise, at the very latest, in the Wuerm [Glacial] Period, we should examine where the corresponding climate prevailed at the time.

In the Wuerm Period, a cool and moist climate in Europe was only in the extreme west, in Ireland, England, on the western coast of Jutland, and in the western parts of France. Only there and at that time, was the climate characteristic of a home for the Nordic race. There, and in the areas further to the north, in the time of the temperature fluctuations of the Wuerm Period, it was free of ice; only this could be the home of the Nordic race. Thus the Nordic race is the child of

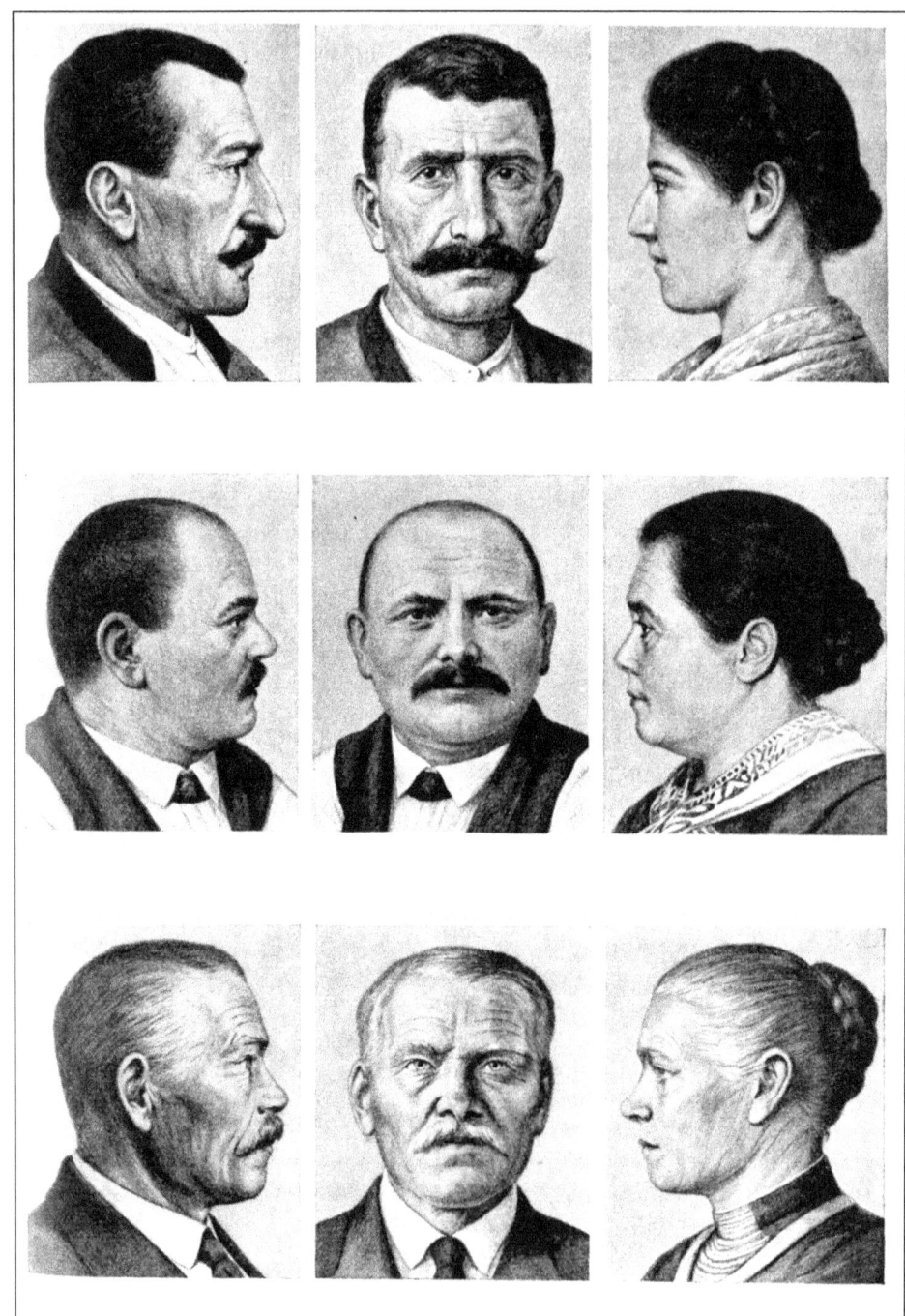

Above: the Basic Racial Types of Europe, according to Otto Reche. Top-Dinaric; Center-Eastern race; Bottom: Eastern-Slavic race.

the West and Northwest Europe of the glacial period; that is, the core population of ancient Europe. No other interpretation of the enumerated racial traits can be assumed."

Otto Reche's monography, *Race and the Home of the Indo-Germans* (1936), is dedicated to the thorough examination of the problem. The scientist dates the rise of the Nordic race to about 100,000 B.C. "In descriptions composed thousands of years ago, we already encounter the Nordic race, with a complete set of qualities as something long established, and not located in a process of formation; these qualities were already so steadily transferred by through heredity, that they could not change more substantially. Light pigmentation was retained on the steppes of Eastern Europe and Asia, and in the sunny climate of the Canary Islands, and in the hot steppes of North Africa, where the Nordic race erected megalithic structures, and had already penetrated Egypt in ancient times. Consequently, these qualities were acquired very long ago. All Indo-European peoples originally arose from tribes that belonged to the Nordic race. These tribes only concluded marriages between themselves, and they had laws against mixing with foreign races, laws which were in force for millennia, but were later forgotten. Only these tribes were Indo-European in the original sense of the word, creators of the Indo-European languages, states, peoples, and cultures. All Indo-European peoples today have a single racial element common to all of them—the genetic pool of the Nordic race."

According to Dr. Reche's opinion, the Nordic race was the creator of not only Neolithic culture, but also the so-called "clay" and "cord" ceramics. By the Neolithic Period, all the Indo-European peoples were already settled farmers, before their exodus from Europe. By the time of horse-breeding, not one primordial Indo-European people existed. It was namely the search for new, arable lands that gave rise to massive migrations. The youth were also often drawn to faraway lands by the thirst for adventures. It is sufficient to take a look at the portraits of the great explorers and seafarers, zoologists, and botanists, in order to certify that they all had a high percentage of Nordic blood. Characteristically, members of that race most often strive for the unknown, for the sake of a pure idea, driven by the thirst for knowledge, and not self-seeking interests. Other races most often gather themselves to journey, [when] pushed by more pragmatic motives. Their characteristic instincts are more often greed, lust, and robbery. For them, money and material interests are a stimulus for advancement, and not any desire to draw a map, or give a name to a sea, or an island, or to discover a new type of butterfly or flower. Only the Nordic race was fascinated by the physics of the Cosmos—the other races were perfectly content with metaphysical discussions about them.

The psychological traits of the Nordic race are best indicated by the fact that it arose in conditions of a severe and constant struggle for life, when a manifestation of will and initiative were regularly compelled, rather than from a life of herd-like habits, in captivitiy of stereotypes. Only the severe north of Europe, with its unruly elements, could definitively sculpt and elevate the spiritual characteristics of the Nordic race.

Ably and rationally, Reche destroyed the false invention of linguists and theologians that "all culture came from the East." First of all, not one other race created an entire system of sea-faring terminology; in the languages of other races, in the best case and on an archaic level, there exists a description of the different conditions of water. And among several peoples of Central Asia, where according to the inventions of culturologists, all civilization should have been born, in general there is not one verb for "swim." Culture, like language, arises from the experience of life, and not from some groundless urges for originality. Dr. Reche wrote: "I consider it a completely mistaken opinion that Northern Europe was "Indo-Europeanized" in some linguistic scheme. I have long emphasized, that any fundamental human race (better "species" as in zoology and botany) in the process of formation in isolation from other species, creates its own language base, which in all regards corresponds to the physical (for example, the racial structure of various organs of speech) and psychological characteristics, and the needs of the race. How could people create a language for themselves that is foreign and uncomfortable to their spirit? Race and language were thus originally tightly interconnected; language was the mirror of hereditary racial traits. Therefore, it is unthinkable that two completely different races could create even similar languages, and impossible that two tribes of one race could create completely [unrelated] languages. As they say, "you just don't jump out of your own skin."

Thus, according to Otto Reche's conclusive base, a variety of ancient European dolichocephalic groups should already have had languages that traced back to one root, to single hereditary traits, characteristics, and needs, which should have developed by similar traits and should have been very similar. All of this group should have had a single language base, later from which, by measure of racial differentiation, they hypothetically separated into Indo-European, Semitic, and Hamitic languages, after the division of the northern and southern Europoids in the last glacial period. The Nordic race arose from the northern Europoids, creating the foundation for all Indo-European languages. Many Indo-European peoples developed gradually, becoming famous in history as a result of the confluence of the most different consecutive waves; therefore, the languages retain many heterogeneous elements, tracing back to this or that dialect, and sometimes to different stages of the development of that language. Namely for this reason, it is incredibly difficult to make reliable conclusions about the formation of separate Indo-European peoples when using such data from linguistics, and not relying on facts from other sciences. It is namely raciology, implementing a productive synthesis, that arranges everything in its place, and shows cause-effect relationships in the process of historical development. "An ancestral people, as it is presented by linguists, which broke into equally valued fraternal peoples at a definite point in time, never existed. The basic race was only capable of reproduction and expansion; at every stage of its development it sent surplus population down the road. Young people gathered often enough in the very inner core of the European home, and using the ceremony of "sacred spring," they set off for some unknown horizon, under the leadership of a selected leader."

This noteworthy book was popular in Germany until the war, and afterwards, by virtue of its clarity and informative conclusive base. In many ways, it has not become outdated. In the context of our narrative, it is worth noting that the German raciologist based the climatological analysis of the conditions of ancient Europe on the methods of Russian scientist V.A. Obruchev, and the racial studies of blood groups on Soviet research: *On the Question of a Biochemical Racial Indicator*,[100] by R.N. Vishnevskiy; and *A New Biochemical Racial Index*,[101] by A.A. Melkikh. It should not be left out, that during this time Otto Reche was a member of the NSDAP. Scientific impartiality and an academic approach enabled him to avoid the post-war persecutions of "de-Nazification" and to peacefully live and work; it follows that no ban on the use of the scientific achievements of an ideological enemy existed in the Third Reich, even with the very "ticklish" racial question.

As we have already become convinced, the studies of V.V. Bunak, G.F. Debets, N.N. Cheboksarov, and other Soviet anthropologists, for their scientific part, did not in the least contradict the postulates of classical racial theory. We emphasize once more, that talk in the given case is only about the conclusions of the academic school, and not para-scientific socialist and National Socialist interpretations, on which historians and political scientists most often rely, when attempting to draw us a picture of that age.

Still another luminary of German science, Gerhard Heberer (1901-1973), Director of the Institute of General Biology and Anthropology in Jena, came to the same conclusions. In a work titled *Racial-Historical Studies on the Ancestral Home of the Indo-Europeans* (1943), he emphasized: "Raciology proves that only Europe could be the center of expansion of the Nordic race, and it is here that it is worth searching for its ancestral home. Race and language have common roots. The Indo-European languages are a creation of the Nordic race and no other. The concept of "Indo-Europeans" and a "Nordic race" in the Neolithic Period entirely agree." In the given work, the scientist substantiated the following important theses:

1) Racial continuity in Central and Northern Europe since Paleolithic times.
2) Racial unity of the Neolithic cultural circle.
3) Racial foundations of the struggle between north and south in Central Europe in the Neolithic Period.

---

[100] *K voprosu o biokhimicheskom rasovom ukazatel'*. 1925.
[101] *Noviy biokhimicheskiy rasoviy ukazatel'*. 1927.

Gerhard Heberer   **Heinz Bruecher**

In the work, *The Theory of the Origin of Species and Modern Biology* (1942), he also pointed out: "Today, a clear enough picture is drawn for us of the history of the Nordic race, since dark times. In the last glacial period, a group of people in a heavy struggle for existence, populated western, northwestern, and central Europe along the edges of the great ice sheet. They developed those qualities which we consider characteristic of the Nordic race. One may follow its history from the Neolithic Period, when the basic Indo-European peoples were formed, through the Mesolithic right down to the Late Paleolithic. Ancient human remains with Nordic traits have an age of no less than 80,000 years. The Nordic race is a product of the last glacial period; we can follow its formation."

And again, it is worth setting apart the characteristic fact that this is no exception, but a rule in the development of the racial theory of the Third Reich, which it appears only the completely lazy have mocked, and on which no one has wiped their dirty shoes.

Gerhard Heberer derived the conclusive base for the development of his concept from the works of paleontologists and specialists in **evolutionary morphology**. Among them the famous Soviet scholar, A.N. Severtsov, and his fundamental theoretical monograph, *Morphological Patterns of Evolution*,[102] occupy first place. And this was at a time, when in Germany's National Socialist period, he himself was one of the major consultants of the SS *Ahnenerbe* [German for *ancestral heritage*] Society, precisely for the question of the geography of the center of origin of the Nordic race. As a conscientious scholar, he composed a map of this zone of Neolithic times, in which the whole northwest region of Russia was included. This means that no sort of theoretically substantiated Russophobia existed in principle in the Third Reich. Science was earmarked for the higher ranks of the regime, but remained politically sterile, even at the height of confrontation on the Soviet-German Front. This cannot help but arouse respect in all right-thinking people.

Gerhard Heberer was also not subjected to the tortuous procedure of de-Nazification, and after the war he wrote a number of academic works on the question of evolutionary biology, and several textbooks on anthropology.

---

[102] *Morfologicheskie zakonomernosti evolyutsii*. Moscow, 1939.

But the following example of history, even in the context of our narrative, could put even the truest lover of political sensation into shock.

Along with craniological, osteological, and biochemical studies for determining the zone of origin of this or that race, many indirect methods of analysis have great effect; for example, the establishment of the origin of different sorts of cultured plants, and breeds of domestic animals. After all, it is completely obvious that members of each race cultivated in their zone of habitation, namely those plants and animals which most correspond to their inherent metabolic processes.

Food prohibitions that have been laid down by several religions for their adepts, do not have some abstract, deific origin, but are dictated by the vital requirements of gastronomical hygiene, since in the specifics of the metabolic processes in each race, some animals, seafoods, and plant products digest better, and others worse, while some are generally harmful. By cultivating certain means of sustenance, a race adapts the environment to its hereditary biology, not the other way around.

In order to clarify the influence of different types of cultured plants on the phylogeny of a race, and also its evolutionary refinement, the leading German biologist and botanist, Untersturmfuehrer der SS, Doctor Heinz Bruecher (1915-1991), was sent to the occupied territories of the USSR in 1943. He led an entire command of Waffen-SS men, who were engaged in the removal of genetic material from Soviet geo-botanical and plant cultivation stations in the Crimea and Ukraine. The framework of our work does not include a defense of the political reputation of this person; however, it is necessary to certify, that despite the fact that he commanded an entire program within the framework of the Ahnenerbe Society, and was subordinate to Heinrich Himmler, he nevertheless always considered the genial Soviet scientist, N.I. Vavilov to be his teacher, something that he wrote honestly about in his scientific works. And such a discovery of the continuity of views is not any kind of an embarrassment. After the war, he settled in Argentina, where he received a position as Professor of Genetics and Botany, later working as a consultant for UNESCO on questions of biology. Remaining an authentic scientist and a principled individual, he struggled against the spread of narcotics plants in Latin America; besides that, using his rich creative work experience as Director of the Institute of Plant Genetics and Experimental Agriculture SS (1943-1945), he developed a pathogenic virus against coca bushes, from which cocaine is made; for this he was killed by the Argentine mafia, at the age of 76. Therefore, in our view, one should use more consideration and accuracy, before placing the "fascist" label on any German scientist of the first half of the 20th Century.

Finally, it is necessary to emphasize that Egon von Eichstadt, Gerhard Heberer, Ilse Schwidetzky, and Karl Zaller were official guests at the 7th International Conference of Anthropological and Ethnographic Sciences in 1964, which was held in Moscow; their works on racial anthropology were highly popular in the Third Reich, and Heberer, as we remember, was a general consultant for the SS Central Racial Bureau, and the Ahnenerbe Society. However, it never came to mind to call them "Fascist scientists," and that automatically liberates us from accusations of being racist propagandists, in connection with any citations of the academic works of the given authors, and those close to them.

Returning to the main theme of our narrative, it is worth a description, although brief, of additional factors that also influence the evolutionary processes of the formation of a focal point of race formation.

Thus, in the monography, *Geographic Pockets of Formation of the Human Races,*[103] V.P. Alekseyev emphasized: "Localization of the very sources of cosmic radiation justifies setting apart cosmic factors of race formation on a particular group. Geochemical factors obviously occupy an independent place of no small importance, in the form of a concentration and thinning out of these or those micro-elements, which immediately influence growth processes, the mineralization of the skeleton, and so on." Besides the astro-biological factor, gravitational effect has important significance in the formation of a race, influencing the setting of the proportions of the body in different races in the process of evolution. The geometry of the structure of the extremities, the bio-mechanics of the joints, the proportion of the organism and its parts, the dynamics of movement, the very aesthetics and ergonomic expediency, are therefore highly diverse among the different races. A connection is observed between the entire complex of the physiological

---

[103] *Geograficheskie ochagi formirovaniya chelovecheskikh ras.* Moscow, 1985.

reactions of the organism of Man, in part between the physico-chemical characteristics of blood flow and the electro-magnetic condition of the surrounding environment, and changes in the intensity of solar radiation.

Also known is the predominant role of calcium and phosphorous in the formation of that solid support for the human body—the skeleton. (Calcium and phosphorous occupy fifth and sixth place, relative to maintenance, after oxygen, carbon, hydrogen, and nitrogen.) Maintenance in different races varies within wide limits.

However, even the introduction of additional informative indicators to the general picture does not sway the very principle of the rise of the Nordic race, with the subsequent propagation of secondary racial generations from it. **The struggle between north and south, on a level of biotypes, constituted a timeless vector in the process of race formation.** Not citing the hundreds, if not thousands of proofs of this from classical sources on racial theory, we now confine ourselves to statements drawn from a foundation of Soviet work.

Thus, in the book, *Paleoanthropology of the USSR*,[104] G.F. Debets points out with all obviousness: "In dependence of the fact that some separate skulls and groups deviate from the original prototype, they may be diagnosed as Northern or Mediterranean."

Thus, a projection of this confrontation in time and space to the present day yielded the palette of light-pigmented racial types of Europe and the territories adjoining it, which we have as a natural scientific and historical fact. According to the system of classification of the basic races of Europe that was fulfilled by Hans F.K. Guenther, modern science defines the following: the Nordic, the Mediterranean, the Dinaric, the Alpine, the Eastern-Baltic, and Pfalzic races.

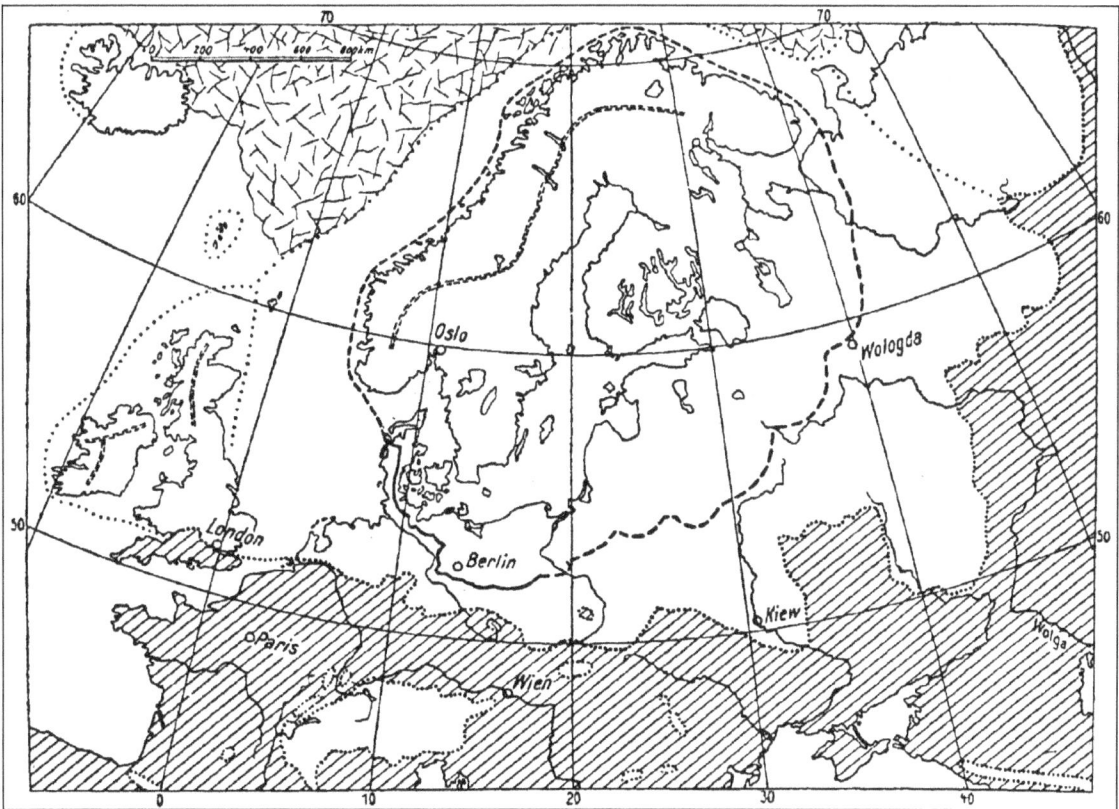

Above: The Cradle of Origin of the Nordic Race (according to Gerhard Heberer).

---

[104] *Paleoantropologiya SSSR*. Moscow, 1948.

107

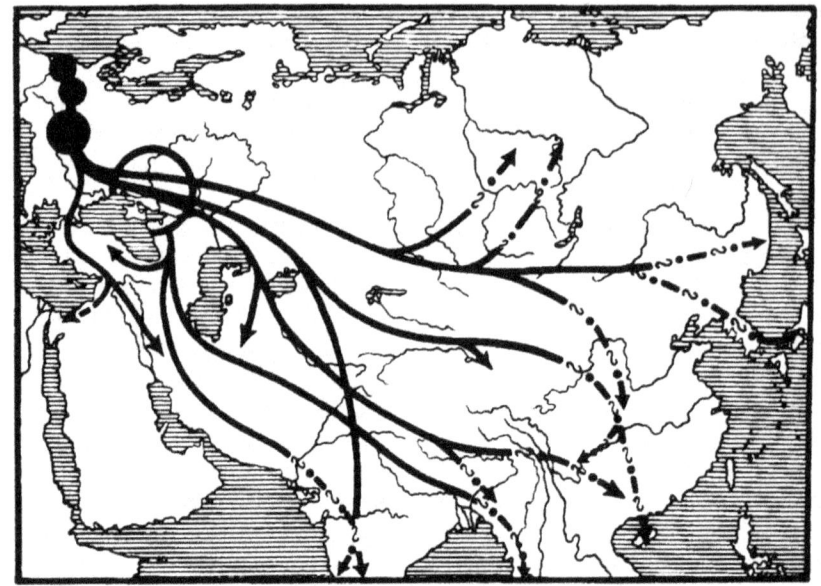

Above: Migration Paths of the Nordic Race into Eurasia (According to Hans F.K. Günther).

THE BIOLOGICAL BASIS OF THE NORDIC WORLDVIEW

"The Nordic Idea is an expression of a worldview,
for which the elevation of Man is a divine precept."
Hans F.K. Guenther

"Today, every clear emphasis of the Nordic
point of view brings profit."
Eugen Fischer

Even people who are not familiar with racial theory have an idea of what the notion **Nordic Race** means. With just that magic combination of words alone, images come to mind of tall, light-blue eyed and light-haired beauties from ancient Greece's Mount Olympus and Scandinavia's sagas, who emit sun-lit energy, unearthly magnificence, and super-human strength.

Ancient Roman patricians, satiated by the awareness of grandeur, the headstrong German knights, the Russian epic of miracle heroes, and the graceful "white swan", Russian fairy-tale beauties, and likewise, the ivory-sculpted faces of SS officers, and finally, reserved English gentlemen—a whole abundance of historical personages from different ages and peoples, are nevertheless characterized namely by the term **Nordic**, meaning the sum total of physical and spiritual characteristics of a people that originated in a single, northern ancestral home.

But these associations, if shifted to facts, draw a completely paradoxical picture. It appears that in the word "Nordic" there lies something ingenious, foreign; something strange that is not understandable to the Russian man of racial theory. Such today is the dominant point of view on the given question, not only in Russia, but abroad, too.

Leading racial theorists of the present time also consider the attribute **Nordic race** to be an inalienable part of science, a customary and consistent scientific category, but there are very few who know that the man who first set down this concept as an indicator of a definite anthropological community—was born in Astrakhan.

The Russian racial theoretician, Iosif Egorovich Deniker (1852-1918), was born of French parents (therefore, the correct accent is on the last syllable of his name) but, being baptized in the Orthodox Church (which is indicated by his patronymic name), by the laws of the Russian Empire, he was automatically recognized as a Russian subject. As a Russian scholar, he is listed in Brokgauz and Ephron's Dictionary, and also in the Great Soviet Encyclopedia of 1955, which recognizes that "Deniker's classification of the races has not become outdated." References to his fundamental work of 1900, *The Human Races*,[105] may be easily found in many works by Soviet academicians on anthropology. One of the leading racial theorists of Weimar Germany—and later, the Third Reich—Hans F.K. Guenther, openly recognized in his own fundamental work *The Nordic Worldview*, that the name of the basic part of German racial doctrine "was first introduced by the Russian racial theorist, Deniker." Another important German authority in the denoted field, Walther Scheidt, named his book on the systematization of terminology, *The History of Anthropology, from Linneus to Deniker*. There is absolutely no evidence at all that he had problems with the political organs of the Reich for mentioning a Russian anthropologist in the title of his book.

---

[105] *Chelovecheskie Rasy.*

Left: Nordic Racial type, ancient Egyptian bas-relief, 30th Century, B.C.

In his book, *Race and the State*, the Austrian racial specialist Erich Fegelin, clearly wrote in his book that "the term **Nordic race** was first introduced by Deniker." Further examples from German literature can be found. In the "democratic" part of the world at that time, the contribution of the Russian scientist is also unconditionally recognized. The American, Otto Klineberg, states in the monography, *Racial Differences*: "Before Deniker, no one created such a racial classification, in which a combination of traits would be used, such as the structure of the hair, the color of the skin, the color of the eyes, the form of the nose and others, that permit the paring down of the number of races to 17, and subraces to 21; previous researchers, basing classification on separate traits, had variously put their number from 3 to 300."

It is striking, but a fact remains a fact: a Russian researcher of French origin was able to achieve general, undisputed recognition of his scientific contribution to the most politicized science of the 20th Century. He was even recognized in Soviet Russia, even though he belonged to foreigners by origin, and to the so-called "old Tsarist specialists". And he was recognized in Hitler's Germany, despite the fact that he personified the hated type "of the Asiatic Bolshevik Hordes." In the "free Anglo-Saxon world," he also won popularity in spite of the fact that they did not very much love to pronounce French names, and regarded the Russians with caution.

In order to better understand the essence of Deniker's innovation, we will examine a brief history of the development of the conceptual basis of racial theory, for without a correct methodology and terminology, not a single science can exist. However, we will give all clarifications in consideration of racial and geographical regional limitations, which are declared in the title of our essay. Besides that we also consider it necessary to specify that we will lead into the history of the development of racial systematization, here and elsewhere, according to the method of Walther Scheidt, with several author supplements; this is in connection with the fact that Soviet classifications, and the list of classification headings that correspond to them, which were worked out by Y.Y. Roginsky and M.G. Levin, do not stand up to criticism. This is the apotheosis of illiterate political prostitution, which unfortunately was taught to several generations of Soviet anthropologists and citizens as a model of "forward science."

In 1672, French ethnographer Francois Bernier was the first in Europe to introduce the term **race** to everyday use. From the start it had a completely ethnographic sense. However, to this day members of the Anglo-Saxon school of science recommend securing primacy in this area for themselves, setting the date of authorship at the very end of the 17th Century.

German philosopher Gottfried Wilhelm Leibnitz introduced the concept of a Europoid race **Europoid race** in 1700, and the Englishmen, James Bradley, applied the more simple and compact variant—**Europids**—in 1721, in order to signify a biological community rooted in the population of the Old World.

In 1735, the ingenious Swedish naturalist, Karl Linneus, was the first to apply the terms **Homo Europaeus** (European Man) and **Homo Albus** (White Man); and in 1746 he created the first racial classification, based on psycho-somatic and physiological traits. It thus appears:

I.   **Americanus rufus**—the American. Reddish hair, choleric, holds himself straight, eager, self-content, submits to tradition.

II. **Europaeus albus**—the European. Blonde, sanguine, muscular, lively, sharp intellect, inventive, submits to the law.

III. **Asiaticus luridus**—the Asian. Yellow face, melancholy, flexible, harsh, stingy, loves luxury, dresses in wide clothes, submits to public opinion.

IV. **Afer niger**—the African. Black color, phlegmatic, relaxed physique, clever, indifferent, slow-moving, oiled-up from fats, submits to despotism.

In 1749, George Buffon insisted on acceptance of the designation **European race**, and John Hunter and Emmanuel Kant simultaneously (in England and Germany, respectively) introduced the concept of a **White race** in 1775.

In 1776, distinguished German scientist Johann Friedrich Blumenbach was the first, in the interests of classification, to use craniometric studies of the skulls of different ethnic groups; in connection with that he came to the conclusion that it was legitimate to use a wider definition—**Caucasian race**—of that same anthropological type.

In 1800, Frenchman George Cuvier used the color of skin as a standard basis for racial classification; therefore, he called the European race *leukodermic*; however, his countryman, Jean Batiste Lamarck, again gave preference to the term "Caucasian race" in 1809.

At the start of the 19th Century, notable German romantics, like August Wilhelm and Friedrich Schlegel, introduced the concept of **Indo-Germans**, on the basis of linguistic research, which many of their followers mistakenly carried over into the area of anthropology. In the 1860s, English philologist Max Mueller was the first to introduce the concept of **Aryans**, relying on data from comparative linguistics and comparative religious studies. Alas, from the start he repeated the same mistake of mixing linguistic and cultural families with racial traits. In the same period, French historian Ernst Renan introduced the **Semites,** the eternal historical antagonists of the **Aryans,** in order to clarify and contrast the content of the latter's designation.

But the start of the creation of **classical racial theory** in the years 1853-1855, is customarily tied with the name of Count Joseph Arthur de Gobineau. He was one of the first who set apart a tall, dolichocephalic, blue-eyed race, within the whole of the white racial prototype. From 1842-1852, his contemporary, Gustav Friedrich Klemm, subdivided humanity into active and passive races; he placed lighter and darker-skinned races among the former.

At the cusp of the 19th and 20th Centuries, the famous German racial theoreticians, Ludwig Woltman and Houston Stewart Chamberlain, used the term **Germanic** for defining the northern racial type—again carrying ethnic and linguistic characteristics over [to the area] of racial traits. Alas, their authority became entangled, later creating unjustified ideological defects in racial theory.

By 1884 the German linguist and historian Otto Schroeder introduced several anthropological correctives to the definition of the term [Germanic]: "The Aryan race at first corresponded to the blonde, northern races, among which the Aryan language and culture developed, grafting itself to other, non-Aryan races in the process of resettlement and cross-breeding. Two important racial theoreticians of this same period—Otto Ammon and Georges Vacher de Lapouge—again preferred to return to the term approved of by Linneus: **Homo europaeus**; but they used it as an anthropological synonym for the term **Aryans**. Both followed de Gobineau in their works, and set apart long-headed and short-headed, light blue-eyed blondes as the basis of the white race, or in other words, its **racial nucleus**.

In 1870, in the interest of greater clarity, English anthropologist Thomas Henry Huxley divided the Europoid race into a lighter xanthochroide race and a darker melanochroide race.

At various times, other anthropologists and naturalists created their own racial classifications, giving obligatory attention to the physical heterogeneity of the population of Europe: Augustin Thierry (1817); Etienne Geoffroy Saint-Hilaire (1818); Boris de Saint Vincent (1827); Amedeus Thierry (1828); James Prichard (1836); Anders Retzius (1842); Robert Knox (1850); Charles

Darwin (1859); Paul Broca (1860); Isidore Geoffroy Saint-Hilaire (1870); Thomas Henry Huxley (1870); and Paul Topinard (1878).

Partly the fault of historians and partly the fault of linguists, the geographical connection to the ancestors of the Aryans was not so clear in the beginning.

Ethnographer Omalius d'Halloy and anthropologist Paul Broca were the first to rebel against the linguists' concept of the origin of the Aryans, created by Max Mueller and other orientalists, which suggested that their ancestral home was in Asia. In 1890, Otto Schroeder 'settled' the ancestors of the Aryans in southern Russia; Isaac Taylor opposed him in 1906, declaring that our ancestors were the "Celtic race from Central Europe." In 1853, talented German anthropologist Karl Penka argued in his book, *The Origin of the Aryans*, that their birthplace was located in Scandinavia: "Pure-blooded Aryans are represented only by the northern Germans and Scandinavians—a most prolific race, denoted by major physique, great muscular strength, energy, and courage. The brilliant natural gifts of the race enabled it to conquer the weaker races of the East, South, and West and impose their language on these peoples." Thus, Karl Penka was the first to point to the disparity in the anthropological factors in the arguments of linguists; he was later supported by another prominent anthropologist, Ludwig Wilser. In the English-speaking scientific world, the first to respond to this trend of "Nordicizing" the birthplace of the Aryans was John Reese, who suggested in 1886 that the Aryans may have originated from somewhere in the limits of the Arctic Circle, in part in northern Finland. And in 1889 Gerald H. Randall gave this definition to the Aryans: "A dolichocephalic race of blondes, originating near the Baltic seacoast. The Aryan presents himself as an intelligent type of man—the basic product of races, in which the basic qualities of dark and light, North and South, emotionality and practicalism, merged and united in a higher and connected state of mind and body."

However, a genuine sensation occurred in Europe with the book written by the Indian Brahmin, Bala Gandakhar Tilak, *The Arctic Home in the Vedas* (1903); based on the encyclopedic knowledge of the Sacred Vedas, he argued that the Aryan race could only have its origin from the north. Some time later, similar data was received in the form of another sacred book of the Aryans, the *Avesti*. Thus, both living ancient Aryan traditions—Hinduism and Zoroastrianism—testified in favor of one and the same theory, which greatly increased its reliability.

Simultaneously, the evolutionary theory of Charles Darwin received further development, and in 1900 the Hereditary Law of Gregory Mendel was rediscovered, in connection with which the concept of "racial purity" was argued on a genetic level. In the same year, [information on] the biologically substantiated existence of different blood groups was released, and several years earlier, the titanic work of the German Anthropological Society was completed. Under the direction of Rudolf Virchow, the Society studied the skulls of modern Europeans, in comparison with excavated materials that were collected by paleoanthropologists. The concept of a Nordic ancestral birthplace was confirmed. At the same time, Thomas Morgan and August Weisman created the *Embryonic Plasma Theory*. And in this same year, the Krupp firm, well-perceiving the state of affairs, announced a competition of scientific works on the influence of Darwin's Theory of Evolution on domestic politics and government legislation.

It was namely at this time that **Racial Theory** took shape as a separate field of science, with names, achievements, and titles. And again in 1900 a large consolidated work of Deniker appeared in the French language, *The Human Races*, in which he was the first to put the synthetic principle of racial classification into scientific practice. "What concerns the classification of races, is that it only takes physical traits into account. By way of the anthropological analysis of each ethnic group, we try to determine the races that compose them. Then, comparing races with one another, we will join races that possess the highest number of similar traits, and set them apart from the other races that are observed to have the greatest differences with them."

By "race" Deniker clearly meant "somatic unity," and thus put an end to any idealism in anthropology. In essence, the entire book was dedicated to separating **ethnography** and **anthropology** from one another. The author considered them phenomenon of different orders: the first is sociological; the second is biological. He wrote: "Several years ago, I proposed the classification of the human races, based solely and singularly on physical traits (skin color, the quality of hair, height, head shape, nose, and so on)."

In essence, Deniker was the first to take the position of strict and consistent **biological determinism** in racial philosophy. In his opinion, the surrounding environment was powerless in the face of racial traits. He wrote: "Racial traits preserve themselves with remarkable stubbornness, despite the mixing of races and changes occasioned by civilizations, loss of a previous language, etc. The only thing that changes is the ratio, in which this or that group comprises part of a given ethnic group."

Substantiating all the accumulated experience of previous researchers, Deniker made a point in the dispute about the Aryans, introducing a new term that had nothing principally in common with the romantic concepts of linguists: "One may call the long-headed, tall, light-haired race **Nordic**, since its members are grouped primarily in northern Europe. Its main traits are: very tall in height, 1.73 meters on average; blonde, wavy hair; light eyes, usually light blue; an oblong head (cranial index 76-79); rosy-whitish skin; the face is long, and the nose is straight and protruding." The terminological mess in racial theory ended, and the term **Aryan** gracefully departed for the sphere of culturology, sociology, and religious studies: "There can be no talk of an Aryan race; we are only able to talk about a family of Aryan languages, and perhaps, of an original Aryan civilization."

The work titled *The Human Races* appeared in 1900, but was synthesized, the term **Nordic race** being brought into use a year earlier. A little later, the leading German racial theorist, Hans F.K. Guenther, who always took the positions of Nordic philosophy, gave an explanation in this spirit in his book, *Racial Elements in the History of Europe*: "In philology, the word *Aryan* signified Indo-European languages; today that term is usually used only with reference to the Indo-Iranian branch of that language family. In the beginning of racial studies, they sometimes called the white or Caucasian race (but not the presently existing one) "Aryan;" later those peoples that spoke in Indo-European languages came to be called "Aryan," and finally the Nordic race was called "Aryan." Today, the term *Aryan* has departed from scientific use and its use is not recommended, particularly from this time, as it has become popular among the profane as a way of contrasting themselves with "Semites." But anthropology also turns away from the term "Semite," since peoples of the most varied racial origin speak Semitic languages."

However, let the Russian reader not be surprised that this information, which is well-mastered in European intellectual circles, is so weakly presented around us. The Russian writer and sociologist, Jakov Aleksandrovich Novikov (1850-1912), an author of dozens of popular publications on racial and ethnic questions, also preferred to write in French, because the Russian intelligentsia, immersed in dreams of ideals in the spirit of a Chekovian "Chaika," refused to fully absorb racial theory. It paid for this in 1917. Alas, but a fact remains a fact: today Iosif Egorovich Deniker and Yakov Aleksandrovich Novikov form the golden collection of racial theory in the French language, but are completely unknown in their own motherland.

In 1912, English researcher Sir Arthur Kent declared that "the political concept of race should be principally understood from a biological point of view." The Nordic Theory began to gather currency, and all the more confidently the view developed, that the outward distinguishing traits of the Nordic race are the result of its biological differences with other races.

Infatuation with the infamous "cranial measurements" began to move to the background, and the **biochemical concept of race** arises. Theoretically, it was created on the basis of numerous summarizations of practical works, by such scientists as Louis Berman (1925), Laurence Hesbrook Snyder (1926), Gilbert Joseph Rich (1928), Wilhelm Cruz (1929), Garrit Smith Miller (1930), Henry Etter Star (1931), Ray Graham Hoskins (1933), Leland Wayman (1935), and William Boyd (1935).

Dr. Otto Reche achieved the most success in the given area, from his numerous works dedicated to racial identification on the basis of blood types. It has been distinctly researched, that the first and second blood groups within the white race are primarily Nordic. Here the percentage of the first blood group always regularly increases in favor of the so-called **racial nucleus**, in which racial traits are expressed with the most uniqueness and distinctness.

American biologists L. Wayman and W. Boyd therefore correctly noticed that "blood groups" have existed longer than the modern races." Their countryman, Otto Klineberg, remarked in this regard that: "The problem of racial differentiation is based, first of all, on the inner metabolism, which for its part is a basis upon which the psychological and mental personality traits depend. Clinical observations have shown with all clarity, the degree of influence of the endocrine system on the human individual. That offers the possibility of using psychological characteristics in the objects of racial classification, and as a whole it enables one to speak already of the discovery of **scientific racial psychology.**"

Nordic Racial Type, according to Herman Lundborg

Nordic Racial Type, according to Herman Lundborg

G.E. Star and D. Rich also pointed out: "The components of human blood, such as hemoglobin, creatinine, phosphates, sugar, calcium, and many others, shed light on the biological basis of human individuality. The basic racial groups are distinctly differentiated by the make-up of these elements."

The German classics of **racial hygiene**, Erwin Bauer, Eugen Fischer, and Fritz Lenz also emphasized in their joint basic work, *Studies in Human Heredity* (1936), that: "Racial differences for the most part depend on the differences of an inner secretion. The constitution of the body, the intellectual and psychological characteristics, and other racial particulars are determined by them."

Thus, racial psychology received a powerful impulse for development. Simultaneously with this, on the basis of simple observations, several researchers began to come to the conclusion that outward racial traits are directly connected with psychological and intellectual capabilities. In 1904, Havelock Ellis established the tie between skin color and the coefficient of intellect (IQ), and then gave new support in favor of the biological superiority of the Nordic race. By studying the National Gallery of Portraits in London, he discovered that the majority of great people were blondes.

In her 1914 essay on racial psychology, Katerina Blackford characterized blondes with positive qualities, such as dynamism, activeness, and initiative, while brunettes [were characterized] by negatives—static, sluggish, lack of initiative, and conservativeness.

Later scholars [like] Donald Paterson (1922), Raymond Pearl (1924), Evelyn Huntington (1924), George Estabrooks (1928), and Catherine Eva Ladgate (1928) conducted numerous studies of racial groups on the basis of the intellectual coefficient (I.Q.), and came to the

114

straightforward conclusion that blondes statistically possess a higher I.Q., in comparison with brunettes, and light blue-eyed [types] also steadily surpass brown-eyed types.

In his book, *The Aryan and his Social Role* (1895), Georges Vacher de Lapouge, not yet knowing anything about future studies, presciently wrote: "The long-headed blondes fulfill the function of brain and nerves in the social organism, while the short-headed brunettes and their mestizos play the role of muscle and skeleton."

In the book, *Pork as a Criterion among Nordic Peoples and Semites* (1933), Richard Walther Darre, the Minister of Agriculture of the Third Reich, developed the original concept of the biological interconnections of totem (taboo) animals, with the racial characteristics of the peoples inclined toward them. Using ancient German and ancient world mythology, he came to the conclusion that the pig always accompanied settled Nordic peoples, namely because the southern nomadic Semites and Turks did not have a liking for it. The ritual prohibition among these peoples against pork consumption is a genetic memory about the shortcomings of the nomadic southerners in comparison with the settled northerners. He writes: "It is a question of the interconnection of defined peoples or human races on the one hand, and defined breeds of domestic animals on the other." The pig is a symbol of Nordic settledness; it is a main biological indicator, and it is namely for this reason that Judaism and Islam so hate it, for this hate best casts light on [their] biological origin.

Later in his book, Richard W. Darre approached an examination of the highly important question, which before that did not receive lengthy attention. "Until now, racial theory has not engaged such a problem, like food and race, but then on the other hand, it is known in animal-breeding that there is an interconnection between food and breed, since similar metabolic reactions flow differently among different breeds of animals. Protein has different value in food...In the process of digestion, protein is broken down into amino acids and then synthesized again into a specific protein for the given organism. Proteins govern metabolism. Proteins are always specific; therefore, protein foods and the digestion of the organism should be compatible, like a key and a lock...Therefore, it turns out that Semites and pigs are physiological antipodes (opposites)."

The correct vitality of an organism depends on harmonious metabolism. In Darre's opinion, in view of this, the members of different races need different foods, according to biochemical make-up. Thus, the pig is an ancient biological indicator of the racial differences of settled peoples, versus the members of nomadic peoples.

Darre reinforced his assertions with an analysis of the bread types preferred by different racial groups, for which he successfully cited the travel notes of Goethe, who traversed the borders of the German and Roman worlds. The ingenious classic of German literature noticed that in southern Europe he saw "black girls with white bread," while in the north of Europe [he saw] "white girls with black bread." From this follows the natural conclusion that not only domestic animals, but cereals which are needed by an individual for food [also] shed light on racial differences. Studying the tiny nuances of the physiology of food, and the geography of agriculture, the author already speaks about racial ecology and makes the conclusion: "The birthplace of the Nordic race is the forested zone of Northern Europe, with its temperate climate."

After the biochemical basis of racial differences, in light of our theme it is worth moving over to an examination of the original assertions of **racial psychology**, for which we will turn to the classic work by Ludwig Ferdinand Clauss, *The Nordic Spirit* (1939).

In it he wrote: "From the point of view of psychology, by race we do not mean a chaotic collection of "characteristics" or "traits," but a general style of experiencing, a defined unity of character...racial psychology is called upon to define those boundaries, which not a single people can violate or open, without destructive consequences for itself. If the Nordic experience is called "centrifugal," then the Eastern experience may be called "centripetal." The outward "coldness" of the Nordic man is explained by his effort to preserve the distance between himself and the surrounding world. The ability to "objectify" the world is a Nordic ability."

Clauss shows us that racial differences superimpose indelible imprints on the whole specific nature of experiences; the landscape of a country only strengthens the genetic given. "Terrain is a material, which the spirit transforms in its style and converts into a landscape...But not every terrain gives similar opportunity for such a transformation...The North Sea is an endless expanse, while in the Mediterranean, the shore is always close. Even if it is not seen, you know that it is close, or you feel its closeness by its evident signs. Here everything is beautiful and bounded by real time, with constant observance of proportion. The northern sky is high, and clouds fly along it all the time, to the southern [horizon]—the hand is extended up and clouds float lazily past it, or they begin a game with each other. The North teaches a man to strive for everything new—new distances—while in the South, the Mediterranean beckons to remain on it eternally: there all is temptation, a happy present time.

The Nordic spirit rushes into the distance, and in part, to the South; but the South for it is like a light to a moth. The degenerative influence of the South reveals itself in the disappearance of the desire to [explore]."

In another of his books, *Spirit and Race* (1940), Clauss pointed out that the members of different races perceive light, space, graceful forms, time, and movement differently. And in this time, while members of the southern races live by its outward effects, by affection and play, the Nordic races live by the essence of inner energy, its constant experience. "Race is a form, a form of living matter that is racial to the degree that it is inherent...Generations change, but not the form. Only if one proceeds from the heart, can he see the regular tie that unites the forms of spirit and body, two parts of one whole. Here we are talking about the form of action of the Nordic man, because that action is a determining value in his hierarchy of values; he perceives the world as something that is opposed to him, something that he should intervene in, in order to create something from it. This is his basic instinct, his inborn position, which determines his method of movement. He cannot [be] any different, because his law of spiritual form orders him to be so. This law is the final explanatory level of authority. There is no answer to the question: *why?*"

Such was the dominant position of racial theorists in the first half of the 20th Century. But modern research in the area of genetics completely supports their general emotional and poetic outlook. The most comical thing is that the Soviet school of anthropology—in its task to officially expose the "mad chimera of racism"—repeated all the basic postulates of classical racial theory in the most unambiguous way—and with a methodological persistence that the Third Reich never dreamed of.

No one ever took it upon himself to study Soviet anthropological works in genetics from the position of classical racial theory. From this perspective, the notorious concept of the "biological superiority of the Nordic race" is simply moved into the academic aura of infallible, "progressive Soviet science," which serves the ideals of "all progressive humankind."

As we showed above, a Russian scientist gave the basic category of German racial philosophy its name—something the Germans themselves repeatedly recognized.

Besides that at the highest point of Nazi-Soviet ideological confrontation, the famous German racial researcher, Ilse Schwidetzky, argued in her book, *Racial Studies of the Ancient Slavs* (1938), that the "eastern and western branches of Slavdom unconditionally belong to the Nordic race." Poland, like Germany, was not numbered among the friends of the Soviet Union at this time; however, her leading anthropologists, Jan Czechanowsky and Karol Stoianowsky, held to the same point of view on the affiliation of the Slavs [with the Nordic race] in their research. American specialists Lathrop Stoddard and Madison Grant characterized the population of northern and central Russia as "Continental Nordics."

Left: Erwin Bauer

Otto Reche, one of the leading specialists in Germany in the area of biochemical analysis of racial traits, conscientiously relied on the works of Soviet scholars B.N. Vishnevskiy, A.A. Melkikh, and V.Y. Rubashkin in his research. Being a member of the NSPAD, nevertheless, as a conscientious scientist, he did not consider it necessary to conceal the achievements of Soviet scientists, and such works as the Kharkov, *Work of the Permanent Commission for Research on Blood Types,*[106] or such Soviet journals as, *Medical Affairs,*[107] and *The New Biochemical Racial Index,*[108] which lay at the foundation of methods for computating of pure-blooded Aryans in the Third Reich.

---

[106] *Trudy postoyannoy komissii po isledovaniyu grupp krovi.*
[107] *Vrachebnoe delo.*

German specialists in the area of **dermatoglyphics**, a method for determining race and nationality with fingerprints, also based their work on the research of Soviet scholars: P.S. Semenovskiy and M.V. Velotskiy. In Germany, the developers of the concept of **racial hygiene** actively published [the works of] N.K. Kol'tsov, U.A. Filipchenko, and B.I. Slovtsov. German racial psychologists greatly honored V.M. Bekhterev, and [German] geneticists greatly esteemed N.V. Timofeyev-Resovskiy and A.S. Serebrovskiy. The famous modern English-speaking writer, Robert N. Proctor, relying on archival documents, brought up some curious facts in his book, *Racial Hygiene* (1988). N.V. Timofeyev-Rezovskiy was sent to Germany under the parameters of a government agreement with the USSR, and became Director of the Institute of Genetics at the Kaiser Wilhelm Institute in Berlin, and later read lessons in courses on increasing the qualifications for SS officers; and in an open 1938 elite party meeting dedicated to contemporary questions of racial policy, he appeared with a report—after the head of the NSPAD racial department, Walter Gross—but before the ideological chief of the Third Reich, Alfred Rosenberg.

Left: Otto Reche

Immediately after the victory over Germany, on the basis of new anthropological and archaeological data, an entire campaign unfolded in Soviet scientific literature, as a corrective to the historical concept of the Russian people. In 1930, the Soviet historian, Y.V. Got'e, wrote in his book, *The Iron Age in Eastern Europe*,[109] that: "The distribution of Slavic tribes on the left bank of the Dneipr in and of itself leads to the thought, that the guides of the earlier Slavic movement to the East and Southeast would have been northerners." P.N. Tretyakov graphically supported this thought in his book, *The Eastern Slavic Tribes*.[110] In publications of the time, a political directive was clearly felt: to prove under the tireless eye of the "father of the peoples", that the Russian is the "older brother in the family of fraternal Soviet peoples"—that particularly, because he is the elder, **he has Nordic origin**. Cinematographers, drawings, and sculptures of that period give graphic support to this [assertion]. The racial ideals of the defeated Third Reich were smoothly transitioned to the Soviet victors, something which is captured in artistic productions. The silhouettes and forms of Josef Thorak and Arno Breker—leading German sculptors—find their reflection in the Mamaev Kurgan in Stalingrad, through the efforts of Soviet sculptors. And they adorn the stations of the Moscow subway, and the main entrance of the Library. To this time, Lenin resembles a façade on the Reichs Chancellory, as closely as two drops of water [resemble each other].

One can still write much more about the Nordic origin of the Slavic world, and bring in numerous evidence, but one single fact is a most eloquent and indisputable proof of this concept. No other people, besides the Russians, by virtue of their ethnic self-awareness, gave definition to an important racial and namely, Nordic trait: light brown hair - or *Rusi*, [in the old tongue].

---

[108] *Noviy biokhimicheskiy rasoviy ukazatel'.*
[109] *Zhelezniy vek v VostochnoyEvrope.*
[110] *Vostochno-slavyanskie plemena.* 1953.

In the article, *Photo-colormetric Determination of Hair Color in Different Ethno-territorial Groups of the USSR*,[111] authors T.I. Alekseyev, V.A. Batsevich, and O.V. Yasina remark that hair color is one of the important racial-diagnostic traits. "In the determination of [race], subjective evaluations are not permissible. In hair colors, particularly ash-colored hues, the retention of phaeomelanin is sharply reduced. Judging by the data of the Russian anthropological expedition, dark brown hair turns out to be a predominant trait in the Russian population, although there is a sufficiently significant frequency of medium brown hues. Among those recently encountered, blonde hair of various hues is the most typical [hair color] for representatives of the northwest territories of the European portion of the USSR."

Thus, the Nordic Theory receives a new biochemical interpretation, for as a result, numerous experiments have proven that retention of phaeomelanin grows by measure of the increase of dark hair in populations, but more important still, it is made clear that "in dark-pigmented Europoid groups, the quantity of phaeomelanin is greater, than in Mongoloids." This says that retention of phaeomelanin is a type of universal, racial diagnostic marker, which unmistakably points to the degree of racial purity within each race. Among the members of the **racial nucleus** of each race, among which unique traits are expressed with all distinctness and uniqueness, the retention of phaeomelanin is less, than on the extreme racial periphery, where the degree of constipation with foreign elements unavoidably increases. A starting postulate of racial theory is that the Nordic race is the racial nucleus of the white race; it is supported in this very notable way: "The members of the southern Europoid race have in their constitution, a great percentage of carriers of phaeomelanin, those who are really representatives of the northern Europoids. Representatives of the Negroid race are characterized by an increased retention of phaeomelanin. In comparison with northern Mongoloids, the frequency of carriers of phaeomelanin among southern Mongoloids will also be increased." Thus, the percentage of a given agent in the hair is directly tied to the percentage of foreign, outside blood. The authors of the article also add: "The variants of reddish-hued hair in the Russian population are very rare."

And recall now, that in Rus', people with red hair were always regarded with distrust and caution, being considered unclean to touch; and during the reign of Peter I, they were even forbidden to testify in court, by official decree. Redheads were also an outside, foreign element, which stood out from the racially pure population. Consequently, the Russian people's distinctive marks are supported by the data of modern geneticists and biochemists.

In the article, *Comparative Studies of the Types of L-Polypeptide Proteins in the Hair of Russians and Yakuts*,[112] [the author], I.S. Afanaseva, also asserts: "In the classification of anthropological types, one of the most important traits is the different characteristics of hair. The form and color of hair on an individual always gives much greater taxonomical significance, such that scientists even assume they are a basis of classification of the human races. In the preceding [19th] Century, it was shown that the races differ also by the depth of hair, by the form of the cross section, size, thickness, and by the distribution of pigmented roots." On the basis of factual material in the given work, it is shown that the Europoids and Mongoloids in the example of the Russians and Yakuts, differ clearly and distinctly by retention of L-polypeptide proteins in the hair, once again canceling out all Eurasian notions of pan-racial blending.

Not only on the basis of phaeomelanin, but also on the basis of another biochemical component—tyrosynase—the concept of a racial nucleus is again supported in favor of the Nordic theory.

In the work, *Modern Representations of Pigmentation in Man*,[113] I.S. Afanaseva writes: "It is revealed that the very lowest average activity of tyrosynase is in light hairs, the shades of which vary from blonde to golden; in this regard, no difference between hue and activity is observed. Of greater significance is the fact that activity of this parameter is discovered in chestnut-brown hair. [There is] greater significance of tyrosynase in black hair; the members of the basic racial groups do not differ in this regard."

---

[111] *Fotokolorimetricheskoe opredelenie tsveta volos v razlichnykh etnoterritorialnykh gruppakh SSSR.* Voprosy antropologii, vyp. 84, 1990. –Questions in Anthropology, Issue 84, 1990.

[112] *Sravnitelnoe izuchenie tipov L-polipeptidov belkov volos u russkikh i yakutov.* Voprosy antropologii, vyp. 69, 1982.

[113] *Sovremennye predstavleniya of pigmentatsii cheloveka.* Voprosy antropologii, Vyp. 82, 1989.

The retention of tyrosynase, just like phaeomelanin, increases in each race in the direction away from its pure, biological nucleus, toward its mixed-race periphery.

Selectiveness is not a metaphor, but a genetic-biochemical given, calculated by a host of independent parameters.

On the extent of all world history, we observe one and the same picture, that within each great race, its biological nucleus attracts a burden of incomplete hybrids to itself.

World history is not only the struggle of great races, but to a still greater degree it is a struggle of racial nuclei with their own genetic peripheries.

The envy of the mongrel toward the pure-bred is the plot in world literature, having a one of a kind popularity. Mozart and Salieri—this was a struggle of biotypes, where the lower one traditionally resorts to dirty tricks.

Namely therefore Swedish anatomist and raciologist, Gaston Bachman, emphasized: "If one measures a civilization, not by the absolute number of creative personalities, but relatively, then the pure-blooded peoples of Europe surpass the peoples of all other countries of Europe."

The German racial theorist, Rudolf Polland, made the following conclusion on the basis of statistics: "The expedient support of the northern racial elements of our population, will undoubtedly contribute to a eugenic mentality, for this coincides with the support of persons, who are gifted above average and are trustworthy in an ethnic regard."

Groups of racial characteristics are always tightly interconnected with one another. The correlation of the data of a single racial diagnostic trait with data results from another, always improves the reliability of the common result, something that I.E. Deniker point out.

In his article, *Diagram of the Correlation of Pigmentation of the Hair and Eyes*,[114] A.I. Dubov showed that the percentage of retention of a component such as melanin increases abreast of eye color: light blue—blue—gray—brown—black. For its part, the color of the eyes is statistically interconnected with the color of the hair. Thus, the possessors of blonde hair at 75% from 100 have Nordic, light-colored irises, and 25% are mixed; those who have varying hues of light-brown hair have light irises in a range of 30% to 65%. Possessors of black hair very, very rarely have eyes of a light hue, and 80% have dark eyes. Besides that it is made clear that the light-blue eyes of the Nordic race have a different structure.

The same applies in relation to skin color. In the beginning of the 1960s, American biologist Kurt Stern proved the inadaptability of skin color in the human races to conditions in the external environment. Besides that he confirmed that the color of a person's skin is a biological phenomena, explained only by the theory of polygenesis; that is, the theory of numerous focal points of race formation. This same point of view is adhered to by the famous African-American racial theorist, Richard A. Goldsby, so there can be no propaganda about racism in the given case, no talk of it whatsoever.

In the collective article, *Primary and Secondary Melanin and Pigmentation on the Skin Covering of Man*,[115] V.K. Vasilevskiy, V.I. Semkin, I.D. Zherebtsov, and I.N. Mikhailov also considered it necessary to emphasize: "Melanin pigmentation is one of the basic factors influencing the color of the skin covering. Diverse variations of skin color depend on it: racial, age, sexual, and individual. It emphasizes the genetic condition of melanin pigmentation."

Outer racial differences, by the most natural means, are interconnected with racial differences on a biochemical level, and Soviet anthropology again confirmed this postulate of classical racial theory. In the article, *On the Correlation of Racial and Morphophysiological Traits*,[116] (Questions in Anthropology, Issue 52, 1976) authors M.G. Abdushelishvili and V.P. Volkov-Dubrovin, relying on factual material, came to the following important thought and confirmation of the Nordic Idea: "A known tie is observed with skin color and several physiological traits. Among the most lightly-colored, a slowed-down bloodstream is observed, and a maximum mineral saturatedness of the bone tissues; and among the darkest, there is a significantly lower mineralization of the skeleton, and a faster bloodstream."

---

[114] *Skhema sootnosheniya pigmentatsii volos i glaz*. Voprosy antropologii. Vyp. 80, 1988.

[115] *Pervichnaya i vtorichnaya melaninovaya pigmentatsiya kozhnogo pokrova cheloveka*. Voprosy antropologii. Vyp. 62, 1980.

[116] *O sootnoshenii rasovykh i morfofiziologicheskikh priznakov*. Voprosy antropologii. Vyp. 52, 1976.

The Salt of the Earth. This is the embodiment of the image of the nucleus of the white race, and the deeply Nordic character of the image is again obvious.

In the article, *Toward a Hypothesis on the "Somewhat Large Adaptation" of Persons with Phenotype O,*[117] E.V. Tikhomirova and E.I. Khrisanfova analyze the statistics of the sickness of people with various blood types, by viral illnesses. On the basis of a multitude of summaries, they make a conclusion about "the somewhat larger adaptation of persons from the first blood group," which reinforces the unique racial-diagnostic observation. "The authors explain the predominance of Type O among donors, not only by its "universality," but also by a healthier quota."

As we recall, the increase of the percentage of carriers of the first blood group is firmly observed in the conditions of the concentration of traits of the white race; that is, when the northern Nordic element in a population increases. The higher biological health of donors of the first blood group also testifies about their affiliation with the Nordic race—the nucleus of the white race.

In the work, *Notes about the Connection of Several Traits of Dermatoglyphics with Groups of Blood of the ABO System,*[118] T.D. Gladkova and L.O. Bitadze analyzed the different unconnected groups of racial traits, such as skin color, blood groups, and fingerprint impressions, and made a strict scientific conclusion: "The frequency of swirls decreases from the darker to the lightly pigmented populations; at the same time, the quota of loops increases in the same direction."

All this means is that in any direction of the consideration of racial-demarcating traits, we did not move; everywhere we will expect one and the same conclusion, made by the first racial theoreticians.

The Nordic race is the nucleus of the great White Race, in which all traits are expressed in the most clear and unique way. The genetic-biochemical differences of the members of the Nordic race are regularly manifested in their psychological and spiritual mentality, which, for its part, finds reflection in the particulars of their religion, culture, aesthetics, and social-political institutions: in a word, in all the specific features of the historical process.

One of the leading Russian specialists in dermatoglyphics, G.L. Khit', wrote in the article, *The End of the Line C/S among Various Racial Groups*[119]: "Dermatoglyphic traits have not had an adaptive character for the entire length of the history of the formation of the Mongoloid and Europoid racial trees."

For their part, A.A. Zubov and I.M. Zolotareva argued in the work, *The Mongols in the World System of Odontological Types,*[120] that on the basis of an analysis of the structure and arrangement of the dental system, Russians, like other central and northern Europeans, differ from the Mongols **by a factor of three**. "The Central Asian Mongoloid racial type is discovered in odontological terms to have sufficient essential variability, with significant deviation away from the side of the "Western complex.""

With the citing of the latter authoritative declarations, we once more want to remove any accusation of racial impurity in the address of Russians, and in the address of the Nordic race in general. As seen, it is still incomprehensible. How, from a genetic point of view, anyone could speak of a Mongol-Tartar yoke? A yoke on whom? The conquered, or the conquerors? Unfortunately, the custom of accusing only the white race of mixing is firmly ingrained.

In 1978, the Scientific Research Institute of Anthropology of the USSR conducted a complex program of study of the sensory systems of Man in different ethno-territorial groups. It was established that members of the Mongoloid race, equally with Mongol-Europoid mixes, possess a different threshold of the olfactory and taste senses. A highly significant sense toward bitter taste was demonstrated.

Now let us not forget the ancient Russian saying: "the bitter bread of the enemy." As is now made clear, this metaphor is an example of a very exact genetic memory of the [Russian] people.

---

[117] *K gipoteze o "neskol'ko bol'shey prisposoblennosti" lits c fenotipom O.* Voprosy atropologii. Vyp. 69, 1982.

[118] *Zametka o Svyazi nekotorykh priznakov dermatoglifiki c gruppami krovi sistemy ABO.* Voprosy antropologii. Vyp. 69, 1982.

[119] *Okonchanie linii C u razlichnykh rasovykh grupp.* Voprosy antropologii. Vyp. 59, 1977.

[120] *Mongoly v mirovoy sistematike odontologicheskikh tipov.* Voprosy antropologii. Vyp. 64, 1981.

During the time of the Mongol and Turkish conquest, the racially pure Europoids—the Russians—had occasion to taste the unfamiliar bitter bread of a foreign race, which had a different threshold of sensitivity. Evidently, this saying was born and became deeply rooted in the popular consciousness, in observance of the inherent racial purity of the arch-type itself. "Bitter bread of the enemy" - this is the taste of detested race-mixing.

Peoples' taste differences are also not only reduced to tricks of the national kitchen, they are genetically conditioned. All sensory organs—without exception—have racial affiliation. We will take into consideration the work of E.R. Sigal and M.I. Potapov, *Property Groups of Saliva in Man.*[121] In it, the authors make the following conclusion: "To the present time, all known genetic systems of saliva show dependence on racial affiliation, but the most noteworthy in this regard is the Rb system. It may be defined as a system that is primarily inherent in Negroids. It is essential also to note the opposing character distribution of the Ra and Db genosystem among Europoids and Negroids."

The classical prescriptions of apartheid, prohibiting people of different races from eating food together, does not have a foundation of prejudice under itself, but rather a biological basis. People of different races have different rates of occurrence of different genes, and therefore become sick from different illnesses. Differences in the activity of immune systems lead to a fit of false democratic ecstasy—the individual of one race can get a disease from another race—to which he has no immunity—through saliva. The Great Age of Exploration is full of examples of the disappearance of entire tribes of natives, who came into contact with the first white settlers. "The old sea wolves" drank an enormous amount of gin, namely with the objective of disinfecting, for the tropical diseases that were not dangerous to the savages, mercilessly decimated [European] ranks. Does anyone ever talk about one kind of food? The authors of the article further write: "Racial differentiation is characterized by serum batches of blood, leukocytes, anomalous hemoglobins, enzymes, tissues, and secretions."

It is namely because of these facts that the science which arose at the turn of the 20th Century, went by the name of **racial hygiene**.

Thus, all the above-enumerated data of the Soviet Academy of Science once again easily agrees with the postulates of classic racial theory, although it is clearly from Richard Walther Darre's discourse on pigs and bread, and their significance in the biological metabolic process of the Nordic race. The traditional Russian kvass drink is based on black bread: and what is this, but another proof of the ancestral Nordic origins of the Russian?

---

[121] *Sobstvennie gruppi slyuni cheloveka.* Voprosi antropologii. Vip 56, 1977.

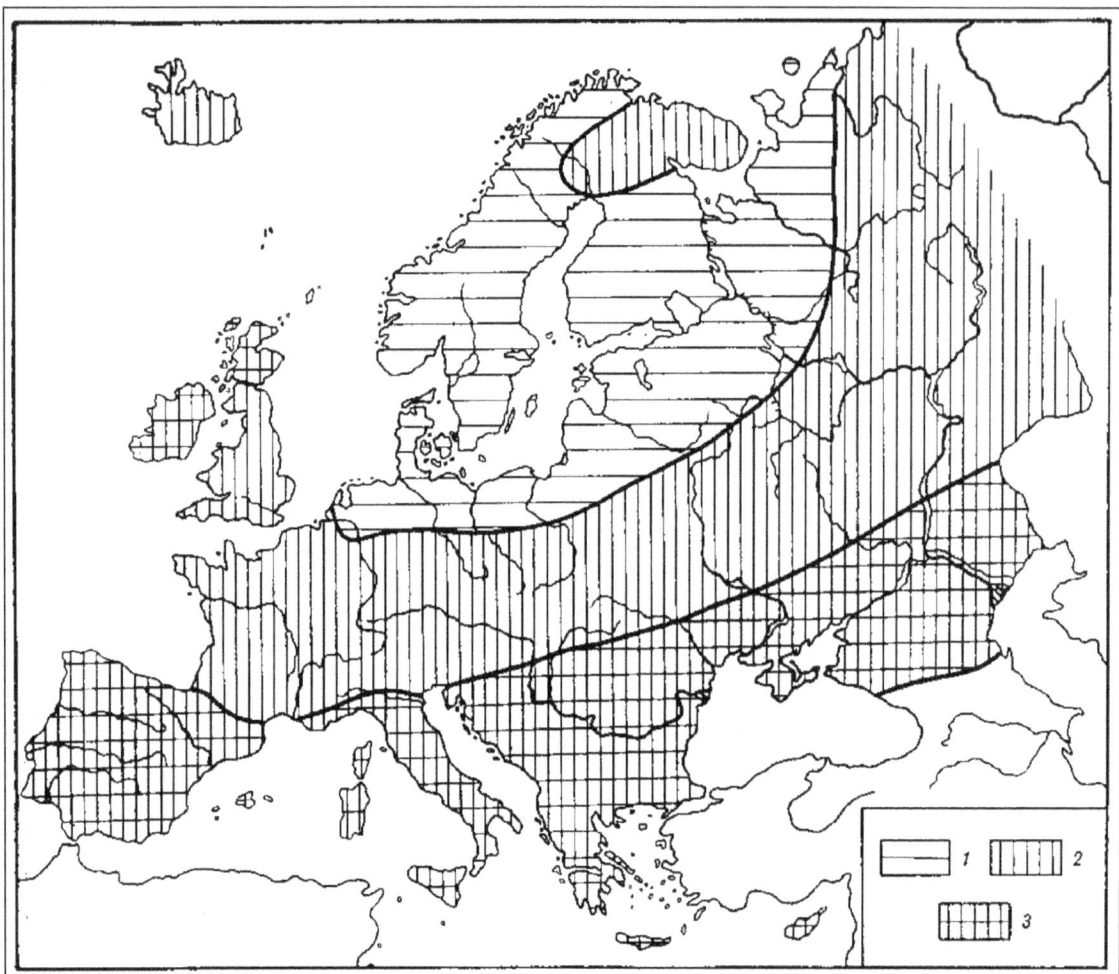

Рис. 2. Географические вариации цвета волос на территории Европы

1 — меньше 50% темных волос (оттенки 27,4—8 по шкале Э. Фишера); *2* — 30—80% темных волос; *3* — больше 80% темных волос

Geographical variation of hair color on the territory of Europe, according to V.P. Alekseyev. 1-less than 50% dark hair (hues 27.4—8 according to E. Fischer's scale); **2**~30-80% dark hair; **3**-more than 80% dark hair.

Soviet race studies, specially created by the communists for the profaning, discrediting, and ridicule of the ideas of racial theory, have on the contrary pointed out to the intellectual elite—in a rather non-trivial manner—that racial theory is the science of sciences, in which the concept of a Nordic race is an integral part. Without this it is impossible to comprehend the sense of world history. As Benjamin Disraeli, Prime Minister of Great Britain, observed in the middle of the 19th Century: "Race is everything; there is not a greater truth anywhere."

V.E. Deryabina's work, *Methods of Statistical Intergroup Analysis of Anthropological Data: and Examination of a Mixed Selection of Traits*,[122] is significant. The given work lays out a method of racial analysis on the basis of which lies a simultaneous examination of the integrated selection of traits, measured in the scales of different systems. Thus, it is a question of a complex, mathematical, multi-dimensional analysis of mixed traits. Summarizing the numerous racial traits of the Russians, the author of the article comes to the following conclusions: "The first and most important conclusion that one may make here, lies in the statement of the significant unity of the

---

[122] *Metodika statisticheskogo mezhgruppovogo analiza antropologicheskikh dannikh: rassmotrenie smeshannogo nabora priznakov.* Voprosi antropologii. Vip. 88, 1995.

Russians on the entire territory considered, and the impossibility of setting apart corresponding regional types, which are distinctly restricted from one another."

On the question of the biological singularity of the Nordic race, in light of the Russian colonization of the north, the author was also unequivocal: "Thus, the existence of a northern sub-type of Russian may be considered a reflection of a definite anthropological community of the population of north and northeastern Europe, preceding the time of Russian colonization of the said territory, and serving as one of the components for the formation of modern Russians on this territory."

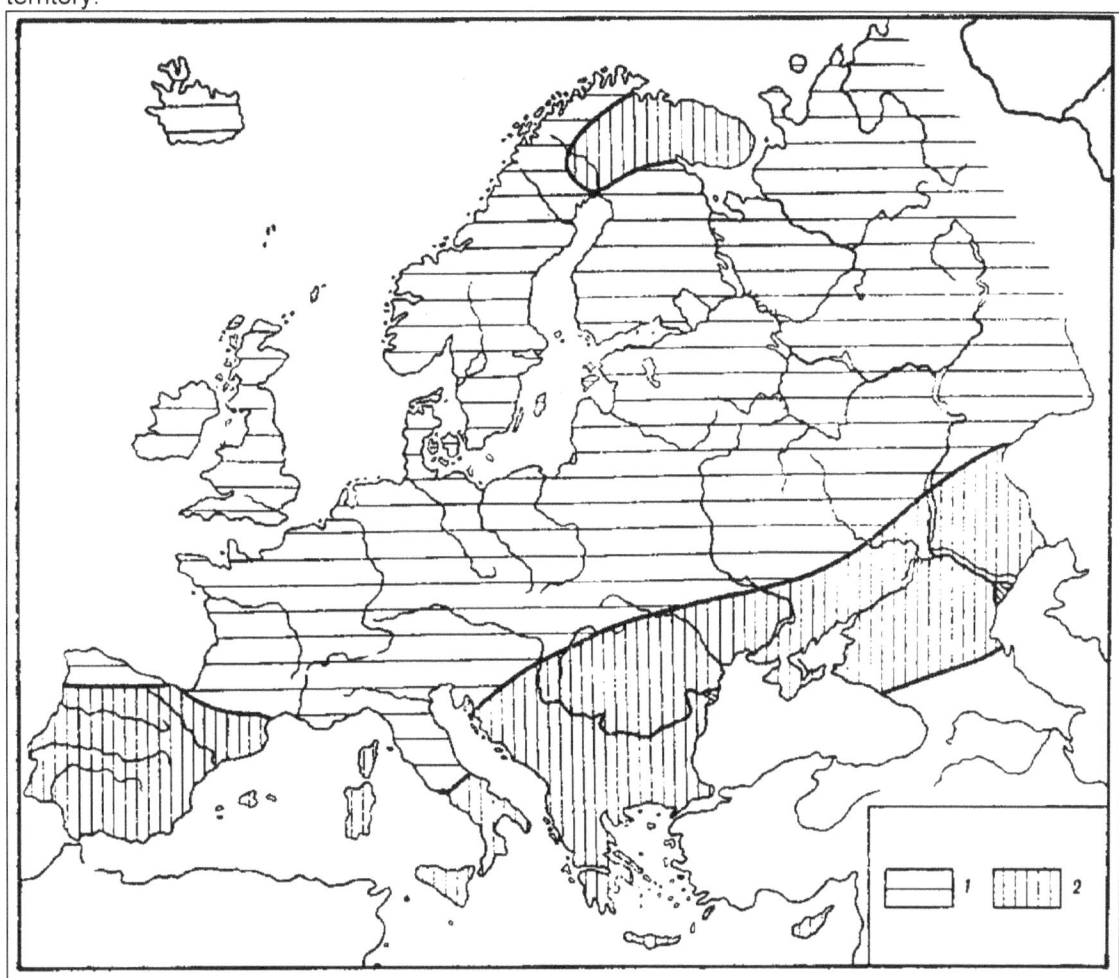

Рис. 1. Географические вариации цвета кожи на территории Европы

1 — очень светлая кожа (подавляющее преобладание оттенков 1—12 по шкале Ф. Лушана);
2 — несколько более темная кожа (довольно часто встречаются оттенки 13—15)

Geographic variation of skin color on the territory of Europe.
1-very light skin; **2** somewhat darker skin.

The given global conclusions are supported again and again by the results of private biochemical studies; for example, in O.V. Irisovaya's article, *Polymorphism of Red Blood Corpuscular Phosphate Acid in Different Population Groups of the Soviet Union*,[123] the author writes: "Noted among the population of Europe is a relatively wide dispersion of three alleles: Ph-a, Ph-b, and Ph-c. As a whole, the rare-frequency gene Ph-c serves as a characteristic trait, marking European populations (0,030-0,070). The Ph-a allele varies in the European population, in the limits of 0.268-0.402. In Negroid populations, the frequency changes to lower parameters:

---

[123] *Polymorfizm eritrotsitarnoy kisloy fosfatazi v razlichnikh gruppakh naseleniya SSSR*. Voprosi antropologii. Vip. 53, 1976.

0.16-0.25. Mongoloids, as was expected, have a very wide spectrum of changeability of alleles Ph-a and Ph-b, but nevertheless, the gene Ph-c is practically absent in them."

As we see, the myth of genetic inheritance from the Mongol-Tartar yoke in Russia again suffers complete collapse, because the Ph-c gene is in Russians, while in Mongols it is practically absent. "Pan-Eurasia" is the result of blood chaos, and the chaos of ideas, skillfully gathered with the goal of choking the genetically inherent arch-type of the White race.

Географические вариации наклона лба на территории Европы и Кавказа (краниологические данные). Мужские черепа:
1—78°,7—81°,4; 2—81°,5—84°,2; 3—84°,3—87°,0; 4 — ареал северных европеоидов; 5 — ареал южных европеоидов

Geographical Variation of the Angle of the Lobe on the Territory of Europe and the Caucasus. (Craniological Data). Male Skulls: 1—78.7-81.4 degrees; **2**—81.5-84.2 degrees; **3**—84.3—87.0 degrees; **4**—range of Northern Europoids; **5**—range of southern Europoids.

Discussing further, the author of the given work again and again supports the basic postulate of the Nordic Idea: "The high thermal stability of phenotypes B and AB testify about the adaptive significance of these variants of phosphate acids in the extreme conditions of heat in a tropical climate." This means that the 3rd and 4th blood groups, by their parameters of thermal stability, graphically testify to their non-Nordic origin. The Nordic race is characterized, as we recall, by a

124

high percentage in namely the 1st and 2nd blood groups. A long way from the nucleus of the white race, in the hot regions of the Mediterranean, the replacement of blood groups 3 and 4 occurred. The viability of these mestizos was lower than that of the pure members of the Nordic race, which, as we said above, is testified to with statistical data from blood transfer centers.

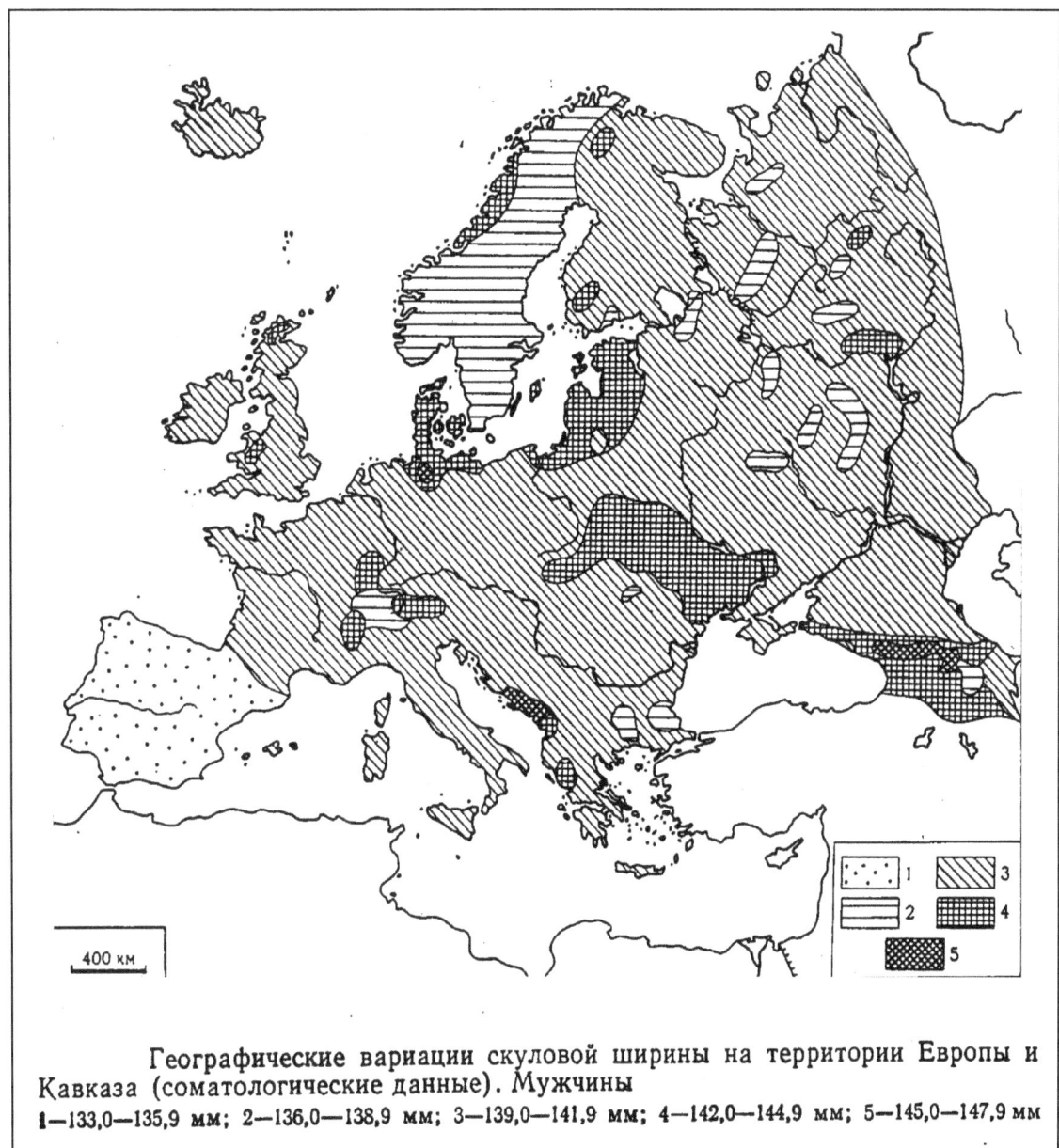

Географические вариации скуловой ширины на территории Европы и Кавказа (соматологические данные). Мужчины
1—133,0—135,9 мм; 2—136,0—138,9 мм; 3—139,0—141,9 мм; 4—142,0—144,9 мм; 5—145,0—147,9 мм

Geographical variation of cheek width on the territory of Europe and the Caucasus (somatic data). Males: 1—133.0-135.9 mm; **2**—136.0-138.9 mm; **3**—139.0-141.9 mm; **4**—142,0-144.9 mm; **5**—145.0-147.9 mm.

The prominent American anthropologist, Ashley Montagu, confirmed the correctness of our assertions in his book, *Introduction to Physical Anthropology* (1951). He wrote: "Today, after numerous studies in the given area, it is completely evident that Blood Group B appeared in Europe, together with the appearance of brachycephalics from Asia. This occurred in the period from the 5th to 15th Centuries, A.D., together with the invasion of Asian armies from the East. Consider it completely proven, that all the basic population on Earth arose in conditions of the absolute dominance of Blood Group O, while the rest are the result of a complex process of hybridization and resultant mutations, on the basis of the later rise of reproductively isolated communities."

The Hottentots and Bushmen of Africa have a very low percentage of Blood Group O, which speaks to their high degree of racial purity. A colony of Gypsies in Hungary has primarily the same picture of distribution of blood groups, as that in Northern India, from whence they came. The Aborigines of Australia also have a low percentage of Blood Group B. "The blood of Asian Mongols is the singular source of Blood Type B in Europoids"—Ashley Montagu summarized. **From this it follows that forms of Asiatic thinking penetrated Europe, together with Blood Type B. Blood types and types of ideas are interconnected.**

The American anthropologist emphasized that genes influence blood groups; therefore, these combinations of traits may be used as indicators.

Taking into consideration the fact that in anthropology correlations between physical and psychological traits have long been sufficiently established, it is possible, using empirical methods, to deduce the definite racial purity of this or that type of world view.

The classical Soviet anthropologist, V.V. Bunak, considered it possible to openly point out in his fundamental monography, *The Species Homo: His Rise and Subsequent Evolution*,[124] that: "The summary indicators of vitality (for example, metabolic processes) differ noticeably in comparisons with Europeans and the peoples of non-European countries, particularly Negro peoples of Africa. In great part, Negroes and many other non-European peoples are inferior to the powerful European indicators."

Below: M.F. Ashley Montagu

It is completely evident, that namely these summary indicators of metabolism render a decisive influence on the development of the principle differences, between types of worldviews, that are genetically present in the different races; the method of perceiving events is always directly associated with the biological indicators of this process [metabolism].

For an illustration of the qualitative differences in worldview types, a completely objective [set of] quantitative characteristics exists. Thus, for example, the retention of cholesterol in the blood strongly fluctuates: among many Mongoloid peoples of the north it is 120mg%; among Turkmens it is 151mg%; among Buryats, 185mg%; and among Russians, 180mg%. For a large number of groups, a correlation has been found in the albumin and globulin in blood plasma. In the distribution of the given indicator, no geographical or other dependent has been observed: among Russians, the average index is equal to 2.0; among Norwegians, it is 1.05; among Buryats, 2.07; among Yakuts, 1.37; among the Chukchai, 1.61; and among Eskimos, 0.89.

The retention of cholesterol in the blood plasma of Negroes is relatively lower—135mg%—than among Europeans—180mg%. A characteristic trait of Negroes and several other non-European groups, is the low retention of calcium secretion in urine. By the sum total of

---

[124] *Rod Homo, ego vozniknovenie i posleduyushchaya evolyutsiya*. Moscow, 1980.

bio-chemical and physiological processes of metabolism—as Bunak points out—all races differ significantly from one another.

Eugen Fischer points out: "The structure of blood serum differs even within the framework of European families." Besides this, among differing racial groups, it fluoresces by different colors. It is possible that this biochemical fact lies at the base of a striking diversity in the attitudes of [different] races toward magic rituals that utilize blood, since there exist tribes and entire racial groups that make use of the blood of strangers, and there are those that use only the blood of fellow tribesmen for [rituals]. Even magic considers blood from the viewpoint of its racial qualities.

The steadiness of this tendency can be traced deep into antiquity. In this regard, Eugen Fischer's opinion is highly indicative, as he points out in his book, *The Rehoboth Bastards and the Problem of Miscegenation among Humans* (1912). He wrote: "It is established [that] among any race, the structure of molecules of blood serum will only become more complex in the future."

Subconsciously at times, the mystical relationship of Man and blood, and its all-powerfulness over fate, is precisely explained by the evolutionary stability of its genetically racial structure.

"The myth of blood", "the call of ancestors," and other similar, generally used forms [of reference] represent nothing else but the lyricized experience of generations, elevated to the rank of super-sensory dogmatism.

Therefore, every person in essence is a biological container, designated for the transportation of a worldview in time and space. And the biologically characteristic packing material is connected with the type of worldview that fills it.

In connection with the above, we do not recommend use of the gastronomic and cosmetic products of other races. The curative mud from the shores of the Dead Sea, which television advertises, is good for applying to Semitic beauties, but not Russian lovelies.

The same applies to fragrant stimulants. During the time of colonization in the West Indies, there was an expression: "The Lord God loves his Negro, and knows him by his smell." Ancient travelers remarked that the Chinese smelled of musk, and Ammianus Marcellinus, a Roman historian of the 4th Century, asserted that the Jews smelled of garlic.

As seen, there is not a body part on an individual, which does not testify to inborn racial differences. According to an article by T.A. Abrazhevich and V.A. Spitsyn, titled, *Genetic Dimorphism of Ear Wax in the Yakuts and the Population of the Baltic Republics,*[125] the portion of Gene d in the earwax of northern Europoids is equal to 0.1815; in Germans it is 0.176; in Yakuts, 0.8939; in Chinese, 0.979; and in Negroes, it is 0.069.

An analogous picture is observed with Gene *w*, for in the Chinese it is equal to 0.021; in Germans it is 0.824; and in Negroes, it is 0.931. Thus, to expose your ears to the gossip of other races is also harmful and un-hygienic. One need only listen to their own [kind].

Besides that the genetically conditioned chemical differences in races also influences the perception of the surrounding world, the development of aesthetic taste, philosophical ideas, and religious preferences.

A.S. Vagina writes in her excellent article, *On the Particulars of the Attitude of Several Ethnic Groups toward Color,*[126] that: "Any population, any ethno-cultural community speaks in its colorful language with a definite selection of "sounds" and their combinations. We perceive that language not only in shades of colorful symbols, but we perceive it by the general color "background" of the national, material culture. A large part of color in traditional culture is obvious. The color of a costume is also closely connected with the outer appearance of the individual, since it usually agrees with the color of her dress, the color of her skin, hair, and eyes. Among people of the Northern race, the preference for black, blue, and also light blue (cool colors of the short wavelength part of the spectrum) is not accidental, for they significantly lighten and decrease the density of the magnitude."

---

[125] *Geneticheskiy dimorfizm ushnoy seri u yakutov i naseleniya pribaltiyskikh respublik.* Voprosi antropologii. Vip. 70, 1983.

[126] *Ob osobennostyakh otnosheniya k tsvetu nekotorikh etnicheskikh grupp.* Voprosi antropologii. Vip. 84, 1990.

Here again it is also precisely explained from a biological standpoint, in I.S. Afanaseva's article, *Modern Representations of the Pigmentation of Man*.[127] In it, she gives a summarization of international scientific information, from which it is made clear, that the eyes of Europoids and Negroids differ in structure, and light-blue eyes differ particularly by the Tindal Effect. In the direct sense of the word, the Nordic man sees the world differently.

And Emmanuel Kant noted in *Analysis of Excellence*, that the "excellent becomes known without any thanks to an idea." Fritz Lenz pointed out: "Already being children, we differentiate beautiful people from the ugly; long before that as we transform the experience of such things, or with the help of comparison, we form an aesthetic sense. We make such differentiations instinctively, to the extent that we carry the appearance of our race in our very selves."

Modern studies in the area of **Biological Aesthetics** argue that each race has its own, genetically conditioned image of Chaos and Order. Konrad Lorenz called this the "inborn model," and Ireneus Ebel-Ebeysfeld wrote in the book, *The Biological Basis of Aesthetics*, that: "Style is the ability to codify imparted information in artistic form." It is namely therefore why the production of art by different races differs in style, for style on a biological level fulfills the function of determining "us and them." The production of modern mass art is specially modeled on the basis of the absence of a bio-password. Style, in a compressed, codified form, expresses the experience of a race. Ebel-Ebeysfeld made a valid conclusion on this basis: "Suggestibility, and readiness to master views, and readiness to accept group values, are to a very high degree, inherent in an individual; they paved the way for racial selection."

Besides that he brings in some curious data from laboratory experiments. It turns out that in rooms painted in red-orange tones, subjects evaluated the measured temperature as 3-4 degrees Celsius higher, than in rooms with blue-green tones. "Warm" tones stimulate the sympathetic nervous system, from which the pulse quickens, and blood pressure increases. "Peaceful" tones, which are called "cold colors" for a reason, belong to the short wavelength part of the spectrum.

Thus, people of the Nordic race prefer cold colors, which correspond to their system of blood donation, that is, their composure; not in the figurative sense, but in the direct sense. Light-blue eyes are an indicator of optimal bio-energetic processes in the organism of Northern man. It is justifiably the same in regard to music and dance, for their rhythm completely relies on the rhythmic structure of the racial archetype. The white race created symphonic, harmonic music, while the black race invented jazz and rap, with their disorderly pile of syncopation. Characteristic in this regard are the harmonious, Russian religious hymns, where the key appears as a genetic unity of harmony, which again confirms the Nordic sources of Russian popular art.

Hans F.K. Guenther wrote in his magnificent book, *Race and Style* (1927), that: "There are no plays in the Semitic languages, and the music of the Arabs was borrowed from the Persians."

Later, Henrich Zollinger contended in the work, *Biological Aspects of Color Vocabulary*, that: "In the linguistics of color concepts, there exists a racial hypothesis, according to which people with a different color of skin can see and call a color differently, because their eyes are affected by differences in pigmentation." Paraphrasing the words of Goethe, Hans F.K. Guenther noted that the Nordic race was created in order to peer into [things], while the Eastern race [was created] for contemplation. This has a very immediate relation toward color, and consequently, toward the structure of the eyes.

The differences also touch on the perception of time and rhythmic forms. This is best of all seen in systems of the different races for the chronological determination of years, and in the predominance of this or that geometric form in their living quarters, something which Oswald Spengler highlighted. Modern researchers like Frederick Terner and Ernst Peppel discovered the presence of racial traits in the organization of poetic form and rhythm. [These traits] are called to work on frequency, by synchronous timing of the frequency of the brain. The timing frequencies of the intellectual bio-processes of different races, also strongly differs, correspondingly to this organization of the processes of instruction. Work with virtual information should be examined in the spirit of racial theory.

It is completely erroneous, therefore, to consider the sweaty, dirty Gypsy camp, with its Asiatic songs and dances, as a symbol of Russian culture. Equally so, the "Quadrat" by Kazimir Malevich: it is a street sign, not a picture—according to the apt definition of Ilya Sergeyevich

---

[127] *Sovremennie predstavleniya o pigmentatsii cheloveka.* Voprosi antropologii. Vip. 82, 1989.

Glazunov; it clearly goes back to sources that are not of the Russian aesthetic worldview. In his excellent book, *Art and Race* (1935), German racial theoretician Paul Schultz-Naumburg expressed the fundamental thought with simple and lucid words: "As the very race itself, so the art."

A modern science—**anthropo-aesthetics**—engages in the study of racial and ethnic canons of human beauty. The planting of foreign racial anthropo-aesthetic canons, by means of television and mass pop-culture is nothing but harmful, and can offer nothing.

Ludwig Ferdinand Clauss, in the book, *The Nordic Spirit* (1939), wrote: "If a person reflects his inborn style, he will not find another, but will seek his won. A spirit which does not live in agreement with its laws, leads a double life: it ends up between its own laws and foreign ones. In secret, it does not feel itself worthy in the face of these laws, or the in the face of others. If the entire time only the foreign is perceived, it leads to the destruction of one's own essence."

100 years ago, I.E. Deniker wrote: "Among several nationalities, the muscles of the face perform such movements that would be highly difficult for foreigners to imitate. For example, moving forward only the upper lip, which is done by Malaysians with an elegance and ease, which would honor any chimpanzee."

Modern beauty contests for girls of different backgrounds produce an impression of the legalized banditry of a slave market, organized on the principle of a "genetic pool", where lewd half-breeds and mongrels always appear in the role of judges. This is a defilement of the race—for race is the highest value we possess.

Hans F.K. Guenther thus described the sacral depths of Northern Man, conditioning his racial biological uniqueness: "If the conduct of Eastern peoples is characterized by an inclination toward pretense, then the behavior of people of the Nordic race differs by reserve, which is perceived by non-Nordic peoples as haughtiness. The Nordic man is not right for preachers. His faith is too solitary, reserved, quiet, and penetrated by reverence. It is characteristic of the preacher to win over a foreign soul and gather disciples around himself—the Near Asian "pathos of community"—which is the opposite of the Nordic "pathos of distance," that was praised by Nietzsche. The sermon among "the non-believers" and the idea of "world religion;" to "go forth and teach all nations"—these are manifestations of the Near Asian spirit; it is foreign to the Nordic spirit."

We read similar thoughts from Ludwig Ferdinand Clauss: "If the Nordic man is religious, then he is religious in this style: he doesn't like to get undressed in front of a crowd, and he doesn't go to a bazaar to pray."

Thus, the entire spiritual world, and all the psychological traits of people of the Nordic race are determined exclusively and solely by their biochemical uniqueness, passed down by a genetic path from generation to generation.

Soviet periodicals, not unlike the official raciology of the Third Reich, came out; one may read journals like *Questions in Anthropology*, *Soviet Ethnography*,[128] *Genetics*,[129] and many others, as supplemental literature. One need only change the criterion for evaluation from "international" to "Nordic," and then everything stands in its place, all on its own.

Created to rebut classical racial theory, Soviet anthropology unwittingly picked up the banner from the toppled German school and operated in the same victorious field, the name of which is **biological determinism.**

The German raciologists were part of the mass of good, decent Catholics or Protestants, who understood race as a unity of experience, a community of style, united in fate. But one of the leading Soviet anthropologists, V.P. Alekseyev, determined that "kinship is a biochemical concept." Such a precise statement of the problem would have shocked even the Head of Ideology of the Third Reich—Alfred Rosenberg.

Now one may daringly say that the business of German racial theory did not go to waste. **The great idea of the Nordic Race has all the more strongly given itself meaning in modern Russia.**

Finally, for the sake of justice, it is also necessary to note that German racial theorists never felt awkward about quoting their Soviet colleagues; therefore, the author of the given essay simply continues the good tradition of academic courtesy, reciprocating with courtesy from the

---

[128] *Sovietskaya Etnografiya.*
[129] *Genetika.*

Russian side. After all, what Hitler never succeeded in doing in his own time, has been done in "modern democratic countries" - that is, conscientious citation of Russian racial theorists. If we take the central work of Fritz Lenz (*Studies in Human Heredity* (1932)), we discover 34 quotes from Soviet scientific works, in which the general number of authors cited is twenty-two. The book is considered canonical in questions of racial hygiene, but no problems arose for the authors from the part of "competent [government] organs," over enthusiasm for the science of ideological enemies.

No problems arose after 1945, either, from the part of the occupation authorities. Not one major racial theorist of the Third Reich went through a single political process about the crimes of Nazism. Moreover, they all retained their positions in the universities, where they taught until the end of their lives. The reliability of this information can be gleaned from the books of two famous anti-fascist writers, who specialized in the theme of the racial theories of the Third Reich—Robert N. Proctor and Stefan Kuhl.

The Nordic Idea is clean, and this is recognized by the Nuremburg Process. No one has accused a Soviet cosmonaut in connection with Stalinist repressions, or the Arms Race. For that matter, A.D. Sakharov, who thought up the [Russian] atomic bomb, has a park named in his honor in Israel. Incidentally, in the same Israel, in the Holocaust Memorial Museum, a memorial stele has been erected for the German racial theorist, Ludwig Ferdinand Clauss, "for saving Jews at the risk of his own life."

In *Racial Elements of World History*, a book that has been published in many languages in the world, and which was written by Hans F.K. Guenther after the war, there is a chapter titled, "The Nordic Ideal." This is the racial-political testament of one of the recognized leaders of the Nordic movement:

"The question is not one of in which measure we, the people living today, are Nordic, but one of whether we have enough courage in order to prepare the world for future generations, which are clean in racial and eugenic terms. The de-Nordicization of Indo-European peoples has always taken centuries; the will of the people with Nordic-thinking should throw a bridge across the centuries. When there is talk of selection, one needs to take into consideration the multitude of generations; modern people with Nordic thinking can expect, for the extent of their lives, only one reward for their labors: the knowledge of their own courage. Racial theory and research in the area of inheritance offers strength to a new aristocracy of youth, which, striving toward high goals, like Faust, will answer to the calls from the spheres, and go out beyond the limits of individual life.

Inasmuch as the movement does not aspire for gain, it will always be the movement of a minority. But the spirit of any Age is always formed by just a minority - including the spirit of the Age of the Masses, in which we live."

DISCUSSIONS ON RACIAL PREJUDICE

*"Natural inequality between separate, distinct tribes and races
is the common principle in an organized world."*
I.I. Mechnikov

*"Clear thought, a strong will, and a deeply-rooted frame of mind,
go parallel with the purity of the blood and juices of an organism,
and stand in strict accordance with them."*
I.A. Sikorskiy

The means of mass information has pounded a multitude of stereotypes, which are similar to heavy boulders into the innermost depths of the public consciousness—stereotypes like "reactionary," "fascism," "totalitarianism," "obscurantism," "nationalism," "chauvinism," "racism," and so on. The trained consciousness of the average person obediently perceives these stereotypes as ponderous monsters from horror films, which have infernal, unearthly origin and submit to their own forbidden, physical laws. Each of the above notions is sufficient enough to trigger alarm in a person and transform him into a reptile, which is wagged by the tail in a constitutional enclosure of political correctness.

Among these monsters of the liberal consciousness, one vividly appears with its unnatural and zoological absurdness. The name of this strange mutant is: "racial prejudice."

One barely attempts to imagine what this is, and immediately runs into the problem of the illogic and unreality of this ideological cliché.

Prejudice is what in the literal sense of the word stands in front of common sense; what precedes it, that is. It is something clearly related to the pre-cultural, pre-social, archaic, and instinctive layers of the consciousness. Prejudice is a psycho-genetic program of the previous experience of our ancestors, which demands blind obedience.

In the case of racial prejudice, about which modern media-humanists from television programs report, the question is one of a hostile and unwelcoming attitude toward a person having racial traits that are different from yours, which are easy to identify by appearance and behavior.

Because of this notion, it is customary today to associate it in a sharply negative sense. When exposing a person as racially prejudice, one is given to understand that he thinks in obsolete and evil categories. This [thinking] is bad, because it is a vestige of the basic ethical sense of a given combination of words, cultivated in the mass consciousness of the average person. Prejudice is something outdated and demands to be made obsolete—such is the present moral-logical context of the use of this expression. That is, by using trite ethical speculation, they changed the assessment of the concept from a positive to a negative.

1. THE NATURAL ESSENCE OF RACIAL PREJUDICE

Racial prejudice is the natural, racial-genetic program of your race, and if it was incorrect, vestigal, and demanding extinction, then you would simply not have come into the world. Racial prejudice is the concentrated evolutionary experience of ancestors, which proves, by the very fact of your existence, the correctness of their life strategy.

If all the boundless natural kingdom decided to interpret, in an instant, the concept of racial prejudice in the spirit of the international dump that is extolled by the troubadors of common human values, then the result of the transgenetic debauchery would surpass the vision of the Apocalypse, with its horrible tapestry. In the spirit of the United Nations Declaration, the hard-working bumblebee would be forced to carry life-giving pollen, not to the noble flower, but to the parasitic weed, and the gibbon would be forced to stare longingly at some macaque. The entire, hierarchically systematized natural kingdom would be transformed into a Broadway show of transvestites in an instant.

In other Indo-European languages, along with Russian, we observe a similar etymology of the word, "prejudice." Correspondingly, the French "prejudge" and the English "prejudice" trace back to the Latin term, *praejuicium*, meaning "a preliminary decision of a judge." In another form, it serves as a warning. The Greek word, *prokrima*, originates from the verb, *prokrion*, which has the sense of "to select," "to give preference to somebody or something." The root, *krima*, also means "decision," "verdict," and "judgement."

The complex German word, *vorurteil*, also means "prejudice." It traces back to the word, *urteil*, which means "opinion," "judgement," but also "verdict," or "decision of the court." But this word divides into *teil*, which means "destiny," "fortune," "fate," "lot," and "portion;" and the universal prefix *ur* is added in cases, when it is a question about something from the beginning, or something long-standing.

As you can see, even a simple linguistic analysis graphically shows that our Indo-European ancestors were united not only by a community language, but by a community of morals; they never thought to mock that word which traces back through them, to an archetypical, original fundamental of existence, which required them to bow both to the verdict of a worldly court, and a higher, ancestral sphere. Legitimately, in a correct explanation of its meaning, the combination of words, "racial prejudice," cannot carry in itself the negative value which the trendoid jugglers of ethical values try to give to it.

If there were no racial prejudice, then the races themselves would not survive.

The racial prejudices of any individual that is completely worth his salt, are based on the fundament of two interconnected, inherent phenomenon of the human psyche. In the first place, on anthropo-aesthetic canons of beauty present in his race, and in the second place, on the awareness of a values criterion that is begat by these canons. Every race carries in itself an idea of what is beautiful and what is ugly, and in connection with this, it works out its scale of values and its criterion of morals. The foreign-born, belonging to a different family, is rejected in the direct sense of the word by the organism of the individual, in accordance with the bio-medical laws of **norm of reaction.** This rejection reaction is characteristic of any healthy organism fighting for life: as with the rejection of transplanted donor organs, so with the rejection of systems of morals and ethics. The immune system of an organism will ceaselessly reject both foreign tissue materials and foreign speculative material.

The stormy growth of natural science in the last two centuries does not contradict the etymology of the word "prejudice." In the middle of the 19th Century, the zoologist and explorer Gustav Radde wrote: "The one who has experienced a certain success in the area of observations in the organic world, begins to fear the word "instinct" and starts to recognize an animal's intellectual reasons, more or less changing with respect to the species." G. Romanes, a naturalist, suggested at the same time that: "Instinct is a reflex action, to which an element of awareness is introduced." Karl Gustav Carus, the famous German psychologist, published a monograph in 1863 titled, *Comparative Psychology*, which in essence laid the basis for the study of the evolutionary psychology of Man and animals. After this, an entire science named **zoopsychology** arose; this science confirmed the fact that the entire animal kingdom exists and develops only on account of racial prejudices. It thus honors, in its illiteracy, its hierarchical genetic cleanliness.

Russian science supports the given view. After 100 years, L.A. Orbeli, a prominent Soviet psychologist, asserted on the basis of new experimental data from the area of evolutionary theory, that the forms of behavior for which our ancestors, who lived millions of years ago, were noted, "were lodged into us under definite conditions, as we emerged onto the stage." Modern biologist V.L. Deglin thinks: "For every reaction of an organism, for every reflex, there exists a definite, coded "model," an anticipating form of effector behavior. The model of an unconditional reflex has an inborn character." In the book, *Physics and Biology*,[130] M.V. Walkenstein states: "A basic physical-chemical principle of evolution concludes in the fact that acquired biological (genetic) information is not lost. At every stage of development, an excess of all possible mutations does not occur accidentally, since the overwhelming portion of mutations does not share in the selection, and the selection goes only between mutants [that are] compatible with the

---

[130] *Fizika i biologiya*. Moscow, 1980.

conditions of existence of already complex organisms. Consequently, evolution "channels" and "accelerates" itself."

The famous neo-Kantian philosopher Herman Cohen (1842-1918) wrote: "We Jews should recognize that racial instinct is in no way a barbarian thing, but only a natural and rightful need, from a national point of view." Konrad Lorenz, a Nobel Prize Laureate for his studies of the biology of behavior, also understood *instinct* to mean "all those psychological mechanisms of behavior, which occurred as a result of evolution, are directed at preservation of the species. They are not individual modifications that are conditioned by training."

The *Modern Dictionary of Sociology*, published in the USA in 1969, also defines: "Prejudice [as] a universal phenomena and an eternal problem of social life." The *Dictionary of Social Sciences*, prepared by UNESCO in 1964, treats "bias" as "a condition, flowing more from the inner processes of the carrier himself, than from factual testing of the characteristics of a group which is in question."

Adolf Bastian (1826-1905), the outstanding German anthropologist and ethnographer, justifiably asserted that the task of anthropology is the necessity of "extracting the initial thought." In accordance with evolutionary biology, psychogenetic and linguistic semantics become obvious; that is, **the primary thought is racial prejudice.**

Inborn aesthetic images, functionally supplying the reproductive procreation of a race, are also located in the riverbed of this "channel of evolution." In the book, *Analysis of Beauty*, Immanuel Kant noted that beauty is recognized for what it is, without the intercession of a concept. Fritz Lenz, the great racial theoretician of the 20th Century, wrote in one of his basic works, *Race as a Fundamental Value Principle*: "Already being children, we differentiate beautiful people from ugly—long before we acquire experience of such things, or with the help of comparison, we form an aesthetic sense. We make such differences instinctively, since we carry the ways of our race within ourselves." With good reason, the philosopher-structuralist, Claude Levi-Strauss (1908-2009), called this "zoological thinking," while Konrad Lorenz [called it] "inborn models." Ireneus Eybl Eybesfeldt, a prominent modern specialist in the area of the biology of behavior, defined identification on the principle of "us" and "them", as an inborn biological filter." The celebrated Russian aesthetic and theoretician of culture, M.M. Bakhtin, also stated: "The physical standards of Man, taken as a biological phenomenon, are pre-cultural."

A priori, Kant wrote about inborn forms of outlook and categories of thought, which exist prior to any experience. He even considered causality a category of thought. Arnold Gehlen (1904-1976), one of the founders of **philosophical anthropology**, suggested that Man derived his "cultural essence" from nature, and that an already phylo-genetically arising program of his behavior is built on the presence of a defined, inborn culture. As Noam Chomsky, the modern philosopher and scientist argued, Man has an inherent program of logical thought and language. At infancy, a child memorizes sounds and combines them, according to an inborn program that reflects his race, and subsequently, his cultural affiliation. Prominent German philosopher Ludwig Klages (1872-1956), who founded a science under the name of **characterology**, accurately determined that: "The soul is older than the spirit." For the spirit of a person always expresses the historical realities of the Age, the time in which it lives, while at the same time, his soul is a reflection of his timeless racial essence. The style of any culture always depends on its racial filling.

In the book, *Homo somatikos: Axiology of the Human Body*,[131] modern author I.M. Bykhovskaya developed the concept of magnifying the independent significance of the human body and its physical appearance in modern culture. "Somatic awareness," passing old speculative templates, itself creates values. The new post-modern age, in the opinion of the author, creates a new, hitherto unheard of doctrine of "the body as a value." The human body does not simply exist in the world alongside other objects, but joins itself to the world, and in a certain sense, creates it." "Corporealness," in the context of reflection and interpretation of activity, is more and more becoming a "universal standard."

According to the accurate expression of Maurice Merleau-Ponty (1908-1961), who was another representative of the philosophical school of anthropology, "the body projects a cultural world around itself."

---

[131] *Homo somatikos: aksiologiya chelovecheskogo tela*. Moscow, 2000.

But saturated with racial traits on all physical levels, the human body should qualitatively anew, intensify the problem of racial values. Abstract truths and incorporeal morals, which acquire corporealness in the context of a new culture, draw into themselves "somatic awareness," the starting point of which, or the "original thought" of which is "racial prejudice," in all its primeval beauty.

Skillfully lathering the optics of the first-born instinct with the foamy substitutes of liberal values, modern mass-culture does not desire to fall into the snare of racial theory, which they started to forget. But the motley, pie-bald crowd has again lifted its gaze to the luxurious attire of the racial archetype, so visibly real and simultaneously transcendentally eternal. In care-free beauty, she is seen on the public sign near the freeway—the eternal blonde princess.

2. ANTHROPO-AESTHETICS

Thus, mass-culture, specially summoned to mock and destroy racial prejudice, by a strange combination of circumstances, lost the beat in a round-dance of cause-effect ties and gave new stimulus to the perfecting of that which they was necessary to throw on the ash-heap of history, like an annoying vestige. In part, by virtue of this, and odd thing arose, and there is now such a modern science as **anthropo-aesthetics.**

N.I. Khaldeyeva, the leading Russian specialist in this area, quite definitively writes in the article, *Comparative Anthopo-Aesthetic Studies in Russia,*[132] that: "It is necessary to say that Man not only perceives the physionomic appearance of another individual, but he also correlates it with his own—and his group's—anthropological assessments. On this basis, the individual and his group may define an area of enlargement of morphological population characteristics, and develop its own scale of phenotypical criterion. Incidentally, the ability to beget the values of an outlook, including in this regard an appearance, is a unique trait of the human consciousness. For an anthropo-aesthetic analysis, the greatest interest is the form of the evaluation of the outer appearance, as an ethnic trait. The particulars of outer appearance may appear as carriers of a defined message, and play the role of signals for the racial and ethnic affiliation of an individual. The equating of physionomic combinations to an entire ethnic characteristic, in the process of perceiving the outer appearance, reflects the anthropological composition of the common mechanism of identification. In the formation of categories of identity, this ingredient is one of the most ancient and stable. Its antiquity is obvious, since the human face was always the most accessible and informative, and on the defining stage of anthropogenesis, vitally important."

This opinion of a recognized specialist again concurs with the results of an etymological analysis of the word "prejudice" in the Indo-European languages. In the time of the stormy development of anthropology, Darwin pointed to the different inborn preferences in the evaluation of types of outer appearance, and as a result, to the existence of inborn, hereditary criterion [that are] characteristic of each race. Each individual carries in himself a characteristic set of "signals" or identifying traits, by which he evaluates the surroundings, in accordance with the biologically based principle of "us-them". A face, from a racial point of view, is a concentration of receptors, of signal information, serving the goals of reproductive or sexual selection. A prospective sexual partner thus evaluates in accordance with an inborn, typical notion of an ideal partner, characteristic for the given ethnic-racial group. N.I. Khaldeyeva further emphasizes: "Racial affiliation has great significance in the process of evaluating outer appearance. The mechanism of mutual human perception is biologically and socially determined; it provides [for] unchanging reproduction."

All these aspects of the anthropo-aesthetic organization of the human psyche, all its architectonics, are encoded in the neural organization of the brain. Original aesthetic canons, characteristic to a given race, are reproduced without distortion in a number of generations; each living generation is directly connected with the ancient archetype. The neurons of the brain seem to fulfill the function of a matrix, the numerical significance of which serves as a corrective and support of the vector of the evolutionary development of a race. This vector of physionomic

---

[132] *Sravnitel'nie antropoesteticheskie issledovaniya v Rossii.* Vestnik antropologii, Vip. 4, Moscow, 1998. (Bulletin of Anthropology, Issue 4, Moscow, 1998.)

idealization is strictly directed immediately to the averaging, that is, toward a single racial standard, in order to provide the maximal reproductive success of an individual, within the limits of his population. Therefore, form and color of the eyes, the nose, hair, lips, and the oval of the face are the most important signs in the ethno-racial characteristic of the perceived appearance, for it is around this combination of racial traits that the genetic information about its carrier most distinctly and concentratedly appears. This for its part, gives the opportunity to maximally use the positive effect of racial prejudice in its evaluation. Information, visibly considered from the face of the evaluated object, is compared with information stored in the brain of the evaluating subject, which brings into action the mechanism of anthropo-aesthetic evaluation. Minimal differences in their genetic matrices encourage reproduction of the race, and the cutting off of harmful mutations, that lead to "chaos of blood."

A summary of the theoretical plan in Khaldeyeva's work has the following character: "The studied material enables one to establish that the ideal, preferred morphotype of a person forms the basis of a real type, according to a priority of gradations of six basic physionomic traits. And being connected with it, is a group characteristic, subject to study by methods of physical anthropology, according to a program of anthropo-aesthetics."

The summary of practical studies absolutely gives particular value to the work of N.I. Khaldeyeva; in the process [of these studies] the **summary indicator of auto-identification** (AI) was introduced. The more consolidated a group is in racial terms, the higher its indicator. Besides this, it was made clear, that within the limits of the samples of Russians from the examined areas, the vector of the male ideal type has a distinctly expressed centripetal character. This quite graphically confirms the presence of a **racial core** among the Russian people, as among any other historically significant community. In the racial core, all racial traits are expressed in the most clear and unique manner. The basic historical load in the process of nation-building and the creation of cultural values falls on the racial core of a people. Among the members of the mestizo periphery, who do not pass [the test of] anthropo-aesthetic selection in accordance with the average racial ideal, the governing instinct is weak, and as a result, legions of all manner of apostates and proselytes are formed from the ranks of the bastards.

The racial core of a people is the hub of the application of historical forces—not some extra-racial, multi-stock slag, which comes out of the sand with the good intentions of leader and religious figures.

"In all Russian groups [that are] outside of a connection with geographical differentiation, relatively low variants of aesthetically preferred beauty, expressed in the summary indicators of anthropological auto-identification, are noted. As a whole, all Russian groups form a common relatively homogeneous, graphic cluster, according to the parameters of anthropo-aesthetics"—N.I. Khaldeyeva summarizes.

Now it is worth emphasizing that these same racial anthropo-aesthetic standards are not a set of unconnected data, arbitrarily interpreted from generation to generation, but namely an archetypical matrix of exact and constant values, written in the neural organization of the brain of every individual. And the cleaner the type of this individual, the more perfect his aesthetic instinct, and consequently, his moral-ethical sense.

Only kind recognizes kind, and therefore generally, only a pedigreed human specimen has a right to judge the traits of a breed. Among mixed peoples, these thoughts, canons, and morals are "freckled."

For an objective evaluation of the standards of racial beauty, modern science has such an exact and impartial method as **anthropological photography.** The work of Nadezhda Nikolayevna Svetkovaya, *Anthropological Photography as a Source for Research in Ethnic Photography*,[133] serves as a graphic and convincing illustration. In it, she writes: "As a result of the analysis of photometric traits, it is clear, that almost all angular proportions of the face possess good boundary characteristics. They have an inter-group range of more than two standards." This means that the quantity of objective racial differences in the build of the face, among the members of various races, consistently surpasses a mistake in measurements.

---

[133] *Antropologicheskaya fotografiya kak istochnik dlya issledovaniya po etnicheskoy fotografii.* Moscow, 1976.

As a whole, racial geometry of the face is such: **Europoids**—according to the data of photometry, have the straightest profile, according to the upper facial angle, the latter among them (83-87 degrees) is always more than the average-faced [individual] (81 degrees); a nose protruding at a relatively small angle to the horizontal (57-63 degrees); a highly strong projection of the nose from the line of the profile (21-27 degrees); and a straight, upper lip (85-91 degrees).

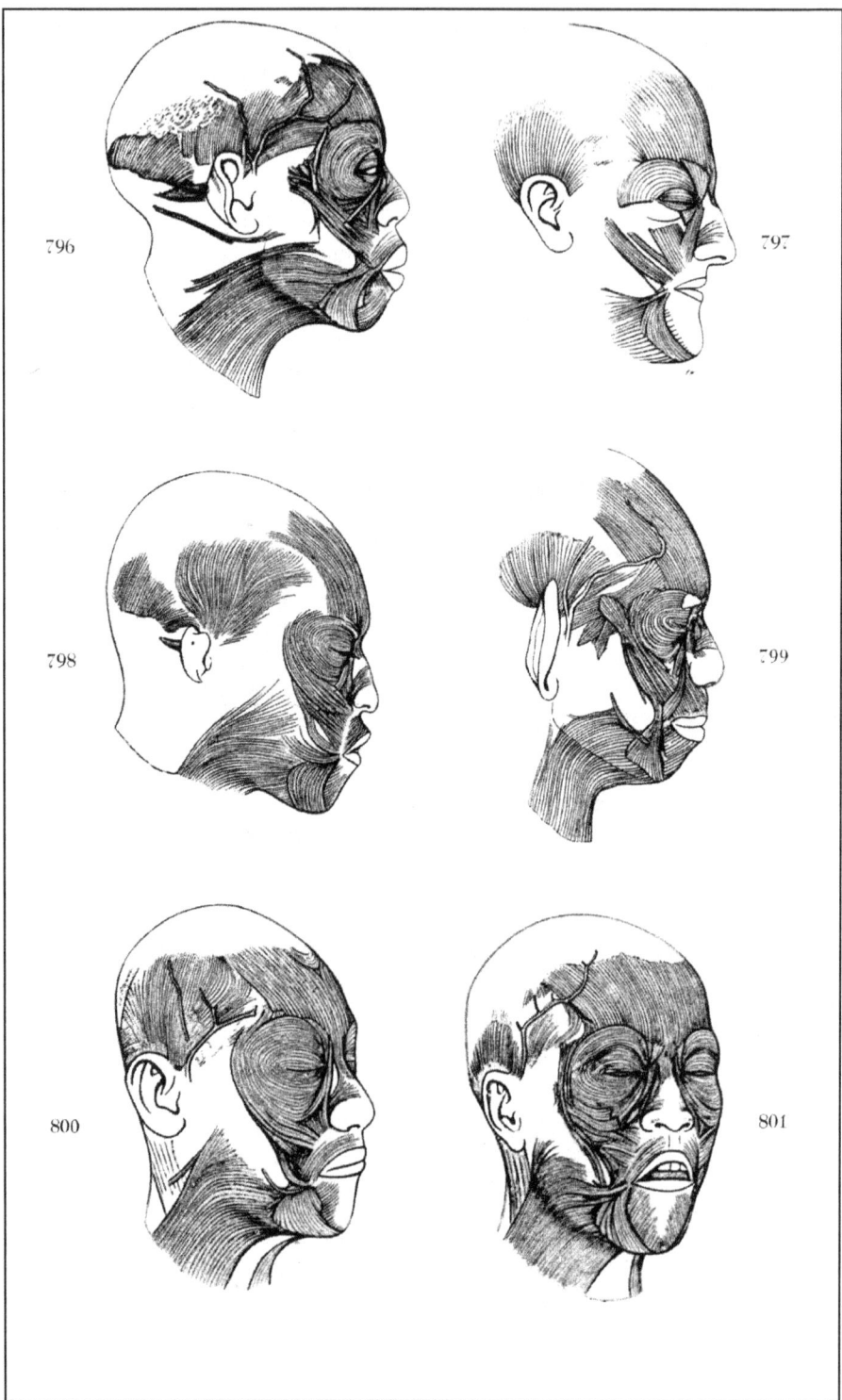

Racial differences in the structure of the facial muscles—according to Edward Lot. Drawing 796-Negro; 797-European; 798-Papuan; 799-Australian Aborigine; 800-Javanese; 801-Chinese

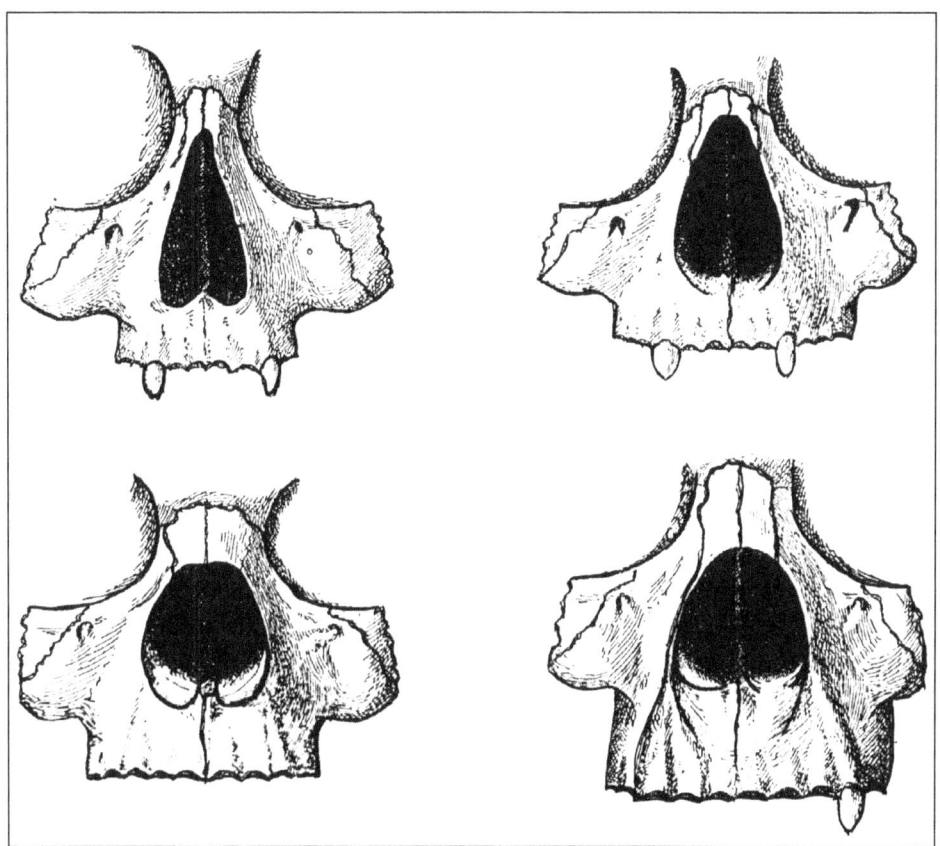

Racial differences in the structure of the nasal opening on the skull—according to Paul Topinard. Clockwise: Europoid, Mongoloid, Australoid, Negroid.

The **Mongoloid** differs by the incline to the mid-jaw by the upper facial angle and the angle of projection of the upper lip (72-82 degrees). Among them, the upper facial angle (82-87 degrees) is always less than the mid-facial angle (83-88 degrees). The angle of the projection of the nose to the horizontal is overall 65-72 degrees, among all groups studied.

Negroids have a sharply protruding jaw, as against the upper and mid-facial angles (73-77 and 76-80 degrees, respectively), and the angle of the protruding upper lip.

One of the prominent Russian anthropologists, V.V. Bunak, concluded in his article, *Photo Portraits as Material for Determining Variations in the Structure of the Head and Face*,[134] that: "It is known that an experienced observer, studying photo portraits, can in many cases and with sufficient accuracy, determine to which ethnic group a portrayed individual is related, and what his anthropological type is. The data cited above establishes that through photographs, one may gain not only the summary and sufficiently general characteristic of an anthropological type, but [also] determine the variants of separate cephaloscopic traits, independently of familiarity with a given type, not by the general impression from the portrait, and on the basis of boundary criterion of a morphological order."

---

[134] *Fotoportreti kak material dlya opredeleniya variatsiy stroeniya golovi i litsa.* Sovetskaya antropologiya N2, 1959.

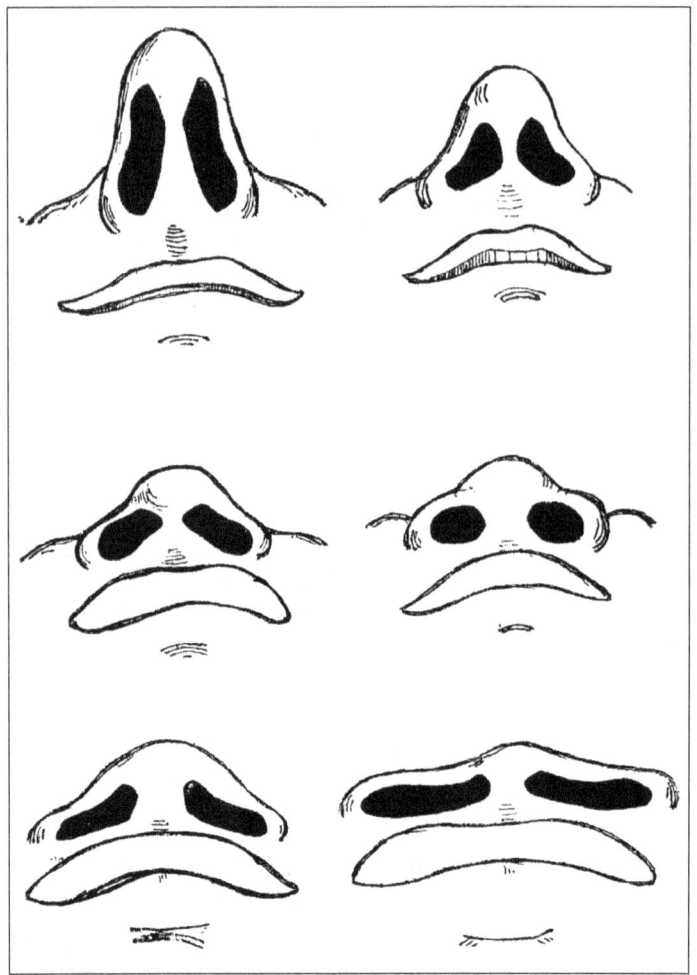

Racial Differences in the Form of the Nose—according to Paul Topinard. Top: Europoid; Center: Mongoloid; Bottom: Negroid.

This again means that a racial and ethnic type is an objective reality, and that it yields to precise measurement, not only as a whole, but also by separate sectors of a portrait. A multitude of folk-sayings in all parts of the world, used to good advantage in ironic form about this or that facial trait, with their obligatory correlation with this or that racial-ethnic type, testify to the powers of popular observation, that is, to the naturalness of racial prejudice.

In the modern anthology of anthropological works, *Problems of Evolutionary Morphology in Man and his Races*,[135] the discussed theme is brought to a qualitatively new level. Thus, in the article, *Prospective Uses of Close Stereophotogrammetrics in Anthropology*,[136] collectively written by L.P. Vinnikov, I.G. Indichenko, I.M. Zolotareva, A.A. Zubov, and G.B. Lebedinskaya, the authors speak about how quality, color photography enables one to reveal all the nuances of the pigmentation of the eyes, the skin, the hair, and also to determine inter-pupil distance and the prominence of the eyeball. In connection with that the authors of the given study think that the method proposed by them "opens wide prospects for extremely detailed research of the surface of the face of a person, and with great success, may be used in ethnic anthropology."

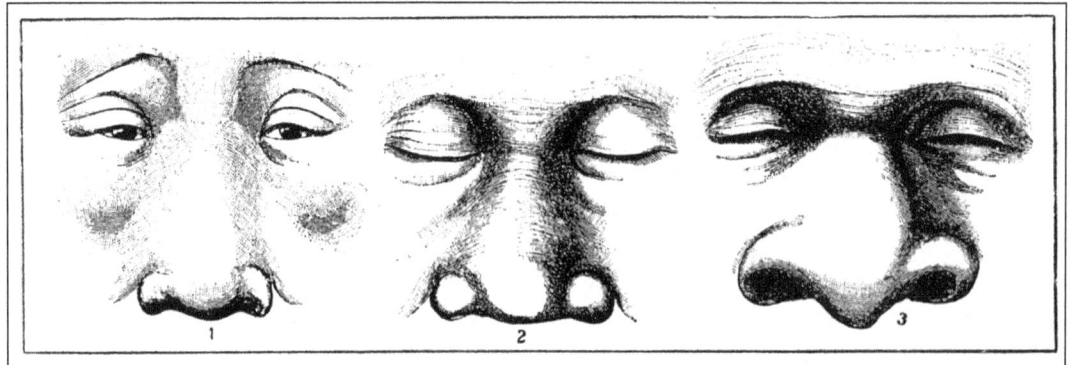

Racial Differences in the Form of the Nose, among "colored" races—according to Paul Topinard: 1-Mongoloid; **2**-Negroid; **3**-Australoid.

---

[135] *Problemi Evolyutsionnoy morfologii cheloveka i ego ras.* Moscow, 1986.
[136] *Perspektivi primeneniya blizhney stereofotogrammetrii v antropologii.*

138

In support of our thesis of the reality of the inborn traits of anthropo-aesthetics, we turn to the monograph of Yuri Kirillovich Chistov, *The Differentiation of the Races of Man according to the Structure of the Median-Sagital Contour of the Skull.*[137] In it, Chistov writes: "The results of the received studies allow one, with sufficient confidence, to speak of the presence of definite characteristics in the form of the sagittal contour of the skull, among "northern" and "southern" populations of modern Man. The intra-racial values of this index reliably differ from the inter-racial; that is, members of certified racial types differ between themselves both by a sum of degrees, and in the linear characteristics of the median sagittal contours. One of the most interesting conclusions is the establishment of the fact that modern craniological series so strongly differ by value of degrees and the linear characteristics of the lobe portion of the contour, as according to the pattern of the occipital section."

Thus, it is completely obvious, that the racial proportions of the face, and the entire head as a whole, [which are] perceived in the process of anthropo-aesthetic evaluation, are a fact of reality. And that fact is not an accidental property of separate personalities, but a fact of an entire people as a whole, preserving its ideals of beauty in its [racial] memory, through a sequence of generations. **Primeval division by the principle of "us-them" is a basic aesthetical category, having a harsh biological foundation.**

Right: Johann Kaspar Lavater

Johann Kaspar Lavater, the founder of **physiognomics,** pointed to the characteristic peculiarities of the structure of the face, among the members of different races and ethnic groups. Friedrich Wilhelm Schadow (1788-1862), the outstanding German painter, created an entire gallery of portraits of national types. But a fundamental, racial-anatomical study of the human face only began with the development of anthropology. In the book, *Man and his Place in Nature,*[138] German anthropologist Karl Voigt (1817-1895) emphasized: "Less depends on the outer conditions of the form of the face. About the well-complicated European, you encounter a harmonious development of three main sectors: the lobe, the nose, and the lower portion of the face; the lobe usually prevails over the rest. Among other races, you see other ratios: the nose or the lower part of the face juts forward, then recedes to the background, furnishing the physiognomy of this or another particular character." Therefore, examining the portraits of members of this or that race, we are persuaded that despite the passing of entire ages, the characteristically tribal characteristics in the structure of the face can be easily seen. "Deviation is always encountered in mixed peoples, whereas the original tribal purity preserves the characteristic form of the tribe's soft parts among all individuals, and as we know, the individuals of a pure tribe resemble each other more, than in mixed tribes."

Not only do the facial sectors of the skull give a face a particular racial mark, but to a significant degree, so do the musculature and soft tissues. A particular section of anthropology, under the name **miology,** studies their racial variations.

In the book, *Race and the Nations of Humanity,*[139] Ferdinand Birkner (1868-1944) emphasized: "The musculature of the face, [which] together with other factors, causes the general

---

[137] *Differentsiatsiya ras cheloveka po stroeniyu medianno-sagital'nogo kontura cherepa.* Moscow, 1983.
[138] Sankt-Peterburg, 1866.
[139] Sankt-Peterburg, 1914.

expression of the face, has general significance from the point of view of racial anatomical characteristics—precisely in those places which are particularly important in the sense of the racial characteristics of living people; they were found [to have] highly noticeable differences between Europeans and Chinese. Direct measurements of the soft tissues of the face of the skull and the contour of the head, furnished by soft tissues, reveal differences; that is, in the area of the skull bones, the soft tissues of the Chinese turn out to be thicker, than in Europeans; the result is that the weakly marked [thicker tissues] significantly increase the flattening of the face on the skull of Mongols.

In the studies of Chinese in comparison with Europeans, the soft tissues are thicker around the root of the nose, the middle of the nose bone, on the bulges of the skull, and also on the very high point of the skull. Papuans display thicker soft tissues only in the area of the most divergence in the skull bulges, and before the ear. In the remaining areas cited here, the soft tissues are not so thick, as in Europeans, and they differ all the more in their thickness among Chinese."

It was also discovered, that among Negroes, in comparison with Europoids, the occipital muscles are better developed, and [there are] few cheek [muscles]; it is also observed that there is not a complete separation of the musculature between the eyes and the mouth, and the chewing muscles are strongly developed. Besides that among negroes there is a strong development and massiveness of surface muscles of the head. Different muscle clusters on the heads of negroes are connected very lightly between themselves, as a result of which, at first glance it creates the impression of the presence of a **continuous muscle membrane** on the faces of negroes.

Generally, the entire musculature of the faces of black-skinned people is far more massive, is more continuous and homogeneous, and the boundaries of the muscles protrude less clearly. Among Papuans, a complete layer of muscle on the face is observed more than is usual among Europeans, and the musculature between the eyes and mouth almost completely lacks fine differentiation.

From there the fundamental conclusion follows, according to which, in the process of evolution all the basic races experienced emotions in completely different ways. Differences in the construction of the facial muscles point to precisely that.

Paolo Mantegazza (1831-1910), an important Italian anthropologist, psychologist, and professor at the University of Florence, specialized precisely in the area of physiognomics and the manifestation of the emotions. In his book, *The Physiology of Love*,[140] he emphasized: "Pleasure, which developed in the middle of the races, differs not only by the degree of what is tasted by them of pleasure, but also by the ability of expression of pleasures." In this essay, and also in the books, *Physiognomy and the Expression of the Senses*[141] and *The Ecstasy of Man*,[142] he developed an original concept under the name **ethnography of pleasure**, in which he paid great attention to the physiological and moral characteristics of the emotional sphere, among the members of the various races.

From a physiognomic point of view, they relate to a number of important racial-diagnostic traits: 1) development of the eyebrows; 2) height of the bridge of the nose; 3) the incline of the lobe; 4) the protrusion of the end of the nose and the middle part of the dorsum of the nose, the lips, and the chin; and 5) the flatness of the face on various levels.

---

[140] Moscow, 1890.

[141] Kiev, 1886.

[142] Sankt-Peterburg, 1890.

Paolo Mantegazza

Soviet anthropologist M.M. Gerasimov, the founder of an entire school of science, emphasized in the introduction of his fundamental monography, *Restoration of the Face around the Skull*,[143] that: "It is known that among the Chinese there are wide, high, cheek-boned faces, in comparison with the faces of Europeans, and many authors note a different thickness of soft tissues in the point of maximal width of the skull bulges. Namely among the Chinese, the thickness at this point is greater, and in Europeans, it is less. For Europeans, according to Guise and Coleman, this size

---

[143] *Vosstanovlenie litsa po cherepu.* Moscow, 1955.

is equal to 6.63mm, while for Chinese, according to Birkner, it is 10.9mm; this fully agrees with our observations."

Baron Egon von Eichstedt (1892-1965), one of the classics of the German school of anthropology, reconciled the traits of the morphology of the face of the human races, with the evolution of their development, in his fundamental monography, *Raciology and the Racial History of Humanity*. In regard to comparative morphology of the soft [facial] tissues, it is worth citing two basic phenomena, which have evolutionary significance. First of all is the presence of Jacobson's organ, which rudimentarily fits in with the blind end in the forward lower part of the nasal septum, fulfilling in lower species a special, functional task. Of further interest are the lateral parts of the posterior cartelage, which in progressive Europoids branches toward the end, while in primitive races, such as the Melanesians, they form a continuous, wide mass. This intermediate form traces back to the man-like apes.

Among primitives, the quadratic muscle, particularly of the dark-skinned races, is also far more compact, than in Europeans, among whom separate fibrils have thus far developed, so that French anatomists consider them as separate muscles. In this, the small lateral tissues of the nasal muscles usually strongly correlate with the general character of the skin covering of the soft tissues. Therefore, their thickness usually also corresponds more strongly with the lowering and large fleshiness of the nostrils of the nose, which is sometimes even encountered among Jews and the pseudo-Jewish type on New Guinea; a great thickness of the folds of the eyelids and the lips. Among Negroes and Paleo-Mongoloids, several can be completely lost in a spongy, connecting tissue. This massiveness produces deep furrows around the alae of the nose, which on flat faces make an almost continuous line from the angle of the eye, through the ala of the nose, to the lower mandible.

If we draw a general evolutionary picture of that then the muscles of the nasal region and the area of the orbit show, that the higher the form, the more differentiation in the muscles. The lips are a characteristic trait on both the individual and racial face; they tell much about the psychological type of an individual. The area of the mouth is the most expressive and indicative, from the point of view of racial physiognomy.

If we also consider the area of the orbits, the nose, and cheeks, the general direction of evolutionary development of the musculature of the human face becomes clear. In all cases, the higher the evolutionary stage, the higher the chance of differentiation of the muscle mass. There are only various forms of expression of a single basic tendency. Thus, we may see and solve through concrete example, the secrets and interconnections of the origin of species and its constructive path.

We may judge about the intermediate stages of human evolution by the atavistic modern forms of primitive races. They all have a muscle mass in the middle part of the face that is thicker and less differentiated. Lack of differentiation in general is considered to be a sign of primitiveness. Massive and multiple interweaving of muscle connections is a characteristic of Mongoloids to this day.

Although bulging of the lips is particularly characteristic for Negroids, it is more or less often encountered in all races, for example, among Eastern Veddoids. The southern Chinese have very thick lips; the primitive Australoids have comparatively thin lips, and the North American Indians have very thin lips. A disproportionately thick lower lip may be an inherited trait for an entire people, for example, the Jews.

The childish mouth, with its indistinct contours, as among European children, is encountered among infantile primitive races. The contours of the upper lip and mouth opening, in the form of a half-moon, is typical among Western Veddoids, particularly among women.

On the Nordic profile, the lips do not protrude, while they protrude among the southern races. With the latter phenomenon, a sagging of the contour of the profile is associated, the concave snout typical for Negroes."

A book by the important German anatomist and doctor, Fritz Lange, titled, *The Language of the Human Face* (1938), is to this day an unsurpassed fountain of information on questions of racial physiognomy.

All these and a number of other talents laid the basis of a science called **habitoscopy**—a part of criminology. The very term derives from the words *habitus* (Latin for "appearance") and *skopeo* (Greek for "examine"). In precise accord with the logic of our previous presentation, it is necessary to note that in habitoscopy they use **objective** and **subjective** images of the appearance of an individual. In the book, *Use of the Traits of Appearance in the Work of the Offices of Internal Affairs,*[144] V.A. Snetkova points out: "The ethno-anthropological type finds reflection in the anatomical outer traits, which can be determined with all accuracy. In criminal practice, the facial type is most often defined in comparison with examples of known groups of a population."

Therefore, it follows that the centuries-old practice of legal-criminal examination of a person of interest is built in precise accord with "racial prejudice," or "inborn anthropo-aesthetic models." Naturally, this should be interest from the point of view of liberal anthropologists, who deny the very fact of the existence of "race." Perhaps they would care to take note of millions of court decisions in all parts of the world, that were made on the basis of the descriptions of accused persons, and [consider them] as judicial fallacies, because of an absence of "ethno-anthropological types" finding "reflection in the anatomical outward traits, that can be determined with all accuracy" ? Then in order to please the "scientific liberals" it would be necessary to bring to sacrifice the identification and firm views of experts, based on group, that is, based namely on ethno-racial traits. In the article, *Ethno-racial bias and Ethnological Science,*[145] V.I. Kozlov legitimately therefore noted: "During contacts with people, essentially differing in anthropological terms, ethnic self-awareness is usually amplified by racial [differences] and becomes a more distinct and stable ethno-racial self-awareness."

A.A. Zubov and N.I. Khaldeyeva—some of the leading Russian anthropologists—concluded in their joint article for the anthology, *Race and Racism. Past and Present,*[146] that: "*Type* means a characteristic sum of genetic and morpho-physiological traits, that mark definite groups within a species; it is a completely real phenomenon, and stands to be worthy of study." And the modern geneticist, J. Nil, declares that in the present time, any individual can be traced to this or that ethnic community, through research, with up to 87% accuracy. The name of a book by A.F. Nazarovaya and S.M. Altukhov, *A Genetic Portrait of Peoples of the World,*[147] speaks for itself, for within it a detailed characteristic of the frequencies of genes is given in all basic, and even in many relic populations of humankind. Finally, the famous *Table of Genetic-Linguistic Differences between Peoples,*[148] by Italian geneticist Luigi Cavalli-Sforza, of the American University of Stanford , conclusively illustrates the objective differences between biotypes. Therefore, in light of the declared theme, it would be fairly useful to turn to the forgotten reservoir of anthropology, which also points to the objective origin of racial prejudice.

3. Fundamentals of Racial Morals

At the turn of the 20th Century, anthropology, not owning the methods of genetic and biochemical control, nevertheless stood on a very high level, namely for its part in the descriptive statistics of outer morphological differences, on which our anthropo-aesthetic and moral-ethical relationships to foreign tribesmen is actually based. The very organization of the task, in the plan of discovery and the description of racial-biological differences between the representatives of various peoples, was more concrete, than in modern ethnic anthropology. The scientific quest of that age was not directed, unlike now—at erasing the differences—but on the contrary, to their strengthening and isolation; that gave maximum concreteness and graphicness to all aspects of racial diagnostics. We emphasize once more, that the descriptive and statistical methods of that age have not become outdated in the least, just as the equations of the celebrated German mathematician, Karl Friedrich Gauss (1777-1855), have not become outdated to this day, and, with the help of which, modern geneticists carry out their calculations.

---

[144] *Ispol'zovanie priznakov vneshnosti v rabote organov vnutrennikh del.* Moscow, 1993.

[145] *Etnorasovie predubezhdeniya i etnologicheskaya nauka.* Rasi i narodi, vip. 23, 1993. (Races and peoples, Issue 23, 1993).

[146] *Rasi i rasism. Istoriya i sovremennost'.* Moscow, 1991.

[147] *Geneticheskiy portret narodov mira.* Moscow, 1999.

[148] From *Genes, Peoples, and Languages.*

Our Russian anthology of the pre-communist era was also no exception to the plan. It founder, by general consensus, is considered to be Anatoliy Petrovich Bogdanov (1834-1896), whom official Marxist science numbered among the stoic "fighters against racism." True, what they based their class convictions on is not completely clear, since Bogdanov completely asserted with the spirit of his time that "the skull, even in its tribal characteristics, represents something constant." Granted this is not racism, but this is far from notorious internationalism.

The goal of one of the main reports by A.P. Bogdanov, *Anthropological Physiognomy*,[149] was precisely to supply a scientific basis for a definition "of the characteristic Russian traits of the face." "For the modern anthropologist-naturalist, the study of Man in general is not an independent undertaking; it is the affair of the anatomist, the physiologist, the psychologist, and the philosopher. Important for it are those variations, which in their form and in their structure, represent tribes; and they are important insofar and inasmuch as they give the opportunity to differentiate and group these tribes; to find in them differences and similarities for the possibility of their natural classification, for reconstruction of the geneaological tree, by which they evolved from each other, under the influence of various causes. In its conclusions and for its goals, anthropological physiognomy sometimes places those traits which are not generally important for the physiognomist, in a significant place—the color of the hair and eyes, for example." Thus, in the opinion of the founder of the Russian school of anthropology, the anthropologist, on a certain level of classification, is first of all a raciologist—the rest is the business of apprentices from the ranks of physiologists and philosophers. The racial-biological priority here is completely evident.

Defining his position with values, Bogdanov is as categorical in questions on the choice of methodology: "Studying a pug or a poodle, for the zoologist the interesting variety is not accidental, originating from these or other outside conditions; and what is more, it is the constant combination, which alone gives him the opportunity to form a representation of a pug or poodle, like the members of natural groups or races. He knows that in genetic theories, traits are not counted, but are weighed according to their significance; they are classified not by their number, but by their clear manifestation, by the revelation of the trait. In the given case, it is important to the zoologist that he gives an indication of the influence of race in every specimen. We have the same in mixed tribes of Man; we encounter the same difficulties and goals in the study of their anthropological characteristics."

The second part of the monograph is dedicated directly to the anthropological physiognomy of the Russian people. A.P. Bogdanov asserts: "We very often use the expressions: 'this is **a pure Russian beauty, this is the very image of a Russian**, a typical Russian face.' Perhaps, in the application of these expressions to private situations, disagreement between observers is encountered, but noticing a number of similar definitions of Russian physiognomy, one is persuaded that something real, not fantastic, lies in that general expression '**Russian physiognomy, Russian beauty.**' This is all the more clearly expressed in negative determinations, in an encounter of physiognomies of those from kinship groups, which historically took shape, or for example, in an encounter with outsiders, in comparing them with Russians. In such cases, '**no, this is not Russian physiognomy**' is decisively heard, and said with great conviction and persuasiveness. In each of us, in the sphere of our "unconsciousness," there exists a sufficiently definite idea about the Russian type, about Russian physiognomy."

As can be seen, this classic of Russian anthropology substantiated all the basic positions, 100 years before the rise of anthropo-aesthetics. In this connection, it would also be apt to cite the words of the Russian ethnographer and historian, N.I. Nadezhdin, spoken by him in 1837: "The physiognomy of the Russian people, in essence Slavic, is etched by the natural touch of the northern climate. The very name *Rusi*, which comes from antiquity, means *light brown hair*."

Further, using the methods of historical ethnography, Bogdanov argued that the colonization of Siberia could not in principle render a disastrous influence on the Russian people. First of all, racial mixing cannot have a place, by reason of the differences in the proportion of the ethnos coming into contact, and also because of the cardinal difference in their biological strategies of survival.

With the start of colonization, huge masses of a racially homogeneous population poured into a territory settled by heterogeneous aborigines, who had neither racial nor political consolidation.

---

[149] *Antropologicheskaya fiziognomika*. Moscow, 1878.

Numerical superiority, coordination of effort, and aggressiveness distinguished the actions of the Russians. Killing the local male population and possessing the indigenous women, the Russian colonizers rolled, wave after wave, along the endless spaces of Eurasia, unavoidably increasing the percentage of Nordic blood in the local population, from generation to generation, in exact accord with Mendel's laws. The administrative and judicial system in the colonized areas, the very character of economic activity, and also the Russian Orthodox Church quantitatively strengthened the process of Russification of the native population, not so much in cultural terms, as namely in anthropological terms. The myth of the "peaceful conquest of Siberia" is a later portrayal of communist propaganda. The list of tribes that disappeared from the face of the earth during the 200-300 years of Russian expansion, is highly imposing. Not one liberal-democratic fabrication has the strength to change the principles of the struggle for existence. Russian chronicles, the travel diaries of merchants, officers, and simply "evil people," preserve testimony of how separate tribes voluntarily surrendered their young women of fertile age, just at the sight of the white conquerors.

While influencing foreign blood, the Russian colonizers guarded their own, since their women and children remained in the home country. Several centuries of such "international peace-making" washed away almost all remnants of racial ethnic distinctions in the indigenous [peoples] from the gigantic territories. "His Majesty's man"—the merchant—and the Orthodox Priest, magnificently complemented each other, coordinating the actions of military detachments, economic factors, and the church; this enabled them to keep control of the local, uncoordinated population. Incidentally, delivery of vodka and tobacco to the Mongoloid tribes of Siberia—for whom this was disastrous—was sanctioned by the Orthodox clergy. Use of the native population, which had a weaker physique, in the mines, tunneling, and pulling boats on the northern rivers, also undermined its racial strength in the confrontation with the Russians. Besides this, **age-old Russian morality** was a cementing factor, making the energetic assimilation of the population of Siberia irreversible. A.P. Bogdanov continues: "Perhaps some married the native girls and settled down, but a majority of the primitive colonists were not such. This was a people of trade, war-like, industrious, concerned with earning a kopek, and then placing things to suit themselves, in accordance with their own ideal of well-being. But for the Russian individual that ideal was absolutely not to lightly tie down his life with some "dreg" of another faith. The Russian will do business with them, he will be kind and friendly with them, enter into friendship with them in everything, except become a member of the family—and introduce a foreign element into his own family. Ordinary Russian people are still firm against this...Often, settlers of different tribes live nearby as neighbors, but marriage between them is rare, although there are often romances. But romances are one-sided: Russian ladies' men with foreign stones, but not the reverse."

Finally, Bogdanov makes the following, highly important conclusions relating to sexual participation in race-mixing: "A woman of relatively high development, of a higher race, rarely lowers herself to the member of a race that she considers beneath her station. The mixing of European women with negroes is extremely rare, and belongs to the accidental—and one could say—eccentric category, but negresses and mestizas have a weakness for European men."

The "lower" the race, the looser its women; this explains modern data of the evolutionary theory on sex and the biology of behavior. In this manner, they simply steal the unobtainable genes from "higher" races. The sense of one's own worth in the sexual sphere is an indicator of biological self-worth.

For example, in this context Russian ethnographer Count A.S. Udarov, relying on personal impressions, spoke extremely negatively about the weakness of the morals of Mordvinian women.

A.P. Bogdanov's outstanding contribution lies in that in 1867, he was the first to compose an *Anthropological Album of the Russian People,*[150] which was demonstrated at international exhibits. Thus, for many years before the turbulent development of anthropo-aesthetics, he substantiated not only its theoretical part, but he set about a systematization of practical material, namely with the goal of revealing "typical Russian faces," which were subjected by him to anthropological analysis, in connection with Russian folk songs. As was expected, the Russian ideal of racial beauty did not make itself hard to find. About the Russian girl: "young, intelligent,

---

[150] *Antopologicheskiy al'bom russkogo naroda.*

without whitener, a white face" or "slender, tall." About the Russian boy: "Smooth Russian girls for a rosy boy. Light brown curls on the shoulders, black eyebrows of sable."

There is no end to the number of similar artistic descriptions from Russian folklore, that once again speak in favor of the objectivity of the conclusion of the modern anthropo-aesthetic of a summary index of auto-identification (AI) for the Russian people, as for any other [people].

The English founder of **eugenics**, Francis Galton (1822-1911), also proposed the creation of summary charts of beauty, according to geographic locations. His initiative dates to 1883, and the German anthropo-aesthetic program only generally arose in 1926.

We emphasize once more, that…pre-revolution Russian anthropology [was] matched with a high Russian civic status, something we almost never see in modern science; it is shamefully shut down by the bugaboo of neutralizing humanism, and arbitrarily converted. Pre-revolution Russian anthropology, just like any other national school, was deeply patriotic and race-oriented, and all the while did not in the least lose scientific objectiveness.

S.I. Lutsenko, in his article, *The Social Environment as a Factor of Development and Beauty of the Human Face*,[151] wrote: "The beauty of any living face depends to a significant degree on this or that form of the skull bones. The skull, like the face, can represent higher or lower forms, beautifulness and ugliness. Morphology of the human face is precisely so connected with the degree of cohesion and civilization of human communities. And its beauty is not created by our fantasies; it has real objective significance, because it is tied by origin with the highest functions of Mankind: the moral and social."

Thus, by following the logic of Russian anthropology, recognizing the self-sufficient aesthetic value of the ethnic standard of beauty, layer by layer, we analyze the morphology of the structure of the facial soft tissues, and then we proceed to a study of the proper "tribal characteristics of the skull." Therefore, "moral and social" values of the beauty of the human face gradually acquire strict anthropological dimension. **Racial-aesthetic morals also have a physical basis.**

4. Craniological Traits of the Evolutionary Worthiness of Races

Continuing our train of thought, we consider it natural to turn to the detailed masterpiece of Russian **craniology**—the work of Dmitry Nikolayevich Anuchin (1843-1923) titled: *On Several Anomalies of the Human Skull and Primarily on their Distribution Among the Races*.[152] Relying on rich, international experience, and also on the results of his own practical observations, he created an interesting scientific study with deep, far-reaching generalizations, the correctness of which we can easily observe in this time.

Anuchin began the presentation of his concept with a description of the **pterion**, a small section of the surface of the skull, where four bones meet: the frontal, the parietal, the temporal, and sphenoidal bones.

It is worth pointing out that we will not tire the reader with the details of a craniological analysis; we will entirely trust the authority of the venerable scientist, and therefore consider it completely appropriate to confine ourselves to conclusions which have a place in this detailed essay. First of all, we start from the position, that the pterion section is a good racial-diagnostic marker, for the different forms of its anomalies among the great races have a variance of 4-8 times, in terms of frequency. Such essential differences graphically demonstrate, that the members of the main human races are highly dissimilar by the tempo of the growth dynamic of the corresponding sections of the skull, and also of the brain itself; for it was discovered by Johann Friedrich Blumenbach's classical school of anthropology, that it is the development of the brain which gives form to a person's skull, not the other way around. A member of that school, Samuel Thomas von Sommering, wrote: "It is necessary to suggest that nature formed the skull bones, in order that they could adjust to the brain, but not vice versa."

---

[151] *Obshchestvennaya sreda, kak factor razvitiya i krasoti chelovecheskogo litsa*. Russkiy antropologicheskiy zhurnal, N1, 1900. (Russian Anthropological Journal, N1, 1900.)

[152] *O nekotorikh anomaliyakh chelovecheskogo cherepa i preimushchestvenno ob ikh rasprostranenii po rasam*. Moscow, 1880.

Left: Samuel Thomas von Sommering

In part, the frontal and temporal bones cover those parts of the brain that are responsible for higher psychological functions and abstract thinking. But among members of the so-called "lower" races, their development completes sooner, than among members of the "higher" races; this finds a corresponding reflection in the premature knitting of these bones. The frequency of these or other anomalies of the pterion, according to Anuchin, stand in direct accordance with the intelligence of a race, as such. An accelerated program of growth development of these fragments of the brain, among the "lower" races, enables the corresponding bones of the skull to close more quickly; this finds reflection in their cultural backwardness.

Of all the remaining anomalies of the brain, of which a significant number are counted, the most interesting in the field of social anthropology is **metopism.**

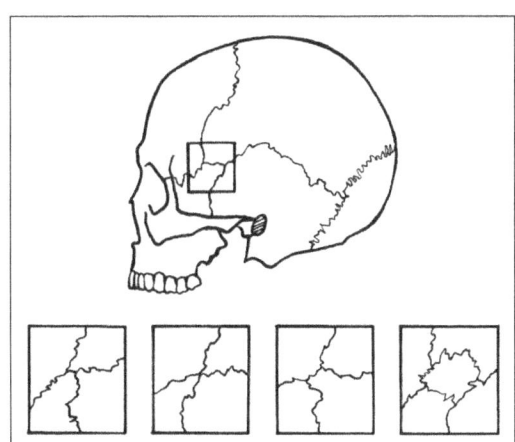

Left: The Pterion (according to Egon von Eichstedt).

By "metopism," we mean the seam that forms at the place of the joining of the two halves of the forehead. This lobe seam closes in the majority of newborn infants, but among several individuals, it is retained throughout life. This anomaly of the skull is an excellent racial diagnostic marker. In some individuals, the lobe portions of the brain that are responsible for higher manifestations of the human psyche and intellect, exert increased pressure on the corresponding sections of the lobe/forehead bones, during the process of the initial growth phase, that moves the bones, which for their part, cause the appearance of the forehead seam that is called "metopism." Many modern, liberal-minded anthropologists futilely attempt to black-out the matter of this sufficiently clear issue, for the development of the fragments of the skull proceed in accordance with the postulates of such a precise engineering discipline as **strength of materials.** No amount of humanitarian speculation can erase the physical boundary separating the "lower" and "higher" races. According to Anuchin's observations, with the metopic, that is, the lobal seam, the skull has a 3-5% higher capacity, in comparison with the usual. Further, analyzing the frequency of metopism among various races and peoples, he made this conclusion: "The table of results for observations shows that among Europeans, the lobe seam is encountered more often, than in other races. As for different series of European skulls, the percentage of metopism is found in variances from 16 to 5, [while] series of skulls of lower races, in the majority of cases [are found] at only 3.5-0.6%. A certain relationship evidently exists between the inclination toward metopism and the intelligence of a race. We see, for example, that in many races, the more intelligent tribes have a higher percentage of metopic seams. Among the higher members of the Mongol and White races, it is expressed by a number that is at the very least, 8-9 times greater, than it is in Australoids and Negroes."

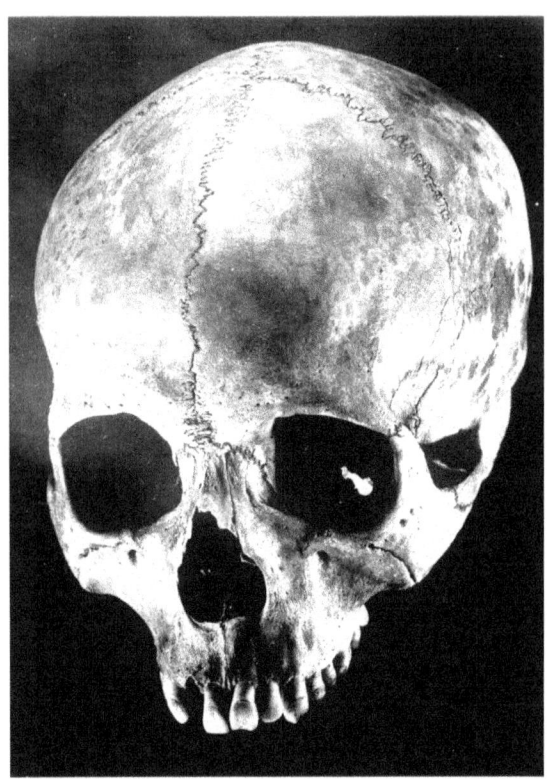

Metopism (from the collection of V.N. Evyagin). Note the metopic seams running down the center and sides of the skull.

The statements by one of the standards of Russian anthropology cannot in any way be regarded in a racist category, for the Institute of Anthropology of the Academy of Sciences of the Russian Federation today proudly carries the name of Dmitry Nikolayevich Anuchin, and the above-cited work is his doctoral dissertation.

Thus, an entire independent **theory of eccentric pressure of the brain** arose in anthropology, to explain the very fact of the inequality of the distribution of the metopic seam in different races, on the basis of the dissimilar nature of their intellectual gifts. Adherents of this concept think that the reason for metopism is a strengthening of pressure from the brain hemispheres on the walls of the skull, in particular on the lobe bone, and that as a result, this creates an obstacle to the timely closure of the lobe seam. On the basis of statistical data, a generalization was made, according to which, individuals with a preserved lobe seam possess a larger brain mass; this enlargement is not only absolute, but it is also relative; that is, it is not tied to an increase in the dimensions of the body. For its part, the retention of the lobe seam speaks to a higher level of mental and intellectual abilities in the given individuals. The most brilliant representatives of this theory of eccentric brain pressure were the major anthropologists, Georges Papillault (1863-1934), George Buschan (1863-1942), and Marciano Limson (1893- ?).

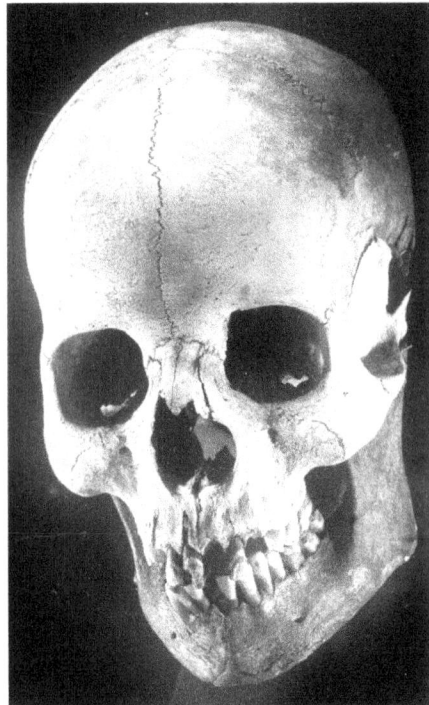
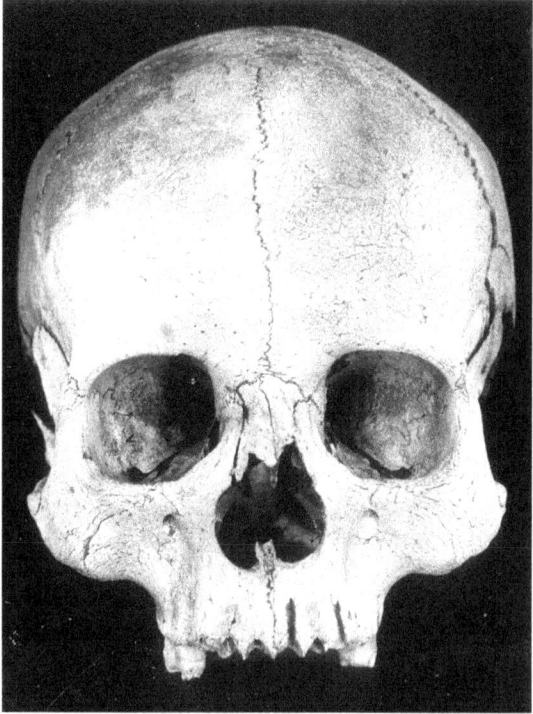

Metopism. Note the metopic seams running down the center and sides of the skull.

148

In the article, *On Metopism*,[153] V.V. Maslovskiy also clearly writes: "Thus, one can look at the phenomenon of retention of the lobe seam, as a phenomenon connected with the improvement of its organization. Such articulation of the skull on the twin lobe bones is a favorable factor for the contents of the skull. Resulting growth occurs in different directions, thanks to the presence of the seams." The pressure of the growing brain, a genetic program that calculates protracted growth, leads to the formation of a lobe seam, called metopism. The brain, developing according to an ingrained program, gives a far smaller probability of its origin. Therefore, it is namely according to this trait that races can be divided into "higher" and "lower".

Among foreign scientists engaged with the anomalies of the skull, it is necessary to include in the context of the given racial systematic, such names as: Wenzel Leopold Gruber (1814-1890); Johann Ranke (1836-1916); Herman Welker (1822-1897); Josef Hirtl (1811-1894); and Paolo Mantegazza (1831-1910).

In his book, *Man, his Origin and Evolutionary Development*, Wilhelm Leche,[154] a famous Swedish anthropologist, anatomist, and professor at Stockholm University, summarized the numerous studies in various countries in the area of anomalies of the seam of the skull. He gave this clear and thorough summation: "The opinion that the level of a culture is connected with the development of the brain, has very strong support—and [so does the opinion] that the [latter is tied] with the development of the brain capsule. Normally paired, the lobe bones on an individual knit together into one bone, in the first 1-2 years of life. More rarely, this knitting comes to a stop at such a stage that the lobe bones remain separated by a lobe seam, for the course of the entire life. Precise research proves that this knitting is delayed by pressure from within, thanks to the growth of the lobe portions of the brain. Thanks to the strong growth of this part of the brain, both lobe bones expand away from each other, and continuing ossification cannot fill the gap between them. One may further consider it proven, that the forward part of the braincase is usually in skulls with a preserved lobe seam, that is, in those in which the lobe bones knit [together]. That preservation of the lobe seam is usually a criterion of intellectual superiority should follow from that; the skull with this characteristic is more often encountered among civilized peoples, than among savages. In connection with this, I want to mention that up to now, not one man-like ape skull has been described as having a preserved lobe seam."

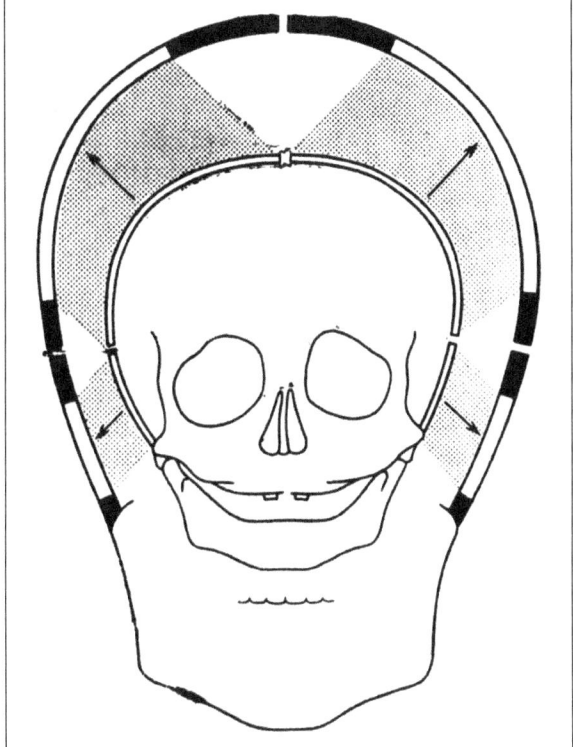

Left: The Basic Mechanics of the Growth of the Skull.

George Buschan, an important German anthropologist of the time, emphasized in his famous book, *The Science of Man*, that: "Metopism constitutes affiliation with high races. The metopic skull possesses more weight, a more complex structure of seams, and a more continuous non-closure of the latter. Lower races yield a lower percentage of such skulls, than do higher, so-called cultured peoples. Besides that metopic skulls absolutely do not express any of the lower traits; rather they express such, which can be considered morphologically higher. We are impelled to the conclusion, that metopism should be recognized as a phenomenon of a higher morphological order, and to see

---

[153] *O metopizme*. Russkiy antropologicheskiy zhurnal. Tom 15, Vip. 1-2, 1926.
[154] 1850-1927

progression in it, rather than regression. The cause for it does not lie in a pathological weakness of the lobe bone, but singularly in great pressure from within, from more developed brain hemispheres, namely their lobe portions. Since greater skull capacity should correspond to an increase in the volume of the brain, and since a larger brain weight—of the lobe portions in particular—usually serves as a sign of higher intellectual development, a further conclusion suggests itself: that the possessor of a metopic skull should stand out by a comparatively greater development of spiritual faculties."

Eugen Fischer, another standard of classical German anthropology, who specialized in the area of comparative morphology, pointed out in his fundamental textbook, *Anthropology* (1923): "Racial differences in the frequency of metopism are connected with varying capacity of the brain. We encounter it, for example, among Germans in 12.5% of cases; on skulls found at Pompeii in 10.5% of cases; among the ancient Egyptians at 7%; among Negroes—in 1% of cases. Among the man-like apes, the preservation of the lobe seam is a very rare phenomenon."

In his dissertation, *Toward the Study of Metopism* (1942), Spanish scientist Juan Comas testified in the same spirit, that: "Anuchin was one of the first to advance the hypothesis of a direct tie between metopism and intellect; that is, the trait is more often encountered among higher races, and consequently, it can be considered a trait of progressive evolution, testifying to the tendency of an organism toward modification of its usual type of skull."

Defining the value of a given craniological trait in its past and present, scientists undertook a daring attempt at substantiating its evolutionary significance in the future. In 1915, Russian anthropologist P.P. Spushkin was the first to express the thought that if not infrequent cases of preservation of the lobe seam among adults is tied to a gradual increase of the dimensions of the lobe portions of the brain, then consequently, Man will possess a very large brain in the future, and the lobe seam will be a normal phenomenon. Besides that he brought attention to the research of his predecessors in the area of distribution of metopism by race. Considering its higher frequency among namely the "higher" races, he emphasized in his work titled, *Are We Changing the Process of Evolution?*[155] that "the frequency of metopism in cultured humanity will grow further. Thus, from the species *Homo sapiens*, the race that stands out with a certain morphological trait is set apart in our eyes."

Precisely so was one of the important anthropological traits of the future superman first described in scientific literature. The given innovative concept in Russian science found continuation in the following works: V.M. Shimkevich's *The Future of Man from the Point of View of the Naturalist* (Berlin, 1923) and A.P. Bystrov's *The Past, Present, and Future of Man*[156] (Leningrad, 1957).

5. Physical Criteria of Intellectual Superiority

At the end of the 19th Century, Rudolf Virchow performed a great work on the territorial distribution of the abundance of craniological information, collected by the German Anthropological Society, as a result of which, the famous *Map of the Distribution of Metopism in Europe* was composed. Bavaria turned out to be one of the regions most heavily populated with this trait.

It is not necessary here to fall into a discussion on narrow German chauvinism, the more so in its Russian rendition. It is easy to guess that the traditionally high level of culture, the well-being, and national self-awareness of this German area is connected to the high frequency of the metopic seam encountered, which for all Bavarians consists of 12%, on average. The famous Soviet anthropologist, G.F. Debets, researching burial sites from the 10th-12th Centuries in Pskov and Novgorod, established the frequency of metopism in them at 14%. And that is the answer to the rise of the first parliament in Europe—the Novgorod Veche. In Novgorod there is a traditionally high level of literacy, and [there is a high level] of patriotism among the Pskovites (from whom, incidentally, elite airborne troops are recruited to this day). For all the inhabitants of the cities that comprised the North European trade and political union—the Hanseatic League—

[155] *Obratim li protsess evolyutsii?* Novie idei v biologii, N8, 1915. (New Ideas in Biology, N8, 1915.)
[156] *Proshloe, nastoyashchee, budushchee cheloveka.* (Leningrad, 1957).

according to the results of excavations, there is a high indicator of this racial-diagnostic trait. Zones exist today in central and northern Europe, in which the frequency sometimes approaches as much as 16%, and for members of the Nordic race, this increase is always significantly higher, than in other Europoids.

The famous Soviet anthropologist, M.I. Uryson, avoided sharp conclusions in his article, *Metopism in Man*,[157] but nevertheless brought up some indeed unique facts that supported the basic postulates of classical anthropology. As a result of excavations, it was established that the frequency of the metopic seam in the craniological series was as follows: 24% for North American Indians, and 26% for Etruscans. There you have an elementary explanation of the reasons for the rise of high Aztec and Mayan culture in the New World, and the core of Roman culture in Europe, laid down by a tribe of Etruscans not large in numbers, but unique in their gifts.

As yourself, in the course of our summaries we are not the least inclined to commit to a narrow European conceit, nor to dishonestly shuffle facts. On the contrary, on the basis of available data, we can certify that at the very least, in the time of their cultural flowering, the native population of America was itself, also a distinct and inarguable example of affiliation with a "high" race.

Thus, with all obviousness, we can conclude that the lobe data of sociology is only the result of biology, not the reverse. Neither environment, nor upbringing can ever make a "metopic" people from a "non-metopic" people. There are races, which from historical and evolutionary points of view, are "laureates of the metopic seam"—and there are those approaching the man-like apes, by the frequency of the presence of this trait.

Finally, even such a luminary of anthropology as V.V. Bunak wrote in his article, *On the Ridges on the Skulls of Primates*,[158] that: "The lobe anomaly seam in Man is observed more often in cultured races, and is tied with an enlargement of the brain, by its growth pressure on the lobe bone."

In general, for the sake of justice, it is necessary to note that in the first years of its existence, Soviet science was not ashamed to be enthusiastic about classic anthropo-social theories. One may see a corresponding position of the issue in the work by S.G. Schmerling, *Toward the Question of National and Social Differences in the Dimensions of the Head*.[159]

In his book, *Man and his Place in Nature*, famous German anthropologist Karl Vogt[160] summarized the data of contemporary science and asserted: "The Negro skull follows a law different than the white man, relative to the knitting of its seam; its forward seams, lobe and crown, as among the apes, knits very early, far earlier than the rear, whereas in the white man, the sequence of the knitting of the seams is completely reversed. But if this is so, then it is not particularly daring to suggest that in the brain of the Negro, there perhaps exists the same simian course of development, which is proven in his skull."

Robert Wiedersheim,[161] another famous German anthropologist, later confirmed this point of view in his book, *The Physique of Man from a Comparative-Anatomical Point of View*.[162] He emphasized: "Grazioli showed that the seams in higher races disappear in a different sequence, than in lower [races]. Among the latter, as among the apes, the process always starts from the front, from the lobe region of the skull; that is, on the boundary of the lobe and parietal bones, and from there proceeds toward the back. This itself reflects the early appearance of the formation of the forward portions of the brain, which can develop further in higher (white) races, when the lobe-parietal seam is completely removed after the occipital-parietal. This should stand in connection with the intellectual differences in tribes.

Much later, the famous Soviet geneticist, Nikolay Petrovich Dubinin, laid down a similar complex of ideas—in politically correct form—in his book, *What is Man?*[163] He wrote: "The brain

[157] *Metopizm u cheloveka*. Sovetskaya Antropologiya, N1, 1959.

[158] *O grebnyakh na cherepe primatov*. Russkiy antropologicheskiy zhurnal, tom 12: kniga 3-4, 1922. (Russian Anthropological Journal, Tome 12, Books 3-4, 1922).

[159] *K voprosu o natsional'nikh i sotsial'nikh razlichiyakh v razmerakh golovi*. Russkiy antropologicheskiy zhurnal, tom 18, kniga 3-4, 1929.

[160] Karl Vogt, 1817-1895

[161] Robert Wiedersheim, 1848-1923

[162] *Stroenie cheloveka co cravnitel'no-anatomicheskoy tochki zreniya*. Moscow, 1900.

[163] *Shto takoe chelovek?* Moscow, 1983?

of Man possess genetically determined traits. For normal development of the brain, a genetic program is needed. It is proven that 5/6 of the brain forms in Man after birth. Unceasing biological continuity for the length of the history of mankind is supplied by the presence of the traits in the genetic program of each individual. Man possesses definite biological characteristics, the specifics manifest at the molecular, cellular, organic, and population levels."

Therefore it follows, that from the point of view of the logic of the development of the science about Man, sciences at the end of the 19th Century, which still did not have the methods and results of the studies of modern molecular biology and neurobiology on the anomalies of the skull, saw regular racial-diagnostic markers, as characterizing the interconnection of the culture of a race and the specifics of the structure of the brain of its members.

Rudolf Virchow, Paul Broca, and Adolf Bastian were the first anthropologists that attempted to place cultural abilities in relation to the volume of the brain. On the basis of these ideas, the school of classical **anthropo-sociology** arose near the turn of the 20th Century. Influenced by the famous scientists Georges Vacher de Lapouge, Otto Ammon, and Ludwig Woltman, they acted with great sweep and daring.

In Woltman's brilliant book, *Political Anthropology*, it was pointed out that: "The fate of the human species is tightly connected with the cubic inches of the brain mass, and the history of humanity is contained in this mass, like a great book full of hieroglyphic signs."

Karl Vogt also thought: "It is found that for the development of intellectual capabilities, it is necessary for a person to have a certain minimal, definite brain weight, at less than which idiocy, limitedness, and feeble-mindedness begins."

Anatoliy Petrovich Bogdanov, the founder of Russian anthropology, noted in 1865: "It is known, for example, that among negroes, ossification and soldering of the seams of the skull occurs far earlier than in whites; that among the latter, soldering occurs more often, beginning with the seams of the rear portion of the skull, whereas among negroes it usually occurs first of all in the forward seams, and then moves to the back. The importance of these traits, having by result an earlier or later cessation of the growth of this or that part of the brain, is obvious for each, in particular if it is taken into consideration that the individual comprises a singular example in a number of beings, among whom the brain continues to grow after youth. If time and order of the succession of ossification of the seams of the skulls varies by race, then it becomes highly probable, that studies of ossification of the ribs or breast cartilage, the cartilage of the larynx, the spine, and even the pelvis, will yield **ethnic differences**."

In his monography, *General Psychology with Physiognomy*,[164] Professor Ivan Alekseyevich Sikorskiy (1842-1919) analogously asserted: "The black race belongs to the least-gifted on the globe. In physique the body of its members visibly has more things in common with the classic ape, than with the other races. The capacity of the skull and the weight of the brain of blacks is less than in other races, and accordingly with this, spiritual abilities are less developed. Negroes never amounted to a great nation, and never played a controlling or outstanding role in history, although in distant times they were far more widespread numerically and territorially than afterwards. The weakest part of the black individual and black race is intellect: on portraits one can always observe a weak contraction of the upper orbital muscles, and even this muscle in negroes is relatively more weakly developed, anatomically, than in whites, but it is a true difference between Man and the animals, amounting to a special human muscle."

George Buschan, mentioned by us above, confirmed all of D.N. Anuchin's conclusions, in regard to the racial differentiations of the Pterion Section. He indicated: " The pterion itself is the zone of the junction of the temporal, lobe, parietal and sphenoidal bones. Usually, the upper edge of the large wing of the sphenoidal bone goes up to the forward lower edge of the parietal bone, separating the temporal bone from the lobe bone; the seams here then form a figure in the shape of an "H"; but now and then the temporal bone projects forward from the forward edge, uniting the seam with the lobe bone. In higher races, this projection is very rarely encountered. Among Europeans, it is [encountered at a frequency] of 1.6%; among Mongoloids and Malaysians at 3.7%; in lower races it is comparatively frequent, so among black races it [is encountered] at up to 13%, among Australoids 15.6%, and among Papuans at up to 8.6% of cases. This relationship says that the junction of the temporal bone by means of an outgrowth, should be considered as a

---

[164] *Vseobshchaya psikhologiya s fiziognomikoy.* Kiev, 1904.

lower (Pithecoid) formation. Moreover, we encounter this constantly in gorillas, chimpanzees, and in the majority of other apes." Eugen Fischer also wrote: "Sometimes between these four bones, which form the pterion area, there is a uniting bone. Among lower races the lobe bone and temporal bones come into contact far more often, than in higher races. We observe this, for example, in Europeans in 1.5% of cases, in Mongols in 3.8% of cases; in Australoids at 9%; among Negroes at 11.8%; in Gibbon monkeys at 13.7%; in orangutans at 33.6%; in chimpanzees at 77%; and in gorillas—in 100% of all cases. Undoubtedly, the presence of the lobe-temporal seam in large measure depends on the relative size of the brain. The stronger the brain bulges in the skull, the more the lobe and temporal bones will separate; more rarely, they will unite into a seam."

In the structure of the face of Man, there are a number of other racial-diagnostic markers, possessing a good discriminating effect. The prominent German anthropologist, Robert Wiedersheim, wrote in regard to this: "The nose bones customarily remain separate, [but] sometimes they grow into one bone, and this is encountered more often in lower races, than in higher ones. Since such a growth is normal for apes, we probably have it in Man as one of the atavistic phenomenon. In chimpanzees, it already appears in the second year of life."

Another Russian anthropologist, Professor A.I. Kryukov, published a work in 1926 (already in Soviet times) under the characteristic title, *On the Degeneration of the Skull*.[165] In it, he indicated: "Engaging in the study of the skull, it struck me to observe how often changes were encountered, like signs of degeneration, primarily in the structure of the skull, but less [often] in the other organs." Seeing a regular connection between the structure of the skull and the brain of an individual, the author of the given work appropriately cited the classical Russian psychiatrist, S.S. Korsakov,[166] who expressed the following thought in his book, *A Course in Psychiatry*:[167] "Although anatomical changes of the skull should never be considered the immediate cause of spiritual illnesses, nevertheless, in the majority of cases they indicate an orientation of physiological processes in the skull, which condition molecular changes in the nerve cells of the cortex."

The logical conclusion in Kryukov's work is therefore simple and valid: "All degenerative changes are evidently tied with premature knitting of the skull seams." It is characteristic that the author does not fear applying the conclusions made by him in his work, in regard to separate individuals, to wider and more complex human groups. In the article, *Anthropological Studies of the Criminal, his Modern Situation and Problems*,[168] V.V. Bunak relies in principle on the same facts and leads us to conclusions, which correspond with the views of Cesar Lombroso,[169] the founder of the school of **criminal anthropology**: hereditary degeneration and the predisposition to commit crimes are interconnected.

6. Social Impact of the Knitting of the Seams of the Skull

However, now having gone through a number of conclusions of the anthropologists, psychiatrists, and criminologist cited by us, we come to a natural summarization: **in a human community comprised of various racial groups, the highest percentage of crimes originate with those, who, by virtue of their hereditary condition, are observed to have a high percentage of premature knitting of all cranial seams.**

The modern, so-called multi-cultural society, which propagandizes the inter-racial International (chaos of blood)—is held as an ideal example; but in the USA and the nations of Western Europe, the maximum percentage of crime originates with the members of the Negroid race and other dark-pigmented racial groups. For ages, the color of crime in the civilized world has not been white, and not one philanthropist has dared to dispute this fact, which has been officially documented in the daily reports of America's FBI.

---

[165] *O degeneratsii cherepa.*
[166] S.S. Korsakov, 1854-1900.
[167] *Kurs psikhiatrii.* Moscow, 1901.
[168] *Antropologicheskoe izuchenie prestupnika, ego sovremennoe polozhenie i zadachi.* Moscow, 1926.
[169] Cesar Lombroso, 1835-1909.

In the given question, the author of these pages does not pretend to be an authority. For these positions of criminal anthropology were introduced in the 1970s, by prominent Russian scientist, Professor Victor Nikolayevich Zvyagin; they are also confirmed today by criminologist Sergei Alekseyevich Nikitin, with [his] collection of the skulls of anti-social elements, among which an inherent, rhomboid deformation of the skull was discovered in a large mass of them, and a very low frequency of the occurrence of the metopic seam (for the Europoid race, 1% in all). In this regard, Swedish anthropologist Wilhelm Leche wrote in his book, *Man, His Origin and Evolutionary Development*, that: "Comparative research of the human skull has established, that all of its component parts originate immediately from those, which the lower vertebrates have."

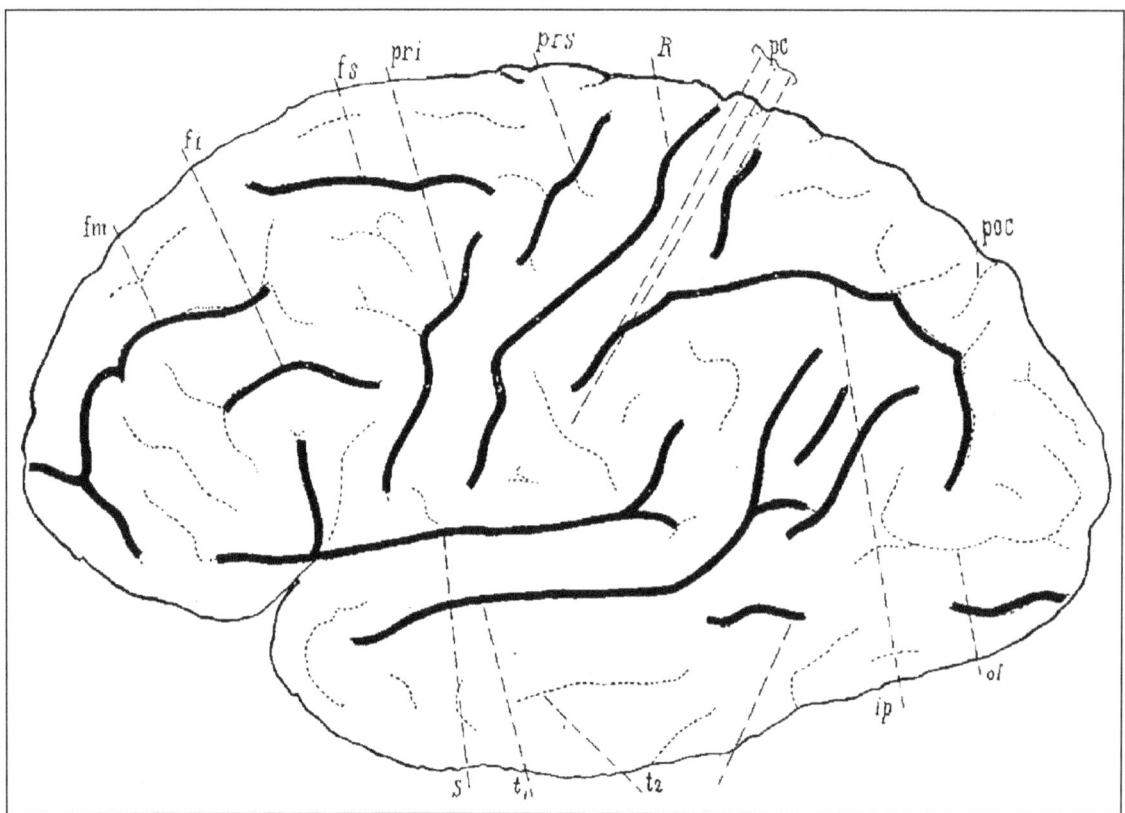

The brain convolutions are divided by fatty lines, according to which racial differences may be determined (According to A.S. Arkin).

Ethnographic confirmations exist. Two English explorers left this characteristic testimony at the start of the 19th Century: "The Hottentots, and the Bushmen in particular, are morally and physically not much different from an orangutan. Africa south of 10 degrees latitude is inhabited only by peoples whose minds are as dark as their skins, and the structure of their skulls makes any hope for their future betterment a Utopian dream."

Prominent Russian raciologist V.A. Moshkov wrote in his monograph, *A New Theory of the Origin of Man and his Degeneration*,[170] that: "By their spiritual capabilities, Negro children do not approach the white child; he is also capable of learning and intelligent, like whites. But just as the fatal period of maturity begins, then together with the knitting of the cranial seams and the forward projection of the mandibles, the same process is observed in them, as it is in apes: the individual becomes incapable of development. The critical period, when the brain starts to tend toward withering, appears far earlier in Negroes, than in whites, namely because of the earlier knitting of the cranial seams in Negroes." In connection with the extremely important specifics of the knitting growth of the cranial seams among members of the different races, and also the graphic and

---

[170] *Novaya Teoriya proiskhozheniya cheloveka i ego vyrozhdeniya*. Warsaw, 1907.

indisputableness of this racial-diagnostic trait in the research of social processes, Professor V.N. Zvyagdin suggested an apt title—**suturology**—for a science of the research of patterns in the cranial seams.

In public life, we observe the confirmation of the following solid rule: **the "lower" a social or racial group from an evolutionary point of view, the sooner the knitting of the cranial seams occurs among its members, and the sooner the programmed development of the brain ceases in them; that is one of the reasons for their anti-social behavior when they end up in the bosom of the distribution of another, "higher" race.**

This rule is confirmed by the statistical research of criminal anthropology, and also completely coincides with the conclusions of neurologists, who attest to the assertions of N.P. Dubinin.

Observing the manifestation of anthropological patterns in socio-criminological patterns, we again easily observe how differences in the physical structure of the races tells on the fate of nations. A.M. Fortunatov's book, *Materials on the Question of the Sequence and Order of the Closing of the Cranial Seams of the non-Russians of Russia*,[171] serves as an excellent testimony. In it, the author writes: "The weight of the brain in high races increases until the age of 40, then remains almost without changes until 50 years, and afterwards begins to decrease. The stronger the brain functions, then the later the knitting of the seams appears. In various races, these cranial seams do not knit together simultaneously. This lack of simultaneity stands in connection with the ability of the brain to develop, and the complexity of the seams. In lower races—the least capable of improvement—the seams are less complex, and smooth out very early; sometimes they more or less completely disappear, by 30 to 40 years of age. Among the more accomplished races, they are preserved longer and they smooth out far later."

According to the author's observations, among blondes the knitting of the cranial seams begins about 40 years of age and later. Along with the time of the knitting of the cranial seams, an important indicator of the general development of a race is the sequence of the knitting of the cranial seams, which is made clear in the very title of Fortunatov's book, in which he wrote: "Among the white tribes, the seams begin to knit from the rear section, whereas in the negro, they close starting from the front portion; the same is observed in idiots belonging to the white race. In the skulls of foreigners [non-Russians] in Russia, the closure of the seams goes in one or the other direction: from the front to the back (in 2/3 of cases), and from the back to the front (in 1/3 of cases)."

On the basis of the above-said, it is not difficult at all to understand why the "multi-national" Russia that is trumpeted to us daily by democratic social scientists, was really built by Russians, and not some other tribe.

The *Russian Empire*, just as *Great [Kievan] Rus'* before it, was founded by the Great Russian tribe, in which, by virtue of its hereditarily conditioned racial traits, the very process and sequence of the knitting of the cranial seams occurs by a model characteristic of "higher" races, at the same time that among the "non-Russian of Russia" a model predominates, that allows one to place them primarily among the "lower" races.

Without difficulty, we can observe this anthropological principle in the history of any great empire and any great civilization. **"Higher" races create—"lower" races destroy.**

The fate of peoples belonging to these base racial types is conditioned by the very hereditary principles of development of their brains, and is not subject to any kind of cultural-educational interference from the outside. In essence, world history is a chemical boiler, carrying out the distillation of the "higher" elements and settling of the "lower" elements.

Proof in favor of this correlation between the specifics of the knitting of the cranial seams among members of the different races, and their ability to create culture, is substantiated by the findings of such anthropologists as: Adolf Frick (1821-1901); Johann Christian Lucae (1814-1885); Josef Engel (1816-1899); Karl Ritter von Edenberg Langer (1819-1887); and Hans Gudden (1866-?).

---

[171] *Materiali k voprosu o posledovatel'nosti i poryadke zakritiya cherepnikh shvov u inorodtsev Rossii.* Sankt-Peterburg, 1889.

From the time of the fall of the Soviet Union, a number of quite different versions of the epochal historical event were advanced. We do not at all intend to engage in polemics over this. From the point of view of the above-said facts, all [their reasons] seem rather trivial. The state-forming population of the USSR—the successor state of the Russian Empire—fell to half of the number of the general population. In the near future, the same lot awaits the United States, where the white nation-building majority will also soon turn out to be in the minority.

Affiliation with a nation-building people is a concept that is not socio-cultural or mystic, but racial-biological, measured by a number of parameters, but most of all reflected in the weight, the complexity of organization, and the evolutionary value of the brain of its members.

There is absolutely nothing "racist" contained in any of these conclusions. Prominent Soviet anthropologist, V.P. Alekseyev, wrote in the monography, *Historical Anthropology and Ethnogenesis*,[172] that: "No kind of painstaking and deep analysis of ethnic mutual relations is possible without an account of the racial situation; no sort of reseach into ethnogenesis and ethnic history can be real and comprehensive, without bringing in anthropological data, that in the final analysis, is the data of the biology of Man. The morphological and physiological characteristics that are studied by anthropologists are genetically conditioned; therefore, the biological traits of a population are closely intertwined with many aspects of their lives, that essentially enrich the picture of human history."

Francis Galton, the founder of **eugenics**, stated his opinion more frankly, suggesting that "conscience, talent, and other clearly human traits, are biologically determined traits, passed down via sexual cells through generations."

A book by the famous Russian anthropologist, Alla Armenovna Movsesyan, belongs among the number of the most modern and significant works on the question being researched. It is titled, *Phenetic Analysis in Paleo-Anthropology*.[173] In the book, the author analyzes extensive material on the racial differentiation of the anomalies of the skull. In the process of numerous studies, it was discovered that they possessed a high degree of heredity. Practically each anomaly in the structure of the skull is an excellent illustration of a trait that differentiates racial groups, since it carries in it a clearly expressed genetic memory of the evolution of a given racial group. "Many findings that speak to what are variations in the structure of the skull, manifest themselves in the process of normal development, and are in great degree, genetically determined. Besides that the stability of the frequencies of traits in populations through time attests in favor of the hereditary nature of the anomalies of the skull, as well as to the corresponding archaeological and historical data results of paleo-population studies"— Movsesyan emphasizes.

In a plan of confirmation of the basic postulates of racial theory, the following conclusions are most important. First, interpreting the historical data of the distribution of the metopic seam, the research comes to the unequivocal conclusion that: "Analysis of the geographical variation of this trait shows that most often, it is encountered in centers of civilization." Secondly, with the help of the aggregate of the anomalies of the skull, the best differentiated of all are the northern and southern race types, for the extent of the entire historical process.

This again attests that world history comes down to the primeval confrontation of the northern and southern racial types; and that centers of civilization do not arise spontaneously from the action of some abstract, refining cultural conditions, but are created by a concentration of individuals of a definite, "higher" racial type, possessing a more progressive construction of the brain, that regularly finds its reflection in the differentiated anomalies of the brain.

A.A. Movsesyan also indicates that anomalies of the skull correspond to the category of so-called descriptive (qualitative) traits, having genetic dependence; in contrast to metrical (quantitative) discrete-variable traits, they are defined according to the criterion of "presence-absence." It is namely this [which] makes cranial anomalies invaluable factors in the plan of racial analysis and the re-establishment of a population representation of history. As an illustration of the given thesis, our modern science legitimately relies on the classical work of D.N. Anuchin: *On*

---

[172] *Istoricheskaya antropologiya i etnogenez*. Moscow, 1989.

[173] *Feneticheskiy Analiz v paleoantropologii*. Moscow, 2005.

Several Anomalies of the Human Skull and Primarily on their Distribution by Race.[174] In it, Anuchin notes: "The detailed study of the anomalous characteristics of races on a significant number of persons, together with the individual deviations within every race, gives the opportunity to not only determine the type of race, and bounds of its variations, but to explain to a certain degree, the genesis of this type, or more exactly, its tie with other racial types."

7. Racial Differences in the Structure of the Pelvis

In all the early years of scientific anthropology near the turn of the 20th Century, the logic of the scientist was nevertheless flawless, for the form of the skull of a child is directly connected with the characteristics of the structure of the mother's pelvis—they should correspond to each other in the goal of eliminating death in childbirth. The mixing of races unavoidably leads to this, because the structure of the pelvis of a mother of a different race does not correspond to the shape of the head of mixed infant, who carries the traits of a father from another race; that leads to complications during childbirth, and affects the viability of the descendents of both original races, reducing the number of mestizos.

"Of all the parts of the skeleton, the pelvis most corresponds to the skull; one can suggest that the pelvis, sooner than any other part, will yield findings of the characteristic traits of the races. The pelvis of a negro differs by narrowness and elongation in length. All the diameters of the lesser pelvis, through which the baby's head passes during childbirth, are far shorter in negroes; the larger diameter is shortened, in particular. This fact in itself is not surprising: it should have been expected, because the head of the infant negro already carries the imprint of its race in childbirth; it is long and narrow, and the small pelvis of the negress has either a wedge-shape or a tube-shape, in conformity with this [fact]"—Karl Vogt wrote. Here, nature acts in strict accord with simple mechanics. The shape of the skull of the infant, according to racial indicators, should line up with the racial indicators of the mother's pelvis, like a nut and a bolt. Any mismatch weakens the so-called "mechanical strength" and "durability" of the race, thread by thread, turn by turn, from generation to generation. Therefore, the pureness of race is the first and main condition of reproduction, and the mixing of races unavoidably leads to degeneration.

It is ideal to bring up M.I. Lutokhin's, *Historical Observation of Literature on Racial Differences of the Pelvis,*[175] from the classical Russian works on this theme. In the beginning, the author cites the opinions of famous anthropologists like Paul Broca, Paul Topinard, and Samuel Thomas von Sommering, who compared the pelvises of lower races with the pelvis of the ape. Franz Pruner-Bey (1808-1882), by virtue of the graphicness and exactness of the traits, he generally suggested giving up the classification of races according to the structure of the skull, and to switch to classification of the races according to the form of the pelvis. The anthropological division which engages in the study of racial differences in the pelvis is called **pelvimetry.**

In conclusion, Lutokhin wrote: "In this essay I mentioned the views of authors, on the cause of very sharp differences in the structure of female pelvises of different races, as a result of adaptation (to a certain degree) of the pelvic ring to the head of the new-born. There is much data in favor of [the statement], that miscegenation of races leads to far more difficult, and sometimes impossible [childbirth].

Similar conclusions were supported by the magnificent Russian raciologist, Vladimir Alexandrovich Moshkov, in his monography, *A New Theory on the Origin of Man and his Degeneration.*[176] Moshkov wrote: The act of childbirth, completely natural for every creature of pure nature, should be just the same for Man; that is, painless, like all other physiological functions. Women of lower races endure births very easily, sometimes even without any pain, and only in highly rare cases do they die from childbirth. But this can never be said of women of lower races, who birth the children of white fathers. Thus, they report on American Indian women, that they often die in childbirth from pregnancies with a child of mixed blood from a white father,

---

[174] *O nekotorikh anomaliyakh chelovecheskogo cherepa i preimushchestvenno ob ix rasprostranenii po rasam.* Moscow, 1980.

[175] *Istoricheskiy obzor literature o rasovikh razlichiyakh taza.* Moscow, 1899.

[176] *Novaya Teoriya proiskhozhdeniya cheloveka i ego vyrozhdeniya.* Warsaw, 1907.

whereas pure-blooded children within them are easily born. Many Indian women know well the dangers [associated with] a pregnancy from a white man, and therefore, they prefer a timely elimination of the consequence of cross-breeding by means of fetal expulsion, in avoidance of it."

In this regard, the famous Russian ethnographer and anthropologist, Octavius Vasilyevich Milchevskiy, emphasized in his essay, *The Foundation of the Science of Anthropo-ethnology*,[177] that: "The form of the pelvis of correspondingly different tribes was satisfactorily and reliably studied by Weber. The pelvis of the Hottentot woman, or the Botokud woman, [being] more elongated in form, more vertical and high by the ilium, [and] more narrow and high by the sacrum, closely approximates the pelvis of animals. The reader knows that when being born, Man passes between the pelvic outlet of the mother, and as midwife obstetric practice shows, the infant passes with difficulty, such that in particular, the head snugly contacts with these bones and even significantly spreads them. If the form [of the head] is the same, then in this time the bones of the newborn are so soft and spongy, that they may be given a favorable form by a simple and easy grip in the hands (which midwives do); and of course, it should not surprise the reader a bit, that with the bones of the mother being very hard, it is ideal that the skull [of the infant] passing through the pelvis should absolutely accept the form of the vent. And actually, these measurements and others have shown that a very close correlation between these two forms (the mother's pelvis and the infants head) exists. Professor Weber even divides people into four classes, observed according to the different forms of their pelvises: oval (Europeans); circular (American Indians); rectangular (Mongoloids); and wedge (Black races)."

The given branch of anthropology later received stable scientific development. The highest flower of **racial pelvimetry** was achieved in the works of Egon von Eichstedt; in part, he indicated: "Racial differences in the size of the pelvis are significant, and are not explained by only the dimensions of the body, but are conditioned by racial variations of heredity. Thus, the pelvises of Veddoids, Negroes, and Paleo-Mongoloids (in Japan) are absolutely and relatively smaller, than in European women. Negro pelvises differ by small dimensions, by narrowness and height, whereas in Europeans the lateral and forward edges of the outlet bones are widely divergent. The transverse oval form predominates in Europoids; the circular in Negroids. Among Chinese, the forms vary, but among the southern brachycephalics, the transverse-oval form predominates. The incline of the pelvis also goes into the number of racial differences. In the Japanese, [the incline] is not great."

For the quantitative measurement of racial differences, there is **William Terner's Source Index.**

Theodor Mollison (1874-1952), a prominent German anthropologist-evolutionist and Director of the Anthropological Institute of the University of the City of Munich, also gave attention to this question in his textbook, *Anthropology* (1923). He wrote: "The opening of the pelvis noticeably varies in peoples, particularly in its transverse diameter. In lower races, the form of the opening of the pelvis is closer to the primitive; that is, to what we observe in the man-like apes. If you look at the pelvis from above, we see a depression, the lateral walls of which form arrangements after the articular surface parts of the *nadvzdoshnikh* [mid-hip section] bones, and the bottom—the sacrum, is far deeper in Europeans, than in members of lower races. The upper opening of this depression in Europeans is more narrow than its bottom; in Papuans, it is wider. These changes in form are connected with the fact that the sacrum bones in Europeans embrace the transverse more, than in lower races."

In Millison's opinion, all these differences have a particularly evolutionary origin. The coefficient of the transverse diameter gives an index of the opening of the pelvis, and graphically shows the magnitude of racial differences: Europeans—80; Ayni—85; Japanese—87; Veddi—88; Australoids—93; and Andamans—99.

This substantial spread in size shows that the process of racial mixing is not desirous, and is even dangerous.

Martin Stemmler, a famous German raciologist, emphasized in *Disease and Race* (1935), that: In places where there are often mixed marriages, women experience difficulties in childbirth. For example, long-headed Nordic children pass through the narrow pelvis of Mediterranean women

---

[177] *Osnovaniya nauki antropoetnologii*. Moscow, 1868.

with difficulty. Among various races, the pelvic bones form differently, and in an unfavorable combination, harmful dislocations of the pelvic hip joints is possible."

In the final analysis, free love propaganda does not offer satisfaction: it destroys races. Any quest for biological diversity-sensitivity is fraught with evolutionary irresponsibility.

8. Racial Differences in the Structure of the Brain

At the end of the 19th Century, Rudolf Virchow, the founder and director of the German Anthropological Society, evaluated—like a true scientist—the problem as a whole. He thought that all the data—including circumstantial—that was found during the study of the skull, could have essential significance, only inasmuch as they are found to be in definite dependence of these or other traits of the structure of the brain.

Charles Darwin argued in his fundamental essay, *The Origin of Man and Sexual Selection* (1871), that: "The conviction, that in Man there is a connection between the volume of the brain and the degree of development of the intellectual faculties, is based on the comparison of the skulls of savage and civilized racees, ancient and modern peoples, as well as on analogies from the whole family of vertebrates."

The outstanding Soviet anthropologist, G.F. Debets, considered the Cro-Magnon type as the proto-European, but in the Cro-Magnons the average weight of the brain was 1,800 grams. The distinguished French anthropologist, Paul Topinard, developed the Darwinist position in his book, *Anthropology* (1879), in the following way: "Lower races have a lower cranial capacity than higher races. In this regard, the worst of all are the Australoids, since they have an average [cranial capacity] of 1224 cubic centimeters; by our measurements, the cranial capacity of the American Indian is also small; it increases in the yellow races, and reaches its greatest size in white peoples."

When Paul Topinard weighed the brain of one Hottentot woman, he found that its weight just exceeded 800 grams. The cranial capacity of the Indonesian Pithecanthropus was 930 grams. Therefore, according to the calculations of Eugen Fischer, the weight of the brain of the Bushmen, the Hottentots, the Veddoids, and the aborigines of Central Australia, fluctuated within a range of 900 to 1200 grams.

N.V. Gilchenko's fundamental work, *The Weight of the Brain and its Several Parts, in the Different Tribes Populating Russia,*[178] is also committed to resolution of this global problem. The clarity and conclusiveness of the position—abundant in statistical material—makes this essay relevant today, in many ways. From the title it is already obvious that the author thought completely in the spirit of racial theory, for on the basis of experimental data, it was proven that among the members of different races, the corresponding parts of the brain have different tempos of growth, and as a result, they are not identical in weight. For its part, this is supported by variations in the frequency of the occurrence of the anomalous seams of the skull. The science of that time was completely logical and consistent. "The influence of nationality (tribal [affiliation]) on the weight of the brain also undoubtedly exists, along with all the previously and firmly examined influences of growth, age, and so on. Racial and tribal traits do not change from ancestors to descendents. The difference in the weight of the brain, noticeable in separate areas of our vast [Russian] Fatherland, cannot be explained by the influence of growth, nor the influence of age, but exclusively by the influence of national (tribal [affiliation])."

---

[178] *Ves golovnogo mozga i nekotorikh ego chastey u razlichnikh plemen, naselyayushchikh Rossiyu.* Moscow, 1899.

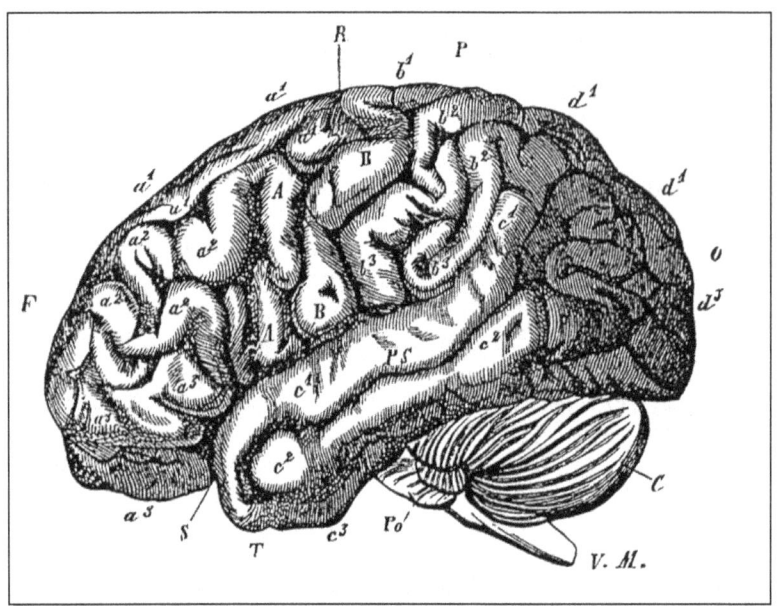

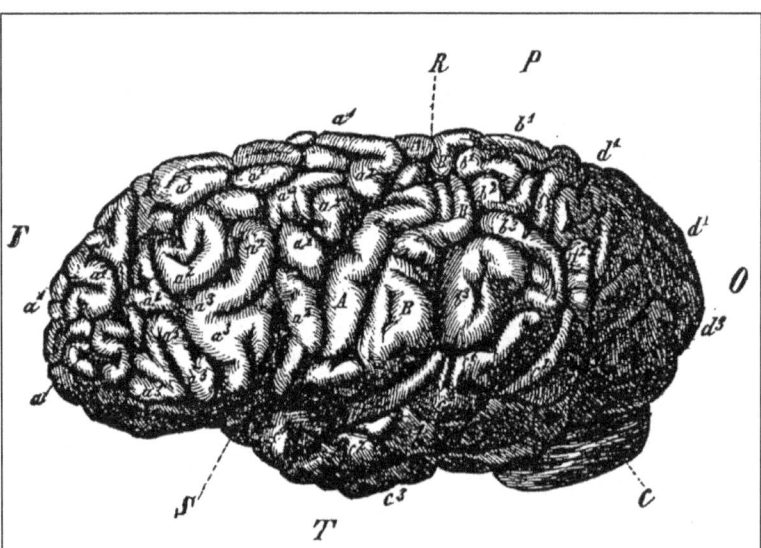

Top: the brain of a negro woman; **bottom:** the brain of German mathematician Gauss (according to Karl Vogt).

Here it is worth mentioning the findings on some specifics of the mentality of peoples living in the mountains. In his book, *Man,* the famous German anthropologist Johann Ranke wrote: "The causes, which in many mountain localities lead to the development of cretinism, often influence the non-cretins in the same area." The insufficiency of iodine in mountain rivers can be added to the number of reasons that cause a high percentage of cretinism among the inhabitants of mountain areas. A saying of the people of Tuscany agrees: "the legs of mountain people are sturdy, but their brains are delicate."

Besides the essential differences in the weight of the brain and its parts, among the members of the great human races, and even separate nationalities, differences in the organization of the convolutions of the brain were discovered.

One of the first racial differences in the structure of the brain was studied by the famous Russian anthropologist, Dmitry Nikolayevich Zernov (1843-1917). His work, with the characteristic title, *The Convolutions of the Brain as a Tribal Trait,*[179] was published in 1873, and in 1877 he

---

[179] *Izvilini mozga, kak plemennoy priznak.*

160

published his fundamental monograph, *Individual Types of Brain Convolutions in Man*.[180] In 1877 the book, *On the Question of the Anatomical Characteristics of the Brain of Intelligent Peoples*,[181] appeared. In all of his essays, there is a distinct morphological description of the structure of the brain of "higher" and "lower" types, not only on the level of separate individuals, but also in great racial-ethnic communities. Besides that this talented scientist was the first to design a device—the **encephalometer**—for determining the differences in the structure of the brain in different individuals. In *The Encephalometer (A Device for Determining the Position of the Parts of the Brain in a Living Person*[182])—a brochure especially dedicated to the description of its construction—Dr. Zernov indicated: "The main goal of the arrangement of the suggested instrument is to give the opportunity to determine the exact raised surface of the brain (fissures or convolutions) in a living individual, independent of the cranial seams, which usually govern in similar determinations."

Professor Wilhelm Waldeyer (1836-1921) was a famous German anatomist and expert of the human brain at the University of Berlin. At the 18th Conference of the German Anthropological Society in Nurnburg in 1877, desiring to characterize the importance of the comparative-racial study of the arrangement of the brain convolutions and fissures, Waldeyer turned to his colleagues with the following famous speech: "I am completely confident that the foundational study of the brain convolutions, from the point of view of their form, their significance, and nomenclature, can be achieved only by the path of a vast and according to possibility, comprehensive comparison between the brains themselves, of all peoples and tribes." In Europe, Gustav Retzius (1842-1919), Jakob Herman Frederick Kohlbrugge (1865- ?), Carlo Giacomini (1840-1898), Aleksander Ekker (1818-1887), Augustin Weisbach (1836-1914), and Gustav Albert Schwalbe (1844-1916) were engaged in the study of the racial differences of the structure of the brain.

R.L. Weinberg, a distinguished Russian specialist of that time, wrote *On the Structure of the Brain of Estonians, Letts, and Poles. Comparative Anatomical Traits*.[183] On the basis of statistical information, he concluded: "Thus we see that although the human brain is arranged relative to its external form, by a single plan [that is] undoubtedly common for the majority of human types, nevertheless it represents an entire range of such traits, which notably differ by their frequency in the different tribes of Mankind, or even are characteristic of only some tribes, being completely absent in others."

Wilhelm Waldeyer

Gustav Retzius

Konrad Lorenz

---

[180] *Individual'nie tipi mozgovikh izvilin u cheloveka.*

[181] *K voprosu ob anatomicheskikh osobennostyakh mozga intelligentnikh lyudey.*

[182] *Entsefalometr (Pribor dlya opredeleniya polozheniya chastey mozga u zhivogo cheloveka).* Moscow, 1892.

[183] *O stroenii mozga u estov, latishey, i polyakov. Sravnitel'no-anatomicheskiy ocherk.* Moscow, 1899.

In his next work, *Toward the Study of the Brain of Man,*[184] Weinberg emphasized in the spirit of the program declarations of the time, that theoretical medicine and anthropology should equally subject the racial differences of the brain to a comprehensive study. Proceeding from a feeling of civic obligation, scientific objectivity, and also tribal solidarity that was customary for those times, the author thought it necessary to emphasize: "After an entire series of works, published in the last three decades on the somatology of the Jews, it is unlikely that any doubts remain of the existence among them of a **basic physical type**, expressed not only in distinctive traits, or so-called Jewish "physiognomy", but [also] in the structure of the skeleton, in the proportions of the skull to the torso, and in the particulars of outer hair covering. The **psychological** traits of the Jewish race appear sharper than the physical characteristics. These and others, primarily the [psychological traits] reflect, as is known, on the development of the central nervous system, or more precisely speaking, are an outward expression of the individual arrangement of the central organ of the mental and physical life of the given tribe."

More was revealed on the traits of the organization of the fissures and convolutions [of the brains] in Jews. Among racial-diagnostic traits are the Roland and Sylvius fissures—they are a feature between the lobe and parietal portions. The numerous breaks and bridges between the neighboring convolutions comprise a tribal trait in the structure of Jewish brains, that is expressed in their enhanced social adaptability and individual situational instincts, which are usually absent in Russians. The great Russian explorer, N.N. Miklukho-Maklay, pointed to this same aggregate of morphological differences, as characteristic racial traits, when he compiled experiences in the Papuas. Karl Vogt wrote: "The Sylvius fissure in negroes has a more vertical direction, as does in equal manner, the Roland fissure."

Describing a specific feature in the structure of the Jewish brain, R.L. Weinberg analogously emphasized: "Thus, in this case we encounter a series of such characteristic patterns of the brain's surface, which, according to the observations of our authors and others, undoubtedly belong to that category of rarely observed variants in the brain convolutions; therefore, [they] should not be avoided with silence in the comparative-racial studies of the human brain." It is namely in Jews that an anomaly in the growth of the Roland and Sylvius fissures is most often observed.

In his article, *On the Racial Traits in the Structure of the Brain Hemispheres of Man,*[185] A.S. Arkin was even more frank and effective. Besides the above-indicated racial traits, he introduced new ones: "The Middle fissure of the brain is a fissure, which to greater degree than the other fissures of the cerebrum, is subjected to changes, and has different contours in the members of different races." Besides that based on an enormous amount of foreign materials, Arkin spoke in the length of his article about "brains, rich in convolutions, which, as is known, are considered more completely organized."

Paul Topinard, the greatest French anthropologist, also wrote in his fundamental book, *Anthropology*, that: "Convolutions are thicker, wider, and less complex in lower races. The nerves of negroes, and primarily the nerves at the base of the brain, are fatter, and the substance of the brain is not so white, as in Europeans." Possessing a thicker cranium bone, of which ancient Greek historian Herodotus wrote, the members of the Negroid race therefore naturally have a lower threshold for detecting pain. In the second half of the 19th Century, the Association of Boxers pointed to this neurophysiological fact, turning down the participation of black sportsmen in competitions, on the basis that they were less sensitive to pain than whites.

Undeniably, black athletes demonstrate their superiority on the sports field. But where, allow me to ask, are the black Nobel Prize winners? Where are the black scientists, philosophers, and classical composers? The White Man never denied that the members of other races, and some animals, can run and jump better than him; he always saw a different goal for existence for himself in this world.

---

[184] *K ucheniyu o forme mozga cheloveka.* Russkiy antropologicheskiy zhurnal, N4, 1902.
[185] *O rasovikh osobennostyakh v stroenii mozgovikh polushariy cheloveka.* Zhurnal nevropatologii i psikhiatrii, imena S.S. Korsakova, kniga 3-4, 1909. (S.S. Korsakov Journal of Neurology and Psychiatry, Books 3-4, 1909.) *Or* "Journal of Neurology and Psychiatry called S.S. Korsakov".

In the given article, Arkin's main finding is the conclusion that "the most characteristic racial differences are noted in the area of the **associative centers.**" These centers have comparatively later development in comparison with other sectors of the brain. It is also easier to calculate the morphological differences of the structure of the brain between members of "higher" and "lower" races. The comprehension of foreign, and in equal measure, the creation of his very own culture, is very closely tied with the development of these associative centers. The language of a concrete culture, its style, its known refinement, or on the other hand, its barbaric rudeness, and the depth and purity of experiences characteristic to it, thus have clear, physical contours.

The conclusion in Arkin's work is simple and persuasive: "The favorite fissures and convolutions have racial differences in the structure of the cerebrum, where they manifest more often in relief [contours]."

There is still one more index of the racial diagnostic and applied culturology, that is completely forgotten by modern anthropologists, namely: "The substance of the Negro brain is incomparably more dense and solid, than in the white [brain]"—Karl Vogt declared.

Weinberg and Arkin, two of the leading above-named Russian [Imperial] specialists in questions of the structure of the brain contours, were Jewish by nationality, which automatically frees us from any possible accusations of racist or anti-semitic propaganda. Their work, on an equal footing with others, made the golden collection of the Russian Academy of Anthropology, in the address of whom no such accusation was ever put forth. In general, for the sake of justice, it is necessary to note that the Russian anthropology of that time, besides vast theoretical problems, also successfully resolved highly specialized tasks, which one can observe in the article by N.I. Balaban and A.I. Molochek, *The Structure of the Body in Schizophrenic Tartars of the Crimea.*[186]

Now it will be a natural assertion in evolutionary terms, that in members of all races, beauty of the face is interconnected with the shape of the skull and the facial bones - and that for its part, reflects the perfection of the structure of the brain. One is a diagnostic marker of the other, enabling the detached observer to quickly—and with a great degree of probability—judge the inner world of any individual, by his out appearance. But if from an anthropo-aesthetic point of view, racial beauty is subjective and relative, then the structure of the skull and brain enables an objective and unbiased judgement on the degree of perfection and worthiness of their possessor, for they carry in themselves qualitative, measurable parameters.

M.I. Uryson, the famous Soviet anthropologist, wrote in his work, *Interconnections of the Basic Morphological Traits of the Human Skull in the Process of Anthropogenesis,*[187] that: "Proceeding from an examination of the skull as a total skeletal structure, one can suggest, that the progressive development of the brain renders its influence not only on the formation of the brain case, but also on the rebuilding of the facial section through its development. Consequently, it is a question of the mutual influence of the brain case and facial section of the skull, and also of the dependent factors of their change, in the process of the evolution of the skull."

It is hardly necessary to clarify that in this interconnection, a key role is fulfilled by that which in common speech is referred to as "racial prejudice;" it fulfills the function of an evolutionary biological filter. Namely therefore, prominent Soviet anthropologist V.P. Alekseyev wrote in his fundamental work, *Geography of the Human Races,*[188] that: "Studies on the spatial regularities of variations of the human organism comprise one of the main chapters of the biology of Man." For our part, we add that subjective, inborn evaluations of these regularities by all the individuals of a given racial group, is also important for the objective realization of the racial existence of humanity.

---

[186] *Stroenie tela shizofrennikh tatar Krima.* Russkiy antropologicheskiy zhurnal, vip. 3-4, 1925.

[187] *Vzaimosvyazi osnovnikh morfologicheskikh osobennostey cherepa cheloveka v protsesse antropogeneza.* Moscow, 1964.

[188] *Geografiya cheloveskikh ras.* Moscow, 1974.

9. Basics of the Biological Hierarchy of Culture

Paul Topinard indicated in this regard: "Impulses present in the cerebral substance, are so firm, despite education and civilization, that they are still preserved after cross-breeding and miscegenation and help to identify the latter...Then the question of distinguishing traits in the human races, depending on their brain organization, is significantly simplified, and it can be said at the same time, that the method of functioning of the brain renders distinguishing traits, such as the form of the skull or the characteristics of hair...It is sufficient to say that ideas of morality can form physiological differences between races. Comparing fables and allegories, which lie at the base of mythology, science dates them [back] to the knowledge of the mutual contact in which peoples found themselves, and consequently they distinguished acquired traits from their own racial characteristics...In a still larger expansion of the task, science cuts out the passing stages of intellectual borrowings, which some races made from others...Languages exist that deeply differ from one another and require a particular arrangement of the larynx for conversation in them, and special interpretation for understanding them...

It is also worth turning attention to the different ways of perceiving musical scales in the five parts of the world. That which is harmonious for the listening apparatus of the brain in some races, is not pleasant to the ears of others. Here, upbringing is nothing, since the very fact initially has an anatomical basis.

The same applies to differences in systems of calculation. Peoples that are called Aryans understand and in general differ by their abilities toward mathematics. Other lower peoples do not know how to count higher than 2, 3, or 5; after that in their opinions, the infinite follows, the unknown, and the non-understood; sometimes, despite all efforts, it is impossible to convey the concept of a large number to them...

The ability to draw also differs. There are tribes that only know how to draw circles and sticks; several of their members do not even know how to distinguish a drawing of a head from a tree or a ship...

Races differ quite deeply by way of life and understanding of social conditions. There are peoples, like the Gypsies, the Jews, and the Arabs, that seem destined for an eternal, nomadic life."

The somewhat sharp in form, but in essence valid summary by the French anthropologist is confirmed by the modern Austrian ethnologist, Konrad Lorenz, in the book, *The Other Side of the Mirror*. In it, Lorenz writes: "The *higher-lower* scale of worthiness is completely identically applied to animal species, cultures, and artworks created and produced by Man...The social conduct of Man also has instinctive content, that is not subject to change by means of cultural activity...Standards of behavior are also as reliable as physical characteristics, as signs of kinship among groups...In recent times, biochemistry has shown that the chemical codes of individually acquired information in molecular chains are impossible [to acquire] by momentary causes. At birth, the organism is given information about "biologically correct" situations, and about the means that enable him to cope with such situations...Therefore, the attributes of "lower" and "higher" are strikingly and uniformly applied to living beings, and to cultures; this justified evaluation immediately corresponds to the [instincts] in these living systems, whether to conscious or unconscious knowledge—independently of whether this knowledge was created by selection, training, or heredity, whether it is retained in the genome of the individual or in the tradition of a culture...Many types of grass are distinctly "constructed" with the expectation that they will be constantly shortened and trampled by large hooves; it happens that now the lawnmower does the same...One can consider it a firmly established scientific fact, that the species *homo sapiens* possesses highly differentiated forms of behavior, serving for the eradication of parasitic threats to the community, and acting completely analogously as a system of formations of antibodies in the cells of the State.

Also, built-in mechanisms lie at the basis of value judgments that protect humanity from threats of quite a different phenomenon, which is degeneracy. In order for some dangers to appear to threaten Mankind with the extinction of hereditary instincts, it is necessary to understand, that to the exclusion of our inborn feelings towards these values in conditions of modern civilization, not a single realized factor is selected on the basis of simple kindness and decency."

The conclusions of anthropology and etology have at last found substantial confirmation in new achievements of neurology. Sergei Vyacheslavovich Savelev's fundamental work, *The Stereoscopic Atlas of the Human Brain of Man*,[189] is a brilliant example. The author writes: "The human cerebrum possesses significant changeability. It differs in men and women, among different races, ethnic groups, and even within a family. These differences are highly persistent. They are retained from generation to generation, and can be an important characteristic of the variability of the brain of Man as a biological species. One of the most interesting indicators of the nervous system of Man is its ethnic changeability. The weight, size, and the organization of the fissures and convolutions of the brain of different races and peoples is thoroughly researched." It is namely in the hereditarily conditioned specific features of the structure of the brain, that one needs to look for the entire depth of the differences in the moral, social, economic, and political development of peoples and races.

S.V. Savelev makes an important conclusion of an anthropo-sociological character: "It is important to note the extreme significance of the brain, which does reflect on the intellectual capabilities of the studied peoples. According to various sources, and the results of repeated weighing [of brains], 900 grams is around the minimal human brain mass, which does not impinge on social behavior. In lower amounts, a full-fledged life in human society becomes impossible."

Simulation of a large-sized brain, with the help of a head of hair, by members of small-brained races. (according to Paul Topinard).

And now, respected reader, let us remember a characteristic fact of recent history: a demonstration by Aborigines of central Australia, which sabotaged the 2000 Olympic Games in Sidney. Television news agencies, delivering this news to the whole world, kept silent about the fact that the average weight of the brain mass of these "people" is around 750 grams. They simply did not have the critical minimum brain mass necessary for understanding that these were Olympic Games. Therefore, in our view, the expectation of a different reaction from them is a high form of impatience and chauvinism.

Renegadism and social anarchism are phenomena that are of a deeply anthropological character, rather than ideological.

In support of these words, one can cite a corresponding quote by Savelev: "In ancient Egypt, religious tradition demanded mummification of the deceased. An analysis of the skulls of mummies allows comparison of the volume of the brain, at the time of the burial of a definite individual. During the time of the flowering of Egyptian culture, the cranial capacity of the Egyptian was 44.5 cubic centimeters higher, than in the period of prolonged collapse. This is the answer to

---

[189] *Stereoskopicheskiy atlas mozga cheloveka.* Moscow, 1996.

the question of why a country, once a cradle of world civilization, has turned into a place for cheap tourism today.

In personal discussion with the author, the remarkable Russian scientist demonstrated laboratory plates of brain slices of different races and ethnic groups. It is not difficult to get a clear idea, that all history is reflected in the intricate pattern of the neural connections of the cerebrums of its players. S.V. Savelev asserts that the quality of connections between neurons serves as a real reflection of the degree of the intellectual and cultural development of racial and ethnic groups, which differ within ranges of 10 to 10,000 (!). Besides that the mass of the brain responsible for this or that function—for example, abstract thought or mathematical ability—may differ by up to 5 times in the members of various races (!). This means that the members of this or that "lower" race will never be able to adopt the culture of a "higher" race and understand it, because **the very process of instruction cannot increase the mass of this or that center of the brain, and it cannot create new neural connections. You cannot develop that for which there are no pre-conditions [for development].** This is the reason why all philanthropist programs for the compensatory instruction of negroes in the USA and the South African Republic constantly suffer failure. You cannot teach something to someone, who doesn't have the physical potential to learn it.

German anthropologist George Buschman considered it necessary to emphasize in his book, *The Science of Man* (1911), that: "Races that stand on a spiritually lower level possess less brain weight, than cultured peoples. If on the one hand, an increase of spiritual abilities entails an increase in brain mass, then conversely, a decrease [in brain mass], that is, a fall in its weight is observed in the disappearance of spiritual capabilities. Lower mammals do not possess the same developed centers of feeling; on the other hand, in apes it is already well-developed, and in several species it occupies almost the same surface area as the spheres of the senses. In Man, the latter only takes up 1/3 of the surface area of the brain. True intellect and thought are tied with the grey cortex of **the centers of association.** In a highly-developed intellect, this cortex is usually strongly developed, but for purely mechanical reasons does not have so much depth to length; that is, the grey cortex mass needs to widen in surface, gathering into folds as a result of the exertion of strong spiritual work. The surface area of the brain therefore shows convolutions, divided between themselves by fissures. This explains why in lower vertebrates, in which the thought process is insufficiently developed, the surface of the brain remains largely smooth, but the more we ascend to higher forms, this surface becomes more complex and rich in convolutions; however, in lower races, these convolutions are simpler, or have a straighter direction, and the fissures between them are more open and less deep. In higher races, these convolutions are usually wider, more curved, and drawn tightly together; their fissures are deeper and narrower."

Another prominent German anthropologist, Robert Wiedersheim (1848-1923), wrote in his book, *The Physique of Man from a Comparative-Anatomical Perspective*, that: "In men of the white race, the cranial capacity is 1,500 cubic centimeters on average, and the weight of the brain averages 1,400 grams As far as the cranial capacity of lower human tribes is concerned, of particular interest in this regard are observations of the Sri Lankan Veddoids, by the Brothers Sarazin. They not only have a delicate skull, but all parts of the skeleton are highly remarkable by their delicateness; that in the words of Virchow, characterizes a whole series of savage tribes that populate the eastern islands.

In the skull, this is reflected by the fact that it weighs 200 grams less than the average European. The skull is very small, and its cranial capacity in pure-blooded Veddoid men is not higher than 1,200 cubic centimeters, and in women it is 150 cubic centimeters less.

According to the volume of the Veddoid cranium, without a doubt it belongs to the lower human tribes, and this stands in full agreement with their low culture. The Andamans are close to them, whereas the Bushmen and Australoids rank somewhat higher. If, as indicated above, the average volume of the cranium of North and Central European men is taken as 1,500 cubic centimeters, then the skull of the Veddoids lags behind it by 250 cubic centimeters, and even as much as 500 cubic centimeters.

Pierre Grazioli (1815-1865), a major authority in the area of racial neurology, said that the normal brain of the Hottentot would indicate idiotism in a European. Another specialist in the said area, Friedrich Tiedemann (1781-1861), asserted that the brain of a Hottentot has few convolutions and they are more irregular, than in a European [brain], and more closely resemble a chimpanzee brain.

Summarizing the above, in the area of racial differences in the structure of the brain, one can assert with confidence, that for the development of higher cultures, a certain weight of the brain and a certain amount of morphological complexity is necessary. Both of these conditions may be met only by means of sexual selection and racial isolation. No sort of sociological project by humanists and culturologists has had any success [in this], and will not meet with success in the future.

This conclusion can in no way be categorized as racist, for one of the luminaries of Soviet anthropology—B.C. Zhukov—wrote with the same candor and spirit, in the essay, *The Origin of Man*,[190] that: "The great difference between the terse language—similar to clicks and whistles—of some black inhabitants of the forests of Central Africa, and the melodious speech of the Frenchman or Italian, is a huge abyss between the intellectual development of the one and the other. And this difference in the intellectual capabilities of the cultured man and the savage, primitive man, stands in close connection with the dissimilarity of the development in them of those organs, in which these abilities are concentrated—that is, in the brain. The member of the higher human race (in regard to intellectual development) has a heavier, more voluminous brain, than the savage, and this difference in the development of the brain consists not only in a greater quantity of brain matter in the first, but also in a more complex structure of the brain itself, and in the particulars of nourishment for the brain through blood carried by blood vessels."

Classical German anthropologist Johann Ranke was one of the first to substantiate the evolutionary significance of the quantitative characteristic: the relationship of the cerebrum to the spinal cord in members of various races. The "higher" in intellectual thought a race is, then the greater this indicator. The given rule is valid for all members of the Animal Kingdom, without exception, for the more primitive an organism, the more peripherally the nervous system in it is developed, to the detriment of the cerebrum. Therefore, it is namely individuals that are related to the so-called low-cultured tribes, who differ on the strong side, by development of purely physical and physiological qualities in comparison with Europeans; but they unavoidably lose to [Europeans] in the intensiveness and complexity of the processes present in the spiritual and intellectual spheres. And in quantitative and qualitative terms, the "animals" with a characteristically lower evolutionary nervous system than the "lower" races, develop better than the members of the "higher" [races]. The evolutionary development of intellect displaces all criterion in a functioning nervous system, the simplest of its functions being gradually replaced with evolutionary more "advanced" [functions]. An aptitude toward analysis and synthesis gradually displaces intuition.

In the fundamental monography, *Raciology and the Racial History of Mankind* (1937-1943), German raciologist Baron Egon von Eichstedt definitively formulated [the following] in his introduction: "Morphological orientation attracts psychological orientation in its wake. We can and should search for the racial differences of the brain in the macroscopic structure (absolute size of the brain as a whole and its parts, the layout of the convolutions and fissures) and in its microscopic structure (type, number, extent and connections of nerve cells). But the opportunity for research should not remain outside of our attention: the material productive strength of various sectors of the brain, their chemistry with all its movements, tenseness and weakness, with thermal and electrical phenomena and rhythms—in short, the very creative force of the brain [should be studied]."

In accord with his conclusions, Eichstedt pointed to the "definite infantile primitiveness of the Chinese brain," which correlates with an entire number of infantile characteristics, like the structure of the body, and the psyche of the Chinese. Of different racial groups, he definitively said: "The brains of the Bushmen turn out to be extremely primitive, not having anything in common with the brains of Europeans."

---

[190] *Proiskhozhdenie cheloveka.* Moscow, 1928.

Ludwik Krzywicki, the noted Polish anthropologist, gave the following conclusion in the monograph, *Anthropology* (1901), on the basis of racial differences in the structure of the nervous system: "The nature of Negroes or Bushmen shows, that the true anthropological spiritual traits are a brake on the manifestation of higher forms of technology, and the more complex social relationships that arise from that. Savages are hunters in an emotional regard, not able to live a settled existence, or better to say, a regular way of life. They would sooner die than get used to a settled culture, to which they do not feel the least bit attracted.

Consequently, racial traits bar them from a path to cultural-technical progress, and the social evolution tied with it. The causes for the differences in levels of development in civilizations are to be sought, first of all, in the racial-anthropological nature of each people. There is a certain spiritual fingerprint on social evolution; it gives some tribes more war-like, enterprising, or artistic traits, and gives others peaceful, marginally-useful, and cowardly traits."

John Randall Baker (1900-1984) was one of the prominent biologists and physical anthropologist of the 20th Century, and a professor at Oxford University; he set aside 23 evaluative factors of the independence of this or that civilization, in his fundamental monography, *Race* (1974). With the help of those factors, he arrived at **total biological determinism** as an explanation of the causes of cultural evolution. Self-standing civilizations always and everywhere were created by definite racial types, the psyche of whose cognitive styles was superimposed everywhere, as a stamp of cultural distinctiveness. In accordance with this, John R. Baker for example, rejected placing the ancient societies of the Inca and the Maya in the category of self-standing civilizations, for despite all their refinement in the areas of calendar-making, higher mathematics, and astronomy, they were nevertheless unable to invent the wheel, in the most literal sense of the word.

10. Racial Odontology

Returning to the stated them, we now turn to a very important trait of facial morphology—the teeth. Besides a distinctly anthropological significance, teeth are also unique evidence of all the nuances of the biological evolution of this or that tribe. **Racial Odontology** is a science specially dedicated to the given problem.

Left: Peter Camper

The great Dutch anthropologist, Peter Camper (1722-1789), was the first to suggest creation of a classification of races, by the angle of protrusion of the lower jaw, which is called **prognatism.** Of the measurement of this facial angle, he wrote: "If I slightly lift the facial line more forward, I get the head of an ancient human; if I fold it back, I get the head of a negro; still further back, a chimpanzee; further back, a dog; and finally, a snipe." George Buschman, whom we cite, [stated] more precisely: "The average value of the angle (angle of the profile) in the white race reaches 80-76.5 degrees; in the yellow [race it reaches] 77-68.5 degrees; in the black [race] 69-59.5 degrees. Prognatism is most rarely encountered in European races, particularly in the Nordic race; [it occurs] more often in Mongoloids, and most often in Negroes. Therefore, one absolutely cannot take into account the light degree of prognatism in the white race." Besides that many anthropologists have noted, that among the members of the different races, there are extreme differences, relatively and absolutely, in the index of the weight of the lower jaw. The "lower" the race—the heavier the weight of the lower jaw.

Also, Robert Wiedersheim, whom we have cited many times, points to the distinct racial differences in the structure of the bone of the hard palate, remarking: "The palate of the negro is a transitional [form] between the palate of the Europoid and the orangutan."

168

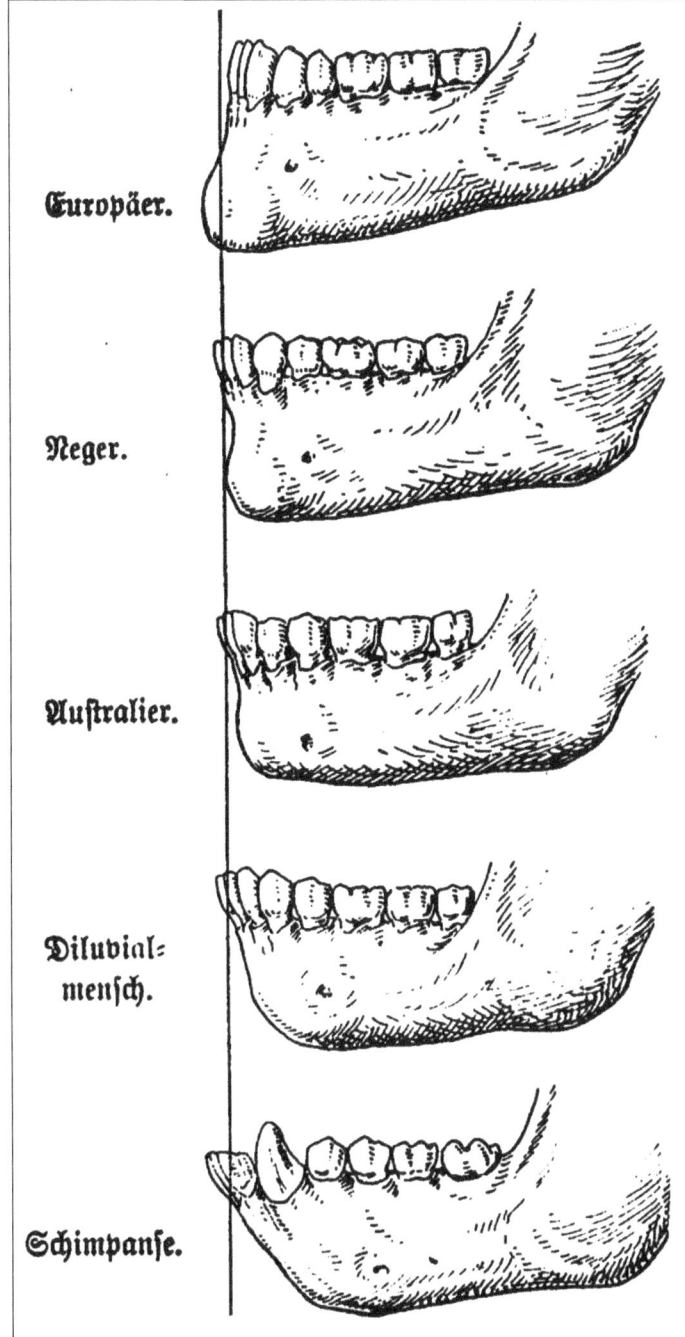

Left: Racial Differences in the Structure of the Lower Jaw (According to Herman Klaatsch). From top to bottom: Europoid, Negroid, Australoid, Neanderthal, Chimpanzee.

For his part, prominent Soviet scientist B.S. Zhukov emphasized: "The form of the skeletal bone forming the upper palate, that is, the so-called hard palate, stands in close dependence on the number and distribution of the teeth, which are dissimilar in the higher apes and in the members of the human races. Among modern Europeans, and in several other peoples, the hard palate has rounded characteristics, and the whole row of teeth in the upper jaw, if you look at it from inside of the cavity of the mouth, reminds one of a horseshoe by its traits. However, in Negroes the hard palate is somewhat more elongated in length, and the dental row, by its form, recalls an arch. Contrarily in members of the lower races, like the Australoids for example, the "wisdom teeth" reach greater development than the rest of the molars, and in tooth structure, this makes the Australoids resemble higher apes somewhat."

In his work, *The Facial Skeleton, and Factors Determining Variations in its Structure,*[191] prominent Soviet anthropologist V.V. Bunak calculated that the surface of the palate bone in European skulls amounted to 1,670 square millimeters on average, while in Papuans it is 1,990 square millimeters, and in Mongoloids, it is 2,020 square millimeters.

G.I. Vil'ga wrote in the article, *Teeth in an Anthropological Regard,*[192] that: "One of the organs of the human body occupying a prominent place in the formation of a type, is the teeth, which in their structure represent not only racial, but significant individual variation." Summarizing a rich, historical literature, the author of the article begins his analysis from a subdivision of races, according to a reciprocal arrangement of upper and lower incisors, into straight-jawed and [prognathous] protruded-jaw. The given discovery is credited to the famous French anthropologist, Etienne Geoffroy de Sainte Hillaire (1772-1844).

---

[191] *Litsevoy skelet i faktori, opredelyayushchie variatsii ego stroeniya.* Moscow, 1960.

[192] *Zubi v antropologicheskom otnoshenii.* Russkiy antropologicheskiy zhurnal, N2, 1903.

In this regard, G.I. Vil'ga pointed out: "The white race is straight-jawed; prognatism is encountered in colored races: the black and the yellow. It is expressed to a stronger degree in Bushmen." Karl Vogt also remarked: "The degree of development of the mandibles is found to be in direct relation with the culture of peoples, and with their aptitude towards culture: slanted teeth are only encountered in the lower races of the human species." Paul Broca and Paul Mantagazza discovered that racial differences according to the size of the teeth have great significance.

Vil'ga continued: "Large teeth in civilized races gradually become smaller in their volume, displaying an inclination toward disappearance; in races with low culture, they are very developed. Besides that the size of the molar teeth [in higher races] decreases from front to back; in lower races, the Australoids and New Caledonians for example, and in apes always, they increase in size; this is called an **ape-like trait.**"

The given trait was discovered through the efforts of Richard Owen (1804-1892) and Franz Pruner-Bey (1808-1882). William Henry Flower (1831-1899) was the first for an evaluation of quantitative differences between races; he suggested calculating a so-called **dental index**. Thus, for Europoids, it amounts to 41; for Mongoloids, 42; for Negroes 44; for Australoids 46; for chimpanzees 48; for gorillas 54; and for orangutans the dental index is 55. As one can see, there is no distinct boundary between Man and the [primates], but between races there is.

Thomas Scott Lambert (1819-1897), on the basis of his own studies, came to the conclusion that on the whole, all dental systems depict racial differences. On the basis of the morphologically outstanding traits of the teeth, he set apart three large human groups: the white, the yellow, and the black. He found sharper differences in the black race. Its incisors are larger, than in the yellow or white races. The canines project more over the level of the neighboring teeth, than in the white race.

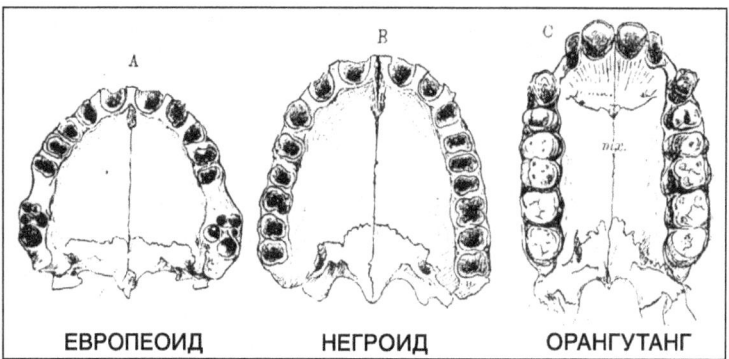

ЕВРОПЕОИД   НЕГРОИД   ОРАНГУТАНГ

Above: Differences in the Structure of the Palate Bone (according to Robert Wiedersheim). From left to right: Europoid, Negroid, Orangutang.

Continuing this thought, Vil'ga wrote: "The incisors of Man are sharper, than in the lower human races. The relative width of the crowns of the large molar teeth in lower races is greater, than it is in the higher [races]." In civilized peoples, the teeth on the right side are more dense and strong, than the teeth on the left; the result of this is that in [civilized peoples] the right side takes a greater part in the act of chewing. This difference is not observed in savage peoples." The author based these conclusions on his own observations and the works of such authors as Julius Emile Joseph Renier (1873-?), Emile Magitot (1833-1897), and Victor Galippe (1848-1922). For their part, Johann Friedrich von Brandt (1802-1879) and Karl Rose (1864- ?) noted that the inclination towards illnesses is not identical among the different races. It was discovered that with the development of culture, the size and strength of the teeth and jaws decreases.

The article *Teeth among the Different Human Races,*[193] by distinguished Russian anthropologist A. Ivanovskiy, is highly informative and illustrative. It contains natural scientific facts in it, that are still relevant today. The scientist testified: "Besides differences in mutual distribution, the teeth of Man depict racial differences by their size. These differences, particularly

---

[193] *Zubi u razlichnikh chelovecheskikh ras*. Russkiy antropologicheskiy zhurnal, 1901, N3.

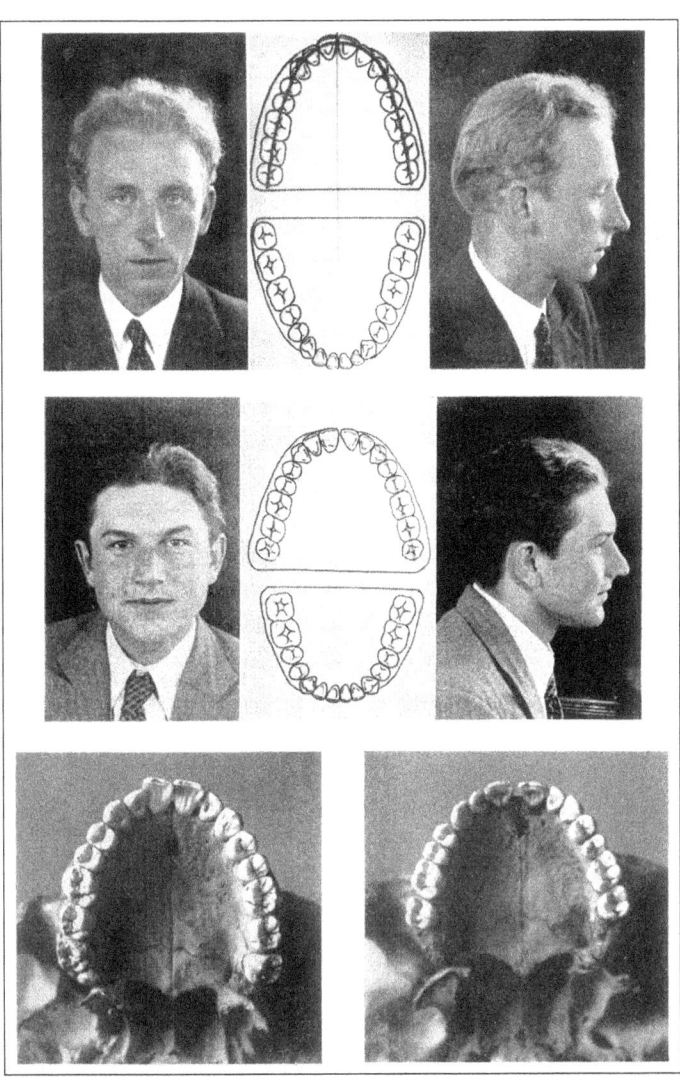

in the incisors and molars, are sometimes very significant. Lower races are characterized by enormous size of the incisors and molars; the latter in them are the equivalent of canines, like those we find in the man-like apes. At the same time that the size of the molar teeth in a man of the white race decrease from front to rear, in lower races (Australoids, New Caledonians) and in apes, it goes in the reverse direction. The teeth of Man, as equally as the jaws, generally decrease with the development of culture, since cultural methods of food preparation now make that energy use by the teeth and jaws, which was necessary for primitive, coarse food—superfluous. The jaws decrease in size in greater measure, than the teeth; thanks to this, the jaws have a longer period of development, than the teeth.

The various races differ between themselves by the form of their teeth. The incisors of Man show the distinctive feature, that their tips are sharper than in a lower race. This points to the fact that in uncultured peoples there is not a difference between the teeth on the two sides of the mandibles, which is observed in civilized peoples, and which is expressed by the teeth on the right side being more dense, than those on the left. Since the majority of Europeans work food with the teeth of the mouth cavity, they chew more, and sometimes exclusively, on the right side. The upper incisors of Malaysians, besides their prognathous condition, are also apish in form, with a bulging front and a slightly concave, rearward surface. This is undoubtedly a simian trait that is constantly encountered in apes."

Specialists in this area of race identification have not turned their attention to an important question like the different tribes who ritually deform teeth in order to satisfy their anthropo-aesthetic canons. To this day, the pulling out of teeth, filing, breaking off of tooth fragments, and the insertion of foreign objects, is widespread among the members of the "lower" races. In Africa and Polynesia, the disfiguring of teeth is extremely widespread. In Senegal for example, a bride, together with the groom, goes to the blacksmith, who files her teeth; upon this, the wedding is concluded. And in Australia they extract the incisors upon sexual maturity, or upon entering into marriage. Many tribes practice the custom of extracting teeth when mourning for deceased relatives; and after epidemics, droughts, and famines, the surviving inhabitants turn out to be almost half-toothless. Also widespread is the practice of inserting gold or precious stones into teeth, with the object of demonstrating the height of one's tribal social status. And there is also the practice of ritually coloring the teeth black, by the Indians of Central America, and coloring them red, by the inhabitants of Burma. It is known that at the high point of colonial politics, many English firms literally became rich from preparing false black teeth for the inhabitants of Senegal, and light-blue teeth for the Chinese, as well as simple devices, with the object of indulging the native leaders of the most remote regions of the boundless British Empire. The English merchants easily recognized the entire prospect of the accessories market, serving to satisfaction the exotic anthropo-aesthetic tastes of the new subjects of Her Britannic Majesty.

Russian scholar Aleksandr Aleksandrovich Zubov, a prominent modern specialist in the area of odontology, summarizing a world experience of studies, certifies that according to several parameters of the structure of the dental system, the differences between the major races are in tens, and sometimes even hundreds of percentages. In the book, *Odontology in Modern Anthropology*,[194] Zubov and N.I. Khaldeyeva state: "An odontological type is a basic classification unit in odontology, the formation of which—according to key traits of dental morphology—coincides with phenomena of the human races. The odontological type unifies specific structural complexes, characteristic for a group or several groups, and consists of many morphological particularities, including race. Variants that are included in an odontological type have a hierarchy.

According to data collected from the territory populated by Russians, it is established that the basic odontological type of the Russians is a well-expressed, satisfactorily pure Western complex. The base of the mass of Russian groups stands out for homogeneity in the distribution of the odontological traits of the Central European type. As a whole, the summary odontological characteristic of the Russians is placed within the framework of the range of indicators of the Central European odontological type."

In a later joint work from an anthology characteristically titled, *Race and Racism. Past and Present*,[195] Zubov and Khaldeyeva state that the differences in the structure of the dental system allow one to speak of a separation of the races, around one million years ago. Zubov estimated nine independent odontological traits that set apart racial differences.

In his article, *Several Findings of Odontology on the Problem of the Evolution of Man and his Races*,[196] Zubov delineated the following important classification traits:

The **Mongoloid Race:** relatively short roots with a high crown. Paliformed incisors.
The **Europoid Race:** small absolute size of the teeth.
The **Negroid Race:** large absolute size of the teeth.

Besides that A.A. Zubov, in the spirit of the postulates of classical racial theory, allowed himself to talk about "primitive" and "Pithecoid" traits in the structure of the morphology of the dental system of some races, and "progressive" [traits] in others. The distribution of "lower" and "higher" types among the basic races is quite easy to surmise.

In the book, *Peoples, Races, and Cultures*,[197] N.N. Cheboksarov and I.A. Cheboksarova emphasized that paliformed incisors are encountered in Mongoloids at a rate of 60-100%, whereas in Europeans and African negroes, their share falls to lower than 15%. Papuans, Australoids, Andamans, and the Veddoids of eastern India, are closer to Mongoloids, by the frequency of paliformed incisors (30-60%).

Another trait of the teeth, like the "Karibelli nodule" (an additional nodule in the upper molars) for example, is very characteristic for Europoids (40-70%), and is encountered in Mongoloids far more rarely.

[194] *Odontologiya v sovremennoy antropologii*. Moscow, 1980.

[195] *Rasi i rasizm. Istoriya i sovremennost'*. Moscow, 1991.

[196] *Nekotorie Dannie odontologii k probleme evolyutsii cheloveka i ego ras*. Problemi evolyutsii cheloveka i ego ras. (Problems of theEvolutoin of Man and his Races). Moscow, 1968.

[197] *Narodi, Rasi, Kulturi*. Moscow, 1985.

11. Pigmentation of the Skin and the "Mongolian" Birthmark

Skin color as a racial trait was set down as a basis of classification by the first natural scientists of antiquity. This condition of affairs did not change, and with the beginning of the Age of Exploration, the hereditary ability of skin to make color was recognized as a decisive factor in the classification of tribes of the human species, by such scientists as Francois Bernier (1684); Leibnitz (1728); Bradley (1739); Linneaus (1735); F. Mueller (1773); Immanuel Kant (1775); and Hunter (1775). In 1793, von Breitenbuch was the first to compose a "Universal Chart of the Color of Peoples."

Jeanne-Joesph Virey (1801), Georges Cuvier (1817), Charles Pickering (1848), Joseph Arthur de Gobineau (1853), and d'Omalius (1869), simply divided humanity into black, white, and yellow races. In the studies of Thomas Henry Huxley, Karl Stratz, and I.E. Deniker, the color of skin was again advanced to the first rank of diagnostic traits, by which examination they are subjected to tones and hues.

It is worth noting that the ancient Chinese philosophers divided people into white and red, the Chinese considering themselves white, and the red as "foreign devils"—the Europeans, that is. And indeed, European skin, even in its lightest variants, can never be considered white: it has a dull, rosy tone, and even in corpses it is yellow. The skin of Mongoloids, even the typical Tungusi and northern Chinese, is not yellow; it is strongly mixed with chestnut and reddish tones. The skin of Negroes is not black, but dark-brown, reddish-brown, or gray-black.

It is necessary to turn attention particularly to the genitalia, which at birth are already more strongly pigmented in Negroes and their hybrids. This serves as an excellent diagnostic marker in the case of miscegenation of skin colors. The second graphic trait is the yellow pigmentation of the skin under the fingernails, and particularly under the thumb and toe nails.

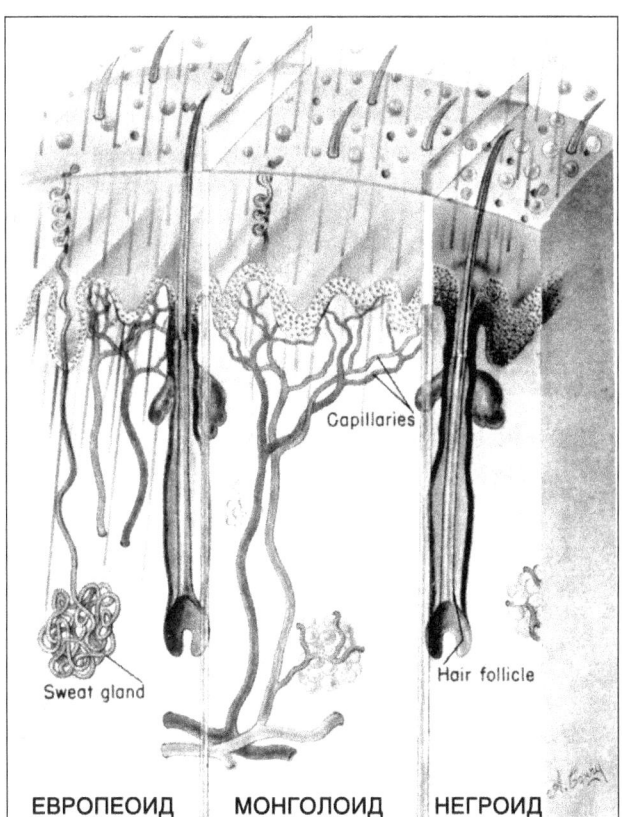

ЕВРОПЕОИД   МОНГОЛОИД   НЕГРОИД

The given trait was a favorite means of slave traders and plantation owners for identifying half-breeds. The strongest pigmentation in dark-skinned individuals is reached in the third and fourth decades of life. In completely black Sudanese Negroes and southern Bantus, the pigment reaches to the uppermost cells of the epidermis, and in Europeans melanin is distributed in the depths of the skin covering.

Left: Racial Differences in the Structure of the Skin Covering (according to Carlton S.Coon). From left to right: Europoid, Mongoloid, Negroid.

But in order to perceive all the hues of the skin, it is necessary to resort to the help of similar statistical methods, and in every way to avoid poetic comparison, such as: blood with milk (Nordic girls), coffee with cream (Ethiopians), old elephant ivory (Vietnamese), red wood (Indians), chocolate (Australoids and Negroes), and so on.

In Russian works, K.A. Bari's *On the Color of the Skin of Man*,[198] and V.G. Stefko's *On the Morphology of Skin Pigment in Man*,[199] also emphasize the importance of skin color in the process of anthropo-aesthetic auto-identification. It

---

[198] *O tsvete kozhi cheloveka.* Russkiy antropologicheskiy zhurnal, N1, 1912.

173

is found to be an essential reflection in the canons of popular graphic art, songs, stories, rituals, and so on.

The famous Russian anthropologist, Eduard Yuliyevich Petri (1854-1899) gave a digression on the history of the question, in his monography, *Anthropology*:[200] "The color of skin has long been a popular symptom for classification. Actually, the color of skin is a trait, and one could say an elementary trait. Not for nothing, the word **varna** simultaneously signified color and caste among the ancient Hindus. Not for nothing in the present time is the slightest suspicion of dark skin color a proof of lower origin, and a proof of disfranchisement for [racially] mixed Americans. Accepting Indians into their midst, American authorities would say, "Let him think he is white."

Today **coloredness** in different races is determined according to Felix von Luschan's Table; the measurements are taken from the bend in the forearm.

The lightest tones in Europoids are encountered in the zone of the blonde races, where the Nordic race, with its rosy, transluscent skin, properly stands in first place. In Eastern Europeans, the skin has a pale, gray hue. Among the peoples of Siberia, the skin has a yellow-brownish hue, but still retains some of its rosiness. Such a portrayal is observed right up to the zone of distribution of the Eskimos, Yakuts, and Ainu.

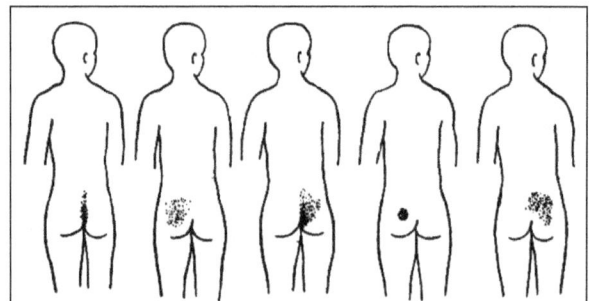

"The Mongolian Birth Mark" (according to Egon von Eichstedt).

In the Europoid group, the belt of the Alpine and Dinaric mountain races present a transitional zone, toward a zone of dark, swarthy races. The Mediterraneans stand in first place here, with light, but already noticeable swarthiness, strong, dull red on the cheeks, and a common predominance of olive tones. If on the northern Mediterranean Coast we have indicators of 10-12 on Luschan's Scale, then on the southern shore we have 12-15. Further to the east, the olive play of colors disappears, but swarthiness increases from 12-15 in the Oriental race and in Turanians. In Armenians, the index is equal to 13. In India we have a greater variety of hues, from a light, almost European light-brown in the Kashmiris and upper castes of the north (14-15), to the typical north Indian of the Punjab (17) and India Proper (21-23). Further to the east, the belt of swarthy races extends from retaining the yellowish hues of the paleo-Mongoloids (14-16), to the swarthy Polynesians with their light-brown skin (16-17).

In the Veddoids, a dark brown tone (24) predominates; the Khmers and Thai are lighter, with reddish hues (14-18).

Among the Asiatic Mongoloids, a light—often light yellow—wheat tone is encountered among northern Chinese (7-10), and further to the south the swarthy index increases (7-15), but the southernmost Vietnamese, who belong to the southern Chinese type, differ from the dark, eastern Veddoids of Cambodia; thanks to their golden hues, they look lighter. The yellow hue enables the Lao (14) to stand out from the Thai (15); and the Malaysians are set apart by a light-brown background (18).

The American Indians of the Pacific Coast and forest zones have light-brown skin, with very weak yellow hues, but with a play of honey colors. The Indians of Central America appear to have a stronger again yellow hue. The Indians of this place belong to races whose covered body is darker than the face.

[199] *K morfologii kozhnogo pigmenta u cheloveka.* Russkiy antropologicheskiy zhurnal, vip. 3-4, 1925.
[200] *Antropologiya.* Sankt-Peterburg, 1890.

Among African negroids, the Sudanese have very dark skin (30-33); the southern Bantus differ from them by very little. But the maximum saturation of pigment (33-35) is encountered among the Nile River Valley inhabitants, and in several groups of the Saharan transitional belt. Mixed groups of the lower social strata of the Tuaregs, like the Ethiopians, have [skin color] indicators of 28-30. The Pygmies have reddish-brown skin (23), and the Kaysan group, with its light skin color (4-5) and ashen hues, completely departs from typical African coloring. The Tamils of southern India have dark skin, like that of Negroes. Here, like in the Saharan transitional zone, dark skin combines with European facial characteristics. The Negroid dwarfs of the Andaman Islands have very dark skin (34-35). The Melanesians are notably lighter (27-30), but their skin always has reddish hues. Lighter still are the Negroid dwarfs of Malacca and Luzon (at 25-26 on Luschan's Scale). Finally, velvety dark-brown skin (26-27) predominates in the Australoids.

Left: Erwin Beltz

Anthropologists Erwin Beltz (1849-1913), Otto Finch (1839-1917), and Buntaro Adakhi (1865-1945), were the first to turn attention to the formation of blue spots on the sacrums of the new-born infants of the colored races. By virtue of the frequency of incidences in the yellow race, the blue spots received the name "Mongolian birthmark." Their manifestation in other colored races is observed more rarely, and in white Europoids, they are generally not encountered. In the book, *Man as a Productive Force*,[201] the famous Russian anthropologist also wrote: "Obstreticians, surgeon's assistants, and doctors have easily observed newborns and noted in them the frequency of a phenomenon, which is mainly on the sacrum: distinctive blue-black blemishes. This trait is considered characteristic for newborns of the Mongolian race. Similar "Mongolian birthmarks" do not remain for life, but soon disappear." The given phenomonen does not have any adaptive or physiological significance, but on the other hand may without a doubt shed some light on the secrets of the evolution of the white race, for it is one of the pathological traits in the embryonic development of its members.

Beltz asserted: "The presence of *Mongolian birthmarks* is a very narrow marker for the differentiation of the white race from other races." And really, the presence of dark blue birth marks is observed not only in newborns of the Mongolian race (Japanese, Chinese, Koreans, Siamese) but also in members of the other "colored" races, like the Ayni, the Indonesians, Eskimos, and Indians. For their part, the presence of these marks on the sacrums of Europoids is a highly precise testimony of the presence of this or that racial admixture. The color of the "Mongolian birthmark" is most often dark-blue or blue-black, and in Indians it is green-black.

Baron Egon von Eichstedt also wrote in this regard: "The Mongolian birthmark is also called the coccyx or sacral blemish. It is a streak of a dark blue color accumulation of pigment, in the region of the sacrum. Its density and form can be quite varied; it can be large or it can be coin or saucer-sized, and have cog-like extensions, which end partially on the spine, partially on the rear. The color varies from light-gray to dark-blue. It undoubtedly has a hereditary nature. In Ecuador and Brazil, they perceive such a circumstance as unpleasant, if the child of a Brazilian mother of European origin has, as they say there, "a medal on the butt." If an adult Ecuadorian woman is discovered to be painstakingly hiding the secret, they say she has a "green ass." On the other hand, among Eskimos and Polynesians, the sacral blemish is considered a sign of pure breeding." The given Mongolian birthmark is present in the majority of apes. From this fact Eichstedt made a valid conclusion about an atavistic, simian origin for this trait, in the morphology of different colored races.

In 1901, English scientist Sir Herbert Reesely, by order of the colonial authorities, conducted a racial examination of the population of India. He established that in numerous regions of this country, 80-90% of newborns have the given trait. The worship of apes among the lower castes of India thus finds an extremely simple, naturalist explanation.

---

[201] *Chelovek, kak proizvoditel'naya sila*. Leningrad, 1925.

Left: Buntaro Adakhi

The author's given conclusion can in no way be categorized as exotic or extremist, since the prominent German anthropologist, Eugen Fischer, emphasized in his book, *Anthropology* (1923), that: "Pigment most strongly appears in apes on the back, in the region of the sacrum, so that this part of the body becomes dark blue, and in people it manifests as symmetrically distributed color spots. Such blue spots on the lower part of the back are [found] in all Mongoloids. It appears shortly before birth, and remains for 5-10 years. Japanese children of one year of age have them in 99.5% of cases; up to 5 years of age—in 62% of cases; and up to 10 years of age—in 6% of cases. There are such birthmarks on Eskimos, Indians, and more rarely on Malaysians, Samoans, and mulattoes of South America; it is sporadically encountered in European children (in 0.6% of cases in Bulgaria) and always and whenever there is Mongoloid admixture (in Hungary, Moravia, and so on). In general, skin color is inherited as a racial trait."

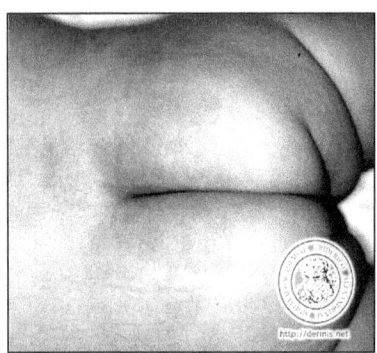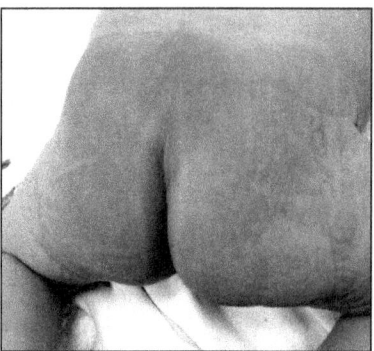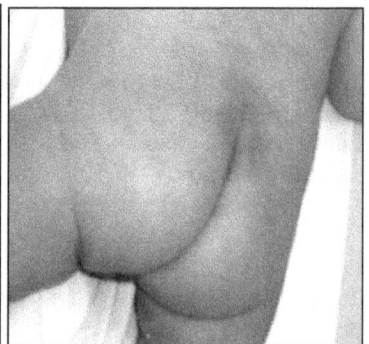

Photos: The "Mongolian Birthmark"

Members of Cesare Lombroso's school of criminal anthropology very plainly pointed to the racial-evolutionary aspects of the origin of the given anomaly. In part, N. Pende wrote in the book, *Deficiency of the Constitution*,[202] that: "Concerning degenerative stigmata and pigmentation, in regard to Mongolian birthmarks, they are almost normal for the yellow race; in Europeans they indicate cross-breeding with Mongolian blood, and are observed mainly in Mongoloid idiots."

---

[202] *Nedostatochnost' konstitutsii.* (Moscow-Leningrad, 1930)

Modern studies in the area of biochemistry also testify about the hereditary origin of the given factor of race formation. American anthropologist M.F. Ashley Montagu correlated the given anthropological observations with genetic blood groups, in his fundamental textbook, *An Introduction to Physical Anthropology* (1951). He indicated: "Almost all Bushmen and Hottentot infants are born with the so-called Mongolian birthmark."

12. Hair as a Racial Trait

The pigmentation and texture of hair is found to be closely linked with the pigmentation of the skin. In 1864, Franz Pruner-Bey created the first extensive classification of races on the basis of these traits. He divided humanity into two groups: wooly-haired and smooth-haired. The first group was comprised to two sub-groups: tufted, wispy hair and fleecy hair. The Hottentots and Papuans belong to the tufted-hair group, and Kaffirs and Negroes belong to the fleecy-haired group. The smooth-haired [group] is sub-divided into straight-haired and curly-haired [types]. Australoids, Malaysians, and the Mongols of northern Eurasia belong to the first group, while Europoids, Nubians, and Dravidians belong to the second group. Besides that he also indicated the importance of such traits as the cross-section of hair, its structure, and the depth of the position of the pigment, which gives hair its color.

In his article, *Hair in an Anthropological Respect*,[203] P.A. Minakov noted with respect to this: "The study of the cross-section of hair deserves the particular attention of anthropologists. For each race the form of the cross-section always has significantly predominant characteristics."

It is not the color, but the structure of hair which points to the racial origin of an individual. George Buschan noted in this regard: "In mixing Negroes with other races, the curly form of their hair is retained on the covered parts of the body." From ancient esoteric tractates, it is known that during acceptance into most secret societies, neophytes were subjected to illumination of the sexual organs. Similar methods for revealing undesirable racial admixtures were worked out by the ideological departments of the Ku Klux Klan in the USA, and the racial-political department of the Third Reich's NSDAP. Therefore, since ancient times, hair was considered an important criterion in racial typology.

In 1825, Bory de Saint-Vincent was the first to introduce "smooth" and "curly" hair for use in determining race, in the interests of graphic classification. In 1860, Etienne Geoffroy de Saint-Hillaire gave a more detailed description of races according to hair, and in 1873 Friedrich Mueller widened and perfected the scheme of classification. Besides that in 1863 Franz Pruner-Bey established that various cross-sections of hair exist among the different races, and that they correlate with form.

Paul Topinard, the distinguished French anthropologist, wrote in his monography, *Anthropology* (1879): "Fleecy-formed hair is thin or hard, and comes in different types. They are long and hang down in the form of a fringe, as among the Tasmanians; or they are long and go to all sides, forming a round mass, which sticks out to 30 centimeters from each side; it is encountered in the Papuans and Kaffirs; or they are very short, in the shape of small rolls, giving the hair the form of peppercorns, like among the Hottentots. The placement of the hair [in the scalp] also produces several of these differences. Normally, hair is placed at an incline. In the Hottentots and several other negro [types], it is placed perpendicular to the skin. Usually, hair is also distributed uniformly on the surface of the head, or irregularly, or by certain straight or curved lines; on the Hottentots and Papuans it grows in short patches, separated by bare spaces. In such cases, when the hair is shortly trimmed, it gives the head the look of a brush with its patches of bristles."

Finally, in 1924 prominent English anthropologist Alfred Heddon again made hair a high principle in classification of the races. German scientist Baron von Eichstedt, in confirmation of the importance of this trait, pointed out: "Mongoloids stand in first place for the length of the hair; after them come the Europoids, and then after a large gap come the Negroids. Haircuts have no influence in this. In all Mongoloid groups, hair can easily reach an extremely long length, and in the men, too. This is especially characteristic for Paleo-Mongoloid groups, and also for Indians. In

---

[203] *Volosi v antropologicheskom otnoshenii*. Russkiy antropologicheskiy zhurnal, N1, 1900.

second place come the Europoids. The curly of African negroids is rarely longer than 15 cm; therefore, both sexes usually cut theirs shortly. Among Melanesians, hair is not curly in the narrow sense of the word: among them it swirls and is twice as long, but close in form to the hair of the Ethiopian race. According to the thickness of hair, races are distributed in the same sequence. The hair of Mongoloids has particularly sturdy roots. Naturally, the races are also distributed in the same sequence, according to the weight of the hair.

Neither length nor thickness, but general form prompted several authors to turn attention to hair as a criterion of race affiliation. Modern classification of the forms of hair are as follows: **1) smooth-haired**, with the subgroups: a) straight; b) smooth; and c) slightly wavy hair; **2) wavy hair**, with subgroups: a) widely wavy hair; b) narrowly wavy hair; and c) curly; **3) curly**, with subgroups: a) twisting; b) locks; c) densely twisting; d) peppercorn; and e) spiral."

However, it is necessary to note that Aristotle and Galen knew that the hair of Negroes becomes curly, because the pores from which they grow are curved.

The density of the sprouting hair also has great significance; in Europoids, it grows in groups of 2-4; in redheads in groups of 5; in Negroes in groups of 2-3, and at an angle of 20-70 degrees; and in Mongoloids at an angle of 90 degrees, always fairly regularly. Among smooth-haired races, hair joins in strands; in curly-haired peoples it joins in bushes or in the form of pepper grains, and only in Melanesians does hair grow in various directions. In real curliness, one typical element of style appears, which Pruner-Bey was the first to draw attention to: namely, the cross-section of curly hair has the shape of a bean or an ellipse. These differences between the long and short diameters were used for substantiating such a quantitative characteristic as **Frederick's Cross-Section Index.** Among curly-haired Negroids, the index is 60; among wavy-haired Europoids, it is around 75, and in straight-haired Mongoloids it is around 90. In mixed-race groups, this size occupies intermediate significance, in comparison with races of pure origin. Besides that the concentration and distribution of pigment spots has huge importance in the racial characteristics of hair.

The measurement of **hair-coloredness**, with the objective of racial identification, is realized today using Eugen Fischer's scale. As a great expert on the question being considered, he wrote in his fundamental monograph, *Anthropology* 1923, that: "Hair has particular significance for anthropology in many respects: on the one hand, there is interest in the hair covering as a whole, that sets man apart from primates, and on the other hand, in the form and color of hair, as distinctive racial traits. From the end of the 5th month, the fetus is covered by an embryonic fluff—the so-called "down." Toward the moment of birth, this fluff in large measure disappears. It is preserved in the children of Australoids and African pygmies. A secondary hair covering appears on separate sectors of the body in the period of puberty. It develops particularly strongly in Europeans, Australoids, and the Aynu. On the opposite side of the pole are Indians, among whom there are no sexual differences in hair cover. The thickness of hair is also different among the races. Races differ by the natural length of the hair on the head. In Bushmen and Hottentots, the length is usually equal to 10-15 centimeters; among Negroes, it is somewhat longer, but not as long as it is with Europoids; but then among the Indians of North America it is not uncommon for hair to reach lengths of 0.75-1.0 meters. Full-fledged beards are only characteristic of a few races. In Veddoids and others, a weak beard only grows under the chin.

Straight hair that retains its straightness for its entire length is most characteristic in the Mongol race. Wavy hair, which forms wavy curves, is encountered particularly often in Europeans. Finally, curly hair that forms tight spiral bends is typical for black-skinned types; in the majority of cases among them, the hair forms tufts, and between such tufts thare are spaces that are deprived of hair. Particularly important, in the sense of the differences in the hair of different races, is the direction and distribution of hair on the surface of the skin on the head.

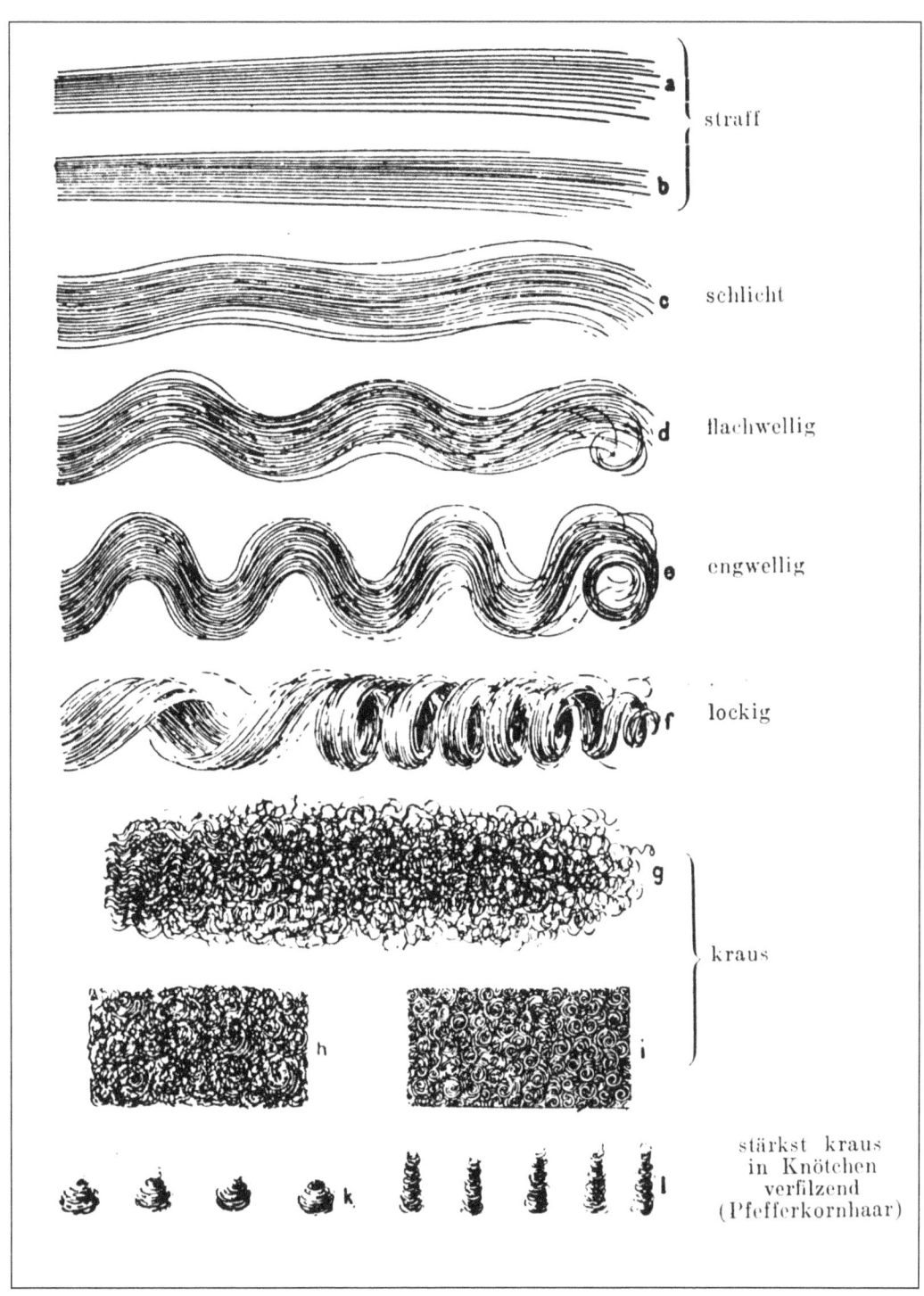

Differences in the Structure of the Hair (according to Rudolf Martin): a-b: straight; **c:** broad and wavy; **d:** narrow and wavy; **e:** curly; **f:** spiral; **g, h, i:** frizzy, fuzzy; **k, l:** screw spiral.

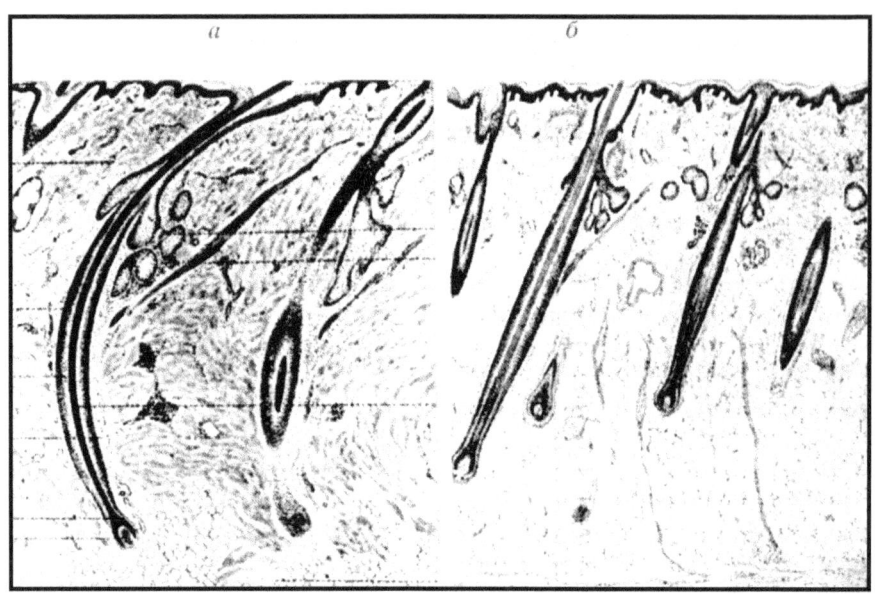

Position of hair follicle shaft on a) Negroes, and b) Europeans (according to V.V. Bunak).

In races with straight and wavy hair, it is established that they have regular distribution according to definite lines of direction, and besides such, the hair is directed toward the crown; in kinky-haired races, the opposite is observed; there is an absence of regularity and a pattern of direction in the hair. Races also differ in the angle of the incline of the follicle bulb, from which hair grows. Races with straight or wavy hair have a strong incline of the hair shaft; for kinky-haired races, it is characterized by a scimitar-shaped curve of the follicle shaft.

Modern German scientist Reiner Flindt indicates in his book, *Biology in Numbers*, that according to recent data, the number of hairs on the heads of blondes is 150,000; on brunettes 100,000; and on readheads the number is 80,000.

Concerning the hairstyles propagandized by the adepts of modern youth culture, which amount to an affiliation with native, tribal "lower" races, and are displayed by models, this cannot be regarded as anything but a sacrilegious and unnatural perversion. For modern young men and girls of the Europoid racial type, under the influence of the ruling "caprices" of fashion, are compelled to pile up a structure on their heads that does not correspond to the structure of European hair, nor the form of the European skull, and also does not answer to European canons of beauty. **Non-European trappings on the head of a European individual are a first sign of degeneracy and the defilement of one's own race.**

13. Racial Somatology

A combination of repetitive analyses of numerous racial diagnostic traits, done according to different measurement scales, raises the accuracy of research. In the process of anthropo-aesthetic perception, each subject is evaluated, and the outer characteristics of the structure of the body of the individual are examined: the proportions of the figure, the grace of movement, its poses. This whole complex of problems is the subject of study of a science called **racial somatology.** Its racial aspects were most fully developed through the work of such scientists as: Joseph Samuel Hepburn (1885-?); William Thompson (1833-1918); Herman Klaatsch (1863-1916); Leon Pierre Manouvrier (1850-1927); Ernst Kretschmar (1888-1964); and Egon von Eichstedt (1892-1965).

Left: Ernst Kretschmar

An article by Russian scientist K.A. Bari, *Variations in the Skeleton of Modern Mankind and their Significance for Deciding the Question on the Origins of the Human Race*,[204] was also indicative in this regard. The author wrote: "The hope, that in the skeletal trunk of several races, one can notice lower traits, turns out to be well-founded. Also, an increase in the number of ribs corresponds to an earlier degree of development, while fewer ribs, and also the number of separate lumbar vertebrae [corresponds to] a later origin." This conclusion was founded on the descriptions of skeletons of various tribes from the ranks of the "lower" races, on the skeletons of which the number of ribs reached 15(!). A difference in the number of vertebrae is also observed, the structure of the collarbone, the shoulder-blade, a noticeable deviation in the curve of the tibia bones, and also an established increase in the number of incisors in the jaws of several wild tribes. Racial differences in the shoulders were known long ago. It is worth mentioning that although [there are] different positions of the neck and shoulders, which in Australoids and negro races is formed further back, than in Europeans. In Europeans, the axis of the shoulders forms an open, outward sharp angle with the axis of the elbow joint."

There are also differences in the proportions between the upper and lower extremities, and in the structure of the hands and forearms. "This applies to the predominance of long, lower extremities over the upper extremities in European races. From this point of view, the significantly long arms in the Australoid, Veddoid, and Negroid races is a primitive stage. In Europeans, this primitive stage is only recalled in newborns"—Bari summarized.

Left: Leon Pierre Manouvrier

Concerning statistics in the area of the racial variation of the number of ribs and constructive traits in the structure of the skeleton, distinguished Russian anthropologist Octavian Vasiliyevich Milchevskiy focused attention on the following eloquent facts, in his monography, *Foundations of the Science of Anthropology*:[205] "For the most part, in negroes there are seven pairs of true ribs, but far more often than in Europeans, eight ribs are encountered, just as there are in apes. Among Bushmen and the extinct "guankhov" there is a similarity in the front and rear apertures, like in the Sumatran orangutan, and many carnivores, like dogs, for example."

Karl Vogt—a famous expert of the human body—expressed similar views in this regard: "In negroes the trunk is smaller, relative to the extremities, in particular the arms, which in a perpendicular position to the body, always reaches lower than the middle of the hips. The majority of negroes, without bending or leaning, can easily scratch the

[204] *Variatsii v skelete sovremennogo chelovechestva i ikh znachenie dlya resheniya voprosa o proiskhozhdenii i obrazovanii ras.* Russkiy antropologicheskiy zhurnal, N1, 1903.
[205] *Osnovaniya nauki antropologii.* Moscow, 1868.

skin under the kneecap. In the formation of the back of the neck, negroes have several noticeable similarities with the gorilla."

And George Buschan emphasized: "The pelvis of the European is absolutely and relatively wider, than in non-European races; this also shows distinctly in the downwards narrowing. Its incline is greater, than in non-Europeans. Besidest that the lower races possess a longer forearm, than higher races; in this respect they come closer to the anthropoid type."

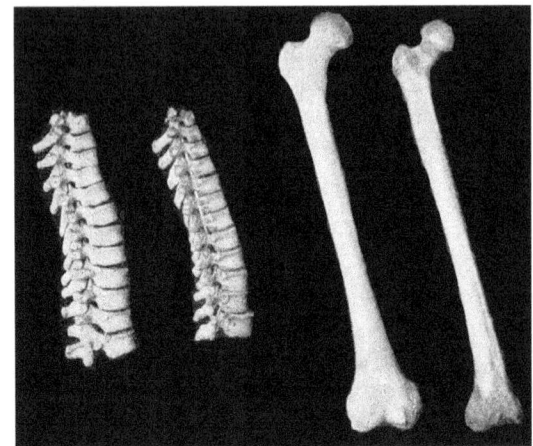

Left: Comparison of the thoracic section of the vertebraeand the femur of a European and an Australian of approximately the same age.

Cited by us many times, Eugen Fischer emphasized that for determining racial differences in the proportions of the body with the eye, it is best to examine the individual from the rear, since in evaluating his appearance from the front, we will involuntarily and subconsciously give maximum attention to the face, and fail to notice other obvious traits of the [individual's] constitution. In his textbook supplement, *Anthropology* (1923), he emphasized: "In comparison with Europeans, Negroes have proportionally longer arms and legs, particularly the forearm and shins; in Mongoloids, they have very long trunks and short extremities. In Europeans, the shoulders and ribs can be called wide, in comparison with the long trunk, and in Negroes they are narrower; therefore, the Negro appears stenothoracic and well-proportioned, while the average European looks more thickset, and many Indians, Malaysians, and Mongols look even more thickset and awkward."

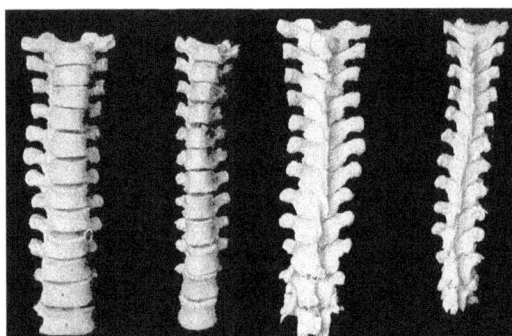

Left: Comparison of the thoracic section of the Vertebrae (front and back) of a European and an Australian aborigine (according to Herman Klaatsch).

In the book, *The Geographical Traits of the Structure of the Body of the Population of the USSR*,[206] A.L. Purunjan and V.E. Deryabin, both prominent modern Russian specialists in the given area, focus attention on the fact that from an evolutionary standpoint, the skeleton possesses a "genetic memory," enabling it to remain unchanged, in comparison with other groups of organs of the morphological structure of the human body, and to preserve the racial traits of the layout practically unchanged, for the extent of many generations. Therefore, in the process of paleo-anthropological research, the given trait enables one to reconstruct the formation and development of races with maximum accuracy. Besides that it turns out that a basic postulate of racial theory is proven: that of the timeless confrontation between northern and southern biotypes. With all obviousness, the author writes: "The basic vector of somatological changeability is directed from north to south; it more weakly expresses differentiation of groups in the east-west direction."

Among the traits on the skeleton of the "southern" racial type, it is worth noting the somewhat longer body and large, long segments of the arm, and the hands in particular. But the most clearly manifested southern traits are in the combination of the cross-section diameters of the shoulders and pelvis. A characteristic "southern" trait is also the large circumference of the neck; in "northern" groups, the large perimeter of the breasts and lower extremities. Analyzing the

---

[206] *Geograficheskie osobennosti stroeniya tela naseleniya SSSR.* Moscow, 1990.

structure and length of the legs, the authors of the book came to the conclusion of the "centuries-old stability of anthropological zones."

And so we see, that the less industrious "southerners"—and their less able-bodiedness, is expressed in narrow shoulders and thorax, and also an inclination toward laziness in general, with a liking for making money that is reflected in the structure of the segments of the arm, which is excellently fixed in the "genetic memory" of the skeleton. Cesar Lombroso, the criminal anthropologist who taught about the individual body structure, could make a conclusion about a subject's mode of living and his habits.

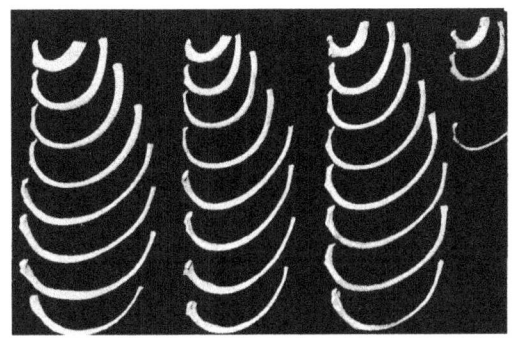

Left: Comparison of the ribs of a European, an Aborigine, an African Negro, and a dwarf Negrito from the Phillipines.

Evolutionary morphology allows one to learn the inside information of entire races. The proportions of the body in one individual reveal generations of soldiers and grain farmers, and in another—cheats and merchants.

We do not have it in mind to insult the members of all "southern" populations, but after all, "southern" ease, together with wanton luxury and idle pastimes have long been the talk of the town. For the majority of "northern" peoples, long hours of sitting at a bazaar is unacceptable from the standpoint of ergonomics. The "genetic memory" of the skeleton itself resists such a mode of daily life, which is a norm for "southern" peoples.

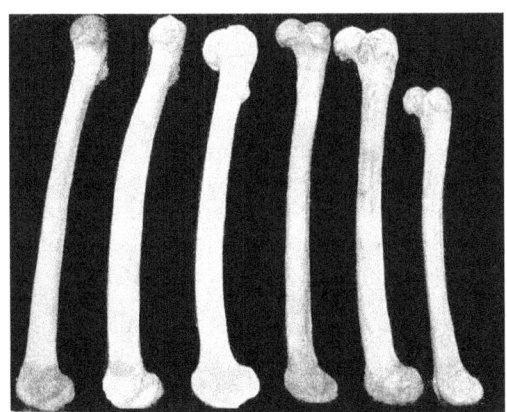

Left: Comparison of the femur of a European, an Aborigine, a Neanderthal, an African Negro, and a dwarfish Negrito of the Phillipine Islands.

In the monography, *Anthropology* (1925), Eugen Fischer spoke on the same subject: "It is worth emphasizing, that all findings about the superiority of Negroes and other coloreds over Europeans in the area of the physiological movement of the body are false. Europeans are better adapted to concentration of muscular energy than any other [people]." This ability of the members of the white race to concentrate muscular energy is a magnificent indicator of the specifics of its origin, tied to the struggle of existence, and consequently, with untiring self-improvement.

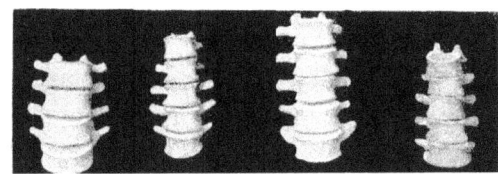

Left: Comparison of the lumbar section of the vertebrae of a European, and Aborigine, an African Negro, and a dwarf Negrito of the Phillipines.

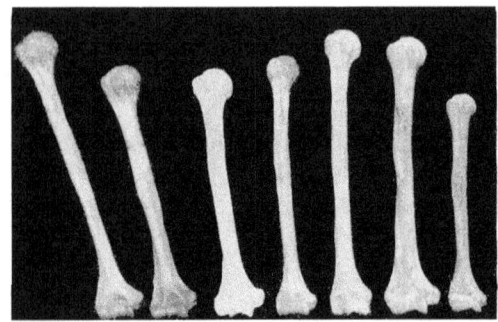

Left: Comparison of the brachial bone of various races, from an inside view: from left to right: two European bones, a Neanderthal bone, two Aborigine bones, one Negro bone, and one Negrito bone.

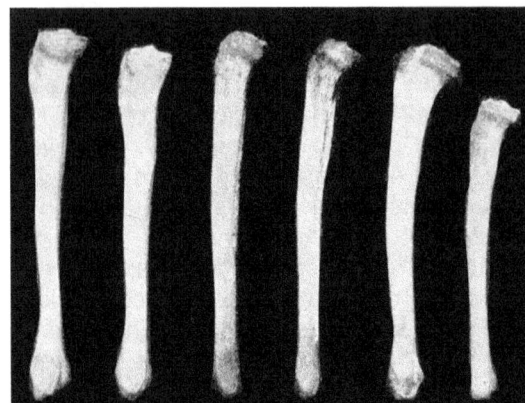

Left: Comparison of the tibia bone among different races, seen from the intererior view: from left to right: two European bones, two Aborigine bones, one African Negro bone, one Negrito bone.

There is still another unique racial diagnostic observation of an especially mundane trait. Bari points out: "Of the relatively lower extremities, it needs to be noted, that to this day there are traits that can be seen in lower races, that indicate several weaknesses of the extremities, since [to attain] a vertical position of the body, it is necessary to acquire strength gradually; and to this day, the inclination toward squatting is widespread in lower races."

Morality, as we already noted above, is found to be closely tied with evolution; therefore, we urgently recommend to all truthseeking fans of heated debate, to clarify the position on the evolutionary ladder of the person being addressed, before the start of a debate, with the help of this racial-physiological test. If the opponent in a debate can squat comfortably, it is better to hold these arguments aside for those who walk upright. From television news broadcasts, one can be easily convinced that many tribes in Africa, Asia, and the Caucasus experience unconcealed pleasure from this pose; this should warn us about their attitude, for morality has a strict physiological basis. The given trait, besides being a racial-ethnic diagnostic indicator, fulfills the function of a marker of the criminal-degenerate elements of society, for squatting is a favorite pastime of prisoners in jails. Besides that it is noticed that Negro women give birth in this pose, like many breeds of animal.

A.P. Bogdanov's notable work, *Physiological Observations*,[207] contains conclusions of a similar character. "Several poses that are very uncomfortable for us, are natural for several other races. Such is squatting, in which the toes are strongly outstretched and planted firmly on the ground, with the buttocks resting on the heels. There are peoples in whom this position takes the place of sitting. We also turn the attention of travelers to the ability of savages to climb trees. Evidently, it is certain that among peoples that are more or less wild and walking barefoot, in particular among those who climb among the trees and rocks, the big toe acquires remarkable flexibility; it not only can bend and straighten out, but it can also draw itself inward and be adjusted by the action of the muscles, in a direction that is parallel to the axis of the leg. Such flexibility of the big toe leads to the suggestion that in several races, [the big toe] is similar to that seen in apes; a foot type that approximates the hand type.

---

[207] *Fiziologicheskie nablyudeniya.* Moscow, 1865.

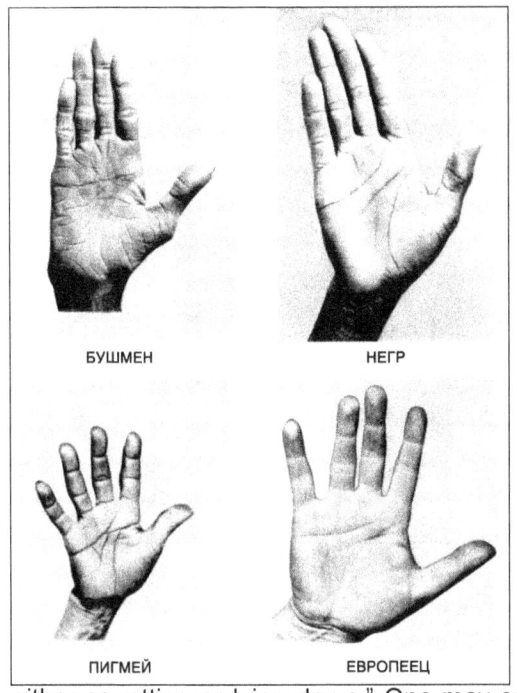

БУШМЕН НЕГР

ПИГМЕЙ ЕВРОПЕЕЦ

Left: Racial Differences in the Structure of the Hand (According to Egon von Eichstedt). Clockwise: Bushman, Negro, European, Pygmy.

In support of this conclusion, I.A. Sikorskiy also pointed out: "Traits continue to show up not only in the structure of the organism, but in the habits of several lower races, of a partial or not fully-matured habit toward a vertical position of the body; this is expressed by the tendency to squat—a tendency from which the European race is already completely freed. The pose itself, it must be admitted, shows that lower races have not completely acquired the constant buoyant muscle tension of the entire body and vertebrae, which is characteristic of whites. As an antithesis to this fact, one may point to the Russian custom of praying in a standing position, and not otherwise—this in particular impresses observers from the East, where prayer is conducted either squatting or lying down." One may say with complete justification that it is the legs which make a man a man. With extreme versatility and surprisingly narrow dependence of all the parts of an animal's body on one another, even the smallest change of one part does not affect the others. Thanks to upright walking, man acquired a trait which physically sets him apart from the ape."

Left: Squatting as a racial-evolutionary trait (according to Eugen Fischer).

Squatting, strong swinging of the arms while walking, and walking on half-bent legs are all undoubtedly atavistic, simian traits, passed down from individual to individual in the process of evolution. Even the modern textbook, *The Morphology of Man*,[208] under the editorship of B.A. Nkityuk and V.P. Chtetsov, clearly points out that "squatting" is a hereditary trait, conditioned by the angle of the neck of the ankle bone. In Europeans this is at 18 degrees; in Mongoloids, it is set at 20 degrees; in Negroids, it is 24 degrees; and in the man-like apes, it is 29 degrees.

If in this case the fantasy came to someone's mind to explain the predilection of different groups for squatting, as some spontaneous cultural influence, then it is worse for them, and best to end any relationship with such a "culture."

---

[208] *Morfologiya cheloveka.*

185

 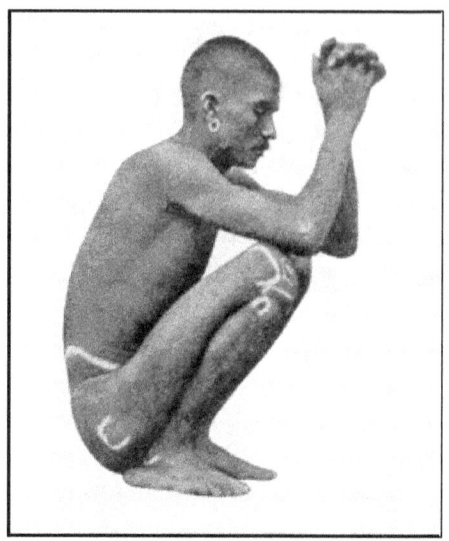

Above: Squatting as a Racial-Evolutionary Trait (according to Eugen Fischer).

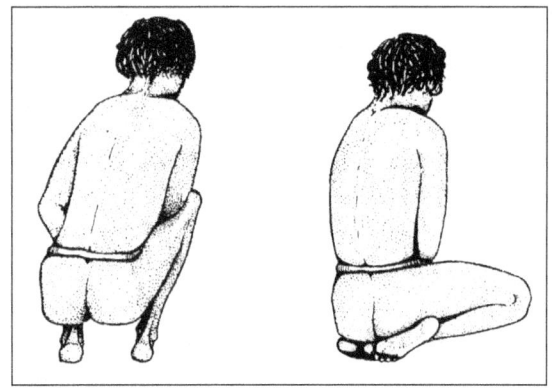

Left: Dominant Pose at Rest among the Peoples of Oceania (left) and the Aborigines of Australia (right). (according to Frederick Vud-Johns).

Bogdanov encourages us to take a closer look at the gait of various peoples, in order to form a more certain opinion about them, for in his opinion, the "gait is as changeable as physiognomy." Much is signified by the method with which people move, as well as any extreme poses of any type, which people adopt while eating, engaging in sex, and necessary body functions. For the attentive observer-analyst, invaluable information about the zoological prehistory and evolutionary worthiness of this or that race, and all its carefully-hidden secret flaws, is contained here.

In the fundamental essay, *Raciology and the Racial History of Mankind*,[209] which to this day is considered a classic of anthropology, Baron Egon von Eichstedt emphasized: "Differences in behavior, gait, and facial expressions, by their pre-conditions have a definite structure of joints and muscles, speed of movement of the nervous impulses, and [standing posture] and gait. Equally, behavior is highly different in different races and peoples. Each has a typical pose at rest, a typical way of walking, typical gesticulation, and manner of squatting, navigation, and prehensile movement."

In agreement with this observation, Pygmies stand on spread legs, like infants of the white race, and the Masai and southern Ethiopians prefer to relax standing on one leg, in a "stork pose." The gait of the Pygmies is ponderous, possibly caused by the disproportion of the legs and body. But at the moment that the Pygmy begins to run, he becomes flexible and easily avoids obstacles. The Pygmy relaxes in a squatting postion, with the foot evenly against the ground, and the hamstring and calf bordering each other for the entire length, in which the upper part of the trunk inclines forward. Whether the Pygmy sits on the ground, a tree trunk, or on sticks, he can squat for hours. For comfort, he will place a log under the buttocks"—Eichstedt wrote.

---

[209] 1937-1943.

Racial differences are easily revealed in the tempo, the swing, and rhythm of movement. Thus, for example, the sweep of the arm while walking is completely different in Negroes, than in Chinese or Indians. Differing physical susceptibility to fatigue, muscle strength, variations in the distribution of the soft tissues of the body, and the center of gravity, make the movement of the members of some races completely dissimilar to the movement of the members of other [races]. Eichstedt specified: "Negroes step firmly; Nordic peoples pace out; Veddoids walk with mincing steps; old people of the oriental race waddle; and Nilotes [walk] as if on stilts. Europeans sit with legs hanging, the Eastern man places his legs underneath himself, the Malayan squats, and Negroes stand on their knees. The European in general cannot correctly squat—this is because of the massive ligaments, thick bones, and the distribution of the muscles. One can become accustomed to this, and learn more or less to skillfully imitate squatting or to sit for some time "like a Turk," but the author is convinced that this is hard work, based on his own experience of attending the night holidays of primitive forest peoples for several hours, while maintaining the appropriate pose. It is known that among various peoples there are different forms of navigation, that among them there are various forms of legs, and that fallen arches is a group characteristic (in Negroes and Jews)."

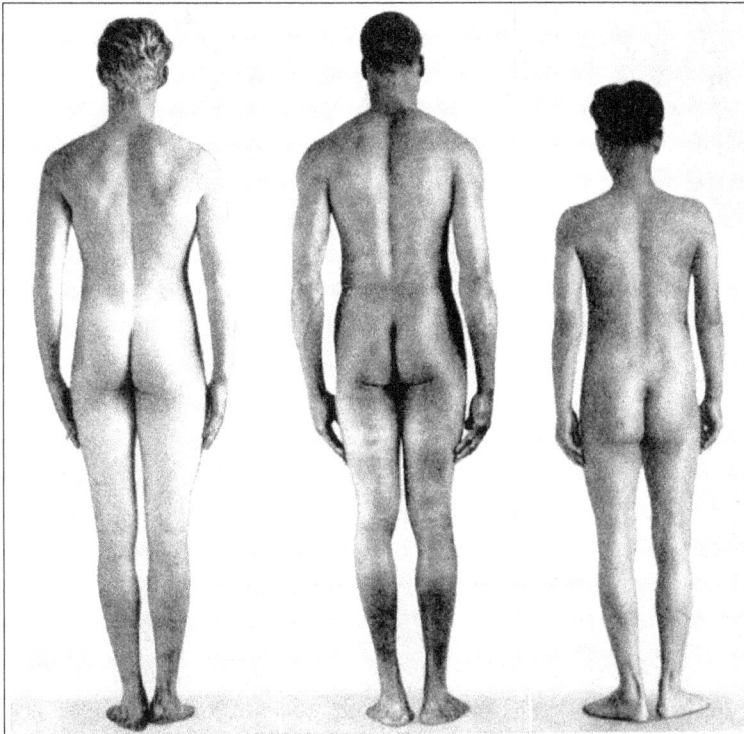

Left: Racial Differences in the Proportions of the Body (according to Bruno K. Schultz).

Eichstedt also pointed out that the characteristics of Japanese miniatures reflect a specific racial structure of the [finger] joints of the maker, the fingers of whom easily flex at right angles to the surface of the hand. Numerous phenomena of everyday life, which culturologists explain with the influence of the environment and upbringing, may have an extremely simple anthropological explanation, rooted in the particularities of the structure of the body of the creators of a culture.

In Russian folktales, one can often encounter this elevated, majestic description of the walk of a Russian beauty: "She steps like a pea hen." This fully corresponds to the psycho-somatic traits of the Nordic race. Therefore, today's propaganda from the adepts of so-called "high fashion," [which promotes] a style of gait, with the required back and forth rock of the hips, is racially foreign, and corresponds to the structure of the body of equatorial races. In studies of folk dances we easily observe, that in people related to the great white race, graceful and rhythmic movement is realized with the arms and legs, while the vertebrae remain straight all the time. Celtic dances, the German zwiefache, the Ukrainian gopak, Russian round dances and kick-dancing, and the Greek sirtaki—everywhere we observe one and the same somatologically similar picture; from an aesthetic and psychological point of view, the white man always considers it incomprehensible to single out the stomach, genitalia, and hips with [body] movements. Further to the south, where the admixture of colored blood increases, dance comes down to a primitive shaking of the genitalia, which is observed in a majority of animals in mating season.

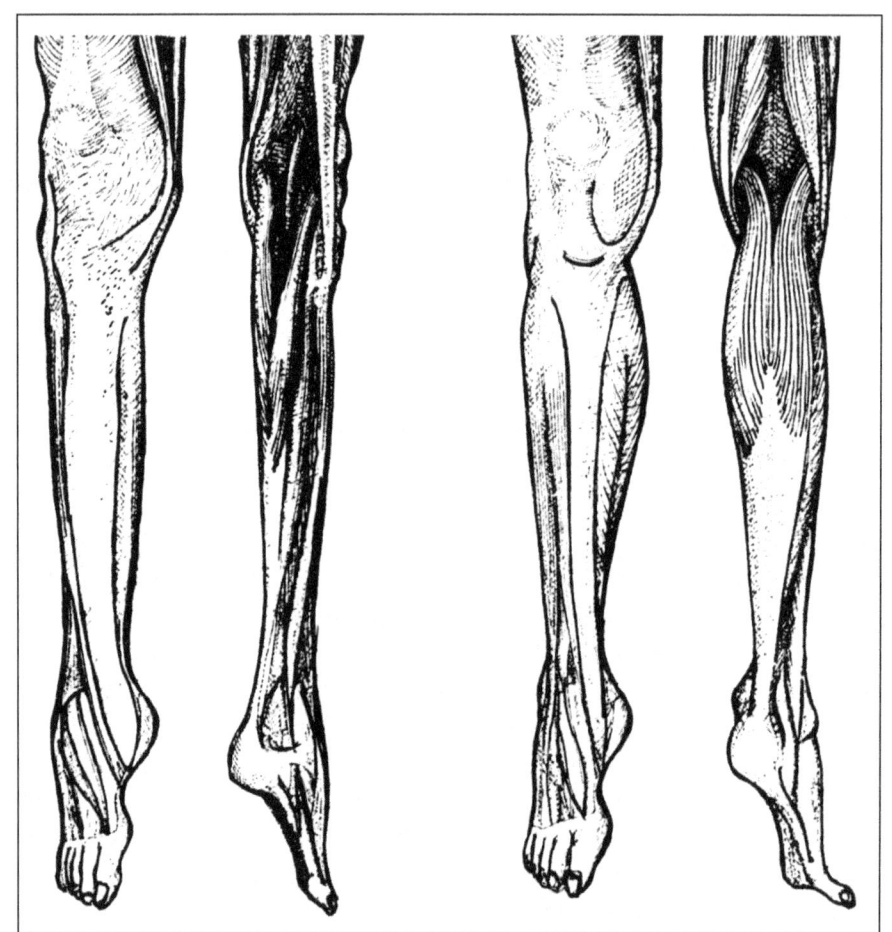

Racial Differences in the Structure of the Leg (according to Edward Lot).

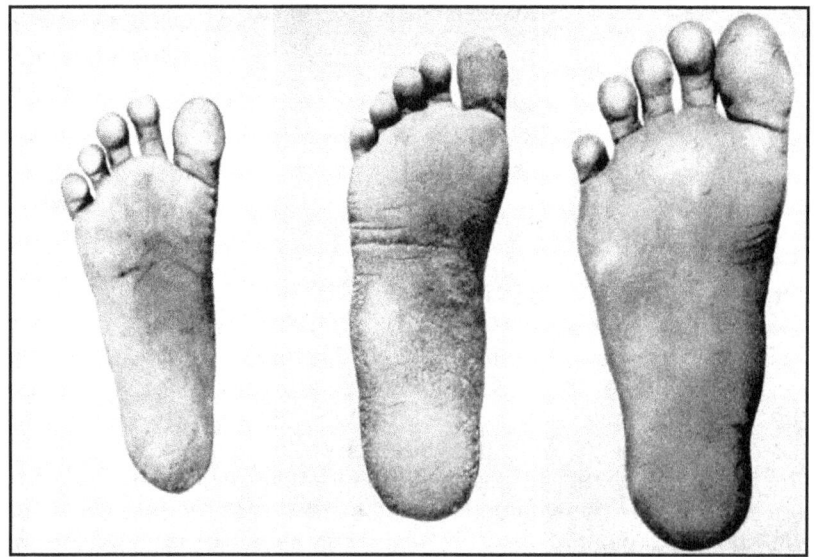

Racial Differences in the Soles of the Feet (according to Egon von Eichstedt).

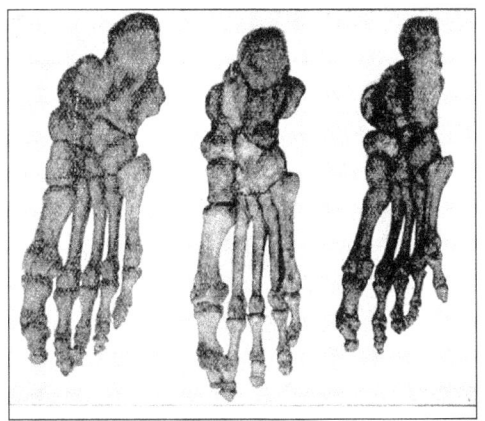

Left: Comparison of the Feet of a Japanese, a European, and an Australoid (according to Herman Klaatsch).

The great Russian explorer, Afanaciy Nikitin, left an excellent description of his outrage at the sight of the notorious "bellydance" in one Asian country, at the time of travel. Originally from the Russian heartland, he experienced genuine physical disgust at the sight of this racially foreign act.

The propaganda about "universal" yoga in the modern European world—which is losing its racial frame of reference—is an open provocation, since all races have different energy chakras. In European races there is greater energy, and consequently the higher chakras of the heart and mind stand out by development, while in the southern, equatorial races the lower chakras of the stomach and genitalia are better developed. This is one of the very main differences in the structure of the races; in all living forms, it firstly manifests itself in the cultural sphere. There exist primordial values and resultant cultural and whole religious systems, that are traced to the mind [chakra], another to the heart [chakra], a third one to the stomach [chakra], and the fourth to the...

No process of leveling with common human values will ever exchange the places of the body and the psycho-energetic accentuations in peoples of different races. "West is west, and east is east, and never shall the twain meet"—this excellent expression is an outstanding geopolitical maxim. From the standpoint of raciology, it would be more precise to say: "North is north, and south is south, and never will they switch places."

The Russian scientist, A.P. Bogdanov, made highly respected observations on the results of race-mixing and miscegenation: "Populations consisting of mestizos present a greater number of idiots, the insane, the congenitally blind, stammerers, and simpletons, compared with the number of such cases that are observed in the communities of the two original races. Thus, in Nicaragua and Peru, although zambosis (a mix of Negroes and Indians), represent a comparatively numerically small class, they nevertheless comprise 4/5 of the prison population."

The positions of classical anthropology are again fully experimentally confirmed by the postions of modern sciences like etology and behavioral biology. Konrad Lorenz wrote in his book, *The Other Side of the Mirror*, that "the positions of the body may be genetically reinforced and become consistent, visible traits of a race. The knights and ladies of the Minnesaenger Age simply could not have sat down on the high, Gothic stools of the time, and walked naturally about Gothic halls, if the style of that time had not already been laid down among them, in the positions of the body."

The given phenomenon of racial somatology was brilliantly explained from an evolutionary viewpoint, by one of the distinguished German anthropologist-polygenists, Hermann Klaatsch. In his fundamental monograph, *The Origin and Evolution of Species*,[210] he wrote: "The lower extremities of the individual were not the least bit simultaneously completed in their transformation into supporting organs, as our animal ancestors became people—quite the opposite; in the time of the spread of Mankind across the globe, the adoption of the lower extremities to a vertical position of the body, and the corresponding gait, occurred completely independently in different groups. From there it is understood, why differences in the character of gait are observed among the different races; it would be a noble task to accurately study the mechanical-physiological relationship to the characteristic gaits which are at first glance noticeable in the Negro and the Japanese. These people walk differently than the European, and the distinctive manner in which Mongoloids keep on their feet is particularly striking.

There cannot be a better confirmation of the view of the relative origins of Man, than the layout of the Australoid foot. It is sooner a tool for climbing, than a support apparatus. By this [example] the foot began already began to change during the spread of Man, and it changed completely

---

[210] *Proiskhozhdenie i razvitie chelovecheskogo roda*. Sankt-Peterburg, 1896.

independently of different races. Therefore, the skeleton presents different traits in Negroes, Europeans, and Mongols. In the Japanese, it is very massive, and it retains many lower traits."

Erwin Belz, a prominent specialist in the area of racial differences in the skeleton, said the following in the work, *On the Characteristics of the Physique of the Japanese* (1885): "The use by the Japanese of the big toe as a thumb, is highly remarkable. They can move them independently of the others, closely press [it] toward the second toe, so that in this state they can firmly hold even shallow objects with them. A sewing woman can take fabric with her toes and turn it as she likes. They even say that Japanese girls can chop with these toes. In general, the leg of a Japanese has preserved very much of its natural flexibility. They can cling to the soil by their footsoles. Therefore, in any kind of exercise, where it is necessary to firmly remain on their feet, like wrestling or fencing, for example, they always remain barefoot. Whoever has seen Japanese working on the roof of a house for the first time, without any [safety] devices, as if they were on the ground, will involuntarily fear for them; but the [Japanese roofers] fear nothing: they do not fall, because their feet are superbly adapted to the slanting form of the roof."

In Europe and America in the 20th Century, an enormous amount of semi-literate fans of Eastern exotica appeared. The Japanese custom of walking barefoot at home was explained by them as their particular way of showing respect to ancestors, the spirits of whom guard the home. But this everyday peculiarity of the Eastern way of life, as we are now persuaded, has a particularly anthropological explanation. The feet of the Japanese are simply not adapted to the wearing of European house shoes. So the white man, yielding to the uncontrolled trends of racially foreign style, fastens sandals on his feet, known among the crowd as "Viet-Namsies", also assaults his own racial constitution, for the big toe is physically not capable of gripping the sandle strap that runs between it and the four remaining toes of the foot.

All the above-said applies in equal degree to the Muslim of the Middle East, for the average European cannot squat for ten minutes in that pose in which Eastern "wise men" are able to sit for hours. The classical Soviet anthropologist, B.S. Zhukov, asserted in the book, *The Origin of Man*,[211] that: "Among the lower races one can find a significant number of examples of a well-developed musculature of the toes. While sewing, Japanese women can hold up and pull material tight with the toes; some Malaysian women use the big toe to help secure the oar in rowing; Negroes, Hottentots, and other peoples use the feet in a bow string, in lifting objects from the ground, and so on."

14. Atavistic (Simian) Traits in the Morphology of the Races

Karl Stratz,[212] one of the well-known specialists in the area of the proportions and morphology of the human body, created a whole system of so-called **pithecoid**, that is, **simian traits**, which attest to a certain "animalness," or evolutionary primitiveness in this or that individual, tribe, and even an entire race.

From their huge number, he set aside the following important ones:

1. Macaque ears—ears with sharped or pointed upper edges.
2. Catarrhines—nostrils that are set close together and opening to the front of the face, and directed frontward or downward.
3. Stenogrotaphia—insignificant development of the temporal area.
4. Inca bone—the interparietal bone, when developed as a separate bone in the skull; grown together with the occipital bone.
5. Torus Occipitalis—an extradordinarily strong development of the transverse elevation of the occipital bone, to which the muscles at the back of the head/nape attach.
6. Significantly long forearm bones.
7. Clawed toes.

---

[211] *Proiskhozhdenie cheloveka.* Moscow, 1928.
[212] 1858-1924

8. Strongly developed eyebrow ridges, with deeply horizontal, widely separated interior corners.
9. Swimming/webbed membranes between the fingers.
10. Four-fingered (simian) flexor folds on the palm.
11. The Mongolian birthmark on the sacrum.

The placement of the eyes is a distinct evolutionary trait, which characterizes the evolutionary situation of an individual; it is simultaneously a **degenerological marker**, and is measured by the **interocular space index.**

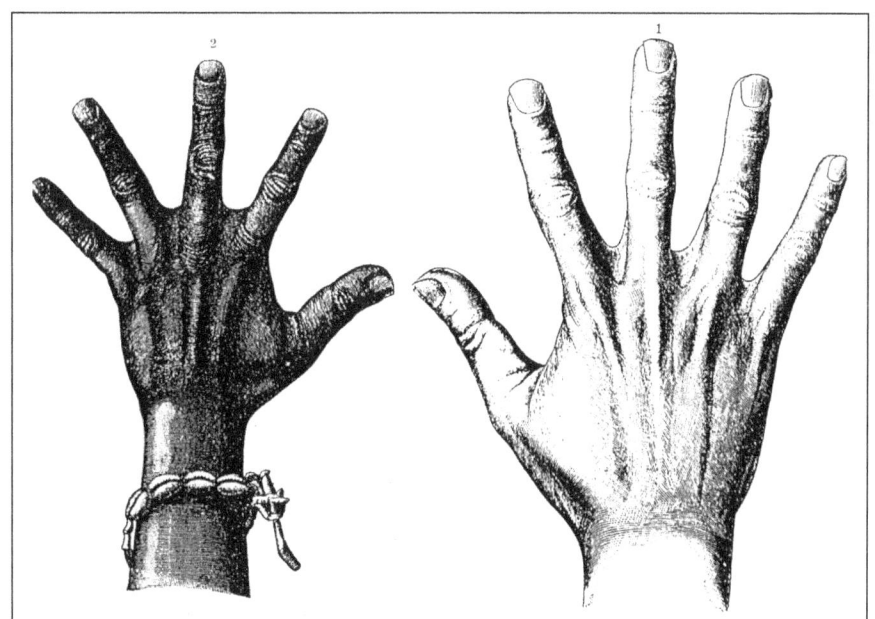

Above: Webbed/swimming membranes between the fingers (according to Karl Stratz).

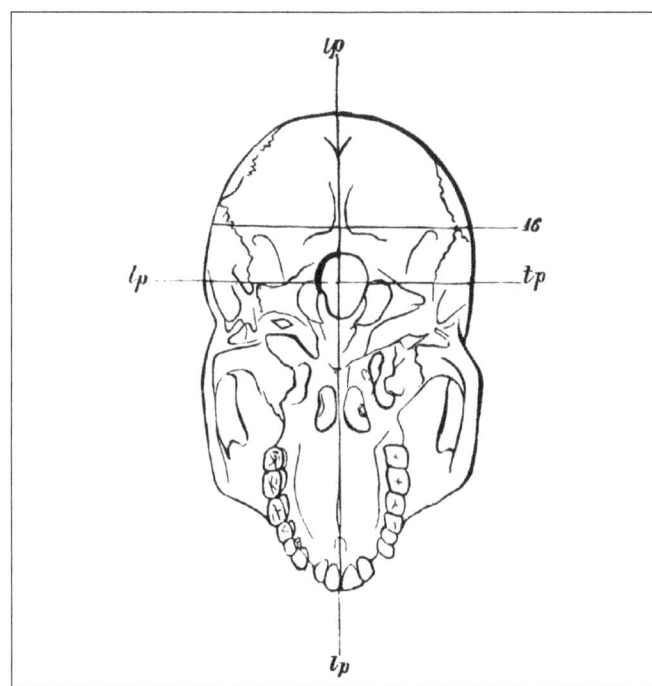

Left: The Base of the Skull of a Kaffir, showing "simian" gaps between the teeth (according to Karl Vogt).

In the book, *The Origin of Mankind*, distinguished anthropologist Hans Weinert wrote: "Which position of the eyes is more ancient—closely placed eyes, or placed at a further distance from one another? Of course, the latter [is more ancient] since eyes placed closely together first appear in lower forms, like the baboon for example, a result of the degeneration of the olfactory organ. A large distance between pupils results in a widening of the field of vision. This leads to a refinement of the stereoscopic portrayal of the surrounding world, and to a more graceful knowledge of space.

According to the findings of anthropological science, the number of degenerative or atavistic traits also includes: excessive ovalness of the face, tail appendages, and besides that an X-shaped form of the legs (genu valgus), which is most often encountered in Jewish mestizos. Finally, all speech defects stemming from incorrectly fused palate bones, were determined to be characteristic traits of degeneracy during the time of the Inquisition. Hans Weinert, who is mentioned above, explained it thus: "The palate mound serves as an excellent example for illustrating the basis of biogenetic law. Besides that this is one of several traits belonging to the soft tissues of the body, which in its development, as in the facial musculature, one can detect a progressive tendency, the direction of which does not upset the data obtained in separate genuses and species."

From an evolutionary point of view, other morphological traits (of which there is a different concentration in all peoples) testify about definite "lowness" of origin. The frequency of the rise of rudimentary traits in this or that population, testifies about its evolutionary situation. Wilhelm Leche emphasized in his book, *Man, his Origin, and Evolutionary Development*, that: "In the inner corner of the eyes there is a small, light-red membrane: the so-called half-moon fold—the conjunctiva. No sort of function can be ascribed to it, nor any use. It is better developed in several wild peoples (Negroes and Malaysians), than in Europeans."

Paul Topinard gave this explanation of the essence of the exanmined phenomenon, in his textbook, *Anthropology*: "There are traits which manifest suddenly in members of all the human races, [but] most often in the so-called low [races]. They are called **rudimentary organs** and **anomalies**. In studies of the transformation of comparatively low forms into more complex and complete beings, many anomalies of the previous [types] receive the name **regressive phenomena** or **reversions**. In them, the expression of blood kinship between two organisms is understood, and why these phenomena touch on the question of the relationship of Man to other mammals.

Gustav Schwalbe (1844-1916), a distinguished German anthropologist, gave much attention to the questions of **evolutionary morphology** and **comparative anatomy**. He defined three main traits which indicate the close kinship of the members of a given race with the Neanderthals: (1) a lower skull dome in relation to its length; (2) an uninterrupted brow ridge; and (3) a strong incline of the forehead, with a small bend of the lobe bone plates. Soviet anthropologist G.F. Debets considered a strong size of the big toe nails to be a certain Neanderthal trait. Besides that he detected a connection between the development of this trait and the characteristic structure of a protruding brow ridge on the skull. All these traits are present in the Aborigines of Australia.

Soviet scientist V.R. Kabo wrote in the book, *The Origin and Early History of the Aborigines of Australia*,[213] that: "The problem of the origins of the Australoids and Tasmanians is one of the important questions in anthropology. It is closely tied to such problems as the formation of the human races (racial genesis), the systematics of race, the characteristics of the racial types of humanity, and their settlement of the Earth. In an anthropological regard, the Aborigines of Australia occupy a special place among the different racial groups of Mankind. Characteristic of them as a whole are wavy black hair, an abundance of tertiary hair covering the body and face (in men), a wide nose with a low or average-height nose bridge, large-sized teeth, average skull diameter, prognatism, above average or tall height, dolichomorphology, and dolichocephalia. The indicated traits are used to connect the Aborigines of Australia with the Australoid race. Evidently there is no disagreement between anthropologists in this regard."

The English evolutionist, Thomas Henry Huxley, and Soviet anthropologist Viktor V. Bunak, also referred to the special Australian race. In accordance with the theory of **polycentrism**, the Australians represent a separate race, which developed in complete isolation from ancient, proto-morphic forms. The given point of view was expressed by Franz Weidenreich (1873-1948), Carlton Steven Coon (1904-1981), and Arthur Keith (1866-1955). The structure of the palate and teeth of the Australoids belongs to extremely primitive forms of evolutionary development, as pointed out by G.F. Debets.

In the book, *Anthropology* (1923), distinguished German anthropologist Eugen Fischer spoke even more decidedly on the issue: "According to many traits, the Neanderthals were even more developed than the Australoids—the latter arose from a more ancient form." At the beginning of

---

[213] *Proiskhozhdenie i rannyaya istoriya aborigenov Avstralii*. Moscow, 1969.

192

the 1930s, he was supported by Soviet scientist I.I. Puzanov, who concluded that the Australoids arose directly from Pithecanthropus; undoubtedly, this is supported by the almost identical weight of the brain in the two of them.

In *The Tasmanians and the Tasmanian Problem*,[214] V.R. Kabo expressed his opinion about Tasmanians being the most primitive people in the world, when they became extinct in the 19th Century. In *The pre-Religious Epoch*,[215] V.F. Zybkovets emphasized that the Tasmanians did not have artifacts of iron, or ironware, and did not engage in fishing at all, even though they lived on the ocean coast; they did not know about ceramic-making. The Tasmanians differed from the Australoids in the form of the hair on their heads, their wider skulls, their wider nasal openings, a more strongly pronounced prognatism, and the exclusively small height of the face and cranium, and the low-placed eye-orbits. The Tasmanians have a wider, but shorter nose and palate, a lower cranial dome, than the Australian aborigines.

Alongside a number of racial complexes, which characterize the similarity of this or that race with the more ancient proto-form ancestors of Man, separate anomalies exist, as we already showed, which indicate a throwback, a reverse move in the development of a species, an evolutionary degradation. In Paul Topinard's opinion, these regressions are most often encountered in the musculature. Among these are **skin muscles**—except for the face and head, which are found beneath the muscles of the shoulder blades, the **pectoral muscles**, and the **buttock and pubic muscles**. "No matter how they explain these facts, in any case they establish a tie between the pattern of organization in Man and animals"—Topinard indicated.

Eugen Fischer also addressed the question of atavism: "The sinewy layers in the forward-most part of the sublingual muscle are encountered in Europeans in 1.5% of cases; in Negroes in 12% of cases; and in Japanese in 14.8% of cases. The Japanese, more than anyone else, have primitive muscle variations, while the Europeans have them least of all. The muscles of the face have been thoroughly studied. In highly-developed forms, these muscles unite in a large, inter-connected complex; indeed, Melanesians, Negroes, and similar others, more often have more primitive forms, which in us [Europeans] are encountered only rarely."

The bifurcated uterus, or so-called **two-horned uterus**, which is a normal phenomenon in rodents, can be added to the number of atavistic traits.

Herman Klaatsch forcefully emphasized in his article, *The Condition of Man in Nature*, that: "Between the known human races and the defined man-like apes, a kinship tie exists not only in general, but even in specifics. Man does not in the least represent a crowning work of art, but is a union of very ancient traits with other, more slowly perfected [traits], and tertiary ones, that appear to be more recently acquired."

Finally, distinguished Soviet anthropologist V.V. Bunak also set apart **anomalous traits** in the process of phylo-genetic evolution, in his fundamental monography, *The Species Homo, his Origin, and Subsequent Evolution*.[216] In accordance with his concept, we applied the given term for defining the variations of a trait, lying at a distance of three standards from average size, or sharply differing from the **typical human structure.**

V.V. Bunak also used knowledge of **structural anomalies**, that is, those which are transferred through heredity, together with different racial traits. From their number, he enumerated the following important ones: first, those anomalies that reflect a disruption in the continuity of the processes in a given racial group, when entire phases of development do not meet, or when it is a question of the appearance of the elements in the structure of a separate organ, in general not analogous with the man-like ape forms. For example, the many anomalies in the knitting of the seams of the skull, the development of additional molar teeth, an occipital crest, and a massive brow ridge. Pathology pertains here, tied with unfinished structural forms, such as split sacral vertebrae, split lips, and so on. Anomalies are tied to the merging, splitting, or mixing of the placement of organs, and all forms of asymmetry, expressed in an incongruent placement, form, and size of the organs on the left and right sides of the body.

Bunak emphasized: "Anomalous structures have an uncomplex genetic structure. The frequency of anomalies of the enumerated types in European groups is lower, than in non-

[214] *Tasmaniytsi i tasmaniyskaya problema*. Moscow, 1975.

[215] *Doreligioznaya epokha*. Moscow, 1959.

[216] *Rod Homo, ego vozniknovenie i posleduyushchaya evolyutsiya*. Moscow, 1980.

European groups; in the skull,anomalies of the Pterion region are encountered more often in Negroes; in Asians—anomalies of the occipital bone. Polar changes are most often [encountered] in Europeans." It is very important to note here, that by **polar anomalies**, Bunak had in mind structural traits that go in the direction of an increase in the specific features of contemporary types of *Homo*; that is, corresponding to an evolutionary vector of maximal humanization. It is namely these traits which characterize the white race. Egon von Eichstedt called them **progressive traits;** he considered the opposite that is witnessed in the evolutionary primitiveness of a structure, as **regressive.**

Like a true scientist, the outstanding English anthropologist, John R. Baker, enumerated and explained the basic anatomical traits that are present in representatives of the so-called "lower" races:

1) **Affenspalte**—"Ape Fissures"; that is, the characteristic morphological trait in the rear portion of the brain, which brings the carrier closer to the categories of anthropoid apes;
2) **Eutebrueste**—"udder-shaped breasts" in women;
3) **Ganzprofilwinkel**—prognatism or strongly forward protruding jaw;
4) **Hottentotfalte**—"the Hottentot fold", the skinfold next to the eyes;
5) **Indianerfalte**—"the Indian fold", the skinfold next to the eyes;
6) **Mongolenfalte**—"the Mongolian fold", the skinfold next to the eyes;
7) **Negerfalte**—"the Negro skinfold" next to the eyes;
8) **Trichterfoermig**—"funnel-shaped" nasal openings.

Left: Sir Grafton Elliot Smith

John R. Baker wrote about the "ape fissure" as a racial diagnostic marker, since in Europeans it is essentially located further back, than in Australian Aborigines and apes. In classical English scientific literature, it received the name of **"monkey cleft"**, and was described by the prominent neurologist, Sir Grafton Elliot Smith (1871-1937). He called it an "impartial, consistent trait, for all those who only saw the brains of an ape and a European in a museum once." In anatomical atlases, the given trait is defined as the **"parietal-occipital fissure."** Psychiatrists tie its presence with evolutionary underdevelopment of the brain and common, hereditary cretinism.

In this regard, it appears completely justified to us, that in his book, *Ethnos, or the Problem of Race*,[217] Sir Arthur Keith (1866-1951) quite accurately and intelligently determined that the characteristic physical traits which clothe our bodies are the **"uniform of race."**

---

[217] 1931

194

15. The Eyes as a Racial Trait

The color of the eyes is one of the important traits by which a person's combined affiliation with a racial-biological circle is determined. In the legends and folklore of all peoples of the Earth, and since ancient times, one may trace the degree of importance of eye color, when identifying others under the "us-them" principle. However, intelligent study of the given important anthropological parameter only began at the end of the 19th Century. Gustav Fritsch (1839-1891) was one of the first to point to racial differences in the eye retinas, and Eugen Fischer discovered a correlation between pigmentation cells in the mucous membranes of animals, and the "lower" races of humanity.

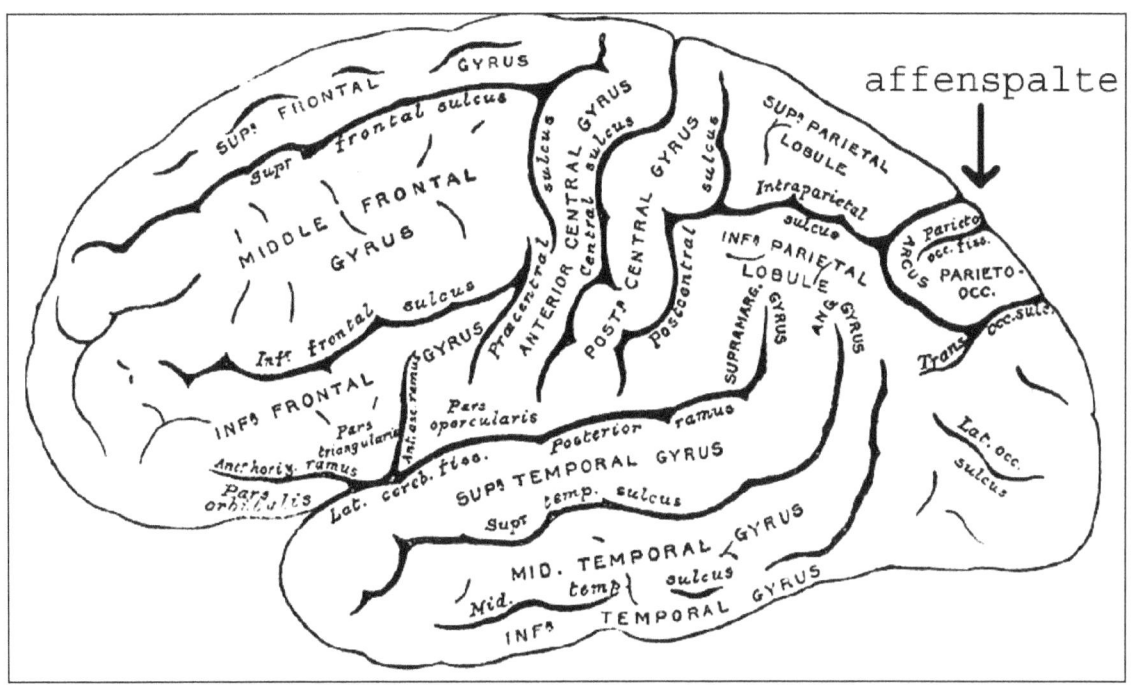

The "Ape Fissure" (according to John R. Baker).

Finally, for his part, Max Wolfgang Hauschild (1883-1924) confirmed the existence of three different types of pigmentation cells in the iris membrane of the black, yellow, and white races, a fact which finds expression in the interpretation of cultural differences. Russian anthropologist P.A. Minakov indicated in his article, *The Significance of Anthropology in Medicine*,[218] that: "Many peoples do not distinguish certain colors of the spectrum. Thus, for example, the Arabs use the words black, green, and brown as synonyms. The Koreans do not distinguish between green and light blue, calling these two colors by one word: *Pehurada*. The Bongo tribe of Central Africa uses one word for black, light blue, and green: *Kamakulutsch*. The color scale of this tribe consists of three colors: black, red, and white.

It should be noted, that with the indicated traits that are characteristic of many savages, they have unusual sharpness of vision and hearing, which enables them see to very distant objects with great detail, and to distinctly hear the most faint of sounds, which are completely inaudible to the ear of a European; on the other hand, the harmonious combination of sounds, colors, and tones [that is found in European music and art], is hardly noticeable to the savage."

From an evolutionary point of view, other morphological traits in the structure of the eyes—the concentration of which is different in all races—also testify to a certain "lowness" of origin. The frequency with which these rudimentary traits springs up in this or that population, are an indication of its evolutionary status. In the book, *Man, his Origin and Evolutionary Development*,

---

[218] *Znachenie antropologii v meditsine*. Russkiy antropologicheskiy zhurnal. N1, 1902.

195

prominent Swedish anthropologist Wilhelm Leche emphasized: "In the inner corner of the eye, there is a small, light-red membrane, the so-called half-moon fold (the conjunctiva), to which no function nor use can be ascribed. It is better developed in several wild peoples (Negroes and Malaysians) than in Europeans." For his part, George Buschan, the well-known German scientist cited by us many times, remarked: "The third eyelid, or the Plica Semilunaris, is a vertically standing membrane joined to the eye tissue, and is a remnant of the blinking membrane in animals. In particular, it is expressed in birds, amphibians, and reptiles. As a reminder of this condition, it is preserved in Man in the form of a small, cartilage-like vestige, which is sometimes encountered rather often in lower races; for example [it is encountered] in Negroes in 75% [of cases], and in whites in only 0.5% [of cases]."

Soviet scientist B.S. Zhukov also wrote: "In the eyes of members of the lower races, the Plica Semilunaris is somewhat more strongly developed, than in the members of European peoples, for example." From this it follows that in the structure of the eyes themselves, and the organs surrounding them, a whole bouquet of morphological traits is enclosed, allowing one to judge with a high degree of probability, the evolutionary value of this or that individual, namely from a racial point of view.

Structural differences in the position of the eyes are no less significant. The lower end of the orbit is very narrow in gorillas; in people, it is wider, particularly in Negroes; in Europoids it is less wide, and it is most narrow in Mongoloids. Baron Egon von Eichstedt wrote in this regard: "One may consider the very wide slit, as in the negro Voot tribe, as an infantile-primitive trait, but only within the bounds of the human series: in people, the structure of the orbit developed in a particular direction. The projection of the frontal maxillary seam on the inner wall of the eye sockets, the expressed projection of the upper jawbone, testifies to this. This is normal in gorillas and chimpanzees, but in people it is very rare, and is only in theromorphic, primitive races." Negritos, Bushmen, and Veddi have the absolute maximum capacity for a similar flatness of the eyesockets, which makes their skulls look more sinister. The form of the orbits is determined with the help of an orbital index. The lower, and usually, the more right-angled the form (as with the Tasmanians, New Caledonians, Tierra del Fuegians, and Guanchi) have an index of around 80, while the rounder and higher forms (in the Chinese, Eskimos, and Polynesians) have an index of around 90. The racial peculiarity of the orbits in Mongoloids is also expressed in the position of the line of the maximum length to width; in Europeans, this line is far more inclined to the horizontal, than in the Japanese, which points to a higher position of the whole inner orbit area in Mongoloids. In them as a whole, the eyeball has a more frontal position. The races differ in the distance between the eyes too, and in the eyes themselves, by the structure of the retina. Egon von Eichstedt indicated: "In chimpanzees, the retina has a very shallow structure; among people, the Bushmen, Veddi, and to a lesser degree—Negroes—come closest to that. The opinion that savages have better eyes is not supported by convincing arguments. The slightly branched cells of the iris membrane are stuffed with pigmentation cells; in Negroids and Mongoloids, the cells are larger; but in Europoids, these cells are smaller and finer. In dark-skinned races this membrane covers the conjunctiva and pupils, as a result of which the latter does not appear white, but yellow, and its edge has a brown border.

New-born Europeans usually have blue or dark-violet eyes, or gray light-blue eyes; [newly-born] Negroes [usually have] brown eyes, and Mongoloids—greenish-brown [eyes]."

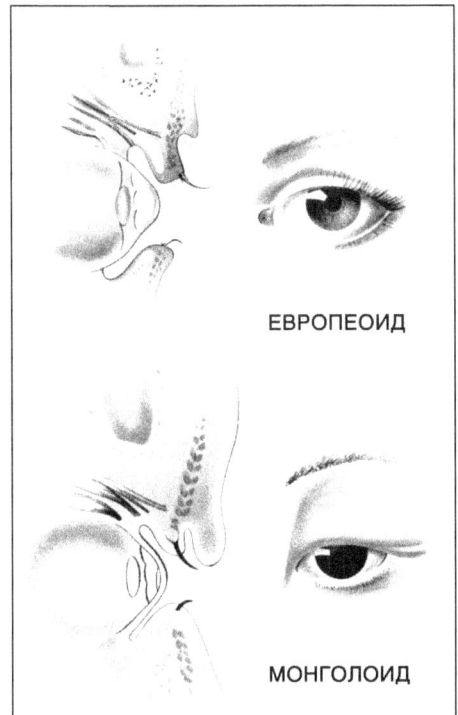

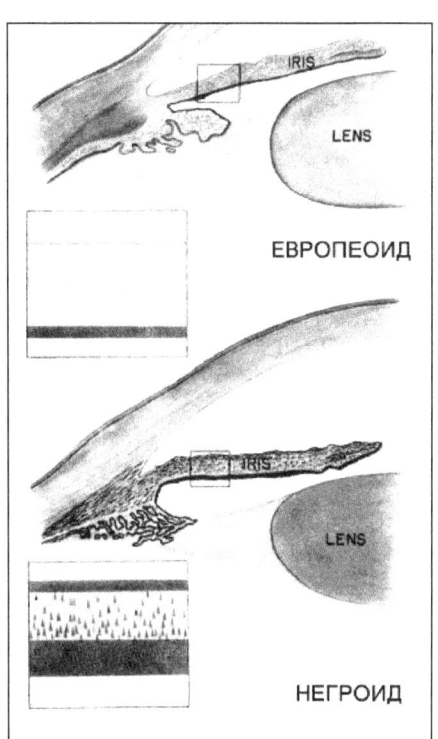

Above: Racial Differences in the Structure of the Eyes (according to Carlton S Coon) Europoid, Mongoloid; Europoid, Negroid.

Left: Rudolf Martin

The muscles for closing the eyelids are also of interest, from an anthropological point of view. In primitive races they are still connected with the muscles of the nasal region. In higher races, they are completely independent. Thus, in Europeans they are distinctly divided into three subgroups. The formation of cartilage on the conjunctiva is an outstanding trait in apes, and is encountered often enough in Negroids, rarely in Mongoloids, and is almost completely absent in Europoids. The given pattern was discovered by Paul Rudolf August Bartels (1874-1914) and Buntaro Adakhi. But Mongoloid eyes have the most notable traits of all. In Mongoloids, the orbits are positioned far higher on the outside, than in Europeans, from which the impression of crossed eyes and goggle-eyes is created in members of the yellow race. But the main trait of Mongoloid eyes is the eyefold, upon which the expression of the face depends in many ways. In northern Europeans, it is usually in the form of a spindle; in peoples of the Oriental race [it is in the shape of] an almond. Egon von Eichstedt pointed to numerous other rudimentary formations in the structure of the eyes in members of the black and yellow races, and their mestizos: epicanthus, the Hottentot fold, the Negro fold, the peach eyelid, the tarsal fold, and the club fold. All these morphological anomalies fall to the

197

portion of the members of the given races, in the inheritance from the first animal ancestors; in a high degree of concentration, they speak of a mutual evolutionary closeness. The coloredness of the eyes of the members of various races today is measured according to the scale of Rudolf Martin (1864-1925).

The famous American anthropologist, Carlton Stevens Coon, discovered that the weight of the ocular orb in Negroes is on average, 8.5 grams; in Europoids 7.9 grams; and in Mongoloids it is 7.4 grams on average.

In the book, *Anthropology* (1923), Eugen Fischer made important observations, such as: "One may determine racial differences in the nervous apparatus, through the eyes. Only in the primates (except lemurs) is the so-called central dimple, or fovea, formed on the retina. It is namely in the area of this dimple, where frequential differences between the races are observed. Thus, the Hottentots and many Indian tribes have particularly shallow and densely distributed cones in the central area of the iris; in East Asians, there are larger zones without rods. The same shallow elements in the optical layer of the iris are found in the Hottentots, then in the Indians and Malaysians; in the Melanesians they are not so shallow, followed further by Europeans.

Far more remarkable are the differences in eye pigmentation, particularly in the iris membrane. But other areas besides the iris have racial differences in pigmentation. In Negroes and a number of Melanesian groups, pigment cells are found in the whites of the eye, and in the optical nerve. Connective membranes also have differences in pigmentation. In all races, except the Nordic race, the epithelial cells are pigmented in the deep layers of this membrane.

There are observable racial differences in the layout of the eyelids. Thus, for example, cartilage on the eyefolds, having the shape of a half-moon, is encountered in Europeans in 0.7% of cases; in Japanese in 20% of cases; in Hottentots in 35% of cases; and in Negroes in 75% of cases. There are differences in the shape of the eyelids themselves. There are very narrow eyefolds (in most northern Asians) and wide eyefolds (in Europeans); in Northern and Central Europe, this fold almost always has the form of a "spindle;" that is, both corners of the eye are approximately identical in size; in Southern Europe, Northern Africa, and particularly in the Near East, almond-shaped eyefolds with a rounded nose corner and sharp outer corners are often encountered. If the line drawn through the corners of the eyes extends horizontally, the position of the eyes is called "straight" (in Europeans and Negroes); if the drawn line forms an angle to the nose, the position is called "slanted" (often in Mongoloids and Indians). Real crossed-eyes do not appear tangled, and the reason for this is the special eyelid folds in the interior corner of the eyes. The profane refer to such a form as "snap eyes." It is a question of an upper eyelid fold, which in the zone of the nose, falls down like a second layer of skin, closes the tear caruncula, and part of the edge of the eyelid from the inner corner, to the middle. The outer [corners] are slanted up. If lifted by the finger, it is immediately evident, that they very edge of the eyelid and the shape of the eyefold, are the same as in anyone else. The formation of this fold is tied with the flat form of the nose and

Carlton Stevens Coon

198

the position of the ocular orbs. This fold, as a rule, is encountered in all Mongoloids, and very often in Eskimos, Malaysians; therefore, it is called the "Mongolian fold." It retains this name, although we know today that the Hottentots and their "bastards" have precisely such a fold, even though they have nothing in common with Mongols. Very rarely is it encountered in individual Europeans."

In the book, *Textbook on the Theory of the Heredity of Man* (1940), prominent German anthropologist Guenther Uest systematized all possible variations in pathological deviations in the structure of the outer coverings of the eye.

Erwin Belz turned attention to one racial trait in the structure of the eyes of members of the Mongoloid race, namely the absence or the weak development of a depression between the eyelid and the edge of the forehead.

With the development of genetics, new and interesting patterns were discovered. Thus, Carlton Stevens Coon made a conclusion about modern evolution, under the influence of the selection of both the color of the skin and the pigmentation of eye retina. People with light blue eyes can better differentiate the ultraviolet part of the spectrum, than those with dark eyes, and therefore better differentiate separate objects in dim lighting, or in fog. Besides that among light blue-eyed subjects, type "O" blood was encountered more often.

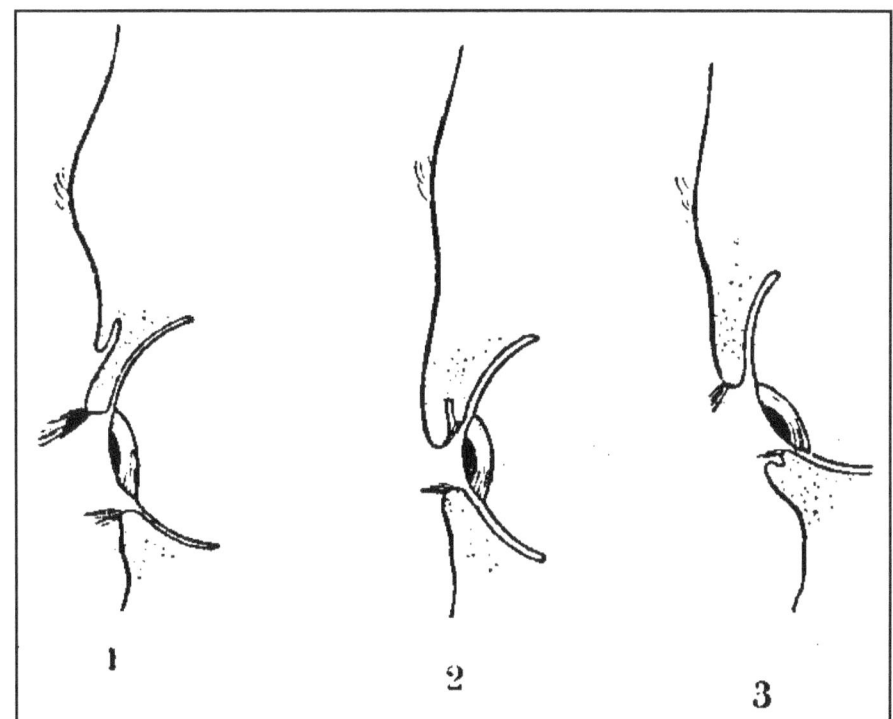

Forms of the Upper Eyefold (according to Erwin Belz): 1 Europoid; **2, 3** Mongoloid.

In the monography, *The Adaptiveness of Beauty*,[219] the famous Russian geneticist, Yuri Ivanovich Novozhenov, emphasized that the color of the eyes of an individual, for the whole extent of evolution, was a key factor in selection, and was in essence, one of the key criterion of racial beauty, on the basis of which prospective mates were selected.

Hauschild came to the conclusion that in the forward section of the eye, pigmentation traits far more strongly act on the common type of eye, while differences in the distribution of pigmentation in the rear section of the eye have less significance. Racial differences are observed in the color and quantity of pigmentation, and in the shape of pigment cells. Their differing types follow:

1) The negroid type of pigmentation cells, which are encountered in Negroes and Melanesians, differ by their clumsy, little-developed pigmentation cell forms in the retina

---

[219] *Adaptivnost' krasoti*. Ekaterinburg, 2005.

membrane, in which the pigment kernel is distributed in a dense mass, such that the cells appear to be formed in a clump. The mass of vessels in the retina membrane are relatively incoherent, and the number of pigmentation cells is less significant, than in Mongoloids and Europoids. The connective tissue is always strongly pigmented, and the granules of pigment, besides the epithelial connective-tissue membrane, also extend around the base of the fold of the cornea, and the upper epithelial cells. The color of the retinal membrane is dark-brown, or dense black-brown. The connective tissue and the *tunica albuginea* appear grey or brown, and not rarely mottled.

2) The Mongoloid-type cells encountered for example, in the Chinese, Japanese, and Koreans, possess a rather thin form of pigment cell in the retina membrane; their appendages are not so coarse or long, and the pigment grains are not so densely distributed. In comparison with the Negroid type, the cells are encountered in large quantity, and uniformly extend to the mass of the retinal membrane. The connective tissue is also strongly pigmented, but only in the lower layers. According to Hauschild's studies, the eyes of Javanese are discovered to be of a Mongoloid type, but with some tendency toward the Negroid. The color of the eyes, as in Negroes, ranges from dark-brown to black-brown; the whites are dingy, brown, and partly mottled.

3) The Europoid type possesses a still thinner form of pigment cell in the retinal membrane, with long, thin, appendages. Pigment granules form irregular clusters, and the relatively small number of pigmentation cells are not large. Connective tissue extends only to the very deepest layers, dispersed around the cell nuclei in small amounts of pigment granules; consequently, they are almost unpigmented. The color of the eyes is light-blue, gray, green, or light brown. Connective membranes look white or bluish.

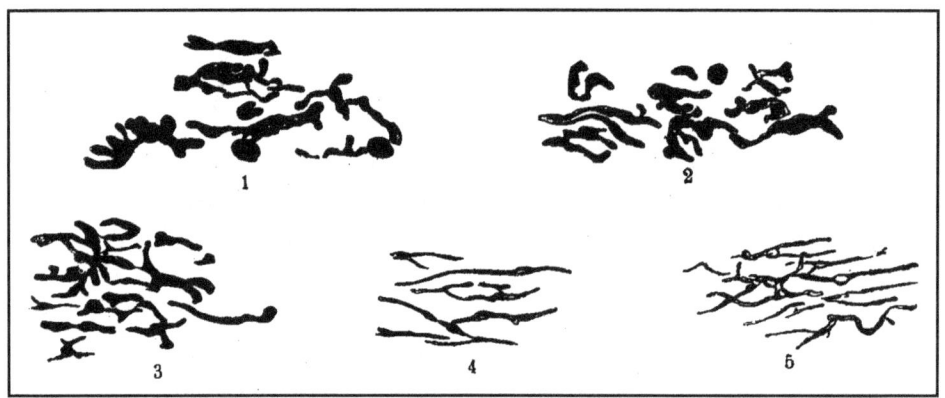

Above: Pigment Cells of the Retinal Membrane of the Eye (according to Max V. Hauschild).
1. Negro; **2.** Melanesian; **3.** Javanese; **4.** Chinese; **5.** European

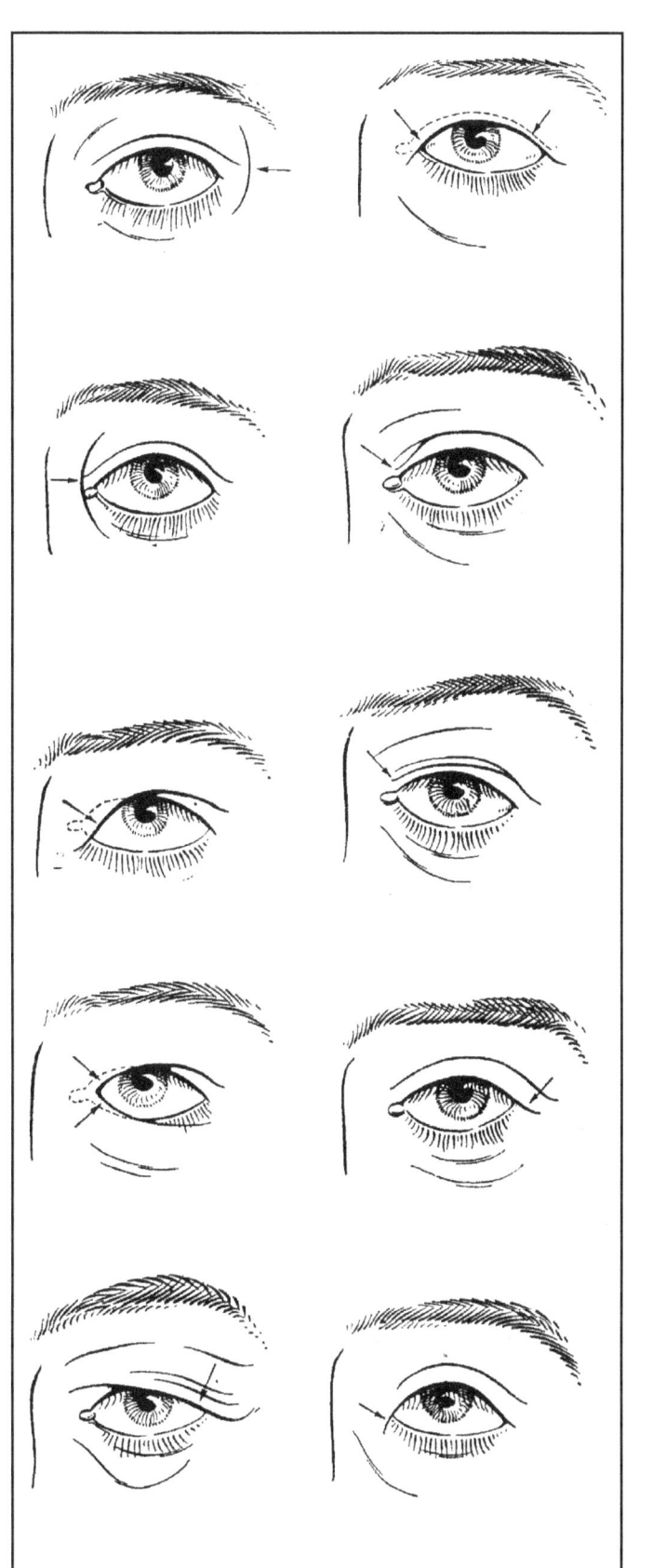

Left: Outer Epicantus
Right: Mongol-Hottentot Fold

Left: Interior Epicantus
Right: Interior Negrito Fold

Left: Mongolian Fold
Right: Tarsal Fold

Left: Double Mongolian Fold
Right: Outer Negrito Fold

Left: Hottentot Fold
Right: Indian Fold

Degenerative Traits in the Structure of the Outer Cover of the Eye (according to Guenther Uest).

Ferdinand Birchner, the distinguished German anthropologist, thus summarily described the given aggregate of structural racial traits.

The long path of the development of anthropology and etology—the investigative sciences of the biological bases of human conduct—continued in the same direction. Many facts of social life that were explained earlier by the influence of abstract cultural differences, received a completely different interpretation. The position of biological determinism essentially became stronger. Morgan Worthy, a distinguished modern American researcher, published a highly noteworthy book in 1974, titled, *Eye Color, Sex, and Race: Keys to Human and Animal Behavior*. Relying on a wealth of statistical material, he explains many core differences in the behaviors of peoples of different races.

It becomes clear that people with dark eyes react more to color, while people with light-colored eyes react more to shape. Dark-eyed individuals are especially sensitive to colors of the long-wavelength portion of the spectrum, since the strong pigmentation partially blocks colors of the short wavelengths. That is namely why southerners love red and yellow colors more, and northerners prefer blue and grey. Besides that light-eyed blondes of the Nordic race differentiate half tones and orient themselves in space better. Dark-eyed people react more to color and are inclined to spontaneous and emotional reactions, while light-eyed types react more to form, and prefer to control their emotions. Dark-eyed people like close contact, while light-eyed people on the contrary build all forms of behavior on distance, considering a reduction of interpersonal space as something bad. In everything, dark-eyed people prefer to follow communitiy standards, while light-eyed people, proceeding from the laws of inner style, work out their life position.

Although the conclusion of Morgan Worthy's work is shockingly simple, it is nevertheless well-argued: "People with light blue eyes first of all perceive shape, and have a scientific mentality, while people with brown eyes perceive colors first, and have an unscientific mentality."

A racial analysis of inventors and rationalists leaves no doubt that genuine science is primarily the brainchild of members of the Nordic race. Consequently, the worldview marked out by the brown-eyed never becomes the genuine heritage of the light-blue eyed man, for the specifics of worldview and eye-color are interconnected.

16. Racial Odors and the Predilections of Cannibals

Doctors and travelers of the ancient world already wrote about the particulars of racial odor; this points to the objectivity and cultivation of the given field of group physiology. In the Middle Ages, missionaries spoke of a "Jewish stench," and the first Europeans that conquered the West Indies heard a phrase from a song by the local negroes: "God loves his negroes, and he knows them by their smell." Another fact is illustrative: in the Europe of the Middle Ages, hair and wig merchants who engaged in the manufacture of wigs, distinguished German hair from French hair by smell, and could even make out Irish, Scottish, English, and Welsh hair. This is how the first Europeans identified the Indians of the New World—by their natural odor, which subsequently received the name *Catinca*.

Johann Blumenbach, one of the founders of anthropological science, substantially discussed the "national traits of the skin." Jean-Joseph Virey (1775-1846), a prominent French anthropologist, doctor, and a member of the Royal Medical Academy, wrote the fundamental three-tome work, *The Natural History of the Human Species* (Paris, 1924). In it, he states: "When the weather is hot for a Negro, his skin becomes oily and blackish; this then soils linen and gives off an extremely unpleasant smell. Negroes stink so strongly, that the places which they pass through remain permeated with this smell for all of a quarter of an hour." And the Russian classical anthropologist, Anatoliy Petrovich Bogdanov, had already pointed out in the middle of the 19th Century that: "Several peoples give off a special odor; thus, it is known for example, that dogs which are used to hunt for runaway slaves, easily differentiate the trail of a Negro from the trail of an Indian."

The **olfactory brain fissure** is also worth adding to the list of specific [racial] traits. Since ancient times, it has been known that all races and tribes have their specific odor, which precedes their origin from the stage of pre-human historical development. Not accidentally then, the sections of the brain that are responsible for smell, have the most ancient origin (from an evolutionary point of view) and their development preceded all forms of mental activity. It is hardly necessary to clarify how great the significance of smell is in the animal world. An astonishing method shows that in the world of people, its significance is also great, although one is not always conscious of it in full measure. **Perfumes, powders, fragrances, and the scents of different peoples also have racial differences, inasmuch as they are used to mitigate the natural odors of their owners.** The tart smell of southerners, which arouses justifiable disgust in members of the Nordic race, is in this regard an outstanding illustration of the biology of the culture-historical genesis of peoples. Karl Vogt pointed out: "The smell of a people is its inalienable historical attribute. Similarly to color, perspiration of the skin has a characteristic trait; in several races, even with the most diligent tidiness, there is always a particular odor. For the Negro a [certain] smell is a pertinent characteristic, as though we did not feed or bathe him."

Only endogenic smell is a subject of study in raciology; that is, that which results from the secretion processes of an organism. In the given case, smell is examined namely as a racial trait. According to the studies of Japanese raciologist Buntaro Adakhi, only 2-3% of Japanese have an underarm smell. In Japanese, the sweat glands are more developed, than in Koreans. And the Chinese give off a distinctive, musky odor. Notions of racial odors are subjective, just like any other racial evaluations, for each party obviously considers itself "the truth in the final instance." Thus, for example, the inhabitants of Melanesia say that "the White Man's smell goes before him." The Indians of the Peruvian forest can distinguish a black man from a white man, at a distance. In general, for the distinct identification of the odors of foreigners, special terminology exists in many languages of the world.

Lord Henry Cames, an English scientist, wrote in his multi-tome report, *Essays on the History of Man* (1774), that Negroes give off "a stench, a skunky smell." Sir Harry Johnson, the famous explorer of Africa, wrote that he identified a thief [who stole] his personal belongings, by his smell, which lingered for a long time in the room, after the unbidden visitors [departed].

H. Ellis, a prominent American anthropologist, wrote the book, *The Psychology of Sex* (1905). He defined the Negro odor as "ammonia and rancid," and wrote that a negro found anywhere smells the same, whether in Haiti or in Washington, D.C. For their part, Buntaro Adakhi and German anthropologist Richard Andree collected numerous information on the natural odor of negroes, pointing out that the inhabitants of Angola, for example, smell stronger than the inhabitants of Senegal; that is, known differences in population variances exist in the activity and structure of the secretion glands of the members of the great Negroid race.

Concerning the Aborigines of Australia, they possess a less intensive smell, in comparison with Negroids; but nevertheless, dogs, horses, and major horned livestock can easily distinguish them from Europeans, and from a significant distance.

It is also worth mentioning the obvious differences in the chemical composition of secretions in different races, and also the correlation of the intensiveness of an odor, with the sizes and shapes of the nasal apertures in the skull.

All this again speaks in favor of the enormous significance of endogenic natural odors in the development of a race, when on the basis of the given trait, prospective mating partners were intuitively and unerringly selected. The division of people into "us" and "them" is not in the least a machination of "fanatical chauvinism" or "zoological xenophobes," as the "politically correct" would try to convince us; it is a basic principle of biochemical evolution. [A] politically correct indistinctness of colors would unavoidably lead to a halt in sexual selection, and as a result, would lead to genetic degeneration. The prospect of a biological certification of the races at this or that stage of development is unavoidable. And it is completely unimportant as to who begins the process. Desire [and motivation] are paramount.

In the middle of the 20th Century, Baron Egon von Eichstedt observed: "Not one explorer said of the black Hindus, that they give off a bad smell. But on the other hand, all explorers testify that the odor given off by black Africans is rotten, and it is preserved even in mestizos, whose great-grandmothers were Negroes. With all obviousness, the Negroids of the South Seas are similar in this [regard]. The Ceylonese smell spicy; the inhabitants of the jungles [smell] smoky; the Chinese [smell of] sweet fat; and many people of the southern latitudes smell of onions and leeks. This author cannot forget the smell of sour fat and burned milk, which permeated all the clothing of subjects of the Toda tribe. Added to that was the smell of perspiration. Unforgettable also are the smells of the small side-streets in Chinese cities, and the sweet smell of the Filipinas. The atrophied olfactory sense of the European does not always distinguish racial smells from the smell of food. Indians and Negroes assert that Europeans smellso bad, that they give off a putrid odor. In Guyana, the Indian women perceive the smell of Negroes as unpleasant, and turn their noses from them. Ethiopians cannot tolerate the smell of the Bantu tribe. Many European explorers assert that Negroes give off a smell of ammonia or goats, and the Chinese [give off a smell] of musk."

Besides differences in the specifics of odor, in the members of different races there is a great difference in the structure of the sweat glands. Japanese raciologist Buntaro Adakhi supported German anthropologist Karl Vogt's opinion, that the sweat glands of the weakly-smelling Japanese are thought to be very small, in comparison with Negro sweat glands. Besides that a direct dependency was discovered between the smell from armpits and the stickiness of earwax. The sweat glands of Negroes in the area of the sex organs essentially exceed the size of the corresponding glands in the Europoid race.

Later, a correlation between smell and taste was revealed. Eichstedt, who had carried out many expeditions, had the opportunity to question cannibals about racial differences. He wrote: "Cannibals particularly like leg muscles. To their taste, whites were tastier than negroes, and the English are tastier than the French." Also, the remarkable anthropologist, Ludwik Krzywicki, relying on rich expeditionary experience, left the following remarks in his monograph, *Anthropology* (1901): "The difference in smell between the white man and the black is so sharp, that it is as sharp as that between a dog and a jackal; in taste (according to the experience of cannibals) the difference is no less significant."

However, Karl Vogt pointed out in this regard: "The color of the meat of the Negro is never such a light-red, as in the European; it is more yellowish, or even brownish."

"Negroes and whites retain their characteristic traits, despite changes in climate. Right down to the inner structure, distinct relationships with orangutans are observed in Negroes. Negroes are a type of man that are different from us", asserts J.J. Virey. As a result of this, for many years until the development of modern embryology, V.A. Moshkov pointed out: "If, according to the law of Gekkel, any living being of pure nature reproduces in its lifetime—the history of its species—there is a question of how history will be reproduced, [if] essentially mixed hybrids are created from two types? The hybrid does not reproduce the history of [just] one species, but two; from the beginning one, and after, a different one. The traits of one parent appear in him at one age—and the traits of the other [parent] in another."

It is namely the mixing of the two different genetic programs that are given to the hybrid in the inheritance of heterogeneous parents, that leads to an imbalance in its whole biological structure, a disruption of psychological moral wholeness. The data from international criminal and psychiatric statistics graphically testify that among racially mixed individuals, there is more criminality, and a higher percentage of nervous and sexual deviations. There is also a greater number of perjurers among hybrids, for the "chaos of blood" is present in their veins. **The hybrid is essentially doomed its whole life to thrash around in the torrents of different bloods.**

17. Major Differences in the Structure of the Inner Organs and Sexual Organs

Deep racial differences are observed in all levels of organization in the human organism, which again allow one to make global conclusions and generalizations. Thus, for example, A.I. Yarkho, a famous Soviet scientist, pointed out in one of his articles,[220] that the spleen of the European is 23% heavier than the spleen of a Negro, and that the liver is 15% heavier. On the other hand, the heart and kidneys in Negroes are on average 12-13% heavier, than those same organs in whites. In the article, *Several Findings on the Morphogenetic Role of the Endocrine*

---

[220] In Russian Anthropological Journal, 1925, issue 1-2. *Russkiy antropologicheskiy zhurnal.*

Glands,[221] (in the same issue of the above-mentioned journal) L.P. Nikolayev developed the idea about essential differences in the structure of the hypophysis in members of the Nordic race, in comparison with Negroids and Mongoloids. Besides that the author came to the unambiguous conviction, that the "endocrine glands play a more decisive, significant [role] in the phylogenetic evolution of Man, and in his gradual development from ape-like ancestors."

Karl Vogt pointed to a simple, obvious fact, the significance of which is underestimated today: "The color of the meat of the Negro is never so light-red, as it is in the European; it is more yellowish, or even brownish." George Buschan, another German anthropologist, stated that: "The intestines of Negroes are shorter than in Europeans; Japanese intestines on the other hand, are longer." For his part, Buschan's eminent countryman, Robert Wiedersheim, noted: "The length of the intestines in whites has an average length equal to 9.6 meters, from which the narrow portions of the intestines account for 8.0 meters, and the thick portions, 1.6 meters. In Negroes, the average length of the intestines is 8.6 meters; that is 1.0 meter shorter [than in whites]. This circumstance, that in Negroes the intestines are shorter, is the result of a shorter length of the narrow intestines, since the large intestine in blacks is longer, than the one in whites. The average weight of the liver in members of the white race is calculated at 1,451 grams, and in blacks at 1,266 grams. The shorter length of the intestines in Negroes agrees perfectly with both the findings of European missionary ethnography, and their national stories, which confirms that the long evolutionary period of this race is tied with cannibalism; for the digestion of human flesh, rich in proteins, one needs an intestine shorter in length, than one for long digestion of plant food. The weight of the liver in members of the black race is also smaller, in comparison with Europoids. This is another testimony in favor of the given fact, for the liver is responsible for filtering the blood.

In the modern textbook, *The Morphology of Man*,[222] written under the editorship of V.A. Nikityuka and V.P. Chtetsov, it is remarked that the average length of the kidneys in Negroes is 11.1cm; in Europeans 12.2cm; and in Fijians—15.0cm; the width of this organ in Negroids is 6cm; in Europoids it is 6.9cm; in Fijians 8.4cm; in Annamites 9.5cm; Indians 10.7cm; and in Arabs—13.2cm. The average mass of the kidney in Malaysians is 210 grams; in Chinese the average weight of the kidney is 275 grams; in Negroes it is 308 grams; and in Europoids it is 313 grams.

It is obvious that a racially uneducated surgeon, might accidentally violate the Hippocratic oath during a transplant procedure, and cause harm to a patient of a different race.

A similar situation may arise in a transplant of the spleen, since in Europeans it weighs 140 grams, in Negroids 115 grams, and in Mongoloids it weighs 90 grams.

Besides that there are big racial differences in the chemistry of the stomach's mucous membrane, and in the branching of the circulatory system.

In *The Arterial System of the Japanese* (1928), Buntaro Adakhi, the prominent Japanese raciologist, outlined **110** traits in the structure of the blood vessels, from which racial differences are easily traced. He indicated: "From an anthropological point of view, it is necessary to create a special anatomy for each race, on the basis of macro and microscopic studies of all the organs. The differences in the soft tissues need to be studied with as much accuracy, as the racial differences in the skull and skeleton. Each anatomist, in the course of his studies, should constantly think about race. The anatomy of Man is the anatomy of the human races."

[221] *Nekotorie dannie o morfogeneticheskoy roli endokrinnikh zhelez v svyazi s voprosom ob izmenenii individual'nikh i rasovikh priznakov*. Russkiy antropologicheskiy zhurnal, vipusk 1-2, 1925.
[222] *Morfologiya cheloveka*. 1990.

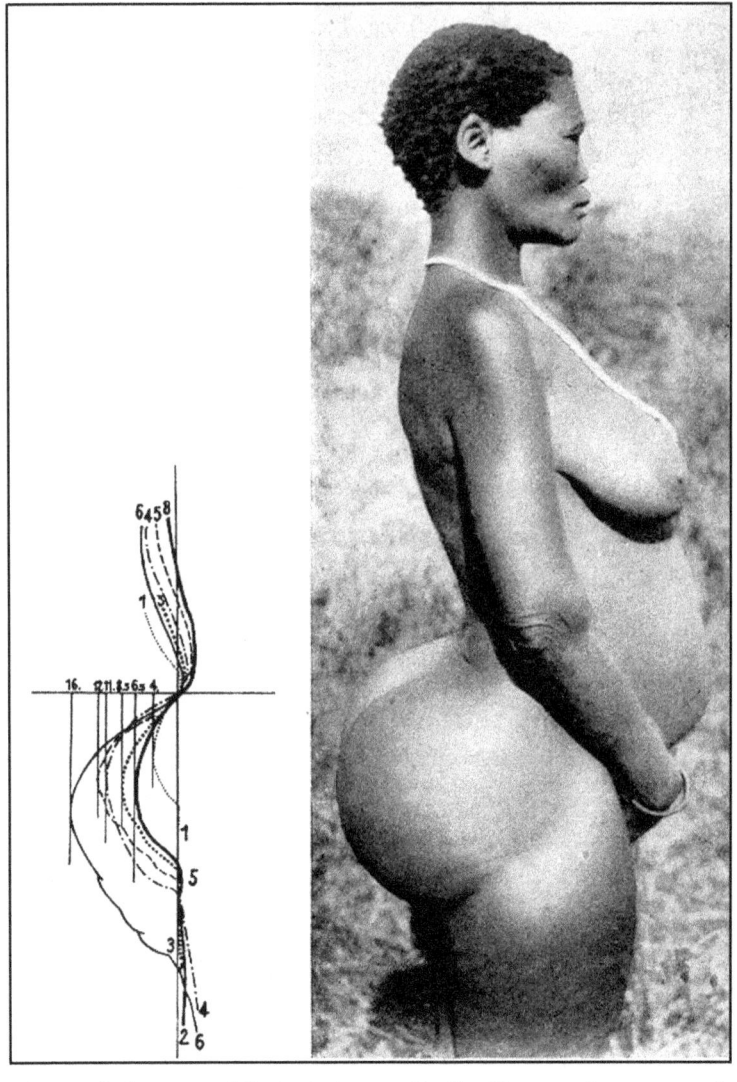

Left: Steatopygia in Bushmen (according to Egon von Eichstedt).

In a dissertation for a doctorate degree in medical sciences,[223] Edward Lot (1884-1944), a Polish anthropologist of German extraction, indicated the convenience of a racial diagnostic, according to a group of traits: "The majority of soft tissues on people are more conservative than the skeleton. On the basis of the observations made, one can show that in the morphological structure of Negroes, their percentage of primitive traits is higher, than in Europeans. From there one may conclude, that by the structure of the muscular system, Negroes are primitive; that is, they are phylogenetically lower. Out of 44 basic muscles on Negroes, 32 are distinctly primitive. Like many other traits, the structure of the facial muscles is an excellent proof of the primitiveness of the muscular structure of Negroes. The structure and function of the muscular system of Negroes is also a distinct proof that phylogenetically, [they are] a more primitive branch of humanity. Besides racial differences, an entire row of variations is discovered in the Negro, which in general are not observed in Europeans. With Mongoloids, the situation is far more complex. Right now, it is not possible to explain why the body of Mongoloids remains more primitive, and only the shins and feet are progressed."

There are also racial differences in the structure of the genitalia—anomalies that cannot be explained by any kind of environmental influences or cultural characteristics. Russian anthropologist Octavian Vasiliyevich Milchevskiy pointed out in 1868: "We notice that in Negroes, the sexual parts stand out in length. Also, the breasts in several women of these tribes are so long, that they can throw them over their shoulders and feed children that they are carrying on their backs."

Josph Hyrtl (1811-1894), the prominent German anthropologist and anatomist, indicated in the book, *Guide to Anatomy* (1887) that: "The springy breasts in girls of the white and yellow races have the singular form of hemispheres; in Negro women in completely identical conditions, they are more oblong, more pointed, and turn down and outward somewhat; in a word, they look more like udders."

Still another luminary of anthropology, Herman Henry Ploss (1829-1885), brought the said issue to a qualitatively different level in his fundamental monograph, *Women in Naturalist and*

---

[223] *Regarding the Anthropology of the Soft Tissues (the Muscle System) of Negroes.* 1912.

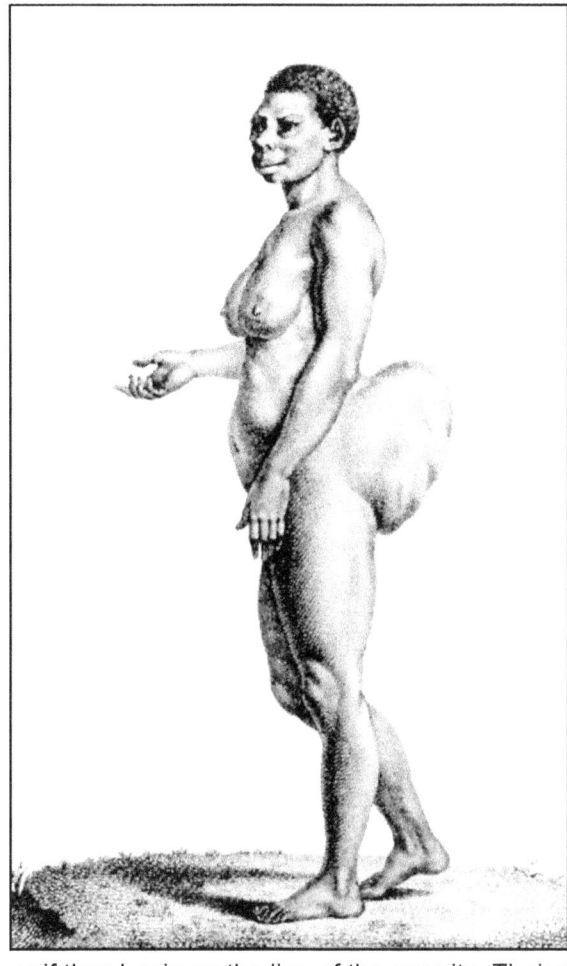

Left: Steatopygia in Bushmen (according to Jeane Joseph Virey).

Social Studies,[224] by using a graphic classification of the trait: "In the structure of the breasts in different races, important differences are observed. The nipples, for example, are small and flat, like buttons, or they are large and have a conical form, a wide base, and a curved peak, or they are large, cylindrical, and equal in size to a finger joint. Like the nipples, the aureola circles are also variously formed. They go from light-colored to brown-colored, or brown, or almost black. There are also differences in their size: they can be small, or extremely large; sometimes they protrude a little from the spherical shape of the breasts, sometimes they form an elevation on it, in the shape of a hemisphere; and in some cases they are distinctly separated by an expressed fissure that surrounds it. Considering the elevation of the breasts themselves, it is necessary to take into account: either they more or less starkly form walls on the surface of the breast, or a layer of fatty cellulose begins to grow from the collarbone, imperceptibly extending to the breasts. It is worth giving attention to the place where the breasts are situated; that is, if they sit higher or lower, closer toward the mid-line, or if they begin on the line of the armpits. Their size, volume, and form have particularly important significance.

It is necessary to differentiate the following forms of breasts:

A) By size:
   1) large or plump;
   2) full;
   3) moderate or average;
   4) small or weakly developed.

B) By density; that is, by less or more springiness:
   1) erect;
   2) hanging down;
   3) loosely-hanging.

It is necessary to remember, that in several cases the breasts may be flaccid, thanks to a certain original form, but they may nevertheless be springy. In general, the form of the breasts may be subdivided into the following groups:

1) cup-shaped;
2) hemispherical;
3) conical;
4) similar to goat udders.

---

[224] *Zhenshchina v estestvovedenii i narodovedenii.* Kiev-Kharkov, 1899, Tomes I and II.

Left: Joseph Hyrtl

Cup-shaped breasts are similar to a half-tangerine in shape: the diameter of the base significantly exceeds its height.

In their size, the hemispherical breasts resemble a half (or three-quarters) of an apple, orange, or coconut; its height is almost equal to the diameter of its base.

The conical breast resembles a pear or a lemon. Its height, that is, the distance between the nipple and the middle of [the breast's] base, significantly exceeds the diameter of the latter; the same may be said about the fourth form of breast."

Egon von Eichstedt brought a classification of races—according to the form of women's breasts—to completion in the 20th Century. He wrote: "Female mammary glands also have great raciological significance. This is in regard to the size, form, and placement of the nipples, areali, and the breasts themselves. In their development, the breasts go through four stages. The formation of a secondary breast gland is worth considering as a progressive trait, since in many colored races, particularly in Negroes, Negritos, and Australoids, the development of the areoli stops at the stage of the first breast gland.

A breast button is very often encountered in these races, even in grown women, while in Europoid and northern Mongloid women, this form is only encountered in pathological cases.

The basic forms of the breast are considered to be a plate or cup shape, semicircular, conical, and udder-shaped (the latter two—the breasts shaped like cones or udder-shaped—are in the shape of goat udders). The first of the named forms is particularly often encountered among Chinese and Paleo-Mongoloids, and quite often in South Indians. The semi-circular form [on average has] a size of 12cm x 12cm x 6cm—usually in young Europoid women, although there are significant differences on hand in frequency in different groups, especially between European and outer-European Europoids. The conical form, usually in the majority of Negroid groups, very often changes, as in the Australoids, in youthful years, to the form of a goat's hanging udder, which is praised by the poets of the Muslim world.

The areoli are quite varied in size and form. Their color fluctuates from a light, often transparent pink in northern Europeans, to a light, dark-brown in colored races. All the named differences are racial differences."

In the Great Age of Exploration, French explorers and naturalists were the first to point to several aspects of morphology, which did not permit one to talk of an evident unity of Mankind—under any conditions. These are the **steatopygia** in Bushmen and the so-called **Hottenton Apron** (Labiorum Minorum). By steatopygia, the excessive development of fat deposits under the buttock muscles in the women of this tribe is meant; this gives them a completely absurd look, increasing the volume of the hips several times. Eugen Fischer revealed the essence of this unusual anthropological phenomena: "Physical parallels exist between the steatopygia of the Bushmen and Hottentot women, and the fat tail breeds of goat and sheep."

The given conclusion of the classical scientist can in no way be considered racist, since Soviet anthropologists expressed a completely analogous [opinion]. Thus for example, in the book, *Geography of the Human Races*,[225] V.P. Alekseyev pointed out: "In the steatopygia of Bushmen and Hottentots, it is difficult not to see traits that are analogous with those morpho-physiological adaptations, such as the humpback of the camel, which normalize the water-saline exchange in animals that are inhabitants of the desert." G.F. Debets, another recognized Russian specialist,

---

[225] *Geografiya chelovecheskikh ras.* Moscow, 1974.

wrote in the article, *Anthropological Data on the Population of Africa*,[226] that: "A suggestion worthy of attention has been expressed, that the structure of the fold of Bushmen may be tied with a periodic abundance and sharp lack of food. In the deep antiquity of the racial types of which we speak, patterns of development which are observed in the animal world, made a conceivable transfer to the human races."

American anthropologist Ashley Montagu also emphasized in the book, *An Introduction to Physical Anthropology* (1951), that the Bushmen and Hottentots in Africa have a very low percentatge of Type "B" blood, and a very high percentage of Type "O" blood, which speaks of their relatively high racial purity; therefore, all the anomalies in the structure of their brains, genitalia, and metabolism cannot in principle be explained by some all-powerful mutation, of which geneticists constantly speak, or some specific trick of the environment, which cultural anthropologists recycle. The anomalous irregular physique of the given racial groups may only be explained by a special path of evolution. "The Bushmen and Hottentot tribes stand apart from the rest of the native inhabitants of Africa. The skin is very dry and thin, to the exclusion of the upper parts of the hips and the berry patch, particularly in women, where the skin is tightly drawn around the vast and thick forms of the buttocks; it is known as a certain type of morphological anomaly called **steatopygia**.

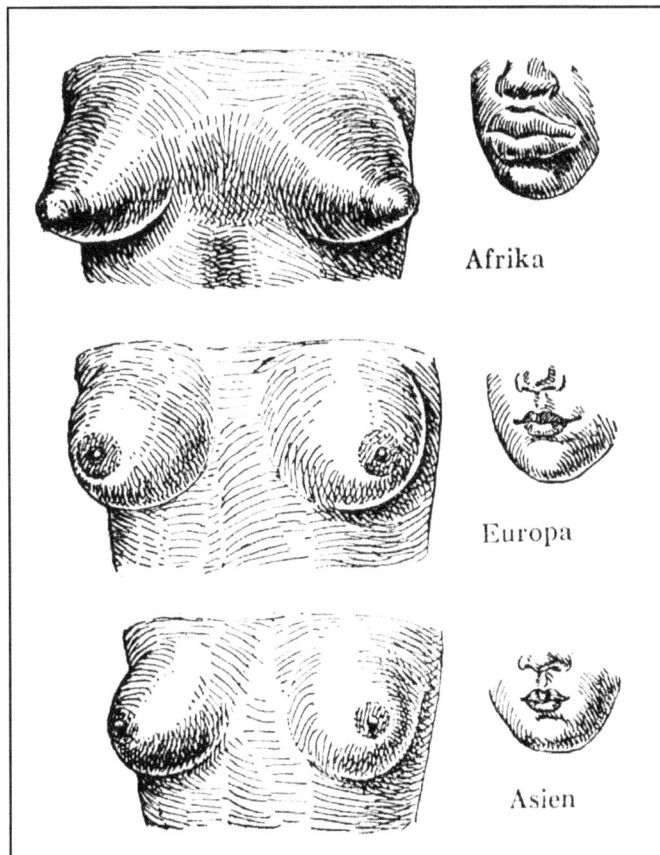

Afrika

Europa

Asien

Left: Corresponding forms of breast according to the structure of the lips, in members of different races (according to Hans Friedenthal).

This surprising characteristic graphically shows the uniqueness of the physique of the given racial group, since the obesity of the lower parts of the body in women are retained between alternating periods of hunger, helping to support life in the extreme conditions of lack of food and moisture."

G.G. Ploss wrote in this regard: "The excessive deposits of fat layers on the buttocks is called steatopygia. This springy mass consists of only fat, from which big branches of connecting fibers penetrate in all directions, highly irregularly criss-crossing each other."

The so-called "Hottenton Venus," described by the great French naturalist Georges Cuvier, possesses buttocks that reach a height of 16.2cm; the brain weighs 812 grams.

Karl Linneus, the founder of zoological systematics, defined the concept of "morphological identity," and "morphological unity." It remains completely incomprehensible, how modern politically correct alchemists, contrive to record all kinds of organisms, with such visible differences, into some "single humankind."

Charles Darwin wrote that he saw a Hottentot woman, who was considered a beauty; the rear end of her body was so huge, that sitting on level ground, this woman could not stand, and had to

---

[226] *Antropologicheskie dannie o zaselenie Afriki*. Sbornik "Proiskhozhdenie cheloveka i drevnee raselenie chelovechestva." Moscow, 1951.

crawl on her haunches to the nearest incline. The men of this tribe, choosing women for themselves, stood them in a row, and preferred the one whose [buttocks] stood out most from the rear. According to the remarks of the scientist, nothing could be more disgusting to the Negro, than an opposite form of the body.

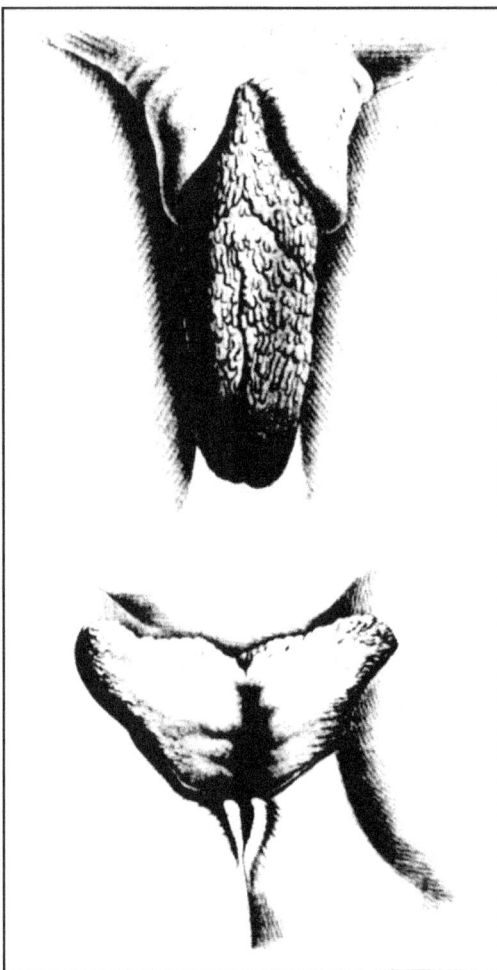

Left: the "Hottentot Apron."

By the term "Hottentot Apron" the unnaturally long sexual lips in the women of the given tribe are meant; they often reach 15-18cm in length, and hang to the knees; in connection with this, since ancient times the custom developed of cutting these off these lips before entering into marriage. In 1554 Jesuit missionaries came to Abyssinia, attempting to bring the Abyssinian Coptic Church into the Catholic fold. In line with this, they tried to prohibit the given gyneacological operation; this immediately sparked a revolt, for the girls who did not undergo this procedure could not find eligible bachelors from their tribes for themselves, since according to the native understanding, the given anatomical phenomenon is disgusting. By special decree of the Pope in Rome, the aborigines were allowed to return to their original custom, in order to not set any obstacles to the spread of Christianity. The Jesuits were expelled in 1633. As a conscientious scientist, prominent French anthropologist Paul Topinard analyzed the given fact from an evolutionary point of view, and came to this conclusion: "We notice that the "apron" does not speak in favor of an immediate kinship of the Hottentots, Bushmen, and apes, since in the female gorilla these lips are completely invisible." Thus, if you relate the given type of racial phenomenon with classical Darwinist science, you are led to unavoidably agree with the presence of a separate, independent branch in the development of the given tribes, their origin coming not from apes, but from some unknown, exotic animals.

Topinard's countryman, Jeane Joseph Virey, wrote in this regard: "Suppose that in Bushmen there is something not unlike a skin apron, hanging from the front and covering the sexual organs. In reality, it is no more than a lengthening of the pudendal labia, to 16cm. They come out on each side of the large pudendal lips, of which there are almost none, and join at the top, forming a hood above the clitoris, and covering the entrance to the vagina. They can be lifted over the front, like two ears. This may be explained by a natural inferiority of the Negro race, in comparison with the white. Therefore, it is more correct to speak of a Negrito species, and not of a race, since this same characteristic structure of the sexual organs is observed in the Coptic and Ethiopian women."

George Schweinfurt (1836-1921), a prominent German anthropologist, ethnographer, and explorer, leaving a history of the science in his brilliant two-tome report, *In the Heart of Africa*, pointed out that in general, in all the southern races the sexual organs of the women lie higher and closer to the front, than in European women. Concerning African women, he wrote: "Only in the orangutan do we find vestiges of large pudendal lips." Other luminaries of anthropology, like Wilhelm Waldeier (1836-1921), Friedrich Tideman (1781-1861), Gustav Fritsch (1838-1891), and William Henry Flower (1804-1899), thought that the construction of the sexual organs in the Bushmen and Hottentot women were morphologically animal in nature, or a so-called theromorphic trait. G.G. Ploss wrote: "The Hottentot apron is a strong hypertrophic on the small, pudendal labia, up to 18cm in length. One cannot fail to recognize an essential similarity of structure in the sexual organs in the labia of the chimpanzee, and Bushmen women."

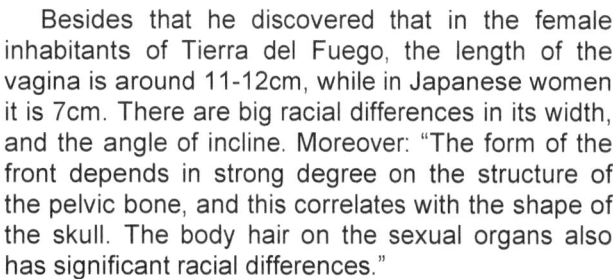

Left: George Schweinfurt

Besides that he discovered that in the female inhabitants of Tierra del Fuego, the length of the vagina is around 11-12cm, while in Japanese women it is 7cm. There are big racial differences in its width, and the angle of incline. Moreover: "The form of the front depends in strong degree on the structure of the pelvic bone, and this correlates with the shape of the skull. The body hair on the sexual organs also has significant racial differences."

In his book, *The Tragic End of the Bushmen* (1956), explorer Victor Ellenburg also emphasized: "The half-erection can be added to the number of distinguishing traits of the physique of the Bushmen; their penis is found to be in a constant state of half-erection. This hereditary trait of the Bushmen race is imprinted on a multitude of Bushmen pictographs." In his classification of races, Karl Linnaeus in general defined the Bushmen and Hottentots as "monozygotic/identical." Besides that in both of these tribal groups, in 80% of cases an under-developed scrotum is observed. And Egon von Eichstedt even developed his own system of classification of the races, according to the form of women's breasts, by the angle of incline of the vagina, and also by the sizes and system of suspension of the male sex organs. He pointed out: "The size of the member of Negroids—which can only be judged in its inactive state—was noted in antiquity and gave rise to superstitious tales about the sexual power of Negroes, which do not appear to be anything special, against the background of the common tirelessness, which is conditioned by a different racial-physiological rhythm. It is true that only African Negroids have such large members; Melanesians cannot brag about this. Among Mongolians, the member is smaller than in Europoids, but there are intra-racial differences: among Koreans, the member is larger, than among the Japanese; and the corresponding size of the head of the member is reversed."

Polish anthropologist Karol Stojanowsky created his own classification, finding a correlation between the racial type, and the form of the head of the penis. Eichstedt turned attention to still another noteworthy fact of morphology: "In the Mongoloid race, the position of the urethra opening [in women] is forward, which is absolutely not so in Europe; everywhere in the East, one may observe the habit of women to answer their need while standing, or walking, thanks to this trait."

Obviously, these morphological differences in the races are not culturally influenced.

Many anthropologists, ethnographers, and psychologists wrote about the influence of anthropological traits on the specifics of the cultural lives of peoples. The intimate sphere is no exception. Ploss wrote: "For anatomical observations, the only people in antiquity possessing at least some kind of exact information about the inner organs of the human body were the ancient Egyptians, in all probability through the custom of embalming corpes at the first convenient opportunity. Easily accessible to the eye and finger, the parts of the female sexual apparatus were very well known to the Talmudists, who had at their disposal an extremely rich nomenclature, and a wealth of synonyms for these organs."

The least "sexually anxious" of all in ancient times were the peoples of the East. For example, the first guide to gyneaocology appeared in Japan only by the 19th Century, while in Europe, one appeared in ancient Rome in the 3rd Century. In the New Age, the influence of the Jews in this area is distinctly visible. The first Institute of Sexual Pathology was opened by Magnus Hirschfeld

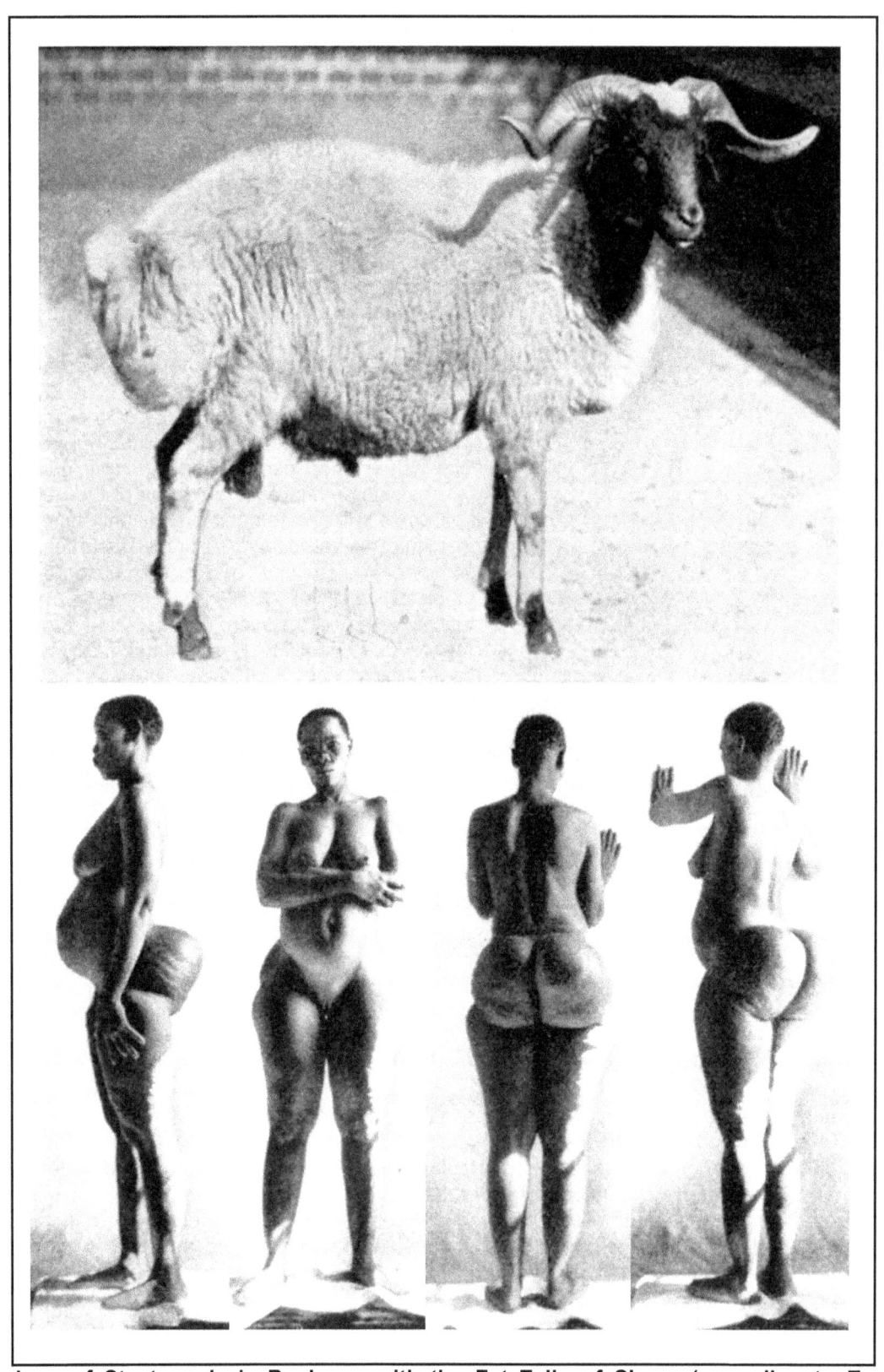

Analogy of Steatopygia in Bushmen with the Fat Tails of Sheep (according to Eugen Fischer).

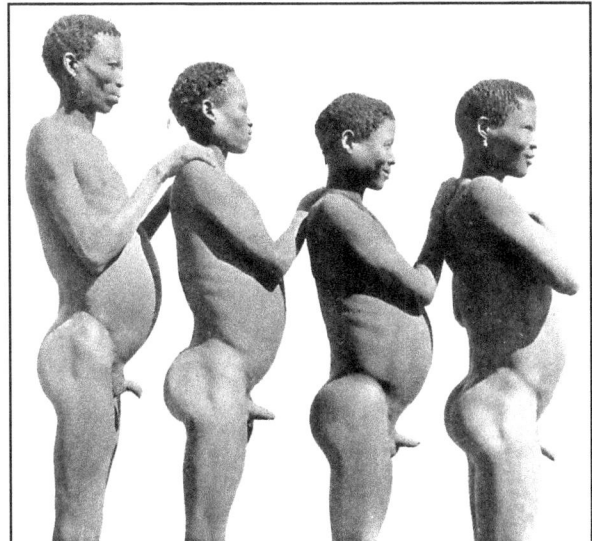

Left: Semi-erection of the Penis in Bushmen (according to Egon von Eichstedt).

In Germany at the beginning of the 20th Century, and scientist Sigmund Freud also rooted all questions of psychology in anatomical specifics. G.G. Ploss turns our attention to the fact that the lower we descend on the evolutionary ladder, the more racial and ethnic diversity in the conduct and canons of sexuality increases. "On the one hand, at one time among the majority of peoples, mainly in the East, the protection of the virginal membrane (hymen) was given extremely high significance, as an external sign of virginity. On the other hand, it was completely ruined in very young girls in China, Africa, and very often in India. In many peoples, there is a rather strange type of custom of trimming away the sexual organs of girls. In the beginning, we knew only of a similar custom in Africa, and considered it peculiar to African tribes only. However, we learned about the existence of such a custom in Asia, namely in Indonesia. This cutting away that is carried out against girls is called **excision**. It is accompanied by ceremonies and festivities, and in great part, is carried out at an early age. Another operation is closely tied with the circumcision of girls—the so-called **infibulation**, or sewing together of the sexual vent. This means of disfigurement does not facilitate conjugal relations. Therefore, among peoples where infibulation is widespread, it is necessary in that case to undo the procedure; the sewn vent is cut open along the scar."

Modern culturologists and sociologists explain such "technical" diversity as a matter of some cultural influences, which is completely absurd in our eyes. Evidently, the structural differences of the sexual apparatus in different races arouse a number of variations around the satisfaction of the basic instinct to perpetuate the species. Culture (if it is appropriate here to use a word that is often slapped on like a label, without analysis, come what may) counts for nothing here, for it is a primary anthropological fact, that all ceremonial, ethnic, and mythological inventions that spring from the different peoples of the earth are only a form of camouflage, with the goal of concealing the anomalies in their own physique, of which they are conscious of in full measure. In classical anthropology, the given phenomenon is called **racial pathology**.

In his fundamental monography, *Secret Libido*,[227] the prominent modern geneticist, Yuriy Ivanovich Novozhenov, writes: "The changeability of the anatomical traits in Man, including the genitalia, is expressed in polymorphism; that is, individual variableness within the limits of one population, and besides that in polytypical (several species) racial differences."

Moreover, he points out that racial differences in the structure of the sexual organs render influence on the psychological aspects of sexual life. Herein lies the secret of the physical modification of the genitalia by the members of different races, for the natural condition does not permit them to achieve desired satisfaction, inasmuch as their biological structure is flawed and insufficient. The classical Russian scientist, Nikolay Nikolayevich Miklukho-Maklay (1846-1888), repeatedly pointed to this fact in his ethnographic works.

---

[227] *Tainstvennoe libido*. Ekaterinburg, 1999.

213

Above: Variation in the Structure of the Breasts in Negroid Women
(According to Egon von Eichstedt).

18. The Chief Differences in the Physiology of the Races

Differences in the physiological and biochemical structure of the members of different races has a huge applied significance; therefore, in the work, *The Significance of Anthropology in Medicine*,[228] P.A. Minakov wrote: "Racial and tribal characteristics, passed down from generation to generation, serve very often as the cause of disease with the help of external factors, which in the subjects of another organization do not usually cause any pathological changes. Entering on a new path of research of the causes of pathological processes, medicine should work on anatomy, physiology, and the pathology of the races, and indicate which anatomical and physical traits are characteristic of pure races and mixed races; and which types of mixed races are more often subject to—or on the other hand—immune to this or that disease."

Despite its origin, the biological science of that time also developed in the river current of full-fledged racial philosophy. In the article, *Biological Reactions and their Significance in the Systematic of Apes and Man*,[229] V.G. Shtefko made a significant conclusion: "Views expressed on the basis of experimental data lead us to an extremely important - and to a high degree - an

---

[228] *Znachenie antropologii v meditsine*. Russkiy antropologicheskiy zhurnal, N1, 1902.
[229] *Biologicheskie reaktsii i ikh znachenie v sistematike obez'yan i cheloveka*. Russkiy antropologicheskiy zhurnal, Tom 12, kniga 1-2, 1922.

interesting conclusion. The cultured races of humanity—for example, the Europeans—have a more complex structure of the protein molecule, than the lower races. Thus, from a biological, or more precisely, a biochemical point of view, they are more complexly organized, than [the lower races]."

In light of our discussions, it is worth remembering again a fundamental work of impressive dimensions: *General Psychology with Physiognomy*,[230] by Professor I.A. Sikorsky. In it, Sikorsky made the valid conclusion that the sum total of anthropological and physiological traits of a race finds its regular continuation in the specific features of its spiritual organization: "The physiological characteristics and attributes of races—like physical types—belong to stable traits, and may be taken as a principle, that the basic spiritual traits of an anthropologically original race endure long and firmly in derivative tribes."

Now it is necessary to turn to the rich reservoir of information which we discovered in the notes and observations of the explorers of preceding times, in the age of the creation of the great colonial empires, when the white man first collided with the aborigines of distant countries, and was shocked by several characteristics of their anthropological and psycho-physiological make-up.

Studied differences in heat exchange in the body, respiration, and pulse in the members of different races: John Devi (1790-1868); Pierre Fouassac (1801-1896); and James Pritchard (1786-1848).

Studied differences in the time of sexual puberty: Louis armand de Lom d'Arce de baron de La Hontan (1666-1715); George Henry Loskiel (1740-1814); and Jean Marie Keating (1852-1893).

Analyzed the types of racial deformities: Benjamin Ellis (1798-1831) and Charles Pickering (1805-1878).

Drew attention to the characteristics of the racial layout of the digestive organs: Le Comte Stanislas d'Escayrac de Lauture (1826-1868); Martin Dobrizhoffer (1717-1791); and Karl Friedrich Heusinger (1792-1883).

Studied questions on racial aging: Filippo Luigi Gilli (1721-1789); William John Burchell (1782-1863); Johann Jakob von Tschudi (1818-1889); Alcide Dessalines d' vedi Orbigny (1802-1857); and John Marie Keating.

Implemented measurement of racial traits in the musculature: John Tanner (1780-1847); Michael Alberti (1682-1857); William John Burchell; and Lambert Adolphe Jacques Quetelet (1796-1874).

Studied racial lice and parasites: Friedrich Martin Duttenhofer (1818-1859) and Jeane Joseph Virey (1775-1846).

Subjected the ability to turn red from embarrassment in members of different races to a comprehensive analysis: George Barrington (1775-1804); August Friedrich Ferdinand von Kotzebue (1761-1819); Marcus Jacob Monrad (1816-1897); and Alexis Casimir Dupuy (1774-1849).

Substantiated racial differences in embryology: Karl Ernst von Baer (1792-1876); Albert von Koelliker (1817-1905); and Francis Balfour (1851-1882).

The first to discover differences in the structure of spermatozoa in members of different races: Gustav Retzius (1842-1919).

Besides that Henry Luis Hollard (1801-1866) and Eusebius Francois de Salle (1796-1873) found an unknown specific—a significant percentage of cohesion of the nose bones, and also discovered an elbow dimple on the skeletons of the extinct inhabitants of the Canary Islands—the Guanches.

Theodore Weitz's book, *On the Unity of the Human Species and the Original State of Man*,[231] is unique for its abundance of information of a similar type. From it we know that among the many inhabitants of Africa, there is a tribal custom of butting heads, because in the collision, they do not experience any pain. To this day, among many native inhabitants there is an ability to break sticks in two on a [person's] head; the head acts as a fulcrum. It is also known that the Spanish

---

[230] *Vseobshchaya psikhologiya s fiziognomikoy*. Kiev, 1904.
[231] Moscow, 1867.

Conquistadores who conquered America, complained more than once in Haiti and Cuba that they could not split the skulls of the Indians with a single blow of the sword. Then again, in antiquity Herodotus pointed out differences in thickness in the skulls of different tribes.

Sir Arthur Keith, the prominent English anthropologist, was one of the first to begin to analyze the influence of hormones on racial differences. He came to the conclusion that not only the primitive structure of the Neanderthals, but even the profile of European faces and the anomalies of the deviation in the functioning of the genitalia in Mongoloids and Bushmen, are in principle explained by a species-specific activity of the sexual glands and adrenal gland. Besides that differences in body structure, hair covering, and pigmentation also have [such an] origin. It was established that the thyroid gland in Malaysians is not only absolutely far smaller, but in relation to the height and weight of the body, it is relatively smaller than in Europeans. In the latter there is one gram of the substance of this gland per 1,920 grams of body mass; in Malaysians the ratio is one gram for every 3,751 grams [of body mass]. In Chinese these differences are not so essential, however, the weight of the pituitary gland in them is less than in Europeans. And for their part, in Negroes the weight of the endocrine gland is less than in whites. These hereditary differences allowed Egon von Eichstedt to create a classification of races, according to the weight of the organs of the endocrine system, and also to reconcile these characteristics with the frequency of psychological and sexual disorders, in the members of the different races; in turn, this manifests in all aspects of social life. He wrote: "It was long ago observed, that Negroes stood out for their special resistance to Yellow Fever. Among Negroes, it is rare to encounter rickets, tumors, pernicious anemia, acne, and certain forms of syphilis; on the other hand, they are often subject to tuberculosis, which is characteristic of the majority of colored and primitive races, particularly in the cold climates settled by them. In Negroids, very fat scars form, and they easily heal wounds. They are less sensitive to pain; therefore, in many Negro tribes, scars serve as decorations. Negroes also differ by time of death from various diseases. In the USA, Negroes often die in the summer; whites often die in the winter."

As a factor in hereditary pathology, race-mixing was comprehensively studied, allowing Eichstedt to summarize: "The general subjectivity of Swedish and Lapp hybrids to diseases, the lowering of their spiritual abilities, and the appearance of numerous deformities is a recorded fact. The disharmonious and puffy faces of these hybrids are well-described. Much data exists about the disharmonious and constitutional disruptions in the hybrids of Europeans with Japanese, Europeans with Asians, and also in the hybrids of Europeans with Negroes. It is worth establishing that in many cases, the mixing of distant races causes obvious disharmony and disruption. It is also understandable, why hybrids arouse tension and disharmony within the peoples of their cultural and social groups.

I considered the proof of this on a night express [train] in Java. Opposite me sat an asthenic old Indonesian; in the neighboring compartment was a 15 year-old with a minimum weight of 175 pounds. In Surabae a very handsome Indonesian got on the train. Therein lies the problem of race-mixing: in the changeability of the sub-clinical, hormonal, or physiological shifts caused by it."

By common opinion, English and American researchers are pioneers in the area of the study of racial differences in metabolism and viability. They introduced the term **base metabolic rate**, by which is meant a minimum of metabolic processes necessary for the further functioning of the organs of the body. In principle, this is similar to the condition of a car running in idle. In this state, the body should be in complete rest, and research should be conducted 12 hours after eating. Then the measurable heat transfer of the body will show a continuing expenditure of calories, differences in dependence from the surface of the body, stimulation, time of year, age, sex and race. The very lowest value of base metabolic rate is noted in East-Asian women, and one of the highest indexes is noted in the Mayan Indians; however, they have the faintest pulse. It is more frequent in the peoples of Indo-China and Negritos, than in Europeans. Blood pressure in Mongoloids is even lower than in Europoids or Negroids. The number of blood platelets in the basic races differs by as much as 20%. There are great differences in the size and structure of the circulatory system. The temperature of the body also shows racial fluctuations, since the **average thermo-regulatory value** in Negroes and Indians is lower, than in whites. The first explorers from Europe in the Age of Discovery noted that Negroes and Hindus have cold skin, while Ethiopian women, on the other hand, had very warm skin.

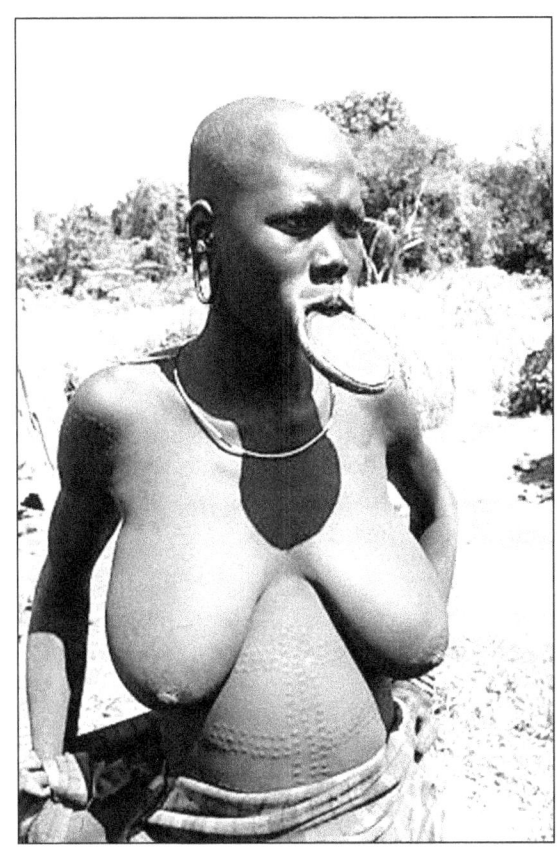

Above and Below: Examples of Ritual Disfigurement

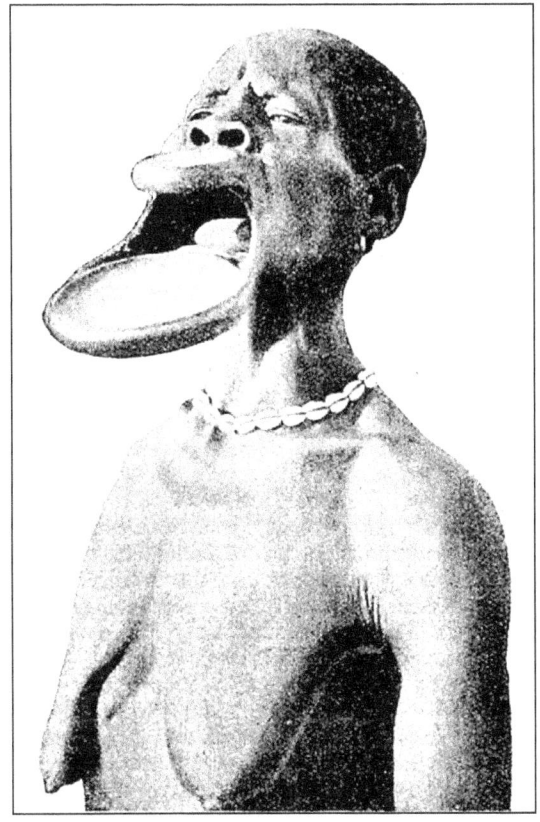

217

19. Styles of Disfigurement in Different Races

And now we return to the theme, with which we started our presentation, namely: anthropo-aesthetics. With the abundance of physical distinctions that exist in the build of the members of the different races, it is completely natural that the criterion of beauty, with which they evaluate countrymen and foreigners, should be different.

G.G. Ploss points out with all clarity in the essay cited by us, that the canons of beauty in all races are highly different: "It is unlikely that we will succeed in [making] the typical form of beauty for each race; what touches on the "eternal canons of beauty," as applied for each case, is that each agrees that they do not exist; every knows that the Negro considers his Negress to be so good, the Kalmyk his Kalmyk woman to be very beautiful, and the European adores the women in Raphael's paintings. A look at the population of the whole world shows us that everywhere, young men strive to possess girls, even in those races, where the girls arouse disgust in us at their very appearance, even in their blossoming years. This unavoidably leads us to the conclusion that the ideal of beauty in different peoples should be highly varied and different."

Charles Darwin, as a great thorough scientist, knew how to critically analyze the biological aspects of beauty: "Negroes do not like the color of our [European] skin; they look at light blue eyes with disgust, and think that our noses are far too long, and our lips far too thin."

Yuriy Ivanovich Novozhenov emphasizes in the book, *The Adaptiveness of Beauty*,[232] that: "The criterion of beauty is worked out by a population, by way of constant selection of such traits and their combinations, which bring adaptiveness to the population, society, and culture."

Konrad Lorenz, the prominent Austrian specialist in the area of etology, wrote: "Our aesthetic perception is distinctly tied with bodily changes." 100 years before him, similar convictions were held by the prominent French anthropologist, Paul Topinard: "People with round heads assert that this form is the most rousing. The Chinese assure us that a flat face and slanted eyes are a pearl of art. In the opinion of the Negro, the most beautiful color is black."

Thus, the value criterion perceptions in peoples of various races are derivative functions of their racial traits.

For a clarification of this thesis, we will turn to historical testimonies.

James Cook (1728-1779), the distinguished English explorer, after discovering Australia, was the first to see the native inhabitants of this continent; he left a description of them that touched on their anthropological preferences.

"One of the main decorations is bone, which they pass through the cartilage that separates one nostril from another. This perversion of taste is considered a decoration! It is outside of human wisdom to think what would spur them to wear similar decorations, and in this to suffer the pain and discomfort necessarily associated with it. This bone is as thick as a thumb and is 5 to 6 inches long; it goes at cross angles to the face, and obstructs both nostrils, so that the savages have to hold their mouths open against their will, in order to have a chance to breath; in efforts to speech, they speak with such a twang, that they are hardly able to understand one another. As a joke, our sailors call this decoration…Actually, this decoration is so strange to see, that before we became accustomed to it, we restrained ourselves from laughing, with difficulty."

It only remains to be suggested, but in light of our discussions, it appears completely logical, that the savages ate Cook namely because of his refusal to accept their essential anthropo-aesthetic canons. Cook called them an "ape-looking nation."

The Great Soviet Encyclopedia speaks of Georges Cuvier (1769-1832) as an outstanding French naturalist, whose contributions are unarguably the creation of an exact method of research and an accumulation of a huge [amount] of factual material." When Cuvier first saw a Bushman woman in Paris, he left this description on the basis of his progressive method: "Her manner of stretching her lips was completely like that of an orangutan. Her abrupt movements and capriciousness recalled the movements of an ape; her lips were fat and formless. Her ears resembled the ears of several apes, by the small size and the weak development of the ear lobe, and the almost complete lack of an outer edge in its forward part. These are all animal traits. I never saw a human head more similar to a chimpanzee, than the head of this woman."

---

[232] *Adaptivnost' krasoti.* 2005.

218

Representations of one's own beauty in the members of the human races have a deep, inborn biological basis; the durability then, of these notions biologically verifies the struggle for existence. In the monograph, *Anthropology*, Polish anthropologist Ludwik Krzywicki left an eloquent observation: "Each group considers itself the most beautiful, when it hasn't encountered another, stronger group. The Tasmanians looked upon their own black skin as the pearl of completion, and advised the arrivals from Europe to cover themselves with coal, in order to hide their ugliness—their white skin, that is. The Negro cannot be "kin" of the White. We could equally search in vain for kinship between an eagle and an ass and a horse."

The book, *Modes of Disfigurement: How it is Expressed in the Customs of Barbarian and Civilized Races*,[233] is food for deep thought on the given question. It was written by the famous English scientist and explorer, William Henry Flower (1831-1899). In it, he analyzes the numerous material on the artificial disfigurement of one's appearance in favor of the local notions of beauty in peoples of different races. The practice of shaving the head disappeared from the customs of savage peoples. They often used stone knives, bits of smashed bone, or shells for this painful and un-hygenic procedure. This is related to the plucking of eyebrows. With the same goal, the native tribesmen on the island of Tahiti liked to scar the shoulders and back with sharp stones. After several days the row of scars are rubbed with wooden ashes or natural dyes, from which large welts form in different colors on the body.

In a visit to the Maiz Islands off Central America, another explorer in the 18th Century left this recollection about the local inhabitants: "They have a custom of cutting an opening in the lips of young boys, alongside the chin itself; in order to prevent the hole from closing, they pass a wooden bolt through, and they do this while the boy has not achieved 14-15 years of age. From this age, they carry a sort of beard in the opening, made from the shell of a turtle. The upper part of the decoration is passed through the opening, so that it is held between the teeth and lip, the lower lip hanging down past the chin. They usually carry this decoration throughout the day, and take it out when they lie down to sleep. There are similar openings in the [earlobes] of men and women, bored into them during youth. As a result of the constant passing through of large bolts through these openings, they reach the size of a 5-shilling piece. They carry pieces of round, smooth-shaped wood in these openings, so that their ears look like wood, tightly hugged by a narrow layer of skin."

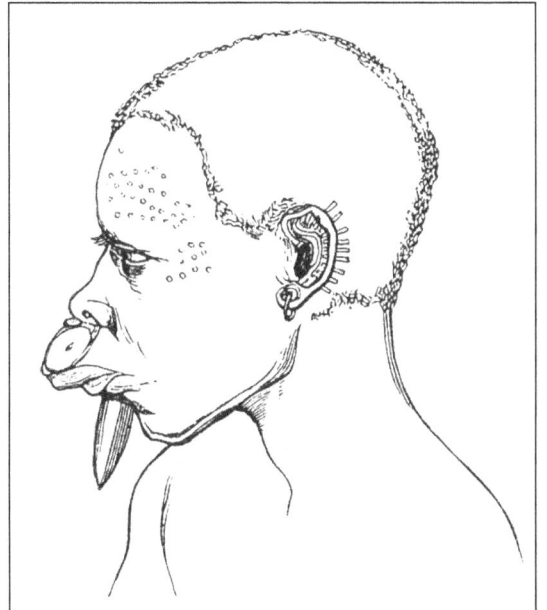 

Right and left: examples of ritual disfigurement.

---

[233] Sankt-Peterburg, 1883.

A similar custom is demonstrated for us by the Botocudo tribe of Central Brazil. Notably, the very name originates from the Portugeuse word "botogue," which means "bolt." The Eskimos of North America pierce the lips at the corners of the mouth, and place a stud with two heads or a dumbbell-shaped pendant of bone, shell, elephant tusk, stone, glass, or wood in each opening. The inhabitants of Alaska pierce their noses and drive all sorts of things into the opening, which have, according to their understanding, an independent aesthetic value. One of the first visitors to Central Africa also wrote: "The outer decoration of the body, with costume, rows, and mutilations, which they subject themselves to, in a word is a common fashion; all have their own distinct character. Most remarkable in women is the disgraceful, unnatural custom of puncturing and disfiguring the lips; they evidently compete with one another with their mutilations, and their vanity in this regard, I think, is not encountered in a like manner in all of Africa. Not satisfied with piercing the lower lip, they also pull out the upper one, for symmetry."

During the first visit of Europeans to New Zealand, to a man the native inhabitants were found to have artificial openings in the ears, into which feathers, bones, finger and toenails, and the teeth of deceased relatives and dead dogs were placed. Some use these openings like pockets in clothing are used. When Zulus were first brought to London, they could not find anything better to imitate than the English gentlemen; so they placed lighted cigars in the artificial openings in their earlobes. Explorer Wilfred Powell reported that on one of the islands of New Guinea, he saw a man in whom the holes in the ears were stretched out to such a degree that one could easily pass a hand through them.

Special attention needs to be given to the custom of disfiguring the skulls of newborn infants, which was noted in a number of peoples in all parts of the world, and was first described by Hippocrates around 400 B.C. In accordance with the notions of these peoples, deformation of the skull should have affected the psychological qualities of the child. Besides that the unnatural form of the head should have spoken of a higher social status for the owner. And to this day, even an abundance of deaths in the process of this painful operation does not stop many peoples from continuing this custom.

In his famous book, *Anthropology*, Paul Topinard also considered it necessary to sharpen the attention of readers on the problem of artificial disfigurement by the members of several races: "From a number of the most original customs, tied with the onset of puberty or carried out on children, we come to the incision in the urinary canal in several Australian tribes; the excising of one testicle among South African [tribes]; the cutting off of one finger/thumb in the women of several Australian tribes and peoples of the African coast, the cauterization of the soles of the feet, and so on. But of all the customs, the most varied are those which relate to the treatment of the deceased. In some cases, the body is burned or simply smoked [like a ham], or the relatives eat it; in different cases, they leave it smouldering or decaying on the branches of a tree, or they leave him to be eaten alive by buzzards."

Russian racial theorist V.A. Moshkov substantiated and provided the real reason for these ceremonies in various parts of the world, which are unusual at first glance:

"When higher races mix with lower ones, the former cannot help but notice the change to a worse direction, arising in the appearance of their descendents. And since the white race is highly valued, it was natural that there would be a desire to come back to it by all means possible. They tried to achieve this by many means, and in particular by selecting brides of a known ideal of beauty, and artificially disfiguring their organs, in order to give them similarity with the organs of the white race. The custom of deforming the skull dates to deep antiquity; according to the words of Hippocrates, among the ancients, deformation of the shape of the skull was considered a sign of nobility. Among the Indians of North America, it was considered a mark of the privileged classes, and was prohibited to slaves. Further, the face of the White Man lost its blush through miscegenation, at first it was pale, then swarthy, and finally, in hot countries—black. Even among our Europeans of fashion, they whiten and rouge their faces as a great habit, but they do this secretly, because there are many women who are not in need of such coloring. Finally, among completely colored peoples, coloring of the face to white and red colors acquires a religious significance. Thus, among the ancient Egyptians, they placed rouge in the tomb of the deceased. Among the Australian aborigines, for example, they whiten the face for dances, or they color it white and red. They color the hair in the same such colors, that is, white and red, in order to portray blondes and redheads. Among many colored races, according to the observations of anthropologists, the earlobes are absent. From this [comes] the desire...to imitate the white race, through elongation of the lobes by hanging weights from them. On Easter Island, where the ears of the natives were artificially stretched to the shoulders, they bowed to the ancient colossal statues, which also have long-hanging ears. Many sculptural portrayals of Bhudda represent him with long, hanging ears and with holes in the earlobes. It is known that lower races differ from the white race by an absence of calves on the legs, and a weak development of the hand muscles; therefore, there is a necessity to mimic these organs. Since baby legs are exclusively affiliated with the white race, in contrast to the lower races, in which [there are] foot soles of prominent size, among Chinese girls we find the custom of disfiguring the feet, in order to make them small. We have enumerated here the main forms of disfigurement, that have as an object of decoration ...the imitation of the white race."

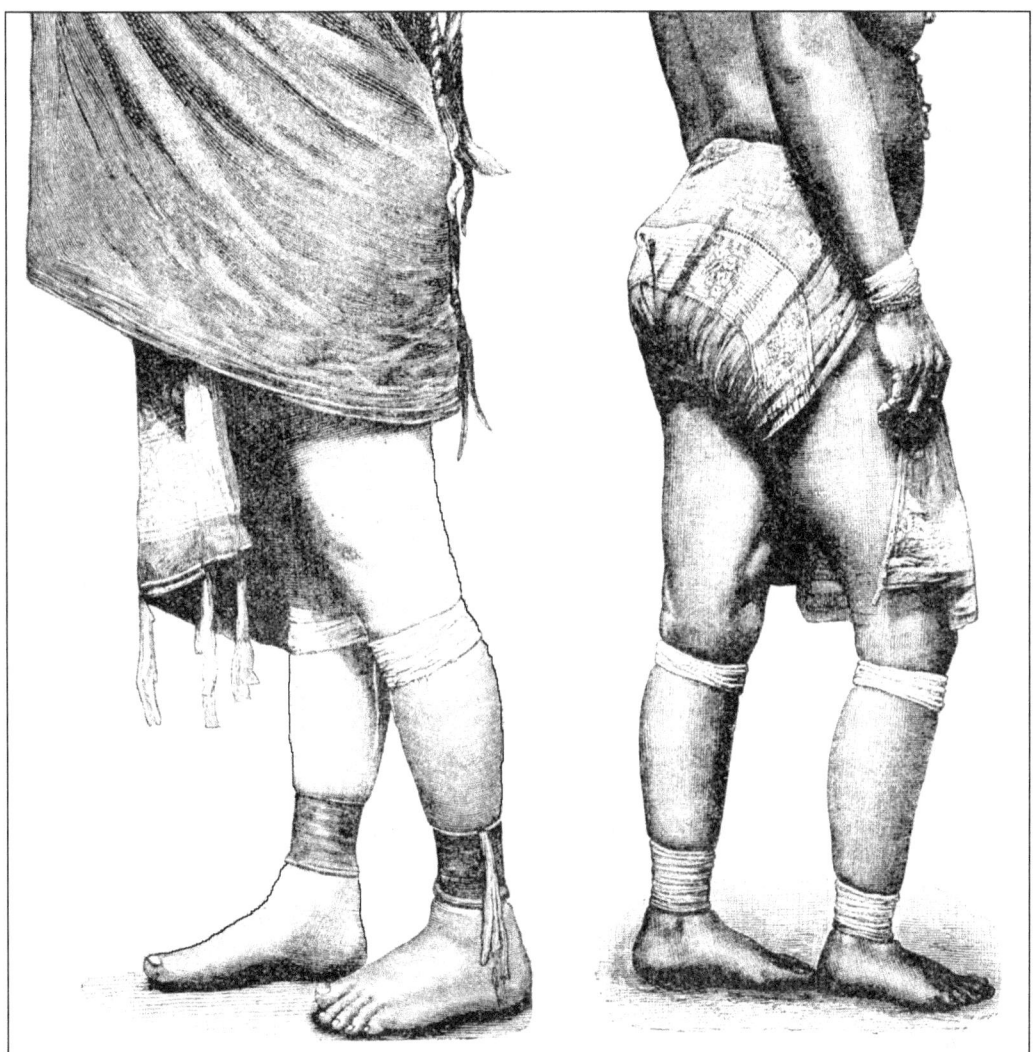

Artificial Enlargement of the Calf Muscles by Women of "Colored" Races (according to Herman G. Ploss).

Left: Example of Ritual Disfigurement

Summarizing all of the above, we consider ourselves justified in forming the following thesis.

Based on anthropological and ethnographic material, one can suppose with a significant degree of probability, that the desire among the members of some ethnoses to artificially disfigure one's body, is tied with a realization by them of the incompleteness of their own physiques. It is namely the subjective, inherent feeling of their own physical defectiveness that compels them to carry out these "modifications" of the natural essence, for nothing that is self-sufficient and self-worthy needs to be remade.

For example, the ancient Aryan religion of Zoroastrianism worshipped human flesh as the peak of Divine creation; under threat of excommunication, ear-piercing, circumcision, and any plastic surgery was forbidden. To alter one's own body, which was given by God, meant to serve the devil. Such was the ancient moral of the ancient Aryans.

All this, for its part, allows us with all inevitability, to again place under question the full biological value of several races, for it is already placed in doubt by their own representatives in the entire history of civilization. We assert nothing. We only advance a supposition.

The given logical conclusion does not contain anything racist, since the classical Russian anthropologist, D.N. Anuchin, expressed it in the same spirit in his work, *How People Decorate and Disfigure Themselves.*[234] "The need to adorn oneself is so characteristic of Man, that it comprises, one can say, one of the characteristic traits of nature—a trait, which is equally characteristic to both civilized peoples and the rudest tribes of Mankind. Among lower tribes, among the so-called savages, this need appears even stronger, than in refined peoples; and many savages, without a moment's hesitation, will make a sacrifice of their means, their comfort, and even their health [for the sake of adorning themselves]. Located at the very lowest stage of civilization, not knowing what permanent housing is, nor farming, nor cattle-raising, and leading an itinerant life of trapping and hunting, the savage is already concerned about the decoration of his body. Not having any kind of equipment, except those made of stone, bones, or antlers, not having any knowledge of metals or precious stones, he decorates himself with shells, bone plates, feathers, beads of clay, or the drilled teeth of beasts; or he paints his body with ochre, chalk, and similar painting ingredients. In hot countries, significantly more attention is turned to decorations, than to clothing, and often the latter is worn as much for protection of the body and covering nakedness, as for vanity and furnishings. The vanity that is characteristic to savages is no less than in cultivated nations, and the demands of fashion, although perhaps not so fickle, are often more overpowering among them, than in modern Europe. In all probability, the motive in the given case was purely aesthetic and similar to other forms and methods of decoration and disfiguration, and was caused mainly by false and barbaric notions of beauty."

Fatal differences in the canons of beauty and self-evaluation, for its part, leads to unavoidable differences in morals, which are also a manifestation of the sum total of racial traits.

Swedish anthropologist Wilhelm Leche wrote in this regard: "Just as among separate individuals physical traits can manifest, which are regarded as a hereditary quality from some very remote ancestor, so a certain individual can commit certain anti-social or immoral acts,

---

[234] *Kak lyudi sebya ukrashayut i uroduyut.* Moscow, 1876.

directed against kin or near relatives, or against all of society; this [behavior] may be regarded as inherited from an ancestor, [who] absolutely did not possess, or weakly possessed, social feelings. Such spiritual defects are not suppressed by a good up-bringing. Similarly, natural selection causes not an absolute, but only a relative perfection of the organism; the notion of morals can achieve a higher or lower development because at different times and among different peoples, the concept of morals was, and even now, is so different. That humanity was at some point liberated from everything that we call "rudimentary organs," however, is unbelievable, because this disharmony is an inseparable travelling companion of each evolutionary process."

20. Racial Prejudices as a Basic Driving Force of Evolution

And now, respected reader, we will clarify our position. We have persistently cited centuries-old works on racial differences, in order to bring attention to this theme, for you will not encounter the facts introduced by us from classical works, in a single textbook on anthropology or criminology. The intoxicating narcotic of "common human values" all the more acquires the characteristic smell of the pyres of the Inquisition of the Middle Ages, and the logic of "political correctness" more clearly assumes the contours of academic obscurantism. It's possible that several of the passages cited by us have shocked the reader with their judgements and sharpness of formulation, but they all belong to the luminaries of world and Russian classical anthropology, who no one can ever accuse of racism; that automatically releases us from any such burden. The persistent immunity of the Russian scientific school [to the charge of racism] is proven by time, and by the bloody ideological battles of the 20th Century. The authority of D.N. Anuchin, I.A. Sikorskiy, A.P. Bogdanov, and dozens of others, remains unshakable, for their actions developed in the common channel of the scientific process, something we graphically demonstrated with references to well-known foreign authors.

The abundance of citations from the German philosopher and naturalist, Karl Vogt, is not accidental. In life, he was a main opponent of Karl Marx in questions of raciology, and his books were very popular in pre-Revolutionary Russia, being distributed in huge printings. By this fact, we want to show that the Russian scientific elite, contrary to common Communist myth, did not in any way surrender uncomplainingly to Marxist propaganda. The famous work titled, *Mister Vogt*,[235] was a disgraceful slander, full of angry insults in the address of a scientist, who had a world name. According to ideological views, Marx repeatedly passed judgement on Vogt, but he was unable to win a single argument, for the logic of the German anthropologist turned out to be unfailingly more strict and persuasive, than the speculation of the German-speaking economist. The names of the classics of Marxism-Leninism are desecrated in modern Russia, and good-natured attitudes toward all their contemporary opponents are automatically summoned. Besides this, Vogt was the best friend of that classic of Russian literature, A.I. Gertsen. In his book, *Bygone Thoughts*, Gersten dedicated an entire chapter to the German philosopher, full of deeply felt and warm lines. Generations of our countrymen studied this essay, within the framework of school programs on literature.

However, it is worth mentioning, that Soviet propaganda managed to deform the philosophical doctrines of many Russian scientists beyond recognition. Thus, for example, *The Great Soviet Encyclopedia*,[236] states in an article dedicated to the great Russian biologist, Ilya Ilyich Mechnikov,[237] that he was a steadfast enemy of the "reactionary philosophy of Nitzsche and idealist trends in biology."

In answer to this, we can refer to I.I. Mechnikov's fundamental work, *The Struggle for Existence in the Wider Sense*,[238] in which he wrote: "The differences between the large human groups, peoples, and races are so prominent and obvious..." Such was the basic conclusion in his work: "...the main condition and motivating factor in the evolution of human society is not the moral, but the intellectual."

---

[235] *Gospodin Fogt.*

[236] 1954, Tome 27.

[237] 1845-1916.

[238] *Bor'ba za syshchestvovanie v obshirnom smisle*. Vestnik Evropi, 1878, Tom IV. (Bulletin of Europe, 1878, Tome IV).

Nietzsche, unknown at the time and living in Orthodox Russia, would have ventured into such articles.

Therefore, we intend to maintain that **racial prejudices are not only a factor, but are namely a basic, motivating force in evolution.**

Arising at the level of the usual outer surface perception, they project onto all the levels of the physical organization of Man. Between traits of the outer appearance of the individual, the shape of the skull, the structure of the convolutions of the brain, the nuances of physiology, the specifics of biochemistry on a cellular level, psychological, behavioral, and the moral, a single vector is traced, tying it to the flails of evolution: as with ancestors, so with decendents. Therefore, this is a sufficiently cursory view, in order to recognize the stranger and the whole tribe. For at the basis of judgement in a given case lies all the huge baggage of the knowledge of our ancestors, and namely their collective will serves as a dominant [factor] in our future conduct. Biological disorientation on the main level of "*us-them*" is tantamount to death, and an interruption of the entire evolutionary chain. Any system of values that disputes that very principle is similar to a deadly virus, whose sole object of existence is the murder of the living system, the [host] body of which it is a parasite. Therefore, our verdict and degree of decisiveness should be firm and crystal-clear.

The moral of the host and the moral of the parasite are not of equal value; therefore, the host and the parasite cannot reach a compromise. For these are two principally different biological programs. Any advice to us to feel pangs of conscience, in the process of honoring our evolutionary heritage, is worth regarding as the usual informational obstacles. Through our actions we will purge not our consciences, but our future.

The famous modern biologist, Ireneus Eibl-Eibesfeldt, writes in the book, *The Biology of Human Behavior* (1997): "Humanity itself is not a single selection. It consists of populations, fiercely competing with each other for limited resources. It cannot be demanded of any single people on earth, to sacrifice themselves for the sake of mankind. We know that we owe almost all which we enjoy today, to the achievements of our ancestors. This knowledge obligates us to think about future generations."

Left: Erich Rudolf Ensch

The distinguished German racial psychologist, Erich Rudolf Ensch, also asserted: "Race and blood; blood and race; [this] lies at the basis of everything. From the structure of the capillary network to the worldview, stretches a single, straight thread. The world of the ideas of Man depends not only on spiritual factors, but on his general being. This common being includes in itself the physical being, as a prerequisite of the progressive world. There is a need to preserve the purity in the general being of the individual, including the physical, in order to preserve the purity of ideas. In order to bring changes to the world of ideas, it is not enough to substitute some ideas for others. It is also insufficient, in order to preserve the purity of new ideas. But the other is sure to demand changes of the common being. Studies of blood and race are the sharpest contrast to all modern forms of idealism, which asserts the primacy of the realization of ideas over being. In reality, the reverse relationship exists. Studies of blood and race show the primacy of the human general being, including its elementary, especially inborn determination, over the world of ideas. Modern psychology proves the same, establishing an indissoluble tie between the most elementary and low psycho-physiological processes, and the highest forms of ideological life. Ideas by themselves are powerless and formless, if they are not tied to physical being. Only in pure flesh and blood can pure ideas be safely developed. A campaign against old, ridiculous and powerless idealism is necessary. Only strong ideas, which can dominate and prevail, are worthy of respect."

The notable position of well-known modern Russian anthropologists E.N. Khrisanfovaya and I.V. Perevozchikov in their joint work, *Anthropology*,[239] can be considered the official point of view of the academic sciences of Russia: "More than more the justified worry about society, in regard to the genie of nationalism, is the connection of defined ethnoses with the false notion of the causal tie of these or those biological characteristics. In several countries, they make a reality of the notion, that if you refuse to recognize the polytypicalness of Man, then the racial problem itself will disappear. The "Straussian" tactic does not take account of the circumstance, that racism and nationalism are tied to fundamental psychological apparatuses [that are built] on recognition of "us-them," which define both the characteristics of the formation of habits, and very ancient devices of group behavior. In this situation, the rejection of the word-symbol, **race**, is no more than a cosmetic solution."

In his fundamental work, *Historical Anthropology and Ethnogenesis*,[240] the outstanding Russian anthropologist, V.P. Alekseyev wrote: "The number of peoples fluctuates, according to various calculations, at around 1,000; the number of racial types in various classifications does not exceed several dozen. From this simple comparison it is obvious, that races in the overwhelming majority of cases correspond to a higher level of unity of anthropological objectives, than do most ethnic groups."

In connection with all of the above-said, we can dare to assert, that the ethnic level of thinking will always be a dead end in the development of an idea, at the same time that the racial level will always be its endless horizon.

---

[239] *Antropologiya.* Moscow, 1999.
[240] *Istoricheskaya antropologiya i etnogenez.* Moscow, 1989.

A New Paradigm in Raciology

"Racial differences sprang up just as Mankind itself appeared."
Baron Egon von Eichstedt

"No one can crawl out of his skin, just as no one can from his soul."
Fritz Lenz

For 100 years, fingerprints have been consistently associated with criminology and forensic medicine. The average consumer of news automatically recalls that the amusing swirls on our palms and feet are lines created by nature itself, and are unique to all people on the planet, for there are not two people in this world with identical fingerprint patterns.

And yet for 100 years, they also "forget" to tell readers that differences in fingerprint patterns, because of which people never resemble each other, are not accidental, but subject to the harsh logic of Nature, which created various peoples and races in the different parts of the globe.

We are not simply different on a level of individuality, but also subject to rigid, genetic differentiation, in accordance with racial, national, and regional affiliation. According to fingerprints, any person on Earth can be identified, with an imperial degree of accuracy.

The first ideas for using fingerprints for the identification of people arose in China in the 12th and 13th Centuries, and were successfully used then in criminal investigations. In Europe, similar thoughts were first expressed by Arthur Coleman in 1883, and in 1888 by veterinarian Wilhelm Eber, who recommended the method to the Prussian police, but without success.

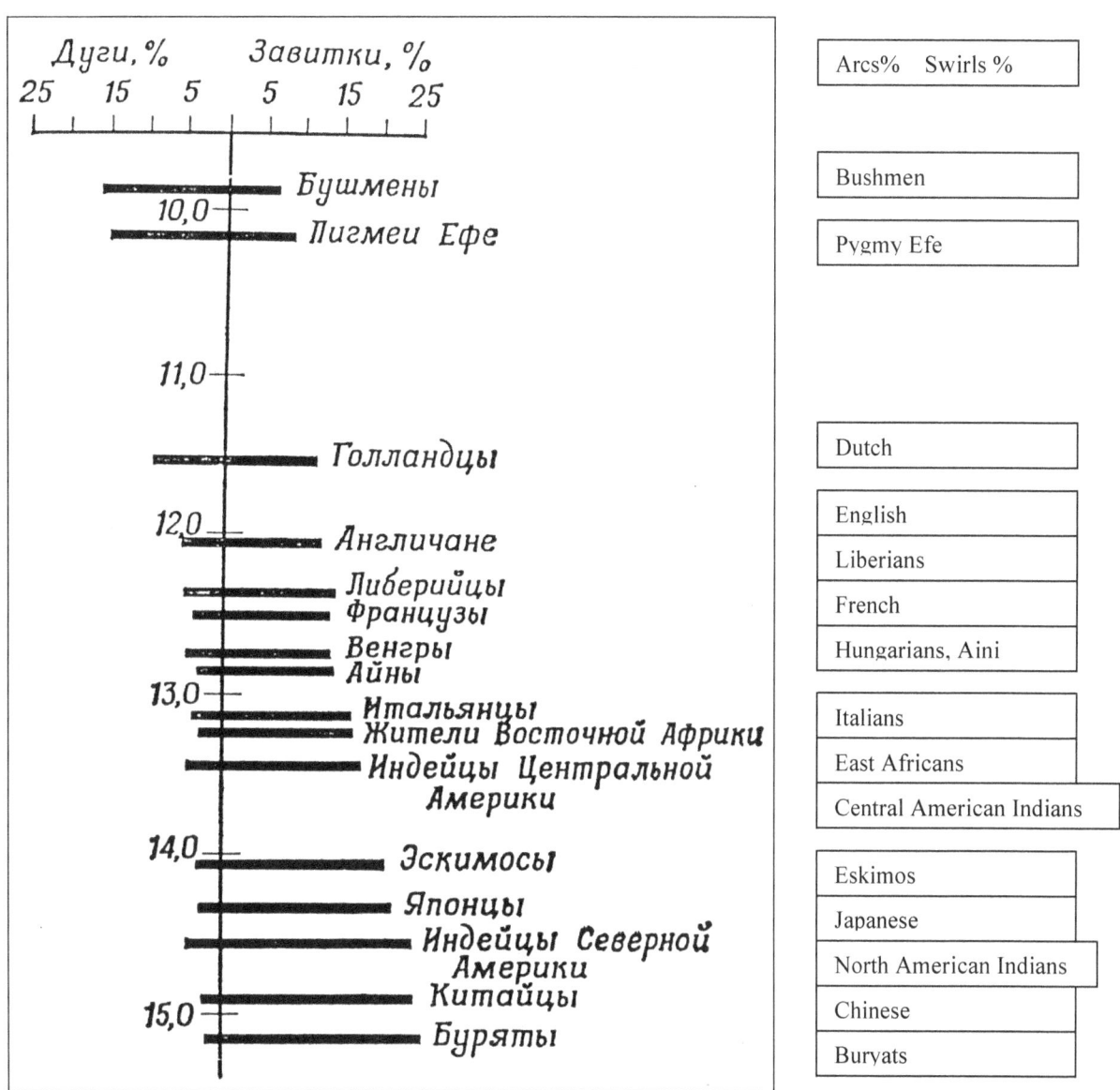

Above: Percentage of Arcs and Vortices [in thumbprints, handprints] in Different Populations (according to Harold Cummins and Charles Midlow).

However, it is known that the very end of the 19th Century marked the stormy blossoming of natural sciences such as anthropology, biology, and psychology; at this juncture sociology participated and political science had a share: **racial theory** arose. At that time, Charles Darwin's cousin, Francis Galton (1822-1911), developed the applied part of racial theory—**eugenics**—the science of the improvement of the human species. Being a talented mathematician, he closely examined the gracefully curving lines in fingerprints, and saw in them graphic portrayals of the mathematical functions of racial traits.

In 1892, Galton was the first to compare the fingerprint patterns of different racial and ethnic types. From that time, alongside solving purely criminal problems, the development of dactiloscopy began to grow in the direction of classical racial theory. Further, Harris Horton Wilder, Harold Cummins, and Charles Midlow made big contributions to the development of a new science, which received the title of **ethnic and racial dermatoglyphics.**

In Russia, dermatoglyphic research only began in full swing in the Soviet era. It is astonishing, but it is a fact, that in a country that took up the cause of internationalism, racial research received official scientific recognition. We refer to the work of P.S. Semenovsky, *The Distribution of the Main Types of Tactile Patterns on the Fingers of Man.*[241] The Institute of Anthropology of Moscow State University organized numerous expeditions to the very corners of our country. Prominent Soviet anthropologists A.I. Yarko, V.P. Alekseyev, and G.F. Debets created a theoretical base of ethnic and racial dermatoglyphics. M.V. Volotskiy, T.A. Trofimova, and N.N. Cheboksarov perfected a methodological basis of research.

From the very beginning, fingerprints begin to differentiate on three levels: racial, ethnic, and territorial; this speaks right away of the accuracy of the method, and the great potential for its development. That is, according to the fingerprints of an individual, not only his race and nationality can be established, but also the geographical region of his origin. Galton's ingenious conjecture at the end of the 19th Century found complete confirmation in the studies of hundreds of ethnic groups, in the most varied places on the earth.

The relatively simple method allows one to achieve staggering accuracy on the first try. They set aside three basic papillary patterns: arcs, loops, and swirls; double loops correspond to the latter. In the table, the proportion of the frequency of swirls, loops, and arcs in several peoples is displayed.

In the article, *Fingerprints as Racial Traits and their Transfer via Heredity,*[242] Doctor Erich Karl, a German specialist in this area, gives this summary to numerous researchers: "Members of the yellow race, along with Eskimos, have the most swirls of all, and the least arcs and loops of all. Among Europeans, the relationship is the reverse: the number of arcs and loops increases in them, at the expense of swirls. Indians are completely adjoined to Asians, and Ainu occupy an intermediate position between the yellow race and whites, while Jews differ strongly from Europeans by the great number of swirls, and the comparatively insignificant number of arcs. Among European peoples, there are more arcs and fewer swirls in Northern Europeans, and in Southern Europeans [it is] the opposite: more swirls and fewer arcs. Among northern Europeans the most arcs and the least swirls in Norwegians; after them come the Germans, English, and Russians."

	SWIRLS	LOOPS	ARCS
Eskimos	72.2%	29.6%	0.8%
Japanese	45.16%	52.76%	1.81%
Jews	42.7%	53.0%	4.2%
Italians	36.46%	58.44%	4.72%
Russians	32.12%	61.3%	6.15%
Germans	26.6%	66.13%	7.27%
Norwegians	25.65%	66.95%	7.4%

It is a remarkable fact that the given article was published in 1936, when Germany had officially adopted racial theory, and ideological opposition toward Soviet Russia as an "Asian country" was marked with all deciseness. However, the findings of E. Karl completely agreed with the data introduced by P.S. Semenovski in 1927; this only speaks to one thing: the struggle of ideologies does not have any relationshjp with the patterns of development of science; and German racial statistics confirm Soviet [figures] - and not in favor of the myth of National Socialism over the "Asiatic hordes of Bolshevism." Research has continued into recent times. The archives of the Anthropological Institute of Moscow State University are growing. As a result, this laborious scientific activity has become the subject of Henrietta Leonidovna Khit's work, *Dermatoglyphics of the Peoples of the USSR.*[243]

A good half of the book consists of numerous tables of comparative analysis of ethnic and regional groups. From the time of the first publications on this theme, the methodological base essentially became more complex, and increased the accuracy of measurements. As a result,

---

[241] *Raspredelenie glavnikh tipov taktil'nikh uzorov na pal'tsakh ruk cheloveka.* Russkiy antropologicheskiy zhurnal, 1927, T. 16, vip. 1-2, s. 47-63.

[242] Published in *Volk und rasse,* 1936, Tome 7.

[243] *Dermatoglifika narodov SSSR.* Moscow, Science, 1983.

their reliability increased. Now the Delta Index, Cummins Index, the axial palm tri-radius t, the pattern of the hypotenar, the supplemental inter-finger tri-radius, the Th/1 pattern, and Eastern Complex are measured.

Within the scientific limits of our article, we will not retell the scientific portion of the book, which is presented by the Institute of Ethnography named N.N. Miklukho-Maklaya. We will focus our attention on the conclusions only, for the civil and scientific courage of G.L. Khit' is expressed in them in full measure, thus raising the worthiness of the given book many times. The purpose of the book is thus described: "Knowledge of the basic patterns of mutation in the traits of any system in an organism on various taxonomic levels, has paramount importance for racial analysis, and the theory of racial genesis as a whole. Racial analysis is based on an examination of geographical variations of traits in territorial groups, revelation of characteristic racial complexes, the typology of ethnic groups and local races, and the determination of the average phenetic distance between taxonomies on various levels."

Left: Harold Cummins

The uniqueness of the method lies in this: a group of diagnostic traits dermatoglyphically occupies a special position, in comparison with defined racial-somatic traits. As was shown by American biologist Arthur Jansen, the racial-somatic appearance of an individual is 80% hereditarily formed, and only about 20% is formed by the surrounding environment. Measurements on the basis of pigmentation of the skin, hair, eyes, and also anthropometric measurements of the head, face, and body, and analysis with the help of psychological tests, carry a certain percentage of error, caused by the influence of the phenotype on the racial-somatic traits of the individual. Fingerprints, and the dermatoglyphic patterns of the palms and feet, are generally not subject to the influence of the environment and are an absolute embodiment of our racial and ethnic essence, which is genetically passed down from generation to generation. Racial-ethnic traits are expressed in them in pure form; therefore, the dermatoglyphic method occupies a special position as the most reliable [method]. G.L. Khit' also made this observation: "The author came to the conclusion that dermatoglyphic traits, being inalterable, are not subject to the action of selection, and are the most stable across time, and more faithfully testify about the preservation of the ancient traits of populations, than do somatic [traits]. The whole system of the traits of skin [contours] is subject to rigid genetic control, which limits the possibility of the variation of each trait in strictly determined limits, and moreover, makes commensurate the degree of intra-intergroup diversity in the traits. As a result of this, a picture forms of the unique stability of the dermatoglyphic complex...The traits of each relief [feature] are retained in the form of controlled genetic systems."

Aware of the accuracy of the method, and using a wealth of bibliographic material, G.L. Khit' set for herself a truly global task: "to trace the dermatoglyphic differentiation of the population of a significant part of the inhabited world,[244] from small component parts (local groups) to basic components (the great races)." Further, summarizing the findings of sexual studies of ethnic groups on the territory of the USSR, the author gives a sternly argued rebuke to the Eurasianists and other adherents of the concept that Russian blood is mixed with Asian blood.

The myth of a multi-national country and...a racially impure [melting pot]...disappears under the influence of the very precise dermatoglyphic racial diagnostic method. The theory that a majority of Russian children are the product of [racially] mixed families, appears to be the ruse of an insignificant, racially mixed part of the population, which desires to bring its sins to the racially pure majority.

"As a whole, the Mongoloids and Europoids of the USSR are well-differentiated on all three levels, from a statistical point of view. Each level significantly differs from the others. In unification of the data, the degrees of intra-racial differentiation increase, and its significance grows to a **maximal threshold.** The ethnic level is the maximum in the intra-racial scale, and reaches 70%

---

[244] *oikoumene*

of the size of the level of differences between the great races. Local races are far more homogeneous, than any of the examined taxonomies."

We once more remind, that race, ethnicity, and territorial traits are meant under the three levels of division. G.L. Khit' emphasizes that even going through the boundaries of a single region, the members of the various races and ethnic groups were not subjected to mixing.

For its persuasiveness and straightforward conclusions on the Russian racial type, it is completely magnificent: "In the European portion of the USSR, the northern branch of the Europoid race and the intermediate types between the northern and southern Europoids are represented. It is established that the Russians are homogeneous in regards to skin relief, and are the carriers of the **most European complex.**"

The myth of genetic pan-mixing among the Russians is only an unscientific provocation, not only against the Russians, but against the entire white race in general.

Finally, G.L. Knit' confirms the basic postulate of racial theory, that any historically significant people has a racial foundation, in which its specific traits are expressed more clearly. It is not abstract social laws that influence history, but namely the racial traits which a people carries within itself. Each historically significant people has a racial base, a racial core, with the help of which it dictates its "rules of the game" to the racial periphery—that is, to racially impure half-breeds. Egypt was created by the Egyptians; the great Roman Empire was created by the Latins; the Chinese Kingdom, by the Chinese, and the Russian Empire was created by Russians. Thus all this talk about multi-national cultures is a rhetorical ploy that conceals an elementary forgery.

Racially pure peoples create empires, and half-breeds destroy these empires. Count Joseph Arthur de Gobineau, the pioneer of Racial Theory, stated: "No one has shown that miscegenation creates higher groups on the social ladder of society." Only a homogeneous type of people, racially pure in its basis, has the desire and opportunity to express itself in historical forms, creating its own statehood, its own religious worldview, and its own cultural-civilizational type.

G.L. Khit' develops this idea in the following manner: "On the level of the great races, the differences should be defined almost wholly on a racial basis. Comparison gives a similar picture: the resemblance of populations of the Europoid group, and the significant distance from Mongoloids. In this distance between members of the Europoid groups, there are fewer regions, than between any of them and the Mongoloids. Resultantly, at the level of the great races, differences between Mongoloids and Europoids are completely dependent on the racial base."

Thus, any talk of the fatal consequences (from a racial-biological point of view) of the Mongol-Tartar yoke is only a harmful myth, without any scientific basis.

Such are the conclusions: **The Russians are homogeneous, racially pure in their foundation, and primarily a Nordic branch of the Europoid race.** The most modern-day, genetic-anthropological analysis gives a completely unambiguous resolution to this question, on the basis of ethnic and racial dermatoglyphics. Those who believe otherwise are mistaken; those who speak otherwise place themselves in opposition to us.

Whoever doubts the veracity of the given information, and considers the position of the author of these lines too biased, can turn to an open, official source - the encyclopedia, *The Peoples of Russia*,[245] where it states in the chapter [titled] *The Racial Composition of the Population of Russia*: "According to an approximate counting, the members of the Europoid race comprise more than 90% of the population of the country, and around 9% fall in the category of members mixed between Europoid and Mongoloid. The number of pure Mongoloids does not exceed 1,000,000 persons."

According to the official international norms of the UN and UNESCO, a country is considered mono-racial, if 66%, or 2/3 of its population belong to one race. Therefore, when democratic social scientists speak of Russia as a multi-national state, it is necessary to add: "Although it is multi-national, [Russia] is mono-racial."

The basic conception of a racial core was formulated in Germany by the outstanding racial anthropologists of the first half of the 20th Century: Alfred Ploetz, Wilhelm Schalmeier, Eugen Fischer, Fritz Lenz, and Hans F.K. Gunther; and it was also confirmed by the **ethnographic and cultural anthropological** findings of Ruth Benedict and Bronislaw Malinowski; and the

---

[245] *Narodi Rossii.* Moscow, 1994.

behavioral biology studies of Konrad Lorenz and Ireneus Eibl-Eibesfeldt. And Ludwig Ferdinand Clauss most fully developed the studies of **racial spirit.** Thus, all talk about multi-national states and multi-cultural communities [is nothing but] a bluff.

Even the above summarizations would be sufficient, in order to add Henrietta Leonidovna Khit' to the number of leading modern racial theoreticians, but her scientific temperament and civic courage do allow her to remain at the halfway point. The monograph contains completely shocking conclusions, and prospects for the development of her scientific method truly have no boundaries. With the help of this book remarkable new opportunities to popularize racial theory have been opened: "The significance of the findings of dermatoglyphics for a resolution of the problems of racial and ethno-genesis as a whole are distinctly and unjustly underestimated. In fundamental summaries of the racial composition of the world, the usual brief and cursory observations, with the usual negative conclusions, are dedicated to this system, touching only on a few aspects. The goal of the author was to show that skin patterns, under the correct methodological approach, can be invaluable sources of historical and biological information. In the enormous volume of material, a confluence of the differentiation of groups and racial complexes, according to the traits of skin contours and racial somatology was discovered. One can assert without exaggeration, that the skin contours of the human races and populations, are the records of the history of their formation. From that flows the possibility of reconstructing the basic stages of race formation/genesis in the scale of all humanity."

It is truly unexpected, but, according to the assurances of Khit "the bibliographic list of works on dermatoglyphics includes several thousand titles."

And now, respected reader, please tell how much you have heard about the given question, with regard to so fundamental a study? It is not difficult to guess, that the concealment of information from the public at large is the result of an intent, which has a definite racial context. The global management of Mankind is realized mainly because of mass-scale racial-biological illiteracy. "He who is forewarned is forearmed," the old saying goes.

Therefore, we will now continue with the chain of logical, reasoned conclusions by the author of the given book.

In light of new discoveries in the area of genetics, the prospect arises of not only the global racial-ethnic identification of humankind, including on a regional level, but also of the detailed recovery of the whole history of mankind on a racial, national, and regional level. Using the unique, unchangeable skin patterns of the fingers, its complete genetic condition, one can penetrate as far as one likes into the historical process, and recover the entire anthropological history of mankind in detail. The rise of any population, tribe, people, or nation can now be traced from the first periods of race-genesis to the present day, on the scale of the entire planet. But what is particularly important, is that the process of forming racial-ethnic communities can now be traced in the mutual connections with each other. It becomes possible to understand which community assimilated with another, and which, on the other hand, dissolved into other tribes. One will be able to take stock of all the social and biological forces, dynamically acting on the racial core of this or that people, as far back in antiquity as one likes.

The biological causes and results of the great migrations of peoples, the birth of cultures, religions, and empires, and also their end, will become the common property of exact sciences, and all dogmatic speculations in the field of world history can be ended once and for all.

The racial ethnic design of any state, independent of its antiquity and degree of study by means of archaeology, can be re-created with the functional reliability and exactness of the technical design of any existing electronic device. On the basis of new methods of dermatoglyphics, it again becomes possible, and now conclusively, to recover the comprehensive political, social, and cultural history of Mankind. It will be possible to enumerate all the peoples that contributed to the formation of great states, and those which took part in their destruction. Now one can learn precisely what to call the descendents of this or that ancient tribe. Every lie and falsification in international relations will cease to have any fertile ground beneath it. The historical method, as a universal method of historical research, should disappear, and **racial historiosophy and revisionism** will have the foundations to go through a second birth. All culture-creating peoples and parasite-peoples will be marked like the elements in Mendelev's Periodic Table of the Elements. The place of population genetics in institutions of higher learning will be taken by **political genetics and population casuistry.**

On exactly the same basis as the modern science of **socio-biology**, one can identify and calculate donor peoples and parasite peoples, and then by very natural means, such sciences as **social parasitology, evolutionary parasitology, and biological culturology**, for example, will spring up. In this regard, a modern researcher from Germany, Jurge Albrecht, quite accurately stated that the prospect now arises of a "genetic inventory of humankind." Henceforth, humanitarians should not develop moral criterion for society, but biological and engineering systemo-technical [criterion].

Henceforth, there should be no more thoughts of scaring one another with the horrors of ethnic purges, which are arising everywhere now. On the basis of new knowledge, one can instantly pull the whole gigantic canvas of the historical process (which includes in itself both the past and the future) like a dust cover, and completely beat all the genetic garbage out of it.

The creation of new states can now be realized, not on the basis of constitutional acts, but on the basis of the ethnic, biological, social, and cultural complimentariness of peoples, and as a result, on the basis of an exact calculation of the racial-ethnic structure for the given territory and the given landscape. Henceforth, specialists in the field of the ethno-functional method should establish borders, rather than political *careerists*.The initial racial-ethnic matrix of the state can be modeled with the ease of selecting building blocks, by shuffling the characteristics of peoples, just like one arranges the colors of the blocks' surfaces.

It is completely clear that on the basis of dermatoglyphics, one will have to completely remake the whole modern theory of the state and civil rights. Why for example, should I, as a member of a state-formed ethnos, pay taxes and submit to peoples with substantially different fingerprint patterns? Why, if our paths separated genetically in the distant Neolithic Age? I should be law-abiding only in regard to my **genetic kind**, and in regard to others, I should build mutual relations [based] on other principles.

My fingerprints are the projection of my racial-ethnic archtype, which was given to me by nature. My ethnicity and racial affiliation are not a chimera, but a scientific fact, which there is absolutely no reason to hide.

In ancient times, people made signatures with their fingerprints or handprints, not because they were illiterate, as modern historians assure us, but because they were aware of the indissoluble connection of fingerprint patterns with their ethnic and racial affiliation. One's own fingerprints under the text of an oath, was perceived as a sign of quality, like a company brand, which the master proudly placed on his creation. People swore by their ethnic and racial affiliation as a genetic mark of quality, before the face of world history, and before the face of their eternal archetype.

The fingerprints of a concrete individual are the projection of the eternal racial-ethnic archtype of one's tribe in time and space; this is a genetically particular record of his indissoluble tie with a genetic constant. This unchanging brand of your family is a mark of quality, through which you touch existence, and enter into history. Vain and eternal, high and low, despicable and heroic, everything that remains of you in time as a record of your stay on Earth, is marked by the indelible brand of your tribe. It is namely this which lays an obligation in front of ancestors and decendents, infusing the inevitability of karma.

The modern "democratic" means of mass information diligently shape public opinion, reinforcing general racial-biological illiteracy. We are assured that there are no differences between nationalities, and that racial differences are an abstraction, a temporary misunderstanding, which is eliminated by factors of socio-cultural evolution. Such is the mantra of the falsifiers of science, among whom the so-called "school of evolutionists" manifest the greatest activity.

For their information, alongside the above-mentioned dermatoglyphics, we introduce a number of modern methods for racial-ethnic diagnostics.

First of all, there is **serology**, which is the science of blood groups; they also have a rigidly conditional ethnic and geographical system of distribution. In recent times, a method of analysis on the basis of the **components of blood and protein combinations** has experienced powerful development; they also differ in the members of various nationalities. Very widely used is the study of the polymorphic systems of blood (PSB).[246] In Russia in the 1920s, a comparatively

---

[246] *PSK.*

232

simple method of determining national affiliation by blood [type] was discovered, with E.O. Manoylov's method. Within the limits of an official government program, Russians and Jews were subjected to an examination. The results of the program showed with a great degree of reliability, that in Jews, the oxidation process in the blood was faster, than in Russians.

In later times, the method of blood analysis were widespread. It was proven that at the level of the great races, and at the level of local subraces, there was a remarkable specific distribution of polygenic, inherent factors in protein serums. In the encyclopedia, *Peoples of Russia*,[247] it states: "The great human races are distinctly set apart, according to the systems of immunoglobins, which provide a protective reaction against various diseases and transferins, and supply a normal circulation of iron ions in the bloodstream."

This shows that people of various races and nationalities differ in the structures of proteins, immune system chemical-biological composition, and also in the electro-magnetic characteristics of the blood. Finally, racial admixtures are comparatively firmly revealed on the basis of as simple an element as earwax.

Traditional methods of analysis are progressing alongside the above-mentioned systems. It is necessary to start with craniometry: that is, that branch of anthropology that specializes in measurements of the skull. One can name two classic works of the given field: Yu. D. Benevolenskaya's *Problems of Ethnic Craniology (The Morphology of the Occipital Region of the Skull of Man)*,[248] and V.P. Alekseyev and G.F. Debets' *Craniometry*.[249]

There are also national traits in the teeth, and **ethnic odontology** engages in their clarification. A.A. Zubov writes in the book, *Ethnic Odontology*,[250] that: "Ethnic odontology places a new method of racial analysis in the hands of the anthropologist, based on the collection of morphological traits in the dental system, through discovering intra-group differences as a result of the divergence of human populations, arising at various times, under the influence of various factors."

Besides teeth, hair, which differs by structure in all races, has been used for 100 years for the purposes of racial analysis.

There is a racial-ethnic diagnostic of the human organism, based on an analysis of the sum total of its biological parameters; it carries the general title of **biometry**.[251]

Research is conducted on the basis of **taxonomy**; that is, on the mutual coordination of taxonomic (systematic) groups of peoples, which stand out by different degrees of relationship.[252] **Somatometrics**[253] engages in the study of the general constitution and build of the body, and its methods are also actual.

A highly effective method of racial verification exists, on the basis of color photographs.[254]

The most accurate and impressive results were successfully achieved in genetics, with the help of so-called **genetic markers**. The accuracy of their racial-ethnic determination is so high, that as proclaimed by mass media, an ethnic weapon was recently created on its basis at the Institute of Biological Research in Nes Zion, in Israel; the weapon only selectively affects Arabs, and is completely harmless to Jews, although they and the Arabs are related to the same Near Asian, Semitic race. A special discipline—**genogeography**—engages in the problems of the study of the territorial range of groups with these or those genes. A detailed genogeographic map of humankind has already been created, and is well-known in the headquarters of the developers of bacteriological and binary weapons.

---

[247] *Narodi Rossii*. Moscow, 1994.

[248] *Problemi etnicheskoy craniologii (Morfologiya zatilochnoy oblasti cherepa cheloveka)*. Leningrad, *Nauka*, 1976.

[249] *Kraniometriya*. Moscow, 1964.

[250] *Etnicheskaya odontologiya*. Moscow, *Nauka*. 1973.

[251] See N.A. Plokhinsky's *Biometry* (Moscow, 1970), and V.Yu. Urbakh's *Biometric Methods* (Moscow, 1964.)Transliterated titles: *Biometriya* and *Biometricheskie metodi*.

[252] See E.S. Smirnov's *Taksonomicheskiy analiz*. Moscow, 1969.

[253] *Somatometriya*.

[254] See O.M. Pavlovskiy's *On the Method of Color Photography in Racial Research and Anthropology*. Sovetskaya antropologiya, 1962, Issue 10. (*O metodike tsvetnoy fotografii pri rasovikh issledovaniyakh v antropologii*.)

On the basis of **psychology, psychoanalysis, comparative racial psychiatry, psychological anthropology, and physiological psychology,** a huge number of tests were created, which identify people on ethnic, social, and cultural levels.

Modern **psycho-genetics** has also achieved impressive successes in the study of the racial differentiation of humankind, on the basis of **IQ tests**, which reveal the hereditary conditions of intellect, that are dependent on racial affiliation. It is necessary to mention the names of such recognized luminaries in this area of science from the USA and Canada, like Arthur Jensen, William Bradford Shockley, and John Philippe Rushton.

And of course, we only give a highly cursory and glancing survey of the methods of racial-ethnic analysis. Our bibliography of books is not even 1/1000th of the known titles on the given subject. But this should be fully sufficient, in order to shake the confidence of those who preach ideas of universal equality.

In the book, *The Racial Differentiation of Mankind*,[255] authors G.L. Khit' and N.A. Dolinova cite a detailed, multi-page system of racial, ethnic, and regional traits of the entire population of the Earth, right down to small, relic tribes. Dermatoglyphic data corresponds with the findings of somatology, genetics, odontology, and other sciences; this allows one to speak of the creation of a systematic, racial-ethnic picture of humanity, similar to Mendelev's *Periodic Table of the Elements*. In the capacity of a universal, differentiating category, it naturally leads to such a concept as a **racial diagnostic marker.**

"The key traits of dermatoglyphics possess a huge reserve of race-demarcating capacity: the difference between the basic races exceeds error by 2 to 40 times. As special research shows, the most distinct differences within a race are observed on the ethnic level: the ethnic barrier is the most powerful factor of differentiation in a population, in the process of its historical and biological development."

Of principal importance is the fact that the conclusions of dermatoglyphics completely agree with the other biologically independent systems of traits, which were enumerated above; this increases the accuracy and reliability of the general picture. In connection with this, the authors of the book talk about a "line of distribution" of the racial diagnostic traits for the whole planet.

Each of the prominent racial groups of Mankind possesses an inherent combination of determined trait frequencies, and their combination is dermatoglyphically unique to each one only. Thus, an analysis of the combinations of traits convinces one that the differences between racial branches have a net configuration/reticular character. This observation is also supported by an analysis of the deviations of each racial group, according to the separate traits in their complex. Europoids and Negroids possess unique combinations. In averaging the size of the differences of each race from the general human sample, according to the sum of traits, we are led to the conviction that the maximally closest to the human norm as a whole are the Mongoloids, then the Australoids, and the furthest from the human norm are the Negroids. As a result, if the hypothesis of the existence of a general human dermatoglyphic complex is accepted, then it follows that we recognize it is most distinctly expressed in Mongoloids and Australoids. Europeans, and Negroids in particular, are most dissimilar to this "general human type." Mutual relationships of the basic racial groups have a durable character. Europoids occupy a more isolated position; Negroids are maximally specialized. Judging by the system of taxonomic differences between races, Negroids are the most ancient and specialized racial branch of humanity, while Europoids are the least ancient, but also a specialized branch."

Christian, then Marxist, and finally, modern liberal-democratic myths about the [biological] specific unity of Mankind should be conclusively discredited as not corresponding to reality. New dermatoglyphic data, and also genetics, graphically show that the concept of **monogenism**, that is, the single-source origin of Man as a biological species, is only a social command of those political structures that rule the world by means of the use of egalitarian technologies, that is, universal equality. And so, the legend of Adam and Eve is vested in the clothes of the "Out of Africa Theory".

---

[255] *Rasovaya differentsiatsiya chelovechestva*. Moscow, *Nauka*, 1990.

Humanity has several centers of race genesis, that are completely isolated and biologically unconnected to one another. Moreover, these centers of origin of the various human races are located apart from one another in time, something that is very important. These basic human racial types are not only divided by kilometers, but by tens of thousands, and possibly hundreds of thousands of millennia. Therefore, any conversation of the biological specific unity of Mankind is an unscientific, provocateur idea, created in the interests of those who desire to control humanity, through a fraud that omits the racial affiliation of peoples.

G.L. Khit' and N.A. Dolinova give the following summary in their book: "The basic racial branches of humankind are orginal, independently formed subdivisions of the human species. The coincidental arrangement of the kinship of the human races is arrived at by genetic markers and other collections of dermatoglyphic traits."

Further developing the methodology of research, G.L. Khit' and N.A. Dolinova conducted an analysis in their new work, on the basis of the following traits: the Delta Index (DL); the Cummins Index (I); the Proximal Palmar Triradius (t); the pattern of the Hypothenar (Hy); the summary percentage of additional triradii (DMT); and the pattern of the thenar of the first inter-finger pad (Th/1). Rapid computerization of all branches of modern science allows us to speak about the development of an independent, applied discipline—**statistical dermatoglyphics.** The uniqueness of the patterns of the skin coverings of the hands are used even before the first year of life by heads of regime secret security apparatuses. The picture of fingerprint patterns of employees in business enterprises are scanned and uploaded to the memory of a computer, in order to henceforth serve as a pass that identifies and individual. In contrast to keys, magnetic cards, codes or passwords, fingerprints are impossible to forge or counterfeit and pass off as one's own. And this in turn shuts out undesirable persons from access to secret information, with nearly a 100% guarantee.

The main advantage of the dermatoglyphic method, besides its accuracy, lies in the subtlety of the moment of the analysis itself. It is impossible to take measurements of a person's skull, analyze his dental system, or analyze his blood, without him noticing. But fingerprints can be taken from anyone, practically without their knowledge. By entering a computer database, and retrieving information about employee fingerprints, one can unerringly determine the racial, ethnic, and even the regional make-up of their origin, without seeing anything of an individual's face. If a certain person agrees that an official application represents the native of a concrete locality, and produces well-forged identification about birth and education, the method of statistical dermatoglyphics instantly exposes the forger; according to fingerprint patterns, one can see for example, if he was born in New York, rather than Moscow, even if he resembles a Muscovite.

We all remember the popular TV series, *17 Instant Springs.*[256] A Soviet agent under cover as SS Officer Stirlitz was discovered namely on the basis of an analysis of his fingerprints, which were taken without notice: "Stirlitz, bring us a glass of water." He succeeded in avoiding exposure, only because his fingerprints were examined at the lowest level of criminal expertise, that is, on compatibility with the images of other fingerprints. If Stirlitz was subjected to a more serious racial-ethnic diagnostic—according to the method of Dr. Erich Karl, for example (which is described in the journal, *Volk und rasse*)—he would not have succeeded in avoiding detection. Although he was described as a "true Aryan" in his personal papers, a dermatoglyphic inspection would have unequivocally attested that this person was not born on German territory. As the outstanding German racial theorist Fritz Lenz so accurately observed: "No one can crawl out of his skin, nor from his soul."

Another example is illustrative.

In modern raciology, the term *PASSING* has been widely adopted; it means that a mestizo, originating from parents of differing racial-ethnic backgrounds, in a move to a new place of residence, uses the simplest methods of masking to pass himself off as a pure-blooded member of one of the races of his parents. Mestizos most often simulate Europeans. Obviously, the popularization of dermatoglyphics would put an end to "passing."

G.L. Khit' and N.A. Dolinova clearly write: "Of all the Europoids, Europeans are the most distinct." This is taken to mean that a pure-blooded European will always be identified, in racial

---

[256] *Semnadtsat' mgnoveniy vesni.*

terms, against a background of similar Europoids. This same thesis then, outstandingly confirms the precept of a "racial core," in the members of which unique racial traits are expressed with all obviousness. Henceforth, this means that any falsifications on a racial ethnic basis will be put to an end, and this will lead to a genuine revolution in all social-economic, religious, and cultural relations. The theory of government and rights will unavoidably be subject to revision, constitutions will collapse, and the "Declaration of the Rights of Man" will once again reshape the political map of the world. But the new state of affairs will not correspond to the interests of merchant myths about equality, but an authentic, natural, and genuine hierarchy of peoples and races. World banking capital based on usurious interest will lose its basis [of existence], and usurious fortunes will fail. There will be a revolution in the worldview of the peoples of the planet, [which will] be principled and in a class by itself. The end of the age of humanitarian values should coincide with the end of the immediate era of the history of Mankind. In agreement with Zoroastrian eschatology, "the Age of Mixing" concludes and is replaced by the "Age of Division." In light of new scientific discoveries, it becomes completely obvious that the predictions of the Persian pagan priests also have an underlying biological basis, for mixing needs to be understood in a deeply racial sense.

Now we move to an examination of the work of still one other outstanding Russian racial theorist: Viktor Alekseyevich Spitsyn. However, certain Soviet ideological stamps of internationalist anthropology do not allow his discoveries to influence the public consciousness in full measure.

V.A. Spitsyn's book, *The Biochemical Polymorphism of Man*,[257] was published in wretched numbers—980 copies in all. With the help of a genetic-biochemical analysis, he not only confirms all the basic conclusions of Khit' and Dolinova's dermatoglyphic method, but he formulates an entire set of new, important conditions in racial theory. In the introduction, the author writes: "As a result of numerous studies, it was established that alternative forms of proteins, found to be under rigid genetic control, are quite unequally distributed among the peoples of the Earth." Using extensive foreign materials and his own works, V.A. Spitsyn speaks in his book "about the global distribution of blood types and other hereditary polymorphisms in Man," and also the "discovery of possible connections between the frequency of the dissemination of defined genetic markers and diseases." Genetic markers are introduced on the basis of mathematical formulas, according to which an account of global migration processes in the history of the world is produced. With a high degree of accuracy, genetic-biological changes in human communities are calculated. From vague, impassioned forms to the structure of impartial scientific definitions, racial differentiation becomes a reality before the eyes, through the efforts of the author. Differences between peoples are determined on a molecular level.

Let us consider facts.

The frequency of **system transferrins (Tf)** in genes increases in the order, "Mongoloid-Europoid-Negroid." Namely with their help, a more detailed differentiation of ethnic groups is produced. In the American city of Seattle, the Central Blood Bank has already worked out and applied **ethnic standards of blood**. In Moscow, it was also established in the course of a research program, that ethnic Russians significantly stand out among other ethnic groups in multi-national Moscow, namely on the basis of system transferrins. The **group specific indicator (Ggl)** is higher in Jews, and distinguishes them from Russians. On the basis of the components of blood serum, ethnic differences are clearly evident, resulting namely from genetic conditions.

The distribution of **genetic frequencies according to Spitsyn's system**, shows that the basic human races differ from one another by 30-40%, on a biological level. Expressed in the language of the Bible, God sculpted us from a separate clay, because between the different breeds of animals, biochemical differences are fewer. "Mankind, with its common human values" is an indefinable phantom. According to this system, it is established that Russians distinctly and clearly differ from Tartars, Jews, and even Ukrainians.

According to a number of biochemical indicators, ear wax differs between the basic races by up to 4-6 times; that allows one to use it as a comfortable and clearly expressed racial diagnostic marker.

---

[257] *Biokhimicheskiy polimorfizm cheloveka*. Moscow, 1985.

Therefore, V.A. Spitsyn came up with a corresponding general **coefficient of genetic differentiation (Gst)**, that enables an empirical calculation of the degree of heterologousness of peoples. He also introduced and substantiated the concept of **genetic distance (D)**, which serves to determine the aggregate of racial differences between peoples. On the basis of these mathematical formulas, the genetic-biological **hierarchy of races and sub-races** is introduced. Next comes the concept of **time of genetic divergence (t)** of peoples and races in the process of world history. Further conclusions and substantiations by the author will again affect the combination of the global scope of the problem, with the mathematical accuracy of its resolution.

"Methods of analysis of the genetic distances and construction of phylo-genetic trees allow resolution of the task of genetic-anthropological classifications, determinations of the degree of divergence of populations, and evaluation of time, arising from the moment of the separation of the corresponding branches of the tree. A more important problem, however, is the attempt to evaluate the duration of the tenure of this or that group of the population of a given territory, and also the immediate determination of the dynamic of the formation of racial-ethnic communities. In this regard, the research of the characteristics of the world distribution of factors of the **system of immunoglobulins (Gm)**, enables one to come to a resolution of the above-stated questions.

The uniqueness of the genetic system Gm lies in the fact that each of the major races possesses a trait, which is characteristic to its Gm genetic complex only, such that the marking haplotypes represent maximal value frequencies." It is completely evident, that on the basis of similar methods, any speculative discussion about "native and non-native peoples," and multi-national cultures", can be decisively put to an end, and the constitutional definition of a "state-forming ethnos" will receive a mathematical definition that excludes mistakes. Formulas for creator-peoples and parasite peoples can also be calculated.

Summarizing international statistical-analytical experience, V.A. Spitsyn gives exact formulas for the frequency of genetic markers in the system of immunoglobulins for the basic human races, corresponding to their maximal frequency:

Europoids—Gm $^{3;\ 5,\ 13,\ 14}$
Mongoloids—Gm $^{1,\ 3;\ 5,\ 13}$
Negroids—Gm $^{1,\ 5;\ 13,\ 14,\ 17}$

An individual familiar with mathematics will easily notice that these three systems of characteristics are completely different and unconnected.

V.A. Spitsyn further emphasizes that "in Europe, the Europoid Gm $^{3;\ 5,\ 13,\ 14}$ is represented with a high frequency in the north of the continent." Thus, it is determined that the most racially pure Europeans are in northern Europe; that corresponds exactly to the geographic origin and range of the Nordic race.

Consequently, notions of racial purity move from the area of journalistic dispute to precise science. From this it becomes clear that **racial purity is not a mythic phenomenon, but a genetic one.** Henceforth, racial theory should base its theses not on poetic form, but on genetic analysis and mathematical formula. The basic postulate of racial theory, that the Nordic, or ksantochroid race, that is, the light-haired, light blue-eyed part of the great Europoid race, is the most genetically pure, and consequently its portion is more worthy; that can be considered scientifically proven.

Another racial parameter testifies in favor of this viewpoint. "A decrease in the concentration of gene **(Pa)**, is traced from the north to the south, within each of the great races."The maximal concentration of gene **(Pa)** is achieved in Europoids, namely according to the measure of the decrease in solar radiation. Therefore, gene **(Pa)** can be called a Nordic gene. Alternative gene **(Pa)** on the other hand, increases in conditions of a hot climate, and namely in members of southern races. Consequently, in the black and yellow races, it is their criterion of genetic purity.

The genetic basis of skin pigmentation also has a highly important racial-demarcation function. In this regard, V.A. Spitsyn writes: "It is known that there is a thick layer of melanin in dark-skinned races, which hinders the penetration of ultraviolet rays deep into the skin, [but] this creates fertile ground for rickets disease. This explains the presence of a compensatory mechanism, [found] in peoples who live in the tropics; they are endowed with an abundance of sweat glands, which are significantly more prominent, than in Europeans."

In Europeans, the frequency of the gene **(Gc)** should never exceed 10%, while at the same time it exceeds 30% in Negroes. It is namely the frequency of this gene that is tied to the characteristic negro odor.

Ancient myths and tales around the world about blonde and white-skinned beauties, as well as their higher value in Asian slave markets, are mostly a result of the genetic envy of the members of coloured races. In this area, mythology, history, and aesthetics, together with genetics, show a very strong resistance to liberal [race] confusion. Racial purity is the first law of Nature itself; mestizos, then, are opposed to Nature's laws.

The Italian biologist Renato Biasutti created a *World Map of Skin Color*, where each hue has a numerical order, according to Luschan's scale. Spitsyn's main conclusion closes with the following: "There is no data on the relationship between climatic-geographical factors, and the distribution of **Gm** factors." This says that racial traits have an unadaptive character, that the environment in general has no influence [on our heredity]. The color of the hair, eyes, skin, and so on, is not the result of a human adaptation to corresponding conditions in the surrounding environment; rather, they are **genetic adornments**, which nature, proceeding from the natural principle, **"to each his own"**, gave to the different races. During the reign of communism, Russian racial theory did not cease to exist, but continued to develop. As a result of this, **the new Russian raciology** is based not only on classical racial theory, but also on the better, independent works of the Soviet era, which enter the honorable world collection of racial thinking.

The Russians are a defined, genetic-biochemical constitution, not those who conceal themselves behind the Russian language and Russian culture. Racial and ethnic differences are not conventionalities or prejudice, but genetic facts, which are calculated with mathematical precision.

The above-cited scientific works are hardly mentioned in Russian scientific and social circles, and the book by Canadian professor John Phillipe Rushton, *Evolution and the Behavior of Races*, produced a real furor in the West. Building on the results of true research methods, the author came to analogous conclusions. This book is destined by fate to be a classical work on racial questions, equal to such masterpieces as *Race*, by John Baker, *Deviations in the Study of Intellect*, by Arthur Jensen, *Sociobiology: the New Synthesis*, by Edward Wilson, and *The Biology of Human Behavior*, by Irenaeus Eibl-Eibesfeldt.

The basic differences between the races are explained by Rushton, on the basis of a strategy for reproduction of one's species, which comprises the biological essence of any living being. It is namely in the sphere of reproduction [that] the differences between the human races are also the most obvious and perceptible. One extreme is strategy "r". It stands out for maximal fertility of an organism, with minimal care about the fate of offspring. A typical example is the oyster. Each year it leaves millions of eggs in the ocean and leaves them to their arbitrary fate. Almost all of them die, and only several become adult specimens. The more primitive a population, the more fertile it is. Among peoples, this rule is also unfailingly observed. Afro-Americans, Gypsies, and several peoples of Asia stand out for fertility, giving extremely little attention or care about the upbringing of their growing offspring. It is namely the strategy for fertility that is the reason for the high child mortality rate in these peoples, and not some "social curse of post-colonialism." Neither the Red Cross, nor other humanitarian organizations can help them, for they are not in a position to alter the fertility strategy of the given ethnoses. Noble sentiments do not influence the physiology of oysters and the like. Appealing to the feelings of compassion in cultured [people] is completely useless. [To] criticize them lacks sense, because they have a principally different fertility strategy—"K". At its basis lies a minimal fertility with a maximum of care by the organism for its offspring. Strategy "K"is characteristic of more highly developed beings.

As a result of the given reproductive strategies in the different human races, the instrumental part, which is employed in reproduction, differs. Unsuspicious humanitarian organizations, which are engaged in the struggle against AIDS, are the first to run into this. They very quickly discover that a one-size condom, in the conditions of an open society and declared equality, does not satisfy the requirements of everyone who wishes to protect himself against the "plague of the 20th Century." The racial problem comes alive again on a qualitatively new level. It is completely obvious that the members of strategy "r" occupy the quantitative side of the question, surpassing the members of strategy "K" in size and in any functional characteristics of the instrumental part in the reproduction of the species. Among Asians the parameters are the least of all. It has long been known that the age of sexual maturity in the different races varies; but now, on the basis of biological behavior, sexology, and psychoanalysis, irrefutable proof has been added to this; [there are] also similar cardinal differences, in regard to sex—including priority poses during the sexual act. The organism surrounds the object of his desire with attention, in direct relation to the strategies "r—K".

The average IQ essentially differs in the members of different races. Thus, in Europeans, it is equal to 100, while in Africans it is around 70. The speed of [mental] reaction differs in approximately the same proportion.

Rushton writes: "Heads of large size (containing more developed brains) are found to be in direct correlation with intellect. Large heads have a tendency to shine with their intellect. This correlation is faithful in relation to various racial groups. At age 7 African children are 16% larger than European children, while the perimeter of their brains is 8% less. In Asians, the perimeter of the brain would be larger than in Europeans, but several of these results manifest after correction, when the fact is taken into account that Asians are smaller than Europeans, on average. This suggests that a small person with the brain of a large person, has a more developed brain, in so much as smaller-sized bodies need less of a brain for the control of their glandular functions.

On average, Negroes have 400 million fewer neurons in the head, than Whites. With a small brain in a large body, they are less gifted intellectually, because a greater part of the brain in Negroes is engaged with life functions, and not with creative thoughts."

Racial differences, by the indicator of which "r—K" appears, are a structure that manifests in the sphere of criminality, the tendency to suicide, and also the demand for narcotics. Whites are more inclined to frustration and thoughts about suicide; Africans, for their part, are predisposed to narcotics abuse, commission of criminal acts, and psychiatric deviation. In Asians it is the opposite; attacks of melancholy are observed more often, and mental diseases are encountered more rarely, than in Europeans and Negroes.

Finally, in the opinion of Rushton, the inclination to altruism and the strategy of social adaptation also has racial properties: 'If altruism is an important characteristic of [strategy] "K", then criminality is parallelly associated with factor "r".' Further, the Canadian scientist ingeniously noticed that "genes also come forward in the role of marital agents," for the members of each race have their own inborn criterion of evaluating beauty and other virtues of the opposite sex. Therefore, racial homogeneity is inarguably good for each society. "Bees and ants offer proof of the unusualness of altruism; they die in defense of their colony, because their system of reproduction leads to this; that is, worker bees have 75% common genes. Monkeys and squirrels can observe genetic differences within their groups, but are more inclined toward cooperation with those that are most genetically close to them."

Dr. Rushton thinks that the sum total of the differences in the structure of the characteristics of "r—K" has a significant degree of genetic origin, and is well-suited for determinations of both class and racial differences. All psychological aspects of the life activities of an individual are also under the strong influence of heredity.

"Even small children recognize racial differences and have corresponding preferences. Ethno-centrism and "racism" are natural mechanisms, contained in the genome of an individual. Just as structures of behavior differ among individuals within a racial group, so they differ between races; they reflect the structure "r—K", and without it, you cannot manage with the formation of any social programs." Therefore, the great German racial theoretician, Baron Egon von Eichstedt, justifiably maintained that peoples of different races and nationalities differ by taste.

Thus, in light of new research in the area of human heredity, it has become completely obvious, that many of the criterion of social values should be subject to decisive revision. A reinterpretation will touch on numerous, modern humanitarian sciences, unavoidably bringing in its wake a structural rebuilding of official public institutions of authority. International relations will acquire a new specific; the role of biological determinism, which will sharply grow into all spheres of life, including art, education, and religion. The very notion of what is human could change; historians will be compelled to begin rewriting history, and philosophers will give their blessings to the man of a **new race.** The degenerative species *homo sapiens* ought to quietly and unnoticeably leave the historical scene, like the Neanderthals and Australopithecines. And again, no one will know why this happened.

But that is in the future, and already now, on the basis of dermatoglyphic methods, serology, and also with the help of genetic markers, one can re-establish the racial purity of our species with a high degree of accuracy. The racial core can be cleaned of foreign admixtures in a short time. Eugenic refinement of the race as a whole ought to now be realized, not with the obsolete methods of cattle-breeding, but on the basis of genetic therapy. Today, in the body of a living person, one can carry out a genetic revision of cells, cleaning them of the racial pollution left by unforeseeing ancestors. Besides that with the use of more accurate methods, one can correct and improve the proportions of the human body.

As we remember, in common anthropometric methods, errors in calculations are higher, in view of the influence of the environment on the somatic structure of the individual. Genetic methods are capable of excluding errors introduced by the environment, thereby helping to create members of a race that are close to an ideal. And with the help of cloning, the number of sanitized, most worthy peoples can be increased to multi-million populations, in the lifetime of a single generation. Cultured peoples, carrying on their shoulders the birth of a world civilization, will no longer look like herds of piebald, outdoor pigeons.

Purity of the breed, strength, virtue, health, and self-sufficiency will again become the norm, a daily phenomenon, rather than nostalgic folk tales forms. Elevating the quality of human material will unavoidably leave in its wake a lifting of all norms of quality of culture and civilization. Hypertrophic inter-ethnic contradictions will disappear. Politicians will no longer shove the peoples of our race into fratricidal civil wars. The racial instinct of the White Man will be restored in his rights, and the racial consciousness will firmly stand above all social, religious, and national associations. Consolidation will again be possible on the basis of regular heredity, and not by virtue of the caprices of a changing environment. The activity of each citizen for the good of the race will stand as a guarantee of his confidence in immortality, which he will acquire through his descendents.

Immortality of the individual within the immortality of race—what other moral imperative can compare with this, by virtue of its simplicity?

For the first time, our Gods have bestowed upon us a chance to change the fate of the [White} race for the better.

Already, today we can model the ideal Nordic man from the days of the dawn of the expansion of the North against the endless spaces of Asia.

We can again show the world a huge number of tall, athletic, well-built, long-headed blondes, conquering the empty deserts, the forests, snows, and native races in their path.

Still another new invention gives us the real opportunity to place the very process of ethnogenesis under full control, and then to actively advance the end of history, which the Old Testament prophets predicted and Francis Fukuyama now predicts. But we can give thought and form to the end of history, according to its judgement. This will only be the sunset of the existence of the species, *homo sapiens*; a new, superhuman race will arise in its place.

The given authentic revolutionary discoveries, the results of which have still not been evaluated in full measure by fantasy and morality writers, are a **genetic wave**.

At its base lies the summary of the findings of medicine and genetics, graphically testifying about the presence of a wave of nature's signals, which govern genetic activity. A sufficiently extensive bibliography on the given problem already exists, but within the bounds of our essay, we will stay with only one work, as it appears to us that it crystal-clearly illustrates our theses. This work is that of Eleonora Nikolayevna Chirkovaya: *The Wave Nature of the Regulation of Genetic Activity. The Living Cell as a Photon Computer.*[258]

It is known that any cells which altogether do not resemble each other within the same multi-cellular organism, retain one and the same collection of genes (DNA fragments), received by the organism from the mother and father, during reproduction. For various parts of the cells, the size of resonance frequencies is established by an empirical method; the frequencies provoke the activation of this or that group of genes, in the desired direction. And this enables [the organism] to regulate the biological clock mechanisms of the cell, giving it this or other parameters of growth and metabolism. As a result, we have an opportunity to regulate the genome of an individual as a

---

[258] *Volnovaya Priroda regulyatsii gennoy aktivnosti. Zhivaya kletka kak fotonnaya vichislitel'naya mashina.* Moscow, 1992.

240

whole, with the help of pathological deviations at a cellular level. This is spoken of in the said article. Continuing our thought, we can also regulate the genetically conditioned racial parameters of an individual on the very same cellular level, without any harm at all to the individual in the process. With the help of genetic markers, the findings of dermatoglyphics, and analysis of the components of blood, we can, without error, establish the racial-ethnic and territorial affiliation of any individual. Affecting the genome of an individual in the assigned ranges (diapasons) of frequency, by means of new, magnetic laser therapy, we can now create new centers of race-genesis and manage the very process of race-genesis in [real] time. Characteristically, the diagnostic methods and management by these processes, can be deeply confidential and inaccessible to wide layers of society.

One must clearly understand that entire global transformations in history are implemented not in the name of Man as such, but in the interests of a concretely dominant bio-type, with the goal of subjugating and suppressing competitors.

In connection with this, our main goal is crystal-clear: the creation of a new, super-perfected White Race, the moral and physical degradation of which has reached its limit. Second, an also highly important task is the creation of a new concept of history, based on new, biogenetic methods.

Both of the tasks will help us work out a **new paradigm in raciology**, which we have attempted to describe in our essay.

THE ANTI-RACIAL, "MELTING POT" MYTH

"God created the White; God created the Black; the Devil created the Mestizo."
Arab proverb.

"The Mestizo cannot love the races: two, three, and twenty, for he would have to defend them."
Albert Fullais

At the end of the 19th Century, the great English naturalist and philosopher, Karl Pearson (1857-1936), prophetically wrote that in his opinion, the historical role of England was to prepare the white race for a good death—euthanasia—by everywhere organizing, creating, and establishing "peace, law, and order", plunging its citizens into a world-wide mediocrity. After 100 years, the strength of his forecast became obvious, for the USA, snatching the initiative from England, persistently prods the white race to biological extinction, everywhere spreading the myth of the so-called "Melting Pot."

This manifests more sharply in democratic [political] science: under the aegis of "political correctness", the very word *race* has generally disappeared from encyclopedias and dictionaries on anthropology in recent years. The harmful and completely unfounded myth of society as a "melting pot", which will remove the racial and national problem once and for all, is touted more and more by mass media. According to the thinking of the creators of this myth, all natural differences between peoples ought to disappear into one legal and socio-cultural space of modern civilization. Besides this, the history of the origin of the human species would disappear, thanks to the shameless and cynical manipulation of the means of population genetics. In the role of a single ancestor of all races, a black-skinned "Eve" from Equatorial Africa has been named. This concept, so indecent and so scientifically absurd, is itself the wonder work of the new **black-skinned Lysenkovites,** in whom we still have the opportunity to be convinced. Concerning the **democratic steelworkers from anthropology**, who invented the notorious myth of the 'melting pot', we urgently would recommend to them to open any book on metallurgy, from which it is clear, that to melt down a single alloy, one can only use metals of a similar crystalline structure. If in the alloy process, one takes and throws elements of varying characteristics into the pot, the 'wonder work' you will get is a heap of metallic garbage with cavities and cracks, which is impossible to refine into a single part or slab.

Besides that even a finished and polished alloy that meets all one's demands, can be separated into its original metals in the reverse process—clean forms without any admixtures. This rule works flawlessly in other areas of science, technics, and even in day-to-day life. As much as you would like to unite the most varied elements in a chemical flask into some unthinkable solution, they will never lose their own characteristics, on account of which all the elements can again be separated into pure form.

Likewise, in biology irremovable physical and chemical traits of matter manifest themselves exactly in the **form of perpetual and indestructible racial traits.**

The different human races are, in essence, different human species. Cross-breeding, about which many modern population geneticists speak, placing an accent on mathematical calculations, does not at all testify about a specific unity of the human species.

In modern liberal-democratic anthropology, in the literal sense of this word that is purged of the races, a traditional viewpoint dominates: that the environment renders a decisive influence on the formation of any living organism, including its inherited traits. Adepts of this theory intentionally confuse constitutional traits, which are actually partly dependent on the environment in the organism's habitat, with racial traits, to which environmental influence does not extend. It is namely here, where a basically dishonest shuffling of facts occurs. This is the stuff of Lamarckism, the evolutionary doctrine of changes in an organism's biological traits, under the influence of the environment, with which the name of French naturalist Jean Baptiste de Lamarck (1744-1828) is connected.

It is hardly necessary to explain further that on the boundary of the 18th and 19th centuries, there was still no system of proofs or scientific methodology; it arose only on the boundary of the 19th and 20th centuries, when August Weisman (1834-1914) and Thomas Morgan (1866-1945) created the concept of "embryonic plasma", on the basis of experimental data, from which it followed that an organism does not change its racial traits under any kind of environmental influences. At the start of the 20th Century, anthropology and biology definitively divided racial and constitutional traits, and in scientific circles the question was resolved; but then big politics interfered. The ideals of equality, and the education in the times of the Great French Revolution, took up Lamarckism in its most speculative and cynical form, and brought it to life again in the doctrines of the European social-democrats, in the second half of the 19th Century. The 20th Century communist experiment in Russia was carried out, namely on the basis of anti-racial conjectures and speculation. Class consciousness was specifically invented with the goal of disabling racial consciousness, since the socially-horizontal fraternization of proletarians in the literal sense of this word was supposed to break the aristocratic vertical line of the "blood myth."

1. "The Black-skinned Persecution"

In the West in recent decades, quite a number of monographs on the phenomenon of "Lysenkoism" and its personal embodiment in Trofim Denisovich Lysenko have appeared; but the state version of his inspired ideologues, as usual, remains behind the scenes among foreign truth-lovers. They transmitted the powerful, influential, and near-scientific field to the stupidity and ignorance of the simple peasant, who was entangled in a jungle of Mitchurinism; but the names of the helpful blind guides were diligently left out of this "routine, regrettable extreme of totalitarian government." However, even a superficial study of the sources clarifies the picture to the maximum.

The young science of Russian anthropology at that time developed in a general context, without any distortions or lagging. The names of D.N. Anuchin, A.P. Bogdanov, and I.P. Pavlov shone on an academic Olympus, like stars of the first magnitude. In 1900, the *Russian Anthropological Journal* was published, and without any upheavals, it survived the Bolshevik Revolution. In the first ten years of Soviet authority in Moscow, Petrograd, and a number of other cities, even a department of The Russian Eugenics Society arose, and the *Russian Eugenics Journal* began to be published. At the same time, despite the fact that the Great War had just ended, contacts between Russian and German scientists did not cease; in the young Soviet republic, they published translated works of world luminaries in the areas of raciology, anthropology, eugenics, racial hygiene, and social biology. However, in the 1930s the situation changed cardinally.

In 1930, the *Russian Anthropological Journal* lost its 'Russian' designation, and the No.3 issue of 1934, Arkady Isaakovich Yarkho wrote the following in his program article, *The Immediate Tasks of Soviet Race Science*[259]: "Anthropological theorists, seeming relatively inoffensive, openly propagandized for 8-10 years in the Soviet press. Until 1930, Soviet race science was completely under the influence of bourgeois racial theories [that are] foreign to us, as well as different bourgeois-idealist currents in the areas of archaeology, ethnography, and linguistics."

Further, Yarkho shows a distinct methodology of distortion of anthropological science: "The struggle against racial theory suggests the presence of completely defined tactics and strategy. Only in the role of a counterweight to the theses of racism, will the concept of historical materialism be shown by us; if we transfer the center of gravity of criticism from the plane of biology to the plane of sociology, our criticism will be effective." Finally, the author of this article does not even consider it necessary to conceal why an entire branch of science in the USSR ought to be so keenly disfigured: "First and foremost—the systematic exposure of the role of the race factor in the historical process."

The entire cohort of the newly-proclaimed luminaries of the young "Soviet science" began a massive attack on classical anthropology, recasting it as in the interests of the party minority, which was undertaking Michurinist manipulations of the Russian people. Some of those individuals are named as follows: Mark Solomonovich Plisetskiy, Mikhail Antonovich Gremyatskiy, Boris Yakovlevich Smulevich, Maksim Grigoryevich Levin, and Yakov Yakovlevich Roginskiy. The last-named particularly stood out with an introduction to everyday science, of special abusive terms addressed to German racial theoreticians: "anthropo-fascists" and "rasovniks". According to the word style here, it is easy to guess the genetic source—children of small-town tailors and tavern-keepers, who flooded the Russian language of that time with proletarian-cult phraseology, and mutant words like "crowd-joker."[260]

The pressure by politicians on the minds of scientists was so unprecedented, that even the outstanding Russian-Soviet anthropologist, Viktor Valerianovich Bunak—whose authority in the world of science was almost absolute, despite the political orientation of this or that regime—was compelled in 1938 to write in favor of the political situation,[261] in a work with the characteristic title: *Race as an Historical Concept*.[262] He literally saved his students and his school from extermination and destruction. The very title of the work contradicted reality, for race is a phenomenon of biological character, not an historical or prehistoric character. In this article he goes so far as to say that "race is an abstract notion" and that "races arise as a result of mutations." Besides this, [he wrote] that "race is not an absolute category, but an historical [one], [with] several stages of formation. Each epoch has its races in their concrete manifestation."

According to the logic of Yarkho and Bunak, the Russian people of the Age of Peter the Great belonged to a race different from ours; Alexander Nevskiy, Dmitry Donskoy, Evpatiy Kolovrat, and so on, [belonged to other races]. And what about us, and our incorrect visual perception, when we look at ancient Greek and Roman statues? *Does it only appear to us, that all these people are of our race, but are totally different, nevertheless?*

But there is more to come. After all, the luminaries of science were making an official declaration. N.A. Bobrinskiy wrote that in biology, "species do not really exist," that "a scheme, an ideal specimen, only exists in our minds." In a word, differences between the Negro and the European are the result of an **aberration** in our vision! M.A. Gremyatskiy believed that the "division into races, of course, is conditional," and that race appears as an abstract as a result of mathematical analysis." It happens that skin color, hair color, and eye color is something not unlike x and y, which one can rearrange on a sheet of paper, to one's liking. A.S. Serebrovskiy also declared that in the notion of race, one unavoidably introduces a moment of subjectivity.

---

[259] *Ocherednie zadachi sovetskogo rasovedeniya.*
[260] "massovik-zateynik"
[261] The Great Purges
[262] *Rasa, kak istoricheskoe ponyatie.* 1938

In support of the anthropologists, they called in official philosophers and historians. Thus, in 1933, the court communist intellectual, V.F. Asmus clearly wrote in the foreword of his book, *Marx and Bourgeois Historianism,*[263] that "..biologism and historism are incompatible."

Also, A.I. Yarkho reported in an article[264] for the *Anthropological Journal*, N1 1932, that in comparison with primitives and other animal races, for the human races "the loss of species (racial) instinct is first of all characteristic."

It is particularly amusing to hear through the mouth of "God's chosen" people, that incontrovertible racial and species solidarity is considered anti-semitism; so too, then, are Jewish theologians. The very principle of Zionism is built on the racial solidarity of the Jews, who among all other peoples, in Yarkho's view, are disappearing in the process of evolution for some reason. Further, the author of this "scientific article" writes that "the greater part that is united within the species *homo*, with the exception of several primitive tribes, is in racial terms, mixed." So then, it follows that commands to observe racial purity—which fill the Old Testament and the Talmud—and on which the citizenship laws of Israel are also based, are no more than a fabrication of a "primitive tribe." And the genetic verification by Israel's Immigration Service of affiliation with the Jewish people—this is the "stupidity of savages." Yarkho further writes: "The entire history of the species *homo* is an example of 'removal', the elimination of biological laws. The new unification of a biological collective unavoidably 'removes' the reality of race. Arising in the process of humanization, productive relations 'remove' the reality of race as such"

There is a more ancient world rule: if you want to check the reliability of some scientific theory, test it on the Jews, and right away, everything will fall into place. In the context of the given narrative, it happens that in all peoples of the Earth, racial traits are abstract and unstable. And besides, with time, all races will generally wash away. But this is for other peoples; in the Jews everything is the opposite: the racial-genetic traits in them are not subject to time. All of the peoples surrounding them are mixed, but the Jews are pure-blooded. And all this despite the fact, that for 19 centuries, they did not have their own state.

Therefore, in accordance with this spirit, Yarkho gives a definition of anthropology: "Anthropology is a science that studies the changes in the biological characteristics of a population in an historical process." Pay attention. Races have "washed away" and "departed", and a people has unnoticeably, but gracefully replaced a population.

At the start of the 20th Century, a famous German racial theoretician, Herman V. Siemens, wisely noticed that "anthropology belongs to a number of rare sciences, which can be completely privatized by several professors." However, the very same Yarkho did not even think it necessary to conceal the true, underlying cause of his ideological thoughts, for he observed: "Great tasks stand before Soviet raciology. The main first [task] on the given stage, is the exposure of any kind of attempts to shift biological laws to society, and the exposure of any false anthropo-sociological or any other imperialistic racial theories; and finally, in the struggle with polygenism, the creation of a Marxist theory of the origin of the races."

In accordance with this political order, Soviet science announced for all to hear, that race does not exist. No terminology—no concept; no concept—no problem. That is the Marxist dialectic.

After the war, and many years after the Stalinist repression, genetics was only restored to its rightful place in the Soviet Union in 1964. As a result, the opportunity again arose to openly discuss the question of the origin of races; however, Marxist blinders continued to disfigure science. The famous Soviet scholar-geneticist, Nikolay Petrovich Dubinin, was again compelled to make a completely groundless, unnatural assertion in his book, *What is Man?*[265] He wrote: "*Homo Sapiens* completed the biological evolution of his ancestors, and one may assume that by a natural path in the future, he will not transform into a new biological form, and he will not acquire the quality of a new biological superman, or on the other hand, of a pre-human. It is a fact that the evolution of Man did not cease on the level of species; it appears to be a contradiction to the dialectical science of the eternal movement of matter. But this is a fact, and it needs to be explained."

[263] *Marks i burzhuazniy istorizm.*
[264] *Protiv idealisticheskikh techeniy v rasovedenii SSSR.* "Against Idealistic Tendencies in Racial Studies in the USSR."
[265] *Shto takoe chelovek?* Moscow, 1983.

And so N.P. Dubinin, who succeeded in peering into the future and establishing the completion of a species, and consequently, the racial evolution of humankind, naturally did not explain it anywhere. Even the Marxist dialectic contradicted his assertion. The scientist further wrote: "In the beginning, races and peoples had the same significance, like [species] among animals." However, whether they "washed away" or "departed", Yarkho did not risk speaking any further. Everyone became quiet.

It is obvious to an elementary school student, that the communist regime could not create an intelligible theory of anthropo-genesis, and subsequently could not solve the race question. Such examples, which go against all laws of logic, can be gleaned from the works of Soviet academic science, as much as one needs.

However, by measure of the weakening of the ideological pressure, the situation began to be corrected, again thanks to the titanic efforts of lone champions, like N.N. Cheboksarov, G.F. Debets, V.P. Alekseyev, T.I. Alekseyeva, G.L. Khit', A.A. Zubov, V.E. Deryabin, A.L. Purundzhan, and A.G. Kozintsev. The works of V.V. Bunak, with the exception of the single one cited by us, are also part of the golden collection of Russian raciology.

2. Evolutionary Stability of Racial Traits

The paradox of the situation, nevertheless, lies in the fact that having hardly recovered from the diktats of the Soviet Lysenkovite repression, Russian anthropology was forced to collide with the Lisenkovites of the liberal West—first of all, American population geneticists, who also declare that there are no races.

With authentic Russian patience and quick good sense, a convincing answer was given to the grandchildren and great-grandchildren of those same tailors and tavern-keepers, who were once again changing the Pale of Settlement for their convictions.

In the collective anthology of authors, *Modern Problems and New Methods in Anthropology*,[266] V.P. Alekseyev, deserving of seniority, opened with his program article, *Problems of Race in Modern Anthropology*.[267] Analyzing the interconnection of the recent achievements of science with public opinion, he emphasized: "Scientific interest toward the question of race in Man, is first of all conditioned by the essential widening of the selection of traits, chiefly morphological, with the help of which races are set apart and characterized. Intensive studies of recent decades have revealed vast aggregates of traits in the morphology of the human body, which with greater or lesser success, began to call for the classification and description of racial variants on a different level."

The progress of science in the area of the increase of the quantity and quality of race-distinguishing traits summoned well-founded criticism by V.P. Alekseyev against vulgar geneticism: "To completely ignore the obvious differences between the geographic groups of humankind in the evaluation of the racial level of group changeability, is in essence a step backward and an attempt at Nihilism." Population genetics has attempted, in the spirit of Soviet command anthropology, "to snip out the problem of race," but race again stubbornly refuses to "depart" or "wash away." "Some traits repeat themselves practically without exception, in all the members of the great races, and consequently, in many cases the determination of a racial type is possible in an individual." The author of the work correctly focuses attention on the fact that adherents of the population approach intentionally remove the problem of racial differences from the framework of research, in view of the weak previous studies on the genetic nature of morphological traits. It is declared that the differences between races are phenotype differences, and there is no need to give them any genetic significance. Improper substitution is present, and this cannot be hidden from the penetrating eyes of the conscientious scientist. In regard to this, Alekseyev writes: "For the modern works of the population field are characterized by an absence of any mention of race, a complex mathematical aid, superfluous faith in the results of a mathematical analysis, and also an excessively developed attention on changeability at the population level." The interest in question represents further an observation of the systematization

---

[266] *Sovremennie problemi i novie metodi v antropologii.* Leningrad, 1980.

[267] *Problema rasi v sovremmenoy antropologii.*

of the populationists, excluding even [the mention] of the skin pigmentation of peoples of various races from the category of traits.

The author explains that this para-scientific speculation, created with the methods of mathematical abstraction, is presented by Western populationists as progressive, democratic science. Finally, he delivers a blow against the theory of Lamarckism and all other concepts of the theory of environment: "The independence of variations in racial traits from the environment, though incomplete, also speaks of their heredity dependence." Racial traits are unadaptable; namely therefore they are called racial, that is, genetic—not phenotypic. The conclusion in the article is simple and persuasive: "Race studies is in need of a new theory, including the practical sides of population developments, and liberation from Nihilism, with regard to the racial level of changeability."

In the anthology's next article, *The Concept of General Likeness in Anthropology*,[268] A.G. Kozintsev subjects the prescribed concept of mongenism to doubt: "If one is to speak on a subspecies level, then often the paths of evolution reflect not a tree, with its diverging branches, but rather a net." In connection with this, it is assumed in the interests of reliability, to build racial classifications, not according to separate traits, but according to their complexes; and to evaluate the total qualitative differences with the help of a special "coefficient of general difference" (CGD).[269]

Further, the author of this article wise reflects in the spirit of the classical postulate logic known as "Occam's Razor", which says: "One should not increase beyond what is necessary, the number of entities required to explain anything. [Or, the simplest explanation is most likely the correct one]." It is namely according to this logic that Kozintsev cuts out the intellectual excess of those who speak of the extreme complexity of evaluating racial differences: "It is easy to establish that the general similarity between Chinese and Koreans is greater, than between each of these groups with Bushmen. But it is complete nonsense to determine which of those two Mongoloid types is closer to the Bushmen."

Sir William Hamilton, another standard of logical thinking, expressed this thought in a similar way: "It is not necessary to admit a large number, nor a great complexity of causes, that is necessary in order to give an account about phenomenon." And Russian anthropologist A.G. Kozintsev, in complete agreement with this rule of logic, made this conclusion: "It is not worth attaching any special significance to the fact, that by the sum of traits, Georgians can prove to be less "Europoid" than Armenians, and Latvians "more Mongoloid" than Estonians. Ethnogenetic conclusions, based on such comparisons, will be no more reliable than the theory of "typologists", who connect people of homogeneous populations to different races."

The next article of the anthology, *The World Distribution of the Occipital-Parietal Indicator*,[270] is dedicated to questions of racial craniology. In the article, Yu. D. Benevolenskaya compares the average value of this indicator for the basic races:

Europoid—91.6%
Mongoloid—96.6%
Negroid—87.3%

The author's conclusion is unambiguous and convincing: "The three racial complexes quite distinctly and completely differ from one another. The occipital-parietal indicator works in very prominent areas, differentiating the great races."

Further, citing the correlation between the occipital-parietal and the vertical-longitudinal indexes, also testifies in favor of racial differences:

among Europoids—0.738
among Mongoloids—0.581
among Negroids—0.706

---

[268] *Kontseptsiya obshchego skhodstva v antropologii.*
[269] *Koeffitsienta obshchego razlichiya.* (KOP)
[270] *Mirovoe raspredelenie zatilochno-temennogo ukazatelya.*

There exists an everyday, unsupported opinion, that Europoids are anthropologically closer to Mongoloids, than to Negroids, but this indicator graphically shows that the depth of difference between the first and second groups [stands at] 27%. More or less, this debunks the notion of the specific unity of Man. "Less disperson, and a tighter inter-group connection with the vertical-longitudinal indicator is observed in the Europoid series, than in Mongoloids." As a whole, this says that in racial terms, Mongoloids are less homogeneous than Europoids.

And now, respected reader, recall how many times in humanitarian literature, you had occasion to encounter the assertions of liberal social scientists, that the white race is extremely mixed, and how this opinion is always given as something that is self-evident and which demands no proof.

In another article, *Toward Taxonomic Evaluation of the Levels of Differentiation, According to Dermatoglyphic Traits*,[271] G.L. Khit', one of the leading Russian authors in the area of determining racial traits according to fingerprints, also testifies that the average genetic distance between Europoids and Mongoloids stands at 21%; this is a testimony of the stability of racial differences.

Still another thematic anthology worthy of attention, *The Historical Dynamic of Racial and Ethnic Differentiation in the Population of Asia*,[272] continues the same general line in the Russian anthropology of recent years, liberated as it finally was from the tutelage of political dogmatism.

In the article, *Differentiation of the Great Mongoloid Race According to Data of Generalized Total Sizes of the Skull Case*,[273] A.P. Pestryakov ingeniously notes that the size of the brain is a "biological birthmark on the body of a race." The author further develops this thought: "It is necessary to spell out a well-known scientific fact: that a so-called *cerebral Rubicon* exists: a minimal, but sufficiently large volume of the brain is necessary...[so] its carrier—Man—can function as a social creature. The average group size of the cranium can be an important parameter in the study of the racial history of humanity."

And really, if we turn attention to the racial history of Mankind, it is easy to observe that the equatorial races, by virtue of their "cerebral Rubicon", have created nothing but sensuality and pleasure. Modern worldwide culture, civilization, science, art—all this is the achievement of the Northern Race. This general idea by the founder of the racial theory of Mankind—the Frenchman, Count Joseph Arthur de Gobineau—has been proven by Russian science with distinction.

Going further, experimental statistical data speaks for itself. Negroids, Papuans, and Veddoids stand out for hypo-craniality; that is, for a small skull, for the conditional volume, "V" of the cranium is found in them to be within the limits of 1,540 to 1,640 cubic centimeters. In Mongoloids, this indicator is an average of 1,714 cubic centimeters. The Europoid race is the most hypo-cranial, for the conditional volume of its cranium is the highest: 1,745 cubic centimeters. Modern data from criminal police in states with so-called "multi-cultural foundations" clearly shows that the highest percentage of criminal, alchoholic, and narcotic predispositions, and also the highest number of AIDS cases and anti-social elements, are among races that are stuck mid-stream, crossing the "cerebral Rubicon."

A.P. Petryakov, on the basis of other material, came to the same conclusion—that according to the sizes of the cranial capsule, Europoids vary the least, and Mongoloids are the most polymorphic; this speaks about their possible "racial heterogeneity." The scattering of traits in the latter is 2.0--2.25 times greater, than in Europoids; and in Negroids and American Indians it is 1.5—2.0 times greater, than in the Europoid race. From that one may make the legitimate conclusion, that of all the great races, the Europoid race is the most homogeneous. "The generalized quantitative characteristics of the cranium studied by us, possess greater stability in time, than the majority of descriptive, racial-morphological traits." From this assertion by the article's author, the conclusion follows that racial traits, particularly one as important as brain size, are actually a "birth mark", [that does not fade at all] in the process of historical development.

"The given parameters may serve as good anthropological markers in the study of ethno-genetic processes. Analysis of the value of generalized parameters in cranialogical series allows

[271] *K taksonomicheskoy otsenke urovney differentsiatsii po priznakam dermatoglifiki.*

[272] *Istoricheskaya dinamika rasovoy i etnicheskoy differentsiatsii naseleniya Azii.* Moscow, 1987.

[273] *Differentsiatsiya bol'shoy mongoloidnoy rasi po dannim generalizovannikh totalnikh razmerov cherepnoy korobki*

one to separate by racial affiliation, and also by foreign affiliation, from a craniological point of view of inclusion."

Yu.D. Benevolenskaya spoke of the "high consolidatedness of the Europoids, in comparison with other races," on the basis of a study of the Lobe-Sagittal Index (LSI, in the article, *Racial Differentiation on the Territory of Asia (according to the Structure of the Lobe Section of the Skull)*[274].

The next anthropological anthology, *Problems of the Evolutionary Morphology of Man and his Races,*[275] again follows the same main line, not leaving any grounds for confusion in this question. In the first article, titled *Several Views on the Dynamic of the Correlations of Relationships in Man, and its Evolutionary Significance,*[276] V.P. Aleksevev writes: "The volume of the interior cavity does not of itself have any adaptive significance. In the terminology of A.N. Severtsev, this trait is endogenic, weakly tied with the environment. In the pair, "brain-cranium", the brain is the leader."

Therefore, in exact accord with racial theory, the hypocranial equatorial races, that is, small-skulled races, can never become adjusted to the culture and civilization of hyper-cranial, prominent-headed Europoids, and can never cross the "cerebral Rubicon," in order to become "social beings" in the European understanding.

In the article, *Racial Variation in the Traits of the Cranial Dome,*[277] Benevolenskaya writes: "Inasmuch as races are not similar, varying qualitatively by type and scale of the intra-racial, race-forming process, a racial diagnostic according to the LSI index looks distinctive in each case. Thus, Europoids are the most consolidated race, and probably therefore the Lobe-Sagittal Index does not give differing racial demarcations within Europoids. The greatest LSI differences are revealed within the bounds of the Mongoloid race."

Still another contributor to the anthology, Yu.K. Chistov, makes the same conclusion in the article, *Racial Differences in the Structure of the Median-Sagittal Contour of the Skull of Man,*[278] on the basis of a different morphological parameter: "The Europoid series differ least between each other by the sum of the linear characteristics of the contour of the skull; equatorial groups differ most."

From a number of modern collective works, it is necessary to mention the anthology, *Unity and Diversity of the Human Species.*[279] In her article, the famous anthropologist, N.I. Khaldeyeva, came to this conclusion: "All populations without exception turn out to be drawn into a single race-genetic process, the basic result of which is the formation of racial systems." But, "formation" is not the "removal of traits", as according to Yarkho, or a "ceasing of the evolution of species," as according to Dubinin. The given statement testifies about the general recovery of Russian anthropology. Besides that the author observes that the processes of polymorphism, that is, of the qualitative differences between the races, have a vector-type trend. Consequently, the concentration of racial traits in a population cannot only decrease, but grow.

Another specialist in this same field, G.A. Aksyanova, remarks: "Polymorphism of those physical traits of modern humanity, which are called racial, exist independently of positive or negative attitudes toward the term, "race". Historical interweaving of this scientific term from the area of biological systematics, with negative social manifestations, does not change its biological essence, as applied to Man. Racial differentiation in the morphology of Man is an objective reality."

Yu.D. Benevolenskaya, already cited by us above, is one of the recognized leading specialists in the field of craniology—a science about the racial differences in the structure of the skull. Her article, included in the given anthology, advances the original concept of the primordial existence of two extreme variants of facial morphology in humankind. "The sums of the analysis lead to the conclusion of the presence of basic racial components. The Europoid type reveals the traits of a

---

[274] *Rasovaya differentsiatsiya na territorii Azii (po stroeniyu lobnogo otdela cherepa).*

[275] *Problemi evolyutsionnoy morfologii cheloveka i ego ras.* Moscow, 1986.

[276] *Nekotorie soobrazheniya o dinamike korrelyatsionnikh otnosheniy u cheloveka i ee evolutsionnom znachenii.*

[277] *Rasovie variatsii priznakov cherepnogo svoda.*

[278] *Rasovie razlichie v stroenii medianno-sagittal'nogo kontura cherepa cheloveka.*

[279] *Edinstvo i mnogoobrazie chelovecheskogo roda.* Moscow, 1997.

trapezoidal morphotype, the eastern [reveals] a quadrilateral [morphotype]. The idea of the existence of these morphotypes finds biological foundation as one of the factors of polymorphism in human populations. Besides that both of these morphotypes reflect the evolutionary phases in the development of the races. Turning to the structure of the morphotypes, we see that the quadratic morphotype is the most characteristic trait in the beginning phase of growth, and the trapezoid [form] is the final phase."

This concept of morphotypes easily reconciles with Alekseyev's theory of the non-adaptive size of the brain; and, in dependence on the size of the cranium, set by the brain in the process "phase of growth," it allows one to scientifically and soundly speak of "higher" and "lower" races. Moreover, the idea of these morphotypes "finds a biological basis", in that one of them belongs to an initial, that is, a lower phase of growth, and the other to an end phase—that is, a higher phase.

Yu.D. Benevolenskaya continues: "These are the 'structural elements,' that is, two morphotypes as the fundamental principle of diversity, not scattered without a trace in a new phase of differentiation of humankind on a level of forming races, but tracing to their foundation." This means that the higher [types] always were, and always will be, higher—and the lower [types] will always be lower: "The hypothesis of dimorphism can be formulated as the phenomenon of parallelism of the races." That is, according to the author's thinking, the difference in types testifies about the mutual independence of their origins."

But after all these statements, the question arises of where one can observe the thesis of the "unity of the human species" in the "initial given morphotypes of the skull" and the "parallelism of the human races"? And if we remember that "the races are not similar, that they are qualitatively diverse by type and scale in the intra-racial, race-forming processes," and moreover, that they are subjected to a ceaseless "formation of racial systems", then in that case, the concept of a "melting pot" is indeed a myth with a clearly expressed ideological direction.

If you consider that these morphotypes are initial components of anthropogenesis, then it is legitimate to study their biological basis in full accordance with the full precept of the last author. In an anthology of theoretical works titled, *Problems of the Evolution of Man and his Races*,[280] the leading Russian specialist in the field of ethnic and racial odontology (the sciences of the differences in the structure of the dental system), A.A. Zubov, distinctly and clearly writes in his article, *Several Findings of Odontology toward the Problem of the Evolution of Man and his Races*:[281] "Dental morphology does not contradict the assumption of the possibility of the independent, parallel development of the races from various local groups of paleo-anthropoids."

Thus, it is time to reconcile with the thought of Miguel Serrano, the ingenious Chilean of racial metaphysics, that there really exists on Earth not one, but several [types] of humankind.

In another anthology, under the characteristic title, *Ethnography, Anthropology, and Related Fields: the Relationship of Subject and Method*,[282] we find a clear and persuasive article by K.M. Kozlovskaya, titled *Experimental Studies of the Epochal Dynamic of the Changeability of Several Physiological Traits*,[283] which gives an unequivocal confirmation of the hypothesis about a primordial existence of two morphotypes and the parallelism of races, on the basis of biochemical processes. The author of the article analyzes biochemical factors that are important to the anthropology of Man, like the mineralization of skeletal bone tissues: "Some information about the level of mineralization of bone tissues shows a rigid, genetic determinancy on the one hand, and on the other hand, a high sensitivity in regard to different outer influences.

It reveals an ability of the organism to support an individual level of mineralization of the skeleton with removal of exterior influencing factors. The geographical condition is not entirely determinant, but corresponds with genetic determinism. Neither morphological, nor local geo-chemical features testify in favor of an increase in the level of mineralization. A comparison of the level of mineralization in the modern Eskimo population of Asia and America, with such traits in Americans of European descent, shows that in all adaptive changes, the mineralization of the skeleton is lower in the first group, than in the latter."

---

[280] *Problemi evolyutsii cheloveka i ego ras.* Moscow, 1968.

[281] *Nekotorie dannie odontologii k probleme evolyutsii cheloveka i ego ras.*

[282] *Etnografiya, antropologiya, i smezhnie distsiplini: sootnoshenie premeta i metoda.* Moscow, 1989.

[283] *Opit izucheniya epokhal'noy dinamiki izmenchivosti nekotorikh fiziologicheskikh priznakov.*

Consequently, the mineralization of bone tissues is also a racial trait, strictly genetically determined. M.V. Kozlovskaya confirms: "The high level of mineralization is not necessarily functional, but recalls genetically determined mechanisms. Concentrations of micro-elements in the bone tissue, is a complex of different indicative traits." From the abundance of elements in this plane, it is particularly necessary to divide the race-demarcating function of zinc, copper, manganese, and lead. In the bone tissues of Mongoloids, there is a markedly lower retention of zinc, which is tied with a higher intensity of oxidation processes in their organism, in comparison with Europoids. For its part, in Europoids the higher retention of lead in the skeleton is caused by use of animal proteins, meat, and milk in the diet. Characteristically, the Europoids of northern Europe precisely differ by the highest gene of assimilability of milk, the frequency of which in Mongoloids and Negroids is significantly lower. Copper also actively appears in the oxidation process of the organism, and consequently, is a good racial diagnostic marker. Manganese testifies to the nearness of the ocean, in the process of the formation of this or that population. Ocean water retains a very low concentration of manganese, in connection with which the bone tissues of island populations are characterized by its reduced amounts.

On the basis of these findings, M.V. Kozlovskaya gave this valid conclusion in her work: "Thus, research of the chemical composition of the mineral component of the skeleton, allows one to observe the preservation over a long period of time, of several traits in the chemical make-up of the bone tissues in members of the ethnos of a given territory."

An authentic masterpiece of modern Russian raciology that is worthy of recognition is A.G. Koznitsev's *Ethnic Cranioscopia. Racial Changeability in the Seams of the Skull in Modern Man.*[284] In contrast to liberal, political-prostitute anthropologists, and their sole pursuits of "laundering" and "removal" of racial traits, Koznitsev sees the direct opposite in his work: "Polymorphism of several traits allows one to change the traditional path of research, reconstructing and creating a morphological scheme with special reckoning of the increase in race-demarcating effectiveness in several cases—in the reductive role of different factors, in particular of [the factor of] age."

Analyzing the results of practical studies, the author of the monograph establishes that the frequency of a race-demarcating trait in the bones of the occipitomastoid seam is on average, 6.4% in Europoids, and 16.6% in Mongoloids. On the basis of the closeness (in morphological terms) on the Occipital Index (OI), racial differences are observed to be more distinct. Thus, for Europoids the frequency of this trait is 8.4%, and for Mongoloids it is 48.5%. The Occipital Index of the second order (OI II), also effectively helps differentiate races: 2.8% for Europoids, and 13.4% for Mongoloids. "In consideration of the value of the Occipital Index (OI) and (OI II), the impression is created that the trait only "works" on the level of the great races. No regularities in the distribution of frequencies within the Europoid and Mongoloid complexes have been brought to light."

Thus, A.G. Kozintsev brilliantly proved the basic postulate of classical race theory, according to which racial and biological traits—that is, pre-social traits—dominate, and not national or social differences.

The cultural and social character of a society is an algorithm of behavior, in which the stylistic, psychological experiences of the individual are set by race. Environment does not form race, but race shapes environment—and racial traits are not pliable plastic materials subject to changes from without, but on the contrary, are a fundamental principle of all socio-cultural transformations in a society. Racial traits are not clay that is obedient to the hands of a sculptor; they are sharp incisors, that create the contours of a being, in conformity with their mechanical durability.

"In all the abundance of suggestions about the role of external factors, it is impossible to present a factor, which could be so well-masked under racial—extremely heterogeneous groups, united here by a racial trait, in all other conceivable terms"—the author continues. Any deformations of the skull lead to differences, which always are significantly fewer than racial [differences]. In connection with this, he makes a conclusion that we have already heard from the mouths of professional Russian anthropologists: "The Europoid race is a more homogeneous

---

[284] *Etnicheskaya kranioskopiya. Rasovaya izmenchivost' shvov cherepa sovremennogo cheloveka.* Leningrad, 1988.

race than the Mongoloid [race]. The horizontal size of the cheekbones is large in Mongoloids and small in Europoids; Negroids occupy as it is, an intermediate place."

Finally, A.G. Kozintsev made a brave, summary conclusion, relative to all depths of the morphological differences between races: "Among Mongoloids the topographical position of the eye is different, than in Europoids. Among African negroes, the eye socket is low, and the eyeball larger, than in Europoids."

And now remember, that Man receives 90% of information by way of sight, but in the above-mentioned facts of the structure of the visual apparatus, it happens that various races have different stereoscopic visual perception, and as a result, in the literal sense of the word, "they see the world differently." The differences in the perception of the world by different races, unavoidably brings in its wake differences in the comprehension and interpretation of moral and behavioral precepts, which in their turn lead to irremovable and unfading genetic differences between cultural and legal norms.

Even portrayals of gods in the religious cultures of various races, or of their absence, testify that the form and its interpretation, from the most simple everyday forms to the most complex metaphysical visions, unavoidably carry in themselves the stamp of race, [each] seeing things its own way. The index of complexity of the suborbital arrangement (ICSA) possesses the highest race-demarcating capacity.

For Europoids, it is 38.0, and for Mongoloids it is 57.9. And this is regularly; after all, the eyes of various races have various attachment devices. This trait also "works" on the level of the great races, rebutting the thesis of A.I. Yarkho about the "removal" of racial traits through historical process.

Peoples, nations, ethnoses, and tribes are really the result of a more recent historical processes, but the gigantic abyss of irremovable racial differences testifies in favor of the prehistoric, and namely, of the biological nature of their origins.

History does not create the racial traits of peoples; on the contrary, it is namely racial traits that influence the history of these peoples.

A.G. Koznitsev writes further: "We analyzed about 30 traits, relating to the seams of the skull, and we set apart six of the basic, most worthy [traits]. The Europoid and Mongoloid races as we remember, differ according to all their basic characteristics." For a summary quantitative evaluation of these differences, the author of the book introduces a special Mongoloid-Europoid Index (MEI). In Europoid populations, it fluctuates between 13 and 39, and Mongoloid populations, from 54 to 82.5.

"For the convenience of calculations, it was necessary beforehand to separate racially pure groups from racially mixed ones." This famous Soviet anthropologist's excellent, sensible thought is worthy of a peer of Count de Gobineau himself—the founder of classical racial theory. Mathematical calculations absolutely need a preliminary demarcation of pure and mixed races. Mathematics is the basis of segregation, for there are no indexes of general human values. In anthropology, the mathematical apparatus is a better remedy for obtrusive, officious humanitarian, utopian [doctrine].

Members of the Nordic race also easily stand apart from the number of different Europoids, with the help of the Northern European Index (NEI). The significance of both indexes (MEI and NEI) among northern Europoids is higher, than in southern [Europoids]." A clear and logical conclusion, given by A.G. Kozintsev in the book, does not leave any room for idle talk or speculation: "Race-demarcating indexes are a simple, but effective means of analysis. Five traits—the occipital index, the wedged-maxillary seam, the posterior cheekbone seam, the index of the complexity of sub-orbital arrangement, and the index of the transverse palatial seam—serve in the first order for demarcation of Europoids and Mongoloids. The combination of traits possess more demarcative strength than other traits. The Mongoloid-Europoid Index (MEI) and the First Main Component (I MC) very effectively separate Mongoloids and Europoids. The differentiation within the Europoid race is traced with the help of the North Eurasian Index (NEAI), and Second Main Component (II MC)."

In the appendix, the author analyzes the skulls of Mongols from the central part of Russia, belonging to the period 4,000-1,000 B.C., and comes to the straightforward conclusion that the ancestors of the Russians in the indicated period were racially pure, northern Europoids.

Therefore, the half-baked, illiterate "Eurasians" and "imaginary luminaries" of the academic world, who assure us that pure races do not exist, and pour streams of lies and dirt on the racial history of the Russian people, need to brought to criminal account for insulting the honor of the blood of our ancestors, for modern "liberal anthropological theories" are in essence, no different than the desecration of graves—a deed that is punishable in both civilized and uncivilized societies. Still another prominent Italian anthropologist, Giuseppe Sergi, made an analogous point in the book, *Species and the Diversity of the Human Genus.*[285] He wrote: "The skull is most important in classification. With one skull, one can differentiate ethnic elements that go into the make-up of mixed groups. From the most ancient ages to our times, no new skull forms have appeared. An important feature in classification is the interior capacity of the cranium; it is directly tied with the form of the skull. The cubic content of the brain does not increase by the measure of the evolution of society. Both the form of skull types and their average capacity remains as before. The increase of the capacity of the human skull is a legend."

With regard to this, the famous German anthropologist, Julius Kohlmann (1834-1918), pointed out: "Despite all anomalies, despite all environmental influences, despite all cross-breeding, the human races and their diverse forms always remain those same ones."

Another luminary of world anthropology, Eugen Fischer, maintained[286] on the basis of extensive statistical material, that: "The form of the skull is a genuine racial trait. One race does not receive full predominance over the other in cross-breeding, but neither does a new race arise as a result; rather, a kaleidoscope of traits forms. New races are created by natural selection. If the choice is no, the mixed population long remains at the level of "the generation of grandchildren." Mendel's Laws clarify the stability of racial traits: the circumstances that shape the heads and faces in Europe today, are the same as in the early Neolithic Period (4,000 B.C.); and from underneath the layer of the conqueror's traits, the appearance of the subjugated, native race also appears in almost pure form.

In his much later book, *Race and the Origin of Races,*[287] he summarized this thesis more distinctly: "There is no merging into a new race. The traits of both races exist in parallel, and not one of them is destroyed. In this case, it is more correct to speak not of a separation of mixtures, but about the re-establishment of an old race."

Many historians and culturologists, who are inclined to explain the rise of pockets of world culture by the action of some exterior causes, are very often placed in an uncomfortable situation, [because] their para-scientific inventions hardly stand up to verification by the findings of the natural sciences. Thus, for example, V.P. Alekseyev emphasized in the book, *Geography of the Human Races,*[288] that: "It is curious to note, that the ancient pockets of human culture did not arise in regions with maximal, primary productive biospheres, that is, in the tropical forests, but to the north of them, in countries with more moderate climates. The old hypothesis abou the co-location of such pockets [of civilization] with large river valleys was repeatedly subject to criticism, but mainly from philosophical positions. Meanwhile, it does not withstand scrutiny from a geographical point of view: the valleys of Siberian rivers in Asia, and the Mississippi and Missouri rivers in North America never became centers for the development of civilization."

And thus we see that no accidental "caprices" of the environment, but the conscious, biological will of a race, which created world history in its best and most visible manifestations.

And in 1928, V.I. Vernadskiy wrote: "In the complex organization of the biosphere, only a regrouping of chemical elements arose within the limits of living matter, not a root change in their composition or number." In the biochemical context of race, this precisely signifies the tirelessness of separately considered racial traits, and only the corresponding change of their concentration in this or that population, in the process of historical development.

I.I. Schmalhausen, an outstanding Soviet biologist, brilliantly developed all the postulates of racial theory, directed against vulgar Lamarckism, in his program book, *Cybernetic Questions in Biology.*[289] He wrote: "The hereditary code is shielded by a nuclear cloud, and the regulatory

[285] Turin, 1900.

[286] In the monograph, *Anthropology.* 1923.

[287] 1927

[288] *Geografiya chelovecheskikh ras.* Moscow, 1974.

[289] *Kiberneticheskie voprosi biologii.* Moscow, 1968.

mechanisms of the cell and the entire organism as a whole [are shielded] from the immediate influence of exterior factors. Hereditary traits, acquired during the lifespan of an individual, are factually impossible, since this "acquisition" only concerns the transformation of information in the given specimen, and dies with it. Hereditary material is not affected by this transformation and remains unchangeable." Here it is emphasized that mutations, about the role of which geneticists love to discuss, actually have an accidental character. Besides that no freedom of mutations exists, for they are only possible within defined boundaries given by hereditary racial traits. In a word, no matter how much you keep a negro in a Russian steam bath in Siberia, and no matter how much you feed him pierogis with vodka, neither he himself, nor his descendents, under conditions of 100% conjugal isolation, will ever start to resemble native Russian Siberians. But geneticists from among the ranks of vulgar populationists attempt to dispute this obvious fact, which is perfectly clear to any individual trained in biology.

I.I. Schmalhausen continues: "A phenotype should be an exact exponent of a genotype. The defense of the individual development of a phenotype from random external influences, and the non-heredity of corresponding changes, are vital conditions for the possibility of the regulated transformation of an organism on the path of its consistent adaptation to common factors in the outer environment, and on the regular path of its evolution in general."

Thus, it is made clear that there is no removal of racial traits as an adaptation in the process of evolution, but rather their strengthening and consolidation. In the process of evolution, racial traits are a distinctive and adaptive instrument, an **"evolutionary sail rigging"** without which, the biological development of a race is not possible. **Racial traits manifest in both the physical and psychological plane, through genetic "nodes of durability", upon which the whole structure of an individual is supported. Without them there is inevitable degeneration and disintergration.**

Schmalhausen asserted: "In the process of evolution, through the influence of Natural Selection in inter-group competition, the means of dissemination of hereditary changes within a population and species as a whole varies, through the organization and use of sexual selection and the limitation of random mating under various forms of isolation. Since in the struggle for existence living populations basically clash side-by-side, genetic differences have a decisive significance in this [process]". Thus, racial conflicts unavoidably lead to revelation, strengthening, and stabilization of racial differences, which for their part, lead to isolation of phenotypes, distinctly fixing differences in the genetic programs of the conflicting races. Racial conflicts improve the health of races and consolidate them, cleansing them of genetically hindering admixtures. "In the process of evolution, under the influence of the stabilizing form of selection, the error-correcting feature of information increases; that is, the structure of the hereditary code becomes complicated."

3. Hereditary Polymorphism

Thus, the evolutionary process promotes not the washing of racial traits, but their consolidation; besides this, in the opinion of I.I. Schmalhausen, "the phenotype reflects the attributes of the genotype at the individual level of the organization." That is, the entire viability of a race is built on the isolation of the phenotypic uniqueness and inimitableness, employed for the maximal realization of its genetic program. The purity of a breed is the main condition of survival, from the point of view of cybernetic biology.

Factors in the capacity of obstacles to the transmission of hereditary information, that hinder the evolution of a race, are divided into **abiotic** (harmful climatic conditions) and **biotic** (microbes, parasites, and predators). These obstacles hinder the viability of a race, and therefore should be excluded from the process of race-genesis by all means. "The struggle with biotic hindrances is never-ending; therefore, for several weeds and vermin, conditions favorable for adaptation to the conditions of culture arose, and they also underwent a fast evolution."

Thus parasite-peoples, and even entire, marginal "lower" races arose, without the talent for independent creation of cultural values, capable only of leeching off the achievements of "higher" races.

Schmalhausen further made the following, highly important conclusion: "As it applies to populations of organisms, entropy is maximal in complete freedom of choice of the variants of a phenotype." Thus, the confinement of sexual selection within the limits of one phenotype reduces entropy in a population to a minimum. Namely therefore, in all tales and legends of the peoples of the world, the ideal wedded pair is always represented by one phenotype—as ideal in racial terms, as a self-sufficient people. The multiplicity of phenotypes, that is, random mating, "chaos of blood", and race-mixing, unavoidably leads to the growth of entropy, and as a result, to the degeneration of the whole population. This is proven by I.I. Schmalhausen, with the methods of cybernetic biology. "The halt of selection, and the disorderly accumulation of mutations, means disintergration—that is, the collapse of the existing organization."

Whatever the influence of the environment on hereditary racial traits, the author of the given monograph makes one simple, logical conclusion: "The quantity of phenotype information is less than the quantity of inherited information." Neither pacifism, nor humanism, nor any other liberal-aesthetic bliss can change the harsh laws of nature, which ruthlessly eliminate the weak, the crippled, and the degenerate from existence. "If the evolution of a population is defined by the competition and selective destruction of its individuals, then the evolution of a species as a whole is determined not by the summary evolution of separate populations, but by the integral effect of inter-population interaction, with competition and exclusion of entire groups—populations, races, and sub-species."

The same capacious and clear conclusion by Schmalhausen in his magnificent program book, is as follows: "Without selective destruction, there is of course, no evolution."

Modern genetics in a correct reading also does not in the least contradict the basics of racial theory. The essay, *Human Individuality: Heredity and Environment*,[290] by prominent American geneticist-populationist Richard Levontin, has already succeeded in becoming a classic. The author explains here that " *changeability*, called **polymorphism**, radically differs from ceaseless changeability, which we observe in such traits as height, weight, form, color, and level of metabolism, or behavior traits; each presents itself on some continuous scale." The famous Russian geneticist, Viktor Alekseyevich Spitsyn, gives an explanation to Levontin's conclusion, proving that racial traits are numbered among the polymorphic, for both carry in themselves namely qualitative, discrete differences.

Today, of 100 studied enzymes in the human populations, approximately 25% are polymorphic; precisely also, 25% of proteins in people are polymorphic. "The fact that molecular polymorphisms are unchanging for the extent of the whole life of an individual, and do not depend on any significant environmental or cultural influences, suggests that they are directly or immediately inherited. Really, using polymorphic groups of blood, we can show the laws of simple heredity"—argues Levontin.

Thus, on the basis of genetic studies, the same picture as a whole is repeated, which we observed on the basis of the biochemistry of the organism, and craniometry of the cranial capsule: racial traits are not adaptive, not sensitive to changes in the environment, and subject to hereditary changes only, which lie at the basis of the unchanging racial differentiation of Mankind, something observed by us for the length of Man's entire history. "There is some polymorphism in which one allele is distributed among Europeans, but is absent in the inhabitants of Western Africa, and vice versa. Allele Fy^2 blood groups of Daffi are encountered with a frequency of around 40% among whites, but as a rule are absent in the inhabitants of Western Africa. Allele Ro, according to the system of blood groups Rh, is rare among whites, but encountered with a frequency of 60% among the inhabitants of Western Africa."

Richard Levontin's explanations are simple and graphic, and also expose the scientific fallacy of the "melting pot" myth. "The basic feature of heredity: determinant traits, encountered in the embryo of Man, again separate when spermatozoids and ovum form. Mixing on a physiological level does not at all mean that hereditary determinants lose their individuality. On the contrary, they remain distinct, and during formation of the reproductive cells, they separate. It is as if, in mixing red and white paint, and getting pink, one could resort the molecules of the new paint, and again get red and white. Hereditary determinants are to a certain extent, quantum, distinct particles, which preserve their individuality in the process of passing to the next generation,

[290] *Chelovecheskaya individual'nost': nasledstvennost' i sreda.* Moscow, 1993.

despite the fact that when they unite during conception, their physiological influences can be joined. Concrete people are only temporary carriers of various combinations of determinants."

In his book, *Physical Anthropology*,[291] prominent Polish anthropologist Ludwik Krzywicki wrote about the stability of the racial traits in different lines, that are united in the organism of hybrid: "One anthropologist, studying the nature of hair, stated that in Mestizos each hair is different in structure; in separate instances, it reconstructs the structure of the hair of one, and then the other parent; in the majority of cases, it is in a completely irregular form. Even such an opinion was expressed, that on the basis of one hair from a Mestizo, one could indicate his racial origin and the intensiveness of the hereditary transmission of each parent.

Brunettes in general differ by having greater intensity in the hereditary transmission of physical traits to offspring, than blondes. Hybrids most strongly resist coloration of the eyes, and then coloration of the skin; hair color is more quickly yields in mestizos.

Parents, of whom one has brown eyes, while the other light blue eyes, produce offspring that stand out by diversity of these two elements. The iris, for example, can have an equal distribution on its entire surface of one color or the other; one color may form a ring around the other, and then again the distribution of one color might interweave with the other, and thus the iris is spotted. Besides that there are cases in which each eye is a different color; in one iris, light [color] might predominate, and in the other, a brown color."

And Krzywicki developed his ideas, emphasizing: "The organism of a hybrid, particularly from types of complete diversity, cannot be a durable, healthy, and fruitful offspring. The harmonious combination of different traits of the pure type is disrupted, and new ones are not firmly established. In this regard, there is the curious spectacle of populations in those countries, where cross-breeding of different types has occurred in large measure. The nose does not correspond to the face; the head does not correspond to the trunk. Also, in cities, one can talk about the so-often encountered phenomenon of eyes that do not match, or that have unequal strength; an individual inherits one eye from the father, the other from the mother."

The outstanding social anthropologist, Georges Vacher de Lapouge, wrote in his fundamental monography, *Social Selection*,[292] that: "Theoretically, for each cell of the organism of a mestizo, the intensity of hereditary transfer differs." Modern studies of the genome of the individual completely confirm this French scientist's intelligent conclusion.

Not having uniform quantum hereditary structure, the racial traits and their complexes are doomed, such that in order to survive, only mating selection and racial hygiene can increase their concentration and restore a pie-bald, liberal society to the racially clean ideals of popular folk tales and legends.

The most integrated and complete expression of the **typological concept** was found in the Polish anthropological school, thanks first of all to the efforts of Jan Czekanowski. The essence of the problem, in his opinion, led to the following: "From the hypothesis of the coupling of hereditary traits in the process of heredity, one can make the conclusion that racial traits are passed through heredity by a whole complex, and that resultantly, the morphological characteristics of an individual coincide with variations of this or that racial type. The study of the anthropological composition of different populations consists of the revelation of its individual racial types and in counting their relative percentage. Tracing the value/size is a basic characteristic of a population and according to it, a comparison of various populations is made in the final analysis. The decisive word in the referral of an individual to this or that racial type, belongs to morphological characteristics, geographical criterion having no essential significance."

The authors of the scientific anthology, *The Problem of Integrity in Modern Biology*,[293] also raised the philosophical problems of the natural sciences, but already on the basis of the systematic method, it turns out that the theory of regulation as applied to living systems supports the classical summarizations of raciology. Czech scientist F. Chizhek emphasizes in his article: "The opinion that the more complex a population is in the genetic sense, the less wholesome it is, is mistaken." And now recall, how many times liberal social scientists and the professors of Marxism-Leninism repeated the myth of a heterogeneous Russian people. With ease, biology

[291] Sankt-Peterburg, 1900.

[292] 1896

[293] *Problema tselostnosti v sovremennoy biologii.* Moscow, 1968.

does away with the problem of this humanitarian speculation. Chizhek argues: "In reality, the differentiation of a species is the reverse side of its integrity. The more complex the phenomenon, the greater its integrity."

Appropriately, the given text recalls the words of Soviet biologist R.L. Berg, who wrote: "Polymorphism is a group adaptation, supplying the chance of life and evolution of the species as a whole." This means that humankind is biologically not able to exist without division into races, and the absurd notion of a "melting pot" contradicts the very laws of the universe.

In the monography, *Genetic Regulating Systems*,[294] V.A. Ratner also emphasized: "Macro-orderliness of a genotype is stabilized by group natural selection, which provides simultaneous genetic isolation of each complex form from others, and a combinant capacity within each form." From this it follows that not only the structure of an individual organism, but the entire population as a whole is oriented toward maximal preservation and retention of the inherent, natural archtype. The given position was successfully confirmed by the famous Russian anthropologists, V.V. Ginsburg and T.A. Trofimova, in the joint work, *Paleoanthropology of Central Asia*.[295] Of gigantic geographical scope and based on extensive material, they came to the same conclusions. Analyzing in detail the graves and burial sites of the given vast territory, which contained within them the remains of skeletons and skulls for the extent of the last 4,000-5,000 years, the scientists wrote: "The basic condition of genetics is the possibility of inheritance of separate racial traits, independent of each other. In other words, racial traits, being outward manifestations—by phenotype—of the collectivity of their hereditary factors (the genotype), like the latter can form different combinations. However, independent inheritance of racial traits is not an alternative, exclusive complex of their transfer, in accord with heredity. A complex transfer creates the conditions for the enduring preservation of a racial type in this or that population. And the concentration on a given territory of not only scattered and varied individual traits, but also a significant number of their complexes, creates racial similarity in separate persons within a population. A complex transfer of similar racial traits speaks of a close and durable genetic composition in a population. And the greater the similar phenotypical complexes in a population, the greater the number within it of racially similar, separate individuals; then the more concentrated (that is, the less mixed) the separate ethnic community is."

Thus, the thesis is born out, that in accordance with the vector nature of race-genesis, the concentration of racial traits in a population in time can increase; in connection with that in exact correlation with the laws of Mendel, it can consequently produce a cleansing of foreign racial admixtures and a return to the original racial type. The great French anthropologist, Paul Topinard, gave this definition in his fundamental monography, *Anthropology*: "By racial type, we understand the average type of a race, accepted as pure."

Consequently, the existence of pure races and pure racial types is not only hypothetical, but practical. Exactly also, as we observe each day in the daily life of dozens of different pure races of domestic poultry, and by demand in the food of animals, birds, fish, berries, vegetables, and fruits of a pure strain. Precisely also the development of a pure race of our Nordic countrymen is not a chimera, but a completely real prospective application of our political will.

The prominent Russian anthropologist, V.P. Alekseyev, wrote in the book, *The formation of Humankind*,[296] that: "Exceptional constancy of the physical type of Man, through time, is attested to by many paleoanthropological studies. This stability of a type is characteristic not only for centuries, but for a millennium, and for several millennia. Secondly, that speaks of the hereditary dependence of a morphological type—the intermediate condition of the anthropological traits of peoples of mixed origin, between original groups."

As we recall, the basic conclusion of the Lamarckists is that in the process of evolution, races supposedly "disappear" and "wash away." But a recognized luminary of Russian science, A.A. Zubov, points out in the article, *Tendencies of the Evolution of Mankind*,[297] that: "In the process of evolution, an organism is emancipated from the factors of the outer surroundings, becoming in its

---

[294] *Geneticheskie upravlyashchie sistemi.* Novosibirsk, 1966.

[295] *Paleoantropologiya Credney Azii.* Moscow, 1972.

[296] *Stanovlenie chelovechestva.* Moscow, 1984.

[297] *Tendentsii evolyutsii chelovechestva.* Rasi i Narodi, Vyp 12, 1982.

life processes and individual development, all the more autonomous." In the article, *The Substantial Principle of Organization in a Living System*,[298] from the anthology already cited by us (*The Problem of Integrity in Modern Biology*), G.A. Yugay analogously emphasized: "An extremely limited concept is the concept of exaggerated influence of the surrounding environment on a living system. The influence of the surrounding environment on an organism was taken to the extreme, to absoluteness, to the assertion that the environment plays a leading determinant role in regard to the organism, and because the environment is supposedly the support of the organism, the organism is only a form of expression of that support. As a result, the interaction of the organism with the environment was not recognized, only the influence of the environment on the organism; that is, the environment, in essence, was considered in isolation from the organism." English biologist Julius Huxley also pointed out: "The tendency of the biological evolutionary process is characterized by an increase of the control of the organism over its surroundings, and growing independence from the changing environment." Therefore, proceeding from the famous postulate of Soviet geneticist N.P. Dubinin, "One genotype-one instinct", we are easily persuaded that history, particularly in its modern phase, presents itself not as a "removal" of the racial problem, but rather, its constant and unchanging growth. The increase in the Earth's population, which has progressed rather markedly in the last decades, compels peoples all the more keenly and actively, to join in a new division of land for resources and spheres of influence. And a constant and steady emancipation of genotypes from the influence of the outer environment, for its part, will lead to inherited differences in instincts, and as a result, [differences] in the systems of values will more and more come to determine the tactics of these genotypes, in the struggle for existence. Racial differences will more and more have their effect on the biology of the behavior of people, in all spheres of life activities.

Modern geneticist-populationists, rejecting the very fact of the existence of stable racial types, and all the diversity of biological processes in the history of Mankind, do not hesitate to explain the influence of mutations. However, the famous American anthropologist, R.A. Fisher, cooled their quick temper with the observation that evolution should produce, "contrary to the onslaught of hostile mutations." The English geneticist, J. Winthrow, also indicated: "One of the particularities of mutations lies in the fact that almost all mutations that have been studied by geneticists are unfavorable." It is obvious that all mutational changes can be produced only within the limits of an existing genotype; no matter how much you influence the environment of cucumber seeds, they will never turn into tomato seeds. The same applies to human races.

This simple rule is supported by the findings of modern biochemistry. Richard Levontin wrote in the book, *Human Individuality: Heredity and Environment*,[299] that: "The presence or absence of blood group antigens is absolutely constant in all environmental conditions. Among people, changes in blood groups have never been observed in the process of life. A blood type is not sensitive to temperature or nourishment." V.P. Alekseyev also testified about this: "The multitude of facts speaks about the selective character of various illnesses, in dependence of this or that blood group; about the various predispositions of a carrier of the Blood Group System ABO toward diseases that are determined by group affiliation."

Racial traits are non-adaptive traits, not in any way dependent on the environment, but completely determined by heredity.

M. Fervorn, a modern researcher, sharply observed once that "a living organism is a monarchical state of cells." We ourselves add that in each concrete instance, the symbols of this monarchy are namely the racial traits of the given organism; this rule extends to the population level.

The symbolism of any state most accurately characterizes the racial essence of its creators, in which connection the racial makeup of the basic foundational part of the population may differ in the process.

---

[298] *Substantsial'niy printsip organizatsii zhivoy sistemi.*
[299] *Chelovecheskaya individual'nost': nasledstvennost' i sreda.* Moscow, 1993.

4. The Problem of Race in Modern Russian Science

The sharpness of the discussion on the problem of race in political and scientific circles has not lost its edge to this day. A highly indicative and even epochal event in this area might be the first international conference, *Race: Myth or Reality?*,[300] which was held in Moscow from 7-9 October, 1998, under the aegis of the Russian Department of the European Anthropological Association, with the support of numerous world and Russian specialized scientific establishments. However, information about this emerging event was confined out there to a narrow circle of the committed, and this despite the fact that the occasion came about with the financial and informational support of the Soros Fund. The reason is simple: the Russian anthropological school did not give in to the presence of stock speculators in science, in order to please the fashion trends of the democratic wave. And they did not reject use of the term, *race*. This brave, principled position by the Russian anthropologists in the international arena was met with openly improper behavior from foreign colleagues, which amounted to the concealment and distortion of scientific facts.

We will dwell briefly on several reports, which graphically characterize the independence of Russian raciological thought.

In the report, *Categorization as a Universal Phenomenon of Awareness of the World (Citing the Racial Differentiation of Man as an Example)*,[301] G.A. Aksyanova declared: 'The refusal to reject the conceptual notion of "race" as an object of study in physical anthropology, in my view, will lead to categorical illusions and mistakes of perception, since the accumulation of material enables one to say that this is an inadequate reflection of the realities studied by us. And the simple rejection of the term "race" does not in general change the objective morphological reality in biology of human populations. The refusal to recognize the reality of the existence of race in Man, works toward the destruction of the group consensus of physical anthropologists...This is a rejection of an effective instrument in the study of the history of human populations. In fact, just the fact of the existence of detectable exterior diversity can serve as the basis for categorization into "us" and "them."'

In the report, *Study of the Human Races in Russian Anthropology*,[302] T.I. Alekseyeva spoke of the deep and thorough tradition of these studies in Russia, uninterrupted even in the conditions of the difficult political and economic situation.

A.N. Bogashev made a report titled, *On the Racial Systematic of the Peoples of Northern Eurasia*,[303] in which he touched on the problematic of research on the pockets of race formation in this region, in connection with which he suggested bringing into use a new taxonomic term, "Western Siberian Race".

In a highly interesting work titled, *Specifics of Common Evolutionary Mechanisms in Anthropogenesis and Race-genesis*,[304] V.Yu. Bakholdina summarized material, according to which the processes of race formation did not go so gracefully, but spasmodically, on the basis of "aesthetic selection." In essence, it is a question of the formation of sexual selection, each time on a new level of development of mental and aesthetic faculties. Art, for example, is considered by her as one of the dominant racial differentiation markers, on the basis of which an evolutionary sifting of the lower, primitive forms of life occurs.

A collective study titled, *A New DNA Marker as a Racial Trait*,[305] analyzes material according to the availability of the new genetic marker CAcf685 on the 19th Chromosome, on the basis of which the value of the genetic distance Gst between Europoids and Mongoloids (in the latter given case, the Chukchi) is evaluated as six-fold. The given marker is recognized as valuable in racial-diagnostic terms.

---

[300] *Rasa: mif ili realnost'?*

[301] *Kategorizatsiya kak universal'noe yavlenie osoznaniya mira (na primere rasovoy differentsiatsii cheloveka).*

[302] *Izuchenie chelovecheskikh ras v rossiyskoy antropologii.*

[303] *O rasovoy sistematike narodov Severnoy Evrazii.*

[304] *Spetsifika obshchikh evolyutsionnikh mekhanizmov antropogeneza i rasogeneza.*

[305] *Noviy DNK marker kak rasovo-diagnosticheskiy priznak*

In the report titled, *The Phenomenon of Harmonization of Craniums*,[306] Yu.D. Benevolenskaya states that on "evolutionarily advanced skulls" a parallel specialization in both evolutionary and racial traits was observed. This says once more that evolution does not "wash away" or "remove" racial traits, but on the contrary, leads to their isolation and strengthening. Each racial trait, just like their sum total, develops together with the general course of evolution, increasing the differences between peoples.

A.G. Gadzhiev touched on this very timely theme in his presentation titled, *Race, Ethnos, and Economics*.[307] The author posed the question of the comparability and likelihood of the influence of racial traits on the economic development and economic way of life of each people: "With a large measure of probability, one may assert that the morphological traits of the pigmentation of skin, hair and measurable differences in pulmonary and skin respiration, and so on, provide this or that race with definitive advantages in the process of labor activity in those regions (geographic range), in which this or that race formed. Achievements in different areas of professional activity, art, physical education, and sports, attest directly and indirectly about the various characteristics (physiological, psycho-physiological, and so on) of the members of different races."

And now, if we remember the conclusions of the luminaries of Russian anthropology, that the size of the brain in the basic human races is not subject to quantitative change, then in light of the discussions of the author of the report, it becomes fatally obvious that it is the fate of several peoples and races, which have no "achievements in different areas of professional activity", to [ever have such achievements]. The basic theses of classical racial theory are again confirmed in the most obvious manner.

In his report, *Several Examples of the Existence of Discrete Racial Variants in the Modern Population*,[308] V.E. Deryabin spoke of the existence of "real racial boundaries". Of enormous interest is A.I. Dubov's speech, *What is Race?*[309] According to his valid observation, to this day, this fundamental term of modern anthropology is distinctly and unequivocally undefined. In this contex, as a supplement of distinctive quality, one can examine the speech of S.G. Efimovaya, *Borders and Opportunities in the Use of Measurable Traits of the Skull in Working out the Concept of Race*.[310]

A.A. Zubov took up the timely theme of the relationship of pure science and the political prostitution of science, in the essay, *The Essence of the "Crisis" of Raciology*.[311] The very word "crisis" in the title speaks of the uncompromising [attitude] of the prominent scientist, emphasizing that only starting with the 1960s did a negative trend take shape in regard to racial classification in race studies. Race is announced to be an "empty category", that is deprived of any biological basis. But Zubov and Alekseyev emphasized that similar discreditation is built on the basis of an intentional mismatching of the data of anthropology and genetics. "All the enumerated positions, on which the rejection of the reality of race is based, are mistaken and insufficiently substantiated, for the biases completely ignore the valuable positive contribution to science of race studies about Man."

The program speech by one of the best Russian geneticists, Yu.G. Richkov, titled, *The Genetic Basis of the Stability and Changeability of Race*,[312] was dedicated to this theme. His report was a summary of many years of theoretical and practical studies. In it, he said that despite the fact that for the last 35 years human geneticists have lacked coordination with anthropology, nevertheless, molecular genetics reveals more and more, "the so-called DNA markers, which can be considered markers of racial differences."

A report by the famous molecular biologist, V.A. Spitsyn, titled, *The Effectiveness of Various Categories of Genetic Markers in the Differentiation of Prominent Anthropological*

---

[306] *Fenomen garmonizatsii cherepnoy korobki.*

[307] *Rasa, etnos, i ekonomika.*

[308] *Neskol'ko primerov sushchestvovaniya diskretnikh rasovikh variantov v sovremennom naselenii.*

[309] *Shto takoe rasa?*

[310] *Granitsi i vozmozhnosti ispol'zovaniya izmeritel'nikh priznakov cherepa v razrabotke kontseptsii rasi.*

[311] *Sushchnost' "krizisa" rasovedeniya.*

[312] *Geneticheskie osnovi ustoychivosti i izmenchivosti ras.*

Commonalities,[313] was dedicated to an analysis of these new racial diagnostic markers. T.V. Tomashevich's speech, *Better to Consider the Differences of Race as Realities,*[314] appears to be completely logical, and therefore, in the context of the uncompromising [attitude] of the presented racial theme; in it, she noted that "..it is namely at the level of the great races that the most essential differences are usually observed: for example, in the distribution of the frequency of the suborbital tertiary cribriform canal in the skull, at the level of Europoid and Mongoloid races, in which mestizo groups occupy an intermediate position. The recognition of the reality of racial differences has a fundamental significance for the development of such areas as physical anthropology and morphology."

The title of A.I. Kozlov's report, *Consideration of Racial Characteristics in Preventative Cardiology,*[315] speaks for itself, for it is a testimony about the deep understanding of the practical, daily significance of racial differences. An unequivocal diagnosis is only found in the morgue. If then, a doctor really desires to heal a patient, he should calculate the **norm of reaction** of racial affiliation, which will unavoidably tell on all metabolic process. The Hippocratic Oath does not contradict racial theory; on the contrary, it justifies it, for to repair injury to a patient means to understand in the first place that the sick of different races are constructed in different ways. Here, the false abstract of universal humanism leads only to the legalization of charlatanism.

In a collective work titled, *Molecular-Genetic Polymorphism in the Study of the Human Population: the Genogeography of Eastern Europe,*[316] S.A. Limborskaya, O.P. Balanovskiy, and S.D. Nurbayev, speak about the great successes achieved in recent times in the deciphering of the human DNA genome. "In the course of this work, a large number of high polymorphic DNA markers was discovered, suitable for population-genetic research. Researching now-living populations with the help of these markers offers the opportunity to gain information about their genetic history, and in a number of cases to date, with this or that probability, [about] important events connected with the origin of Man, his races, and the settlement of Man on a global scale. The results achieved by a complex analysis of the region of Eastern Europe points, in racial terms, to the high-resolution capacity of DNA markers in the analysis of the genetic pool."

The title of T.V. Panasyuk's report, *Race and Sport: Investment in the Physique of Man,*[317] is eloquent. And N.I. Khaldeyeva developed a different, very timely theme in the report, *Racial Components of Anthropo-aesthetic Selection of the Variant of Appearance in Man.*[318] Here the teachings of Charles Darwin are directly proven, that each race has its inborn ideal of beauty. And the modern hypothesis was substantiated about the connection of population aesthetical preferences of both men and women, with racial differentiation. "The anthropological component leads in the complex of different parameters, being the most ancient and stable characteristic in the perception of outer appearance."

According to the results of this conference, a program document was published, titled, *The Problem of Race in Russian Physical Anthropology.*[319] This can be fully considered as the official position of Russian anthropological science. Thus, in part, E.V. Balanovskaya indicated the following in her presentation, which was included in the general version: "The objective classification of individual genotypes according to DNA markers, practically completely corresponds to racial classification." G.L. Khit' supported her in her turn, indicated that each of the major racial groups of humankind possesses inherent combinations of definite, frequential key traits, unique to it only. E.Z. Godina emphasized: "The basic racial differences are to a significant degree already formed in the embryonic period." In the same spirit, N.A. Dubovaya's proofs followed: "Until the present time, there is not one fact of when very dark-pigmentation of the skin, which is characteristic of equatorial groups, would have been marked for individuals, whose

---

[313] *Effektivnost' raznikh kategoriy geneticheskikh markerov v differentsiatsii krupnikh antropologicheskikh obshchnostey.*

[314] *Luche schitat' razlichiya ras real'nimi.*

[315] *Uchet rasovikh osobennostey v preventivnoy kardiologii.*

[316] *Molekulyarno-geneticheskiy polimorfizm v izuchenii narodonaseleniya: genogeografiya Vostochnoy Evropi.*

[317] *Ras i sport: vklad v teloslozhenie cheloveka.*

[318] *Rasovie komponenti antropologicheskogo vibora varianta vneshnosti u cheloveka.*

[319] *Problema rasi v rossiyskoy*

ancestors were not born on the African or Australian continents, or southern Asia. Precisely also, the appearance of a light-skinned, blue-eyed population in Africa or southern Asia is not marked, without the influx having the traits of migrants. If among the members of the Mongoloid race epicantus—one of the most characteristic of its traits—is encountered in 20-100% of cases, then among Europoids this indicator varies from 0-10% in cases. Straight hair is distributed both among Mongoloids, American Indians, and among Europoids, but among classical Negroids, it is never encountered. For Australoids, including Veddoids, wide and narrow wavy hair is characteristic. Mongoloids and American Indians differ significantly from Europeans by the stiffness of straight hair (a trait which is almost never encountered among Europoids)."

And thus, with the help of clear and persuasive presentations by the better Russian anthropologists and geneticists, we are convinced that art, sports, politics, all forms of economic activity, and also mating selection of spouses on the basis of aesthetic preferences, promotes not the "washing out" and "removal" of racial traits in the process of evolution, but rather their revelation, strengthening, and isolation. This refutes the baseless conclusions of A.I. Yarkho, and the other "alchemists" of the "melting pot" theory.

World history is not the removal of the racial question, but its unavoidable and constant intensification.

5. Race as a Higher Value

It stands before us to lift our eyes to the heavens, for there we will see a similar picture. For the majestic mystery of Theogony—the ancient struggle of the Gods—is nothing other than the quintessence of the confrontation of higher racial emanations. Every inch of galactic space is permeated by this struggle at all levels; it is carried out on the level of incorporeal ideas, and on the level of elementary physical particles.

Polygenism in evolutionary theory is a direct result of polytheism in religious studies. One of the founders of **philosophical anthropology**, Arnold Gelen (1904-1976) remarkably and accurately formulated this ecumenical rule, proclaiming: "The sole truth lies in this: that life feeds only on life."

It is appropriate also to turn to Alena de Benoit's program work, *What is Racism?* The greater part of his work is devoted to argument with population geneticists, whose basic conclusion is in favor of the absence of race; as we remember, it concludes on the basis of genetic markers, that it is impossible to establish distinct boundaries between basic racial branches.

In answer to this, de Benoit points out: "If one uses only genetic closeness, it turns out that the "closeness" between Man and the chimpanzee is far more than between several groups of peoples." The essence of it concludes that geneticists arbitrarily use "genetic distances" between genetic population pools, and not between racial groups. This could be expected of them. It turns out, that anthropologists begin the study of race from the usual visual perception of exterior descriptive traits, while population geneticists "construct" populations in their heads, which in general do not correspond to the reality of existing racial boundaries. Thus, they simply ignore the data of **biotypology**, upon which racial classification is in fact based. With such a success then, one may announce that in any river basin, separate rivers do not exist, but only one, since all the tributaries carry water that flows into one ocean. If an analysis of "DNA markers" is not able to distinguish between Malay and Negro, then the problem is in the particular method, not in the science of raciology as such. According to de Benoit's accurate determination, population geneticists have fallen into an "optical illusion", in creating artificial populations, which are models, and not real subjects of physical anthropology. "Geneticists can repeat as much as they like, that genetic pools do not necessarily correspond to phenotypes, or that we do not encounter genetic pools on the street."

If they continue to further deceive us with their phantom "populations", then we will be forced to turn for help to the more ancient and naturalistic methods of racial differentiation.

In his brochure, *Man as a Productive Force*,[320] famous Soviet anthropologist and hematologist B.N. Vishnevskiy wrote: "Insects of the different races display differences conditioned by the

---

[320] *Chelovek, kak proizvoditel'naya sila.* Leningrad, 1925.

261

chemism of the blood of the host. Thus, the external parasites of Japanese stand out from those dwelling on Europeans. The study of insects with the members of different races, helps to understand the question about the kinship connections of tribes."

Lice and other parasites, in contrast to "academic teachings", differ by a great "knowledge of life", only perceiving differences in the quality of the blood which they drink. Finally, with the help of these bloodsuckers, one can resolve the drawn-out discussion of "higher" and "lower" races among peoples with a simple, cursory inspection of prisoners in a cell; one can easily determine the place they occupy in the hierarchy of the criminal world, by their distribution on plank beds. Modern photographic surveying with a high degree of resolution capability, allows one to photograph insects of different types and to present their images to population geneticists with the goal of proving the existence of obvious differences between the races.

Finally, in 1855 Christian Baumgartner came out in favor of the organization of Man from several different pairs of ancestors, based on the existence of a high number of human parasites: "Were those parasites originally on one couple, [that couple] would have died, and with them all of the parasites." And the famous French anthropologist, Jean-Joseph Virey, wrote at the beginning of the 19th Century, that: "It is known that each species of mammal, bird, and so on, has its own insect parasites, which are observed on it alone; so the Negroes have them: they have their lice, [which is] completely different from the lice of the White Man." Negrita lice "have a triangular head, a tuberous body, and a black color: the same as in Negroes."

In the beginning of the 20th Century, Polish anthropologist Ludwik Kryzwicki conducted a survey of cannibals in the most varied parts of the world; all the diners of human meat swore to the investigator, without blinking an eye, that the races differ by taste. Obviously, they were also outraged and vexed by the assertions of population geneticists, for insulting their refined tastes.

In the second half of the 19th Century, the founder of the German Anthropological Society, Rudolf Virchow, said: "Man could just as well have originated from either a swine or an elephant, as from apes"—because, in his opinion, even in a ram or elephant one can observe traits similar to Man. It is worth noting that the "swine theory of anthropogenesis", as expressed by Virchow, now has factual confirmation through the efforts of modern geneticists, because many donor organs "reserved" for people are now taken from transgenation pigs, and not from apes, as one would expect, according to the assurances of traditional evolutionists.

In his article, *The Theory of Anthropogenesis and the Origin of Species*,[321] S.V. Vasilyev was forced to recognize, that to the present day, there is not a commonly accepted "model of the evolution of Man." Besides that it is emphasized in the given work that modern evolutionary theory admits the coexistence of "a hominid, located at various stages of speciation," in the same [space]-time, within the limits of one ecological niche. That is, expressed in the words of classical racial theory, there is nothing unnatural in the coexistence of "higher" and "lower" races at one time, within the limits of one "living space."

In this context, the modern work of G.L. Khit' and B. Keyta, *Dermatoglyphic Divergence of the Basic Racial Branches of Mankind*,[322] is highly significant. In it, the authors unequivocally declare: "Individual profiles of the skin fingerprints of the three racial groups testify to the uniqueness of each of them. The possibility of the apportioning of similar variants is completely excluded. Europoids and Mongoloids have opposite combinations of traits, revealing a certain likeness to a "double spiral." Negroes, without exception, occupy an extreme position, according to all traits. Mongoloids are a less stable race than Europoids. Concerning Negroids and Europoids, unification into one western branch is impossible, according to fingerprint data."

Spitsyn, a prominent Russian authority in the area of the biochemistry of Man, also emphasizes: "Each of the major races possesses a trait characteristic to it only—a genetic complex of gammaglobulins and alkaline phosphate placenta."

Further, G.L. Khit' and B. Keyta made a completely shocking conclusion in their article: "Inasmuch as the traits of the different systems of an organism evolved with different speed, changing with dependence on concrete conditions, the age of the formation of racial, mutually independent systems obviously cannot be identical."

---

[321] *Teoriya antropogeneza i vidoobrazovie.* Vestnik antropologiya N4, 1998.

[322] *Dermatoglificheskaya divergentsiya osnovnikh r asovikh vetvey chelovechestva.* Rasi i narodi. Vip. 11, 1981.

Thus, authoritative scholars confirm that all the basic human races do not match one another in qualitative genetic characteristics, age, and tempo of evolution. Besides that a single evolutionary model for the different races is generally absent, and practical medicine and transplantology shows that pigs have more of a basis to be considered a close relative of Man, than do primates. Incidentally, among all mammals, the pig possesses one of the highest IQs; this again contradicts the research of progressive-humanists, who tie the development of land animal life forms with the complication of higher, nervous-physiological activity.

Therefore, in accordance with the above-said, it seems groundless and improbable to us, to talk about a "certain singular humanity" as a higher value, for that is the same as suggesting the existence in the body of one individual, of fragments of skin from different races, and various biochemical formulas of metabolism, in various parts of the body. The criterion of values did not arise in a vacuum. Ethical values are the quintessence of physical being, and our existence is divided by the timeless boundaries of race, the surmounting of which is not attainable by a single violator of the laws of nature.

Today it seems completely natural and obvious to us, that culturologists, social scientists, and members of the mass media, discuss urgent problems daily, and constantly appeal to a category of values, as if it is a certain truth in the final instance. And in this they most often allude to some "general human values." But in reality it was substantiated comparatively recently—at the end of the 19th Century—as a philosophical category. Besides that it did not relate then to any abstract, "common humanity"; on the contrary, it imparted a concrete, racial-biological significance. Theodoule Ribeau (1839-1916), a prominent French psychologist, pointed out in his book, *The Logic of Feelings*, that: "Since an analysis of values leads us to the same general and elementary manifestations of psycho-physiological life, then it is completely natural to look for their root in biology. A reliable explanation is one which takes the values to ancestral functions of life activities, that is, to a continuous method of reaction to primary elements, to simple processes; which brings out the notion of the value of the principle of organized self-preservation. Since there is a difference in organization between separate individuals, then this explains individual differences in the evaluation of values."

The famous German philosopher, Moritz Eisler (1818-1890), also emphasized that "our mental activities do not create values, they only open before us already-existing values, which have a biological basis." The French philosopher Gabriel Tard (1843-1904), who specially developed the "Theory of Values", wrote: "Values do not exist outside of us, outside of our subjective lives, since each emotion is secured by heredity."

By common recognition, the first in modern times to substantiate the notion of "values" was the German philosopher and naturalist, Rudolf Herman Lotze (1817-1881). Another titan of philosophy, Wilhelm Windelband (1848-1915), who headed the Baden School of neo-Kantism, pointed out: "Since the time Lotze energetically put forward the idea of values, and placed it at the head of both logic and metaphysics, attempts are often made to establish a "Theory of Values", as a new type of philosophical basis of science. Philosophy should not be an image of the world; its task is to bring those norms upon which the value and meaning of any thought depend, to the consciousness of people."

For his part, Max Scheler (1874-1928), one of the founders of modern sociology concepts, developed a phenomenological direction in the "theory of values." According to his concept, one should differentiate values and their carriers. Scheler defined values as objective phenomena, not dependent on the consciousness of the person and the subjects, in which they manifest. But an objective phenomenon that is not dependent on the person or subject, is his racial affiliation. And the differences in the systems of values of the main races of humankind, according to Bibeau's logic, are by natural method conditioned by the differences in the organization of their members.

The excellent neo-Kantist, German philosopher Heinrich Rickert (1863-1936) worked out the conception of differences in the systems of values in the natural and humanitarian sciences; he wrote in his book, *On the System of Values*, that: "Philosophy should rise to the super-historical value derived by it, in existing cultural, material benefits."

Race is such a super-historical value; itself for itself is the reason, the result, and the ethnic yardstick. A racial view of history therefore means subjectivism, but at the same time race is an objective given, in the genetic and anthropological sense. This apparent contradiction is easily removed with the help of a logical definition, which unites the conclusions of the natural and humanitarian sciences. Axiological studies about values uncontradictingly extend to the area of positive anthropology. Heinrich Rickert justifiably noticed that "where there are no values, there is no science."

Therefore, from the point of view of the integration of philosophy, history, and natural knowledge, we make this determination: **race is a super-historical subject of an historical process.**

Rickert's followers developed axiology—the science of values—namely in the 20th Century, when with all sharpness the problem of the ethical evaluation of scientific progress arose. Karl Jaspers wrote: "Scientific knowledge, as such, is not any kind of true value." Philosopher Ralph Barton Perry expressed himself in the same key: "Both pure science and common technology are useful, but [they are] indifferent to the goals they serve." All this in full measure relates to the entire complex of natural and humanitarian sciences, for their achievements are rendered meaningless and useless if they do not serve a higher, objective value, which is experienced and evaluated subjectively by us; **race** is its measuring stick.

Therefore, the philosopher-instrumentalist John Dewey emphasized: "The problem of establishing unity and cooperation between the convictions of Man about the world in which he lives, and his convictions about the values and goals, which should direct his behavior, is the deepest problem of modern science." A statement by the English writer, Oscar Wilde, is very clever and accurate: "If Man is not a gentleman, then excessive knowledge is to his detriment."

Your humble servant has associated with many anthropologists, geneticists, biologists, psychologists, and members of related sciences, to whom notions about values and civic duty were completely alien. For those specialists, the problem of one's own race was of no interest. In the meantime, the great English philosopher and scientist, Karl Pearson, named the introductory chapter of his program essay[323]: *Science and Citizenship.* It was with mention of him that we began our work.

If a scientist considers science to be international, then in the process of scientific works he will unavoidably serve the interests of foreign races, and not his own.

Russian science, being advanced in many respects, nevertheless inherently suffers from **axiologic impotency**, covering its ailment with the academic naturalist's fig leaf of objectivity. "I am free from evaluation." –This slogan has been made fashionable in modern scientific circles. But after all, even the Egyptian vulture, in devouring carrion, is not free from a system of values. Moral sterility is still an illusion.

In connection with this, one may recall Rickert's statement, that "ethical values are first of all, with the will", and also the words of Dewey: "Desire is a spring, which brings the mechanism of evaluation into action." To this one can add an aphorism of Perry: "My interests are I myself in the deepest sense."

The outstanding English philosopher and mathematician, Alfred North Whitehead (1861-1947), also asserted that: "cut off from reality, activeness is separate from value, for only reality represents value." Besides that "by a factor of coercive determinism, which exists in the universum", he said that subjectivity of the comprehension of being is unavoidable, a result of the objective existence of different races, considered by him as "enduring objects."

Again we are persuaded that from the point of view of the theory of knowledge, that race is a super-historical subject of the historical process.

It is namely in this deep subject of racial sense, resulting from will and interest, that Russian anthropology is deficient.

The founder of racial hygiene, Alfred Ploetz (1860-1940), wrote: "Everywhere outside of the individual, where the ethicist seeks a transcendental supporting point of human activities; where the politician struggles for basic, vital interests, the final object, consciously or unconsciously, is always that organic whole of life, represented by race."

---

[323] *Grammar of Science.* 1911.

It was also not an accident that one of the leading racial philosophers of the 20th Century, Fritz Lenz (1887-1976), named his program essay, *Race as a Basic Evaluative Principle*. In it, he remarked: "Race is the carrier of everything—the individual, the state, a people; from it all existence proceeds, and race itself is the essence. It is not an organization, but an organism…Outside of our will toward values, the notion of value loses its significance. The stars of our fate are within us. The substantiation of our higher ideals is in our own essence…Health of race serves both the happiness of separate peoples and the constant, general, basic happiness. A degenerative people is unavoidably unhappy, even if it possesses all the treasures of the world. We do not need race for the sake of happiness, but happiness for the sake of race."

Therefore, from the position of **racial axiology**, that is, namely from a racial system of values, there is still not a completely compiled work of any scientific character, with which to overthrow one of the most grandiose myths of Christian dogmatics and the modern liberal Enlightenment—the myth of the species unity of Mankind. *Homo Sapiens* is a chimera, a collage, an illiterate forgery for the trusting devotees of leveling universalism.

6. Race versus Species

According to the modern findings of **evolutionary theory and systematics**, the general number of species of plants and animals on Earth approaches up to two million. From all this diversity, there are 25,000 species of worm, 70,000 species of mushroom, and 3,500 species of mammal. Of primates—the closest relatives to Man in the Animal Kingdom—there are 101 species. And of all the very diverse "humans", [it is claimed that] there is only one species. Strange, isn't it?

Scientists capable of independent thinking have repeatedly pointed to this obvious absurdity in liberal-utopian systematization.

The famous German naturalist and philosopher, Karl Vogt, wrote in the book, *Man and his Place in Nature*: "If the difference between the Negro and the German is greater, than between the Capuchin monkey and the sayyu, then either the Negro and the German are two different species, like various apes, or these apes should be merged into one [species]." The German naturalist-philosopher Friedrich Gelwald also suggested: "If we just once called Papuans our brothers to us, then we suggest it would that it would not cost great effort, to welcome the chimpanzee as a cousin."

Left: Karl Vogt

The very term "species" entered biology from logic, and the very notion was brought into use by Aristotle; it was used to refer to expressed similarities or the identical essence among a group of individual subjects. In 1693, the English naturalist John Ray substantiated the application of the term for denoting similar individuals that were capable of passing down their distinguishing traits to offspring. But it is namely racial traits that are distinct and passed down by inheritance. The famous Swedish scientist, Karl Linnaeus, wrote in his work, *Philosophy of the Botanist*: "As many species are numbered, as the [number of] different forms created from the very beginning." That is, under species, the founders of the systematic of the Natural Kingdom distinctly and namely understood race. Georges Cuvier, Peter Pallas, and Jeane Baptiste Lamarck stood on these same positions, and understood the word "race" in the sense of "breed", the "descendents of a common ancestor."

Already, at the very height of arguments in the middle of the 19th Century on the problem of species, the founder of Evolutionary Theory, Charles Darwin, explained in his book, *The Origin of Species*, that he considered the term *species* to be "completely arbitrary, devised for the sake of convenience." German paleontologist L. Wuertemberger observed (1880) that it is impossible to point out where a species begins and where it ceases to exist. Then the notion of species loses any natural scientific basis. And in 1881 Russian paleontologist S. Nikitin compared "species" with an arbitrarily cut section of endless ribbon. German histologist Albert Kelliker declared in

265

1872, that: "It is possible that one or another species from different phylogenetic roots, which are so originally different, can strive toward a similar, ultimate phase."

However, in 1901 Russian scientist V.L. Komarov brought clarity to the terminological muddle: "During studies of the basic unity [of a species] one needs to consider not the abstract, standardized notion of "species", but the real race. Race is real, and not species." He also repeatedly wrote about the desirability of replacing the term "species" with the term "race", in all natural scientific literature. "A pure morphological notion of species as a model or type, is naturally abstract, and in the attempt to carry it over to living nature, it easily crosses over into negation of the very species." Being a consistent and decisive scholar, Komarov generally excluded the term "species" from is lexicon.

German biologist Oscar Hartwig also asserted in 1916 that "The notion of species is purely abstract." The famous German evolutionist, Ludwig Plate wrote in 1908 that "The notion of species abstracts Man. Like all concepts, the notion of species is the product of human thinking, and the consequence of this is that it does not represent anything real." Only the prevalence of blood kinship, that is, a community of racial origins, in his opinion, could be supposed as a basis for systematization of the animal world.

In modern science, the argument around the problem of species has not subsided to this day. In 1954, US researcher B. Berma, standing on the position of logical positivism, renewed criticism: "Species as a class stands outside the reality of existence. A highly abstract invention such as species, the basis of which is to set the idea of the reality of evolutionary populations, is only capable of obscuring the understanding of the evolutionary process."

Another luminary of Evolution Theory, Ernst Mayer, wrote in 1949: "What is species? There is not a single viewpoint among systematicians. Disagreement is observed even among specialists in separate groups." A. Kane, a scientist from Oxford, was also very frank: "The boundaries of a biological species in space-time are undetermined, inasmuch as the genetic criterion (the possibility of hybridization) are inapplicable and fit to be satisfied only by the comparative study of morphology, physiology, genetics, and behavior."

Thus, it is a fact that Mankind declares itself to be a single species on the basis of the possibility of free cross-breeding of the members of different races—an unscientific assertion, that is contradictory to the laws of systematic—a science which for more than 300 years already, has engaged in the classification of living organisms. The comparative study of morphology, physiology, genetics, and behavior precisely uncovers the entire depth of differences between members of the different races. Kane thus continues his thought: "A biological species consists of populations that are genetically related between themselves." But after all, under genetically related populations it is necessary to recognize race. No one is going to deny the obvious fact that between Bushmen and Scandinavians there is absolutely no genetic connection. On behalf of classical racial theory, the author of the above-mentioned book sounds the following thought: "Species came into contact only after they acquired their ecological differences." This means that from the very beginning, races were pure, **and mutations were a result, not a cause of their development.**

Left: Hugo de Frise

But then again, similar thoughts were expressed a half-century earlier by Russian scientist V.L. Komarov: "For the rise of a new race, it is necessary that its characteristic traits appeared immediately in all the indivisible inhabitants of a given territory." This is a complete refutation of the mutation theory of the origins of the basic human races.

In zoology and biology, they quite often use such terms as "higher and lower plants." But according to genetic laws of heredity and the rules of systematics, [the terms "higher" and "lower"] should be carried over to the human races. The [following] terms were brought into use as early as the start of the 20th Century, by Dutch zoologist and botanist, Hugo de Frise: "seasonal races", "rubbish-field races", and "parasitic physiological races"—can easily be applied to modern social society, in pursuit of an explanation of many "culturological" phenomena. When visiting *avant garde* art salons and night clubs in racially mixed cities, one can confidently say, using de Frise's terminology, that here is a well-represented "complex, cross-bred character," appearing as a "result of regressive mutations." In the 19th Century, German racial historian Gustav Klemm used classical Latin for the designation of these phenomena—"Bassa Gente"—which means "low people". In modern Russian slang this sounds like "bassahente". One may recall the witty passasge by the great Russian writer, V.V. Nabokov, who spoke of the people in pop-culture as "members of a different sexual flora."

Karl Vogt, cited by us above, make the following logical conclusion: "If one takes a closer look at the definition of race and species, at the differences established between us by custom, it turns out that this difference is extremely relative. Races begin where a certain root is supposed or known, from which they developed; there too, where they are lost in the depth of time, they admit species. As a species, they would not be recognized in real time; one can never admit that the human species consists of several thousand different species, which so differ from one another as much, if not more, than a large part of ape species. If the principles of zoological systematic have general significance, then they should be applied identically and impartially to Man and apes alike."

The opinions of the authors of the collective anthology, *The Biological Evolution of Man*,[324] are indicative in this regard. M.I. Uryson writes in his article, *Toward the Problem of Allocation of Hominid Branches in Evolution*,[325] that: "Insofar as Man by his origins is himself an integral part of the organic world and a higher link in the evolution of primates, the principles of systematics and the rules of zoological nomenclature, which are applied to other groups in the Animal Kingdom, should extend to him. The qualitative differences in Man over animals do not liberate us from the necessity of considering Man as a being that achieves a definite level of biological organization, and is located in general in the river channel of evolution in the organic world."

V.A. Spitsyn's article, *Modern Ideas about the Evolution of the Order of Primates, in Light of New Findings in Molecular Biology*,[326] appeared in the same journal. In light of his research, the modern liberal-democratic concept of racial genesis from a single African root, on the basis of mitochondrial DNA, appears in all its naked repulsiveness. The modern **"Black Eve" Lysenkovites**, of whom we spoke at the beginning, are completely discredited, for according to Spitsyn's argumentation, and the findings of the laboratory experiments cited by him, evolutionary retention of this very mitochondrial DNA is only possible in conditions of "inter-species exchange by females."

In a high measure of indecent para-scientific teachings, artists attempt to name the common maternal ancestor of Mankind, as none other than a "black-skinned Eve." But in light of the findings of molecular biology, it turns out that our 'great-grand ancestress' was a Negress who in turn devoted herself to male primates, half-apes of various species, and that laid the beginnings of the racial evolution of Mankind. A more infamous and absurd rendering of the Biblical Genesis version of sin for "the modern educated public", is impossible to imagine. However, even if the existence of some hypothetical "black Eve" is assumed, with her passion for journeys in sum total, and with [her] craving for sexual variety, "nothing follows from this or that that she needs to be elevated to the rank of "ancestress of Mankind."

In the fundamental anthology, *The Eastern Slavs: an Anthropological and Ethnic History*,[327] for the creation of which the most noted Russian anthropologists, geneticists, and biologists took part, it is clearly indicated that: "It is known among Western European archaeologists that there is an effort by a group of geno-geographers and historians, to tie the modern geography of the genetic pool of the West European population with the Neolithic Farming Revolution, which started in the Near East in the 9th to 7th Millennia, B.C. They even attempt to tie the rise of Indo-European languages to these events, extending their history back almost two times, and radically changing the notion of their historical-geographical ancestral homeland. It turns out, however, that modern molecular genetics, on the basis of findings about polymorphism and the ancientness of mitochondrial DNA, showed an extremely small role (on the order of 5-15%) of Neolithic migrants from Asia Minor, in the formation of the genogeography of modern West Europeans, and

---

[324] *Biologicheskaya evolyutsia i chelovek.* Moscow, 1989.

[325] *K problem videleniya gominidnoy vetvi evolyutsii.*

[326] *Sovremennie predstavleniya ob evolyutsii otryada primatov v svete dannikh molekulyarnoy biologii.*

[327] *Vostochnie slavyane: antropologiya i etnicheskaya istoriya.* Moscow, 1999.

discovered numerous Paleolithic sources from the West European genetic pool." From this it follows that on the path of their possible migration from Equatorial Africa, through Asia Minor and into Europe, neither "Black Eve" nor her descendents enjoyed excited demand on the part of the White aboriginal population; this once more testifies to the fortress of racial consciousness of our Nordic ancestors, and also about their origin from another biological type, other than Negroid or Mongoloid.

Modern political-prostitute population-geneticists, in expanding on the myth of the so-called "African Eve", that is, of the origin of all modern races from Equatorial Africa, from one black-skinned woman approximately 100,000 years ago, on the basis of studies of mitochondrial DNA, complete a disgraceful scientific forgery. In a similar work cited by us, *The Problem of Race in Russian Physical Anthropology*,[328] E.V. Balanovskaya clearly indicates that "unfortunately, mitochondrial genes have no kind of relationship to racial traits."

It so happens, that propagandist ideas about the "oneness of Mankind", which proclaim a common origin for all races, are based on a trait which, in general, carries no racial information within itself. The given primitive cheating is not scientific, and therefore, the question of an "African Eve", may be removed from the discussion, as an example of "black-skinned Lisenkovism [repression]."

The outstanding French naturalist Georges Bouffon (1707-1788) separated Man from the Animal Kingdom, and taught that Man is a product of Heaven, and the animals—products of the Earth: "A strange place for Man! What an unfair assignment, what an erroneous method! To place Man on the same board with simple quadrupeds." The prominent German psychologist and founder of **comparative psychology**, Karl Gustav Carus (1789-1869), asserted: "The cause of the "eternal division" between Man and animal lies not in some single organ, but occurs throughout the whole organization." Robert Hartmann, a German anatomist of the 19th Century, revealed distinct contradictions in the classification of morphological differences: "In general, even the most fanatic defenders of Darwinism are more and more inclined to the conviction, that Man could not have originated from a single one of the now-living anthropoid forms. It is true that one can prove a close, and in many cases, even an extremely close somatic kinship between Man and anthropoid apes, but there is no chance to prove a direct origination of the former from the latter." And in his *Lessons on Man*, the above-mentioned Karl Vogt derived three human races from three various man-like apes: gorillas, chimpanzees, and orangutangs. The famous English naturalist, Thomas Henry Huxley, held the view that the lowest developed apes, from an anatomical point of view, lagged further behind the most developed apes, than the latter lag behind Man.

Finally, the famous Russian biologist-evolutionist, K.M. Zavadskiy, honestly declared in his fundamental work, *Species and the Evolution of Species*,[329] that: "A species does not only not have outer morphological, but [does not have] the anatomical, histological, and cytological traits, which would allow one to mark by them, namely its group and not other groups. Here, systematic ends up in a blind alley." Besides that the author further asserts: "Dissimilar manifestations of different traits among different species is a result of the inequality of species, and the differences in modes of species evolution. The differences testify that a single standard does not exist, by which all could be organized." After this, Zavadskiy made an unexpected stipulation: "Our characteristic of a species can be applied to all races, capable of independently reproducing in nature, and capable of continued existence in evolution." With this small explanation, the below phrase simply sounds deadly: **"Species have unequal worth namely because they stand on different stages of the development of the form of the specific organization of life, or, located on one stage of organization, they have some principle constructive characteristics."**

Remembering that the evolutionary-biological rule of the development of species, in Zavadskiy's opinion, is valid for separately taken races, we now have an excellent definition of the "unequal value of the human races," given by a famous Soviet scientist. Besides that from the very first sentence, he openly declared that his book is dedicated to the problem of the "inventory of species," and in another place he spoke about evolution as a "stairway of creatures." But it is

---

[328] *Problema rasi v rossiyskoy fizicheskoy antropologii.* Moscow, 2002.
[329] *Vid i vidoobrazovanie.* Moscow, 1968.

namely upon all these principles that classical **racial theory**—founded by Count Arthur de Gobineau in the mid-19th Century—rests. His main book is called, *Experiments in the Inequality of the Human Races.*[330]

Besides this, Zavadskiy made a very important conclusion in his work: "The interior absence of a conditional, definite time of existence of a species in nature, is one of the root differences of a species. In this sense, a species may be called 'open' in the time of a system." With this, Zavadskiy once more supported our definition of the term [race]:

A race is a super-historical subject of an historical process.

7. Polygenism

Below: Georges Buffon

Logically relying on more than 300 years of research by evolutionists and systematicians, we are compelled to recognize that a single *homo sapiens* species is fiction. Only races are real.

The outstanding "right-wing" German philosopher Eugene Duering (1833-1921) wrote: "Between the man of one race and the man of another race, there may be as many differences, really, as between a man and an animal." However, we consider it necessary to emphasize that the idea of carrying over the principles of systematics from the notion of "species" to the notion of "race", does not belong to K.M. Zavadskiy first. It is easy to be persuaded of this, if we analyze the very formula for the notion "species", from the given perspective in the works of prominent evolutionists.

The French philosopher-materialist, Jeane Baptiste Robin (1735-1820), wrote: "Under the title "species", naturalists mean the sum total of individuals possessing a sum of differences noticeable to them."

The Swiss anthropologist, Jeane Louis Agassis (1807-1873), subsequently defended the concept of the origin of the human species in several geographical centers, which are unrelated to each other. He indicated: "A species is the final limit of classification, upon which naturalists decide on. On the basis of it are built the least important traits, such as: height, color, and size."

Jeane Baptiste Lamarck (1744-1829)—whose views are so dear to the hearts of all liberal anthropologists for their propaganda teachings about the influence of environment on the hereditary traits of organisms—was forced to recognize: "A species is a sum total of similar individuals, unchanging repeatedly from generation to generation, as long as the exterior conditions themselves do not change [so much] as to change their habits, traits, and forms."

Prominent anthropologist Etienne Geoffroy de Saint-Hilaire (1772-1844) expressed the opinion, according to which: "Species is an aggregate or a number of individuals characterized by a sum of outstanding traits, the transfer of which is natural, regular, and constant in the natural order of things."

---

[330] *Opit o neravenstve chelovecheskikh ras.*

Etienne Geoffroy de Saint-Hilaire   **Georges Cuvier**

The outstanding naturalist, Georges Cuvier (1769-1832) determined that: "A species is an aggregate of organic essences, born one from another, or from common parents and from individuals, as similar to them as they are mutually similar between themselves."

Another famous anthropologist, Armand de Quatrefages de Breau (1810-1892), inserted into the notion of species [the idea of] "a likeness of individuals and their unbroken blood tie, right down to the original group."

Now, if following the logic of Zavadskiy, we substitute the word "species" with the word "race", in all these formulas, then one can easily be persuaded that the essence of the definitions does not suffer a bit from this substitution. Actually, Etienne G. de Saint-Hilaire also concluded, that "race is a chain of individuals, descending one from another, and distinguished by traits that are made constant." And Armand de Quatrefages de Breau emphasized that: "race is a sum total of similar individuals, belonging to one species; of individuals receiving by way of inheritance, and in their turn, passing down the traits of the original version."

Adrienne de Gusier called the races "hereditary versions", and Georges Poucher, following this logic, asserted that "the word 'race' means different natural groups of the human genus and therefore, they are in essence, the same as [different] species." Feeling the vulnerability of his own position in the plan of strictness of the definition of the concept of species, Lamarck himself jokingly called it a "work of art".

Official academic evolutionary science likes very much to cite the authority of Swedish naturalist Karl Linnaeus (1707-1778), but steadily forgets to mention that he divided the entire genus *homo* into three species: *homo sapiens, homo ferus, and homo monstruosus*. Linnaeus subdivided *homo sapiens* into:

"I.   Americanus rufus—American. Chestnut hair, choleric, stands up straight, persistent, complacent, submits to tradition.

II.   Europaeus albus—European. Blonde, excitable, muscular, active, clever, inventive, submits to law.

III.   Asiaticus luridus—Asian. Yellow-faced, melancholy, flexible, cruel, frugal, loves luxury, dresses in wide clothes, submits to public opinion.

IV.   Afer niger—African. Black skin, phlegmatic, sluggish physique, cunning, indifferent, lethargic, oiled up with fats, submits to tyranny."

Besides that the author mentions—*homo ferus*—the "wild man", who is covered in hair and walks on all fours, and also *homo monstruosus*, to which microcephalics and plagiocephalics belong.

And the great Linnaeus suggested that a trait does not exist for distinguishing Man from ape, since both of these types are united by an entire number of intermediate anthropomorphic beings—wild people, pygmies, satyrs, tailed people, troglodytes, and other sub-humans. However, in his understanding, close similarity of species did not in the least way suppose a sure kinship between them, since each species was created separately and remains unchanging from the very beginning of existence. These views were expressed by the scientist in his famous report: *The Systerm of Nature* (1758), and as early as 1760 in a following publication he radicalized these views even more, remarking: "To many, it can seem that the differences between Man and the ape are greater, than between day and night; however, if these same people were to compare the great heroes of Europe and the Hottentots of the Cape of Good Hope, then with difficulty they would assure themselves that these and others have **identical** origin; or if they wanted to compare a noble, aristocratic, educated and cultivated young woman with a wild man, left to his own devices, then they would hardly be at peace with the thought, that both belong to one and the same species."

Georges Buffon and Johann Blumenbach, speaking about the human races, also defined them as "versions." In 1801, anthropologist Jeane Joseph Virey was the first to deliberately come out against this jumble of concepts. He determined that the human genus consisted of two species: white and dark. Jeane Baptiste Boris de Saint Vincent (1778-1846) divided humanity into fifteen species, and Antoine Desmoulins (1796-1828) [divided them] into sixteen species.

Thus arose the philosophical-anthropological field of **polygenism**, which rejected the specific unity of Mankind.

Left: Karl Linnaeus

Jeane Joseph Virey, a Doctor of Medicine and member of the French Royal Medical Academy, wrote in his three-tome work, *Natural History of the Human Genus* (Paris, 1824), that: "Blumenbach and others support the theory of the oneness of Man as a species, with physiological arguments, such as marriages of blacks and whites yield offspring. But horses with donkeys, and wolves with dogs also produce offspring. The human genus as a whole needs to be divided into two species, and those in turn into various races. The first species—a facial angle of 85 degrees. Its races—white, yellow, honey, and dark-skinned. The second species has a facial angle of 75-80 degrees. Its races are black and near black skin (Hottentots and Papuans)."

In the book, *Theory of the Four Movements in Universal Fate,* the famous socialist-utopianist, Charles Fourier (1772-1837), asked forgiveness in an ironic form, from the readers "for the writer's fables" which derive a human genus from a single trunk."

It is especially necessary to emphasize, that the antique world did not know an equality of races, and correspondingly, a specific unity of Mankind. As was to be expected, only the spread of Christianity set down the beginning of monogenetic agitation. Not surprisingly, it was namely a man of mixed racial origin, a mestizo named Augustin, who subsequently received the nickname, "the Blessed." In 415 A.D. he was the first in Europe that began to develop the Judeo-Christian doctrine of the common origin of peoples. However, the intellect of the White Man constantly resisted this preposterous, racially foreign fabrication, for during the reign of the Byzantine Emperor Justinian, a Church council discussed the question, of whether Negroes are descended from Adam, and could they be Christians? In the year 748 A.D., the Monk Virgil, who asserted that another Earth existed, that was populated by peoples different from us, was condemned by Pope Zacharius. In the year 1110 A.D., the philosopher Guillaume de Conche was condemned for writing that Eve was not the only woman created, and that her antipodes lived beyond the ocean. In the year 1450 A.D., the Jew Samuel Sarsa was burned as a heretic, for a theory of the deep ancientness of the origin of the world and Man.

After the discovery of America, it was announced in 1512 that Indians were also descended from Adam and Eve.

Felipe Teofrasto von Hohenheim Paracelso (1493-1541) needs to be considered as an ingenious pioneer in polygenism in Renaissance Europe; in 1520 he restored the ancient ideas about the nature of the origin of races. It is impossible to presume, he said, that the inhabitants of the distant, newly-discovered lands are the children of Adam, and that in them is the same flesh and blood, as in us. Moses was a theologian, not a doctor [of science].

The great Italian scientist and illustrator, Giordano Bruno (1548-1600), expressed similar views in 1584, in his book, *The Expulsion of the Triumphant Beast*. Of the several human races, he wrote: "These from the 'New World' do not in the least bit comprise a portion of the human family, for they are not people, although they strongly resemble them in their members, figures, and brains."

Left: Jeane Baptiste Boris de Saint Vincent

And in 1591, Bruno added that "not one man of the Ethiopians or the Jews, springs from the same protoplasm [as Europeans]." He referred to the Chinese and the rabbis as recognizing the existence of three human breeds. Hardly anyone knows, but in 1600, Giordano Bruno was burned at the stake in The Inquisition—not for the assertion that the world is round, but for propagandizing the ideas of polygenism. In 1616, the same fate befell Vanini. He was sentenced to have his tongue cut out and burned, for in his *Dialogues*, he brought up the old hypotheses about the natural origin of Man, and the "assertions of the atheists, that the Ethiopians arose from apes; that the first peoples walked on all fours, like animals, and that in nature, a sort of hierarchy of beings exists, from lowest to highest."

In 1655, Protestant nobleman Isaac de la Peyrere came out against the Judeo-Christian mono-genetic doctrine, in the published report, *Pre-Adamites*, which was subjected to public burning in Paris. According to de la Peyrere's theory, only white people, tracing their origin from Adam, are people in the proper sense of the word; members of other races are miserable forgeries, sub-human "pre-Adamites."

In London in 1695, a book by an anonymous author came out, in which the question was shifted to a scientific basis. It was dedicated to the American natives. Moses, the book said, was a great lawmaker, like Solon and Lycurgus, and his tales of Creation were oriented to the intellectual level of his listeners. Studying arguments in favor of migration to America by peoples and animals from other continents, the anonymous author came to the conclusion that [native] Americans were indigenous. He also showed that in the distant past, negroes were the same as they are now. Their blackness cannot be explained with a curse from God, nor the influence of the sun. Neither Native Americans, nor Negroes, in the opinion of that author, are descendents of Adam.

In Lyons in 1744, Guillaume Ray published *A Dissertation on the Origin of Negroes*. In it, he pointed out that six species of Man exist, also adding orangutans and seals to people.

The most important book that expounded the arguments of the polygenists of that age, were the two-tomes by Lord Kames, first appearing in publication in 1774. The author, a Protestant-liberal, considered similarity as a criterion of species, and not the fertility of hybrids. In his opinion, God created several human pairs, which were adapted to the climatic zones, in which they lived. Whites and Negroes—these are people of different kinds. Lord Kames' book had a wide response, and his high position in society gave weight to the scientific position of the polygenists as a whole.

In 1785, Samuel Thomas Zommering (1755-1830) of Germany, and Charles White (1728-1813) of England, laid the foundations for comparative anatomy of the races. Both of them came to the conclusion that Negroes, by their build, occupied an intermediate position, between Europeans and apes.

At the start of the 19th Century, a blossoming of activity occurred with outstanding French anthropologist-poygenists like Jeane Joseph Virey, Jeane Baptiste Boris de Saint Vincent, and Antoine Demouslins.

In England in the middle of the 19th Century, the most radical polygenist was Robert Knox. In the book, *The Races of Man*, published in London in 1850, he tried to prove that the European races were different from one another, like the ordinary Negro from the Bushman, the Kaffir from the Hottentot, the red-skinned Indian from the Eskimo, or the latter from the Basques. "Peoples belong to different races. Call them species if you want, or versions: this has no significance. A fact remains a fact: people belong to different races." His ideas were shared by Charles Hamilton Smith, who asserted in the essay, *The Natural History of the Human Species* (1848), that three creations of Man existed, and he considered the origin of one race from another as impossible.

The Swiss naturalist Jeane Louis Agassis also wrote: "Species are not firmly established in their boundaries, and cannot be defined by an exclusive ability of individuals toward fertility only between themselves. So too, do the human races differ, as much as some biological families, genus, and species. They arose independently of each other, in eight different points of the globe."

The outstanding German philosopher and anthropologist, Karl Vogt, emphasized in the book, *Man and his Place in Nature*, that: "If macaques in Senegal, baboons in Gambia, and gibbons on the island of Borneo can develop up to man-like forms, then why deny the same to apes in similar development in America? If man-like apes can develop in different areas of the Earth, and moreover from different groups, then again, why is it reasonable that only one of these different groups further develops into a type of man, and not others? In a word, why not different species of Americans from American apes, and Negroes from African [apes], and finally, Negritos from Asian [apes]?"

Felipe Teofrasto von Hohenheim Paracelso

Samuel George Morton

In America, Samuel George Morton (1799-1851) was a distinguished anthropologist-polygenist, who headed an entire scientific field, and substantiated the legal aspects of slave ownership. His followers, Josiah Clark Nott (1804-1873) and George Robbin Gliddon acted more radically. Their joint essays, *The Types of Man* (1854) and *Indigenous Races of the Earth* (1857), were in essence, an encyclopedia of polygenism, which contained a huge amount of anthropological and ethnographic information, and also knowledge from theological tracts of all the main religions. This imposing body of texts, supported by dozens of historical illustrations from the cultural lives of various peoples of the Earth for added persuasiveness, testified about only one thing—every kind of lengthy discussion about the genetic unity of humanity is an anti-scientific provocation of later origin.

The given point of view cannot be considered as outdated, inasmuch as the essays of Agassisi, Morton, Gliddon, and Nott were recently re-published in an eight-tome academic series, *American Theories of Polygenism*, which came out at the University of Memphis in Tennessee, through the efforts of the prominent academician, Robert Bernasconi. All the tomes are provided with professional scientific commentaries, clearly pointing to a rise in interest towards the given theme—among specialists, first of all.

However, the orginal flowering of polygenism occurred, as we are convinced, at the start of the 19th Century. After some weakening of interest in the theme, as early as the time of Darwin to the second half of the 19th Century, a new boom in the development of this theory began. The first to come forward in support of polygenism were prominent linguists of their time: Max Mueller (1823-1900) and Ernst Renan (1823-1892). The latter asserted: "If the children of Semites and the children of Indo-Europeans were placed separately, and put under the supervision of deaf-mutes, then the former would inevitably start to speak in one of the Semitic languages, and the latter ones in one of the Aryan languages; from this it follows that the type of language does not depend on the will of the individual: it is unavoidably a product of the organization of the brain. These findings are a significant argument in favor of the theory of the origin of the human genus from several pairs. At that time, when Man became Man, by virtue of acquired speech, he was already dispersed across the globe in the form of different groups and races. And in the meantime, it is known that an immense number of such elementary languages—and not including dead languages—have not left a trace of themselves."

The remarkable Polish anthropologist, Ludwik Kryzwicki, later extrapolated the conclusions of comparative linguistics to the data of racial anthropology, and the analytical result obtained

274

allowed him to interpret many facts of culturology anew. In his monograph, *Anthropology*,[331] he wrote: "As if Mankind did not exist for a long time, it was always broken up into a certain number of independent groups, developing independently in cultural and social terms. In support of this, it is sufficient to refer to the fact that human speech arose in several points. In equal measure, other great cultural discoveries, for example, the ability to use fire and the bow, were made in many places around the world, completely independent of each other." For our part, we would like to emphasize, that modern civilization became familiar with such tools as the boomerang, from local natives, after the discovery of Australia. And lo, its functional name, and who invented it, the members of the Australoid race cannot explain or recall. If indeed all races originated from a single pocket of race genesis, then the boomerang would be known to all, and members of all races would at least remember for sure, why it was needed and who was the inventor of the aerodynamic masterpiece.

Precisely so also, when the Spanish of Columbus' expedition landed on the coast of America, the local Indians were shocked, most of all, by the sight and functional effectiveness of firearms, tame horses, and the wheel. They had never heard or seen anything like them.

Kryzwicki developed his logical discussions in the following way: "The existence of racial differences may be so far removed into the past, as far as the collection of paleontological proof to this day will allow. Differences in the methods of forming words attest to the variety of human speech. But the most persuasive proof of the fact that human speech arose in several centers, independent of one another, is found in the structure of languages. And thus, we should presume the existence of racial differences in the distant, prehistoric age. Their existence is even a legacy of half of our human ancestors. Skulls of a completely Cro-Magnon type are encountered today among the Basques, Corsicans, and Berbers."

At the end of the 19th Century (and at the same time) French anthropologists Paul Broca and Armand de Quatrefages de Breau created a long list of races, "which are recognized as pure." A student of Broca, the remarkable anthropologist Paul Topinard, was the first to bring the notion of a **human type** into use. Under that term, he meant "an average norm of traits, which are possessed by a race that is assumed to be pure."

In a three-tome collection of reports published in Paris in 1877, Paul Broca dedicated many articles to harsh and uncompromising criticism of monogenism. He indicated that "modern races come from a direct line, or through the cross-breeding of several species." Besides that he was probably one of the first to expose the obvious trick of the monogenists, asserting that if several races mixed, then consequently this rule of hybridization is justifiable for all races in general. Broca thought that this was not so, for northern Europeans in general don't mix with the native inhabitants of Australia, or the many tribesmen of Africa. The main conclusion in Broca's collection of reports is heard eloquently and persuasively for all, who still daydream with phantoms of political monogenism: "Mankind is distinctly one genus; but if it were one species, it would be the singular exception in all Creation. The human races differ between each other more than do some species of animals, which are divided into several genuses by all naturalists. Being carried to another climate and different living conditions, these races resist any changes whatsoever."

The above-enumerated factors strengthened the theoretical base of polygenism. Besides that the **typological** field of anthropology definitively took shape; it derived the racial differentiation of Mankind on the basis of stable race types. Ernst Gekkel (1834-1919), a student of Charles Darwin and one of the founders of the philosophical-political version of his teachings, (which received the name of **Social Darwinism**), asserted that "not one of the known living apes today, and consequently, none of the indicated man-like apes, could be a distant ancestor of the human genus."

Paul Topinard's monography, *Anthropology*, is an authentic masterpiece that discredits all the unscientific speculation of the monogenists—those adherents of the idea of the origin of all human races from one ancestral pair.

As is known, the basic proof of the monogenists in favor of the specific unity of Mankind, comes down to the possibility of free cross-breeding between members of different races. Topinard counters this with the following: "Let us assume that a rabbit and a hare, a dog and a

---

[331] 1901

275

wolf, and a camel and a dromedary belong to the same species. But a goat and a sheep are much further [apart], and they even belong to different species; meanwhile, it is proven that their cross-breeding successfully [yields offspring]. The Pyrennes goat and the domestic nanny goat also belong to different species, and meanwhile, in the Pyrennes, the yield mongrels. It seems that even the copulation of cows with deer has produced mongrels. Thus, fertility solely within the limits of a species, and their sterility, does not constitute sufficient criterion for a species. Cross-breeding between species is ordinary and fruitful enough, producing sterile mongrels in some cases, for example, horses and mules; and in other cases [it produces] fertile [hybrids], like the hybrids of the rabbit and the hare, the wolf and the dog, the jackal and the fox, two species of camel, horses, zebras, bison, and the European bull, and so on. Undoubtedly now, the boundaries of species do not constitute an absolute obstacle to fertile cross-breeding, and consequently, these same boundaries do not represent anything firm; this gives us full freedom in the discussion about cross-breedings within the human genus. The distance between Europeans and Bushmen is as great, for example, as the distance between the different species of man-like apes, or between the wolf and the dog, or the goat and the sheep."

Finally, Topinard saw the fact that members of different races are adapted to different climates as a basic confirmation in favor of polygenism: "People endure the influence of the environment before our eyes, but they do not pass on the traits acquired this way to their offspring." As far as so-called "free cross-breeding" is concerned (something liberal evolutionists love to discuss), Topinard, being an erudite individual, cited a great number of curious historical facts in his book, which allowed him to seriously have doubts about this "freedom" of cross breeding: "The Australians are cited as one of the proofs in favor of the existence of fertile, inter-species sexual relations within the human genus; however, up until recent times, there is no knowledge of mestizos coming from Aborigines and Europeans. The Egyptian Mamelukes, in the course of 560 years, did not have children from wives who were taken from Georgia, and they could never give durable offspring in the Nile Valley."

Besides this, it becomes clear, that in the former Portugeuse colony of Macao, no Chinese-Portugeuse mestizos took root, and in the former Dutch colony on the island of Java, the Malay-Dutch mestizos produced only sterile daughters by the third generation. All the mestizos produced in Africa from marriages between the Dutch and Hottentots either died out, or returned to the original African type. In the southern states of the United States, and in the countries of Latin America, it has long been observed, that members of the Nordic type of the European race yield mainly sterile offspring in cross-breeding with Negroes—and already in the first and second generations. At the same time, dark-haired and dark-pigmented Europoids yield comparatively durable and vital offspring with Negroes. Among the Arabs during the spread of Islam in Africa, there was an entire terminological hierarchy, designating six steady transitional racial types—from pure Arab to pure-blooded Negro. And across the stretch of centuries it has been noticed, that they all have different fertileness; as a result, their women have different value in slave markets. From this, Topinard made the logical conclusion: "Two good races will produce a better average; two bad races will yield still worse."

Finally, the French anthropologist justly focused attention on racial-anthropological anomalies, which do not allow one to speak of the specific unity of Mankind under any circumstances: namely, **steatopygia** in Bushmen, and the so-called **Hottentot skirt (Laborium Minorum).** By steatopygia, the excessive development of fat deposits under the buttock muscles in the women of that tribe is meant; this gives them a completely absurd look, increasing the volume of the hips several times. By the Hottenton skirt, the unnaturally long genital lips in the women of the given tribe, is meant. They often reach a length of 15-18 centimeters, and hang to the knees; in connection with this, since ancient times a custom developed of cutting off these sexual lips before entering into marriage. When Catholocism was introduced to Abyssinia in the 14th Century, it was prohibited to carry out gyneacological operations; immediately, a revolt flared up, for young girls could not find eligible tribesmen for themselves, since according to their native understanding, the given anatomical phenomenon is repulsive. By special decree of the Pope in Rome, the aborigines were allowed to return to their original custom, in order to not set up obstacles to the spread of Christianity. As a conscientious scientist, Topinard analyzed the given fact from an evolutionary viewpoint: "We note here, that the 'skirt' does not speak in favor of an immediate kinship of the Hottentots, Bushmen, and apes, since on female gorillas these lips are completely invisible."

And so, correlating the given kind of racial phenomenon with classical Darwinist teachings leads unavoidably to recognition of the presence of a separate, independent branch in the development of the given tribes, who derive their origin not from apes, but from some other sort of unknown animals.

Jean-Joseph Virey wrote in this regard: "It is supposed, that among the Bushmen there is something not unlike a skin apron, hanging from the front, which covers the sexual organs. In reality, this is nothing more than an extension of the small labia up to 16 centimeters. They protrude from each side behind the large labia, of which there are almost none, and unite at the top, forming a hood over the clitoris, and covering the entrance to the vagina. They can be lifted over the front, like two ears. By this, one can explain the natural infertility of the Negrito race in comparison with the White. Therefore, it is more correct to speak of a Negrito species, and not a race, since this same trait of the structure of the sexual organs is observed among the Coptic and Ethiopian women."

It remains completely incomprehensible, why many liberal-democratic media, which are engaged in the distribution of erotic and pornographic printed material, and are inclined to propaganda of any kind of sexual perversion, avoid giving attention to the given anthropological phenomenon, which is distinctly able to increase their profits. For the sake of fairness, it is necessary to remark that a similar kind of racially specific feature is observed not only in members of the 'fairer' sex, but also in the 'stronger' half—during full-scale examinations, Virey established that during running, the testicals of Bushmen retract into the abdominal cavity. The given trait is directly tied with hereditary cretinism, and also testifies that in this racial group, all connective tissues in the organism are structured differently, principally.

The Swiss traveller, Viktor Ellenburg, also clearly indicated in his book, *The Tragic End of the Bushmen*, that: "Among the majority of Bushmen, even among the young, the whole body and particularly the face, is covered with folds and wrinkles, as a result of which it creates the impression that the skin of the Bushmen is too much for their bodies, particularly when they are hungry. The Bushman Race strongly differs from all present-day races inhabiting the African continent, and right up to the 19th Century, they lived at a Stone Age level in their development. Among the Bushmen, even the structure of the bones is completely different from the majority of the members of the Negro races. The bones of their extremities have an almost cylindrical form; another characteristic trait of the Bushmen is that their hands and feet are small—and according to the opinion of one traveller—of almost Lilliputian size, hardly more than that of children. The Bushman penis is among the number of distinctive physical traits; among them it is found to be in a constant state of semi-erection. This inherent trait of the Bushman race has been recorded on numerous Bushmen cliff paintings and pictographs."

The prominent German anthropologist and professor of the University of Munich, Ferdinand Bircher (1868-1944), summed up the views of his colleagues in his encyclopedic work, *The Races and Nationalities of Mankind*. He asserted: "At the present time, Mankind is represented in its different races, nationalities, and tribes in a mosaic picture, such that a whole number of researchers adhere to the opinion, that the prominent differences are the result of varied origin; that is, that Mankind arose from several roots and is not one species with several races, but one genus, composed of several different species."

The anthropologist and archaeologist, Herman Klaatsch, analogously indicated in the work, *The Condition of Man in Nature*,[332] that: "Earlier, they gave greater principle significance to the question of whether Man originated from one form or several; expressed in scientific language, this was an argument about a monophyletic or polyphyletic method of origin of the human genus. At the present time, this struggle of opinions appears unnecessary from the theoretical side, since one cannot seriously speak about a strict, monophyletic origination. The origin of each group of

---

[332] From the anthology of articles, *The Evolution of Man*, Moscow, 1925.

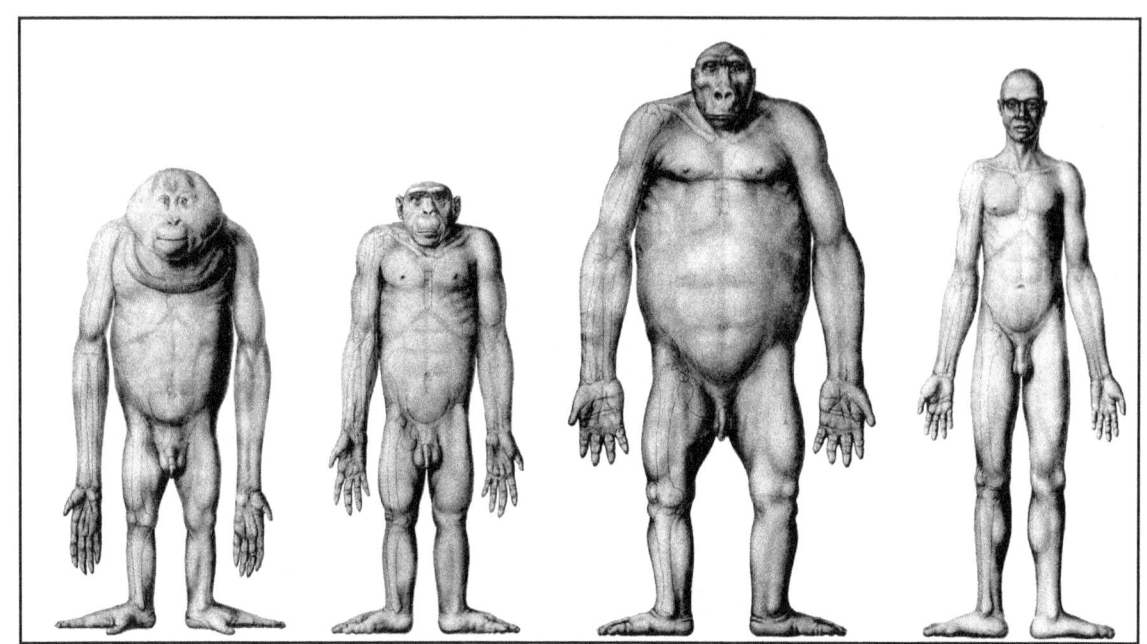

Above: Evolutionary Morphology, according to Herman Klaatsch: orangutang, ape, gorilla, Man.

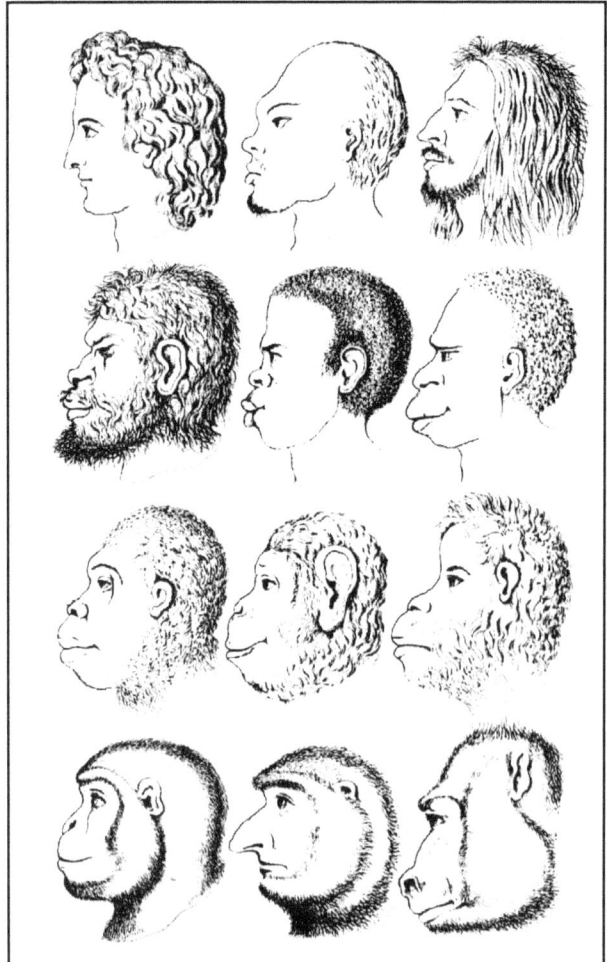

Left: Evolutionary Morphology, according to Ernst Gekkel

animals is polyphyletic. The ancestors of the man-like apes differed from each other, when humanization began. Proceeding from there, one can explain the racial differences of Mankind to a significant degree." In the study, *Race and the Ancestral Homeland of the Indo-Germans*,[333] Otto Reche coorelated the conclusions of paleontologists with the data of blood group distribution in different racial groups, and came to this conclusion: "Man as a species is represented in ancient Europe by local dolichocephalic races; this species separated into the Nordic Pfalzish, and Western races. One may consider the inhabitants of Central Asia, the Australian aborigines, the Bushmen, the Pygmies of Central Africa, Negrito groups, and so on, as other particular species, which also separated into races."

---

[333] 1936

278

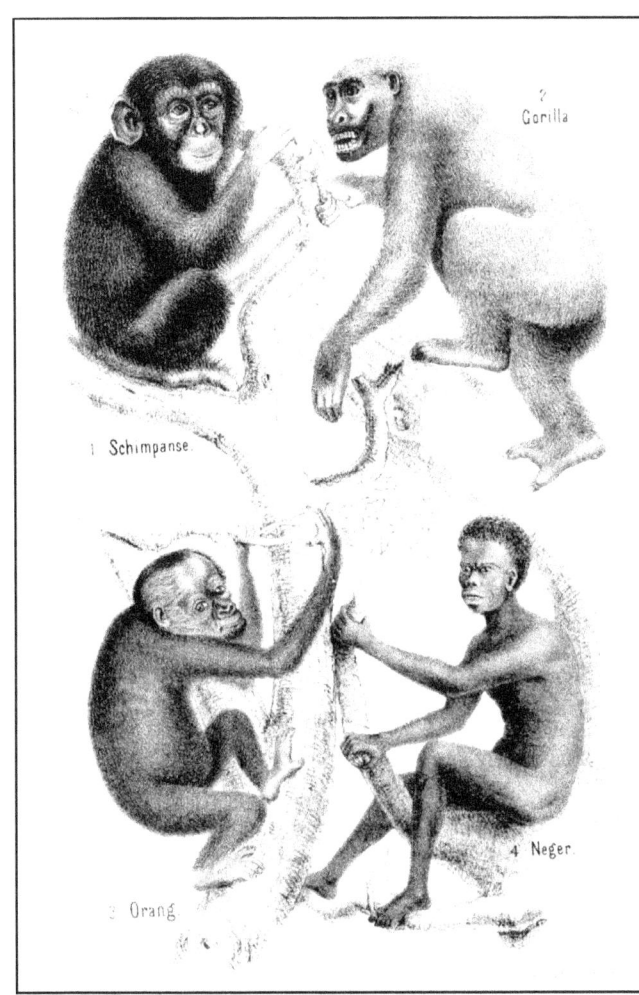

Left: Evolutionary Morphology, according to Ernst Gekkel

Still another recognized authority in the area of anthropology, Franz Weidenreich (1873-1948), formulated his concept of **polycentrism** in a report made in Stockholm in 1938, at the Second International Congress of Anthropological and Ethnographic Sciences; according to his report, Mankind arose in four independent centers—according to the number of modern races—in Southeast Asia, East Asia, Africa, and Europe. The first center served as a zone for the formation of the Australoids; the second, for the Mongoloids; the third for the Negroids; and finally, the fourth and last—[as a zone for the formation] for the Europoids. The Javanese Pithecanthropoids were the initial forms for the Australoids (brain of 930 grams); the Sinanthropoids for the Mongoloids; and the African Neanderthals [were the initial form] for the Negroids. In each separate case, the genetic tie of this or that ancient form with the corresponding modern races, was argued by Weidenreich, with the help of morphological comparisons: the sagittal ridge of the cranium in Australoids was close with [that of] the Javanese Pithecanthropoids (on the skulls of modern Australian Aborigines, this trait is actually encountered more often, than in the members of the remaining races); the Mongoloids resemble Sinanthropus by their flattened faces and spade-shaped incisors; the Europoids, like the European Neanderthals, are characterized by an orthogonal, or lightly forward-projecting profile of the facial skeleton, and strongly projecting nasal bones.

Left: Herman Klaatsch

American anthropologist Carlton Stevens Coon supported Weidenreich, delineating not four, but five pockets of racial genesis, by subdividing the African center into two independent centers. Besides that he pushed back the time of the appearance of racial differences, noting that they traced back even to the stage of development of the Pithecanthropoids. Soviet scientists G.F. Debets and N.N. Cheboksarov developed the concept of polycentrism, depending precisely on the facts of morphology. Within the framework of polycentrism, a theory of multiple transition from Paleoanthropoids to *Homo Sapiens* of the modern type was formulated. This already was an essential radicalization of the given system of views, again enabling one to speak not of polycentric evolution, but of polygenism, in the widest sense of the word. In the brochure, *Stages and Intrastadial Differentiation in the Evolution of Man*,[334] V.P.

---

[334] *Stadii i vnutristadial'naya differentsiatsiya v evolyutsii cheloveka.* Moscow, 1967.

Yamikov also emphasized that: "the acquisition of orthogonalness is a manifestation of a particular path of biological progress." This again speaks of the independent development of the White Race, and consequently, again makes the concept of the origin of Man from a single pocket of race-genesis untenable: "The independent, parallel development of ancient peoples into definite races of *homo sapiens* occurred with different speeds, and was conditioned by inner tendencies toward evolutionary transformation."

Although he came forward as an author of the theory of a single pocket of racial genesis, Ya.Ya. Roginskiy was compelled to recognize, that racial traits in the members of different races are not identically expressed in childhood and in adult years. Negroids are observed to have the greatest difference with Europoids at the stage of adulthood; in childhood years, the differences are mitigated. Children of Mongoloids, on the other hand, strongly differ from the children of Europoids and Negroids.

In the joint article, *Problems of the Settlement of Europe, According to Anthropological Data*,[335] from the anthology, *The Origin of Man and the Ancient Settlement of Mankind*,[336] G.F. Debets, T.A. Trofimova, and N.N. Cheboksarov considered it necessary to emphasize: "The mixing and migrations of the basic races of modern Mankind is relatively easily traced in anthropological material, since the differences between these races is more essential, than the differences between the somatic groups of a single race. Ancient forms of the Europoid, Mongoloid, and Negroid races existed as far back as the Paleolithic Age; any chance of Mongoloid types forming from Negroids or Europoids, and vice versa, is therefore already unlikely and practically excluded."

Left: Sir Arthur Keith

One of the outstanding English anthropologists, Sir Arthur Keith, substantiated the theory of the great ancientness of racial differences in 1950. Sir Keith assumed that as early as the Pleistocene, such differences existed between fossil hominids of various territories—which is characteristic for the great races of a later age. Negroid, Mongoloid, Europoid, and other types of fossil hominids existed; their evolution proceeded in parallel and independently, and led to the manifestation of the modern great races, on those same territories.

And thus, we see that the concept of the origin of the modern races from a single group of animal ancestors, which philosophers, culturologists, and rights advocates so love to propagandize, comes apart at the seams and does not withstand any criticism. **The biological unity of Mankind is an illusion, meant for the convenience of the undiscriminating crowd.** The moralistic assertions of the profane can often be heard: "They are people, too"—and from the natural science point of view: "This is wood also" and "These are rocks, too."

V.P. Alekseyev very astutely formulated the essence of the problem in his monography, *The Geographic Pockets of Formation of the Human Races*.[337] He wrote: "The hierarchy of the races is one of the oldest problems of anthropology. The hierarchy of the races is understood in principle as the hierarchy of the pockets of race genesis."

Prominent German anthropologist Hans Weinert, desiring to correct the conceptual confusion arising in any critical view of anthropogenesis within the limits of all of "humanity", was compelled to state: "The term 'ape' is used in such a general sense and so uncritically, that different objections against an origin from the ape, as well as several deviations from this theory, are often explained by the misunderstanding and incorrect summarization of it. It is difficult to say, how much our "dignity" and inner value factors in this, that we trace our genus not from the ape, but from a lizard or shark. We are in the least bit obligated to believe that our ancient ancestors evolved only in the capacity of a unified, limited group, from anthropomorphic [forms]."

---

[335] *Problemi zaseleniya Evropi po antropologicheskim dannim.*

[336] *Proiskhozhdenie cheloveka i drevnee rasselenie chelovechestva.* Moscow, 1951.

[337] *Geograficheskie ochagi formirovaniya chelovecheskikh ras.* Moscow, 1985.

Returning to Topinard's book, which is listed among the great classics in every textbook on anthropology, one must recognize that despite all the arguments against it, his chief conclusion produces the impression of a destructive bomb, to such a degree [that] the opinion of the student parts ways with the commonly used, everyday stamps about "some single Mankind." Obviously, compilers of textbooks read very little themselves, and are also not familiar with Topinard's work. An old truth is that a classic is something everybody loves, but no one has read. Thus, we render due [credit] to Paul Topinard's authority, and cite the main conclusion from his remarkable book, *Anthropology*:[338]

"The differences are fewer between the species of one genus of the man-like apes, than between the main human races. For example, between the orangutan and gorilla, that is, between two separate genuses, the distance is less than the distance between an Australian aborigine and a Laplander. The blonde-haired Swede, with a rosy-colored face, sky-blue eyes, graceful limbs, and a large cranial capacity, is located further away from the black-as-soot Negro with yellow, schlerotic [eyes], short and wooly hair, protruding jaws, and turned-up lips; or from the Papuan, with wool-like, but long and tufted hair, sometimes disheveled and presenting a spherical mass, which is comparatively as large as the mane on a bison; or from the Bushmen females, with yellow-colored faces, the lips of an orangutan, with vaginal lips reaching to the knees, and the large, deformed rear. Their differences do correspond, of course, to the notion of simple variety, and surpass even the differences that separate many species. Thus, it is necessary to recognize that the distance between the main human [racial] types is greater, really between the varieties in natural history, and so greater than between separate species. Moreover, sometimes this distance is evidently also as great as the distance separating genuses. **The human genus, comprising the first family in the order of primates, is divided into species, or basic human races.**"

Left: Gustav Schwalbe

Of course then, our opponents from the number of prescribed monogenists can object, that the given point of view, although it belongs to the classics, is nevertheless outdated. Then we will bring in a quotation with a similar thought, by the distinguished Soviet scientist, V.V. Bunak. His posthumously published article, *Upper Paleolithic Skull Sungir #1, and its Place among other Upper Paleolithic Skulls*,[339] from the anthology, *Sungir': An Anthropological Study*,[340] the Russian luminary of anthropology emphasized in the spirit of the classical philosophy of polygenism: "In their typical form, natives of the various continents differ from each other by skin coloration, hair form, secondary sexual traits, brain case, the form of the face, nose, lips, and other characteristics, more than some species—the maral and the deer, and many species of rodent. Such forms as the Bushmen also carry differences from the predominant modern type, in the structure of the lumbar part of the skeleton, the position of the pelvis, and the form of the sexual organs." In the twilight of his life and feeling the coming of the end, Bunak decided not to tie himself into the knots of Marxist-Leninist science, and expressed his convictions, which were built on many years of research.

---

[338] Russian anthropologist Ilya Ilyich Mechnikov wrote the foreword to Topinard's *Anthropology*.

[339] *Verkhnepaleoliticheskiy cherep sungir'1 i ego mesto v ryadu drugikh verkhnepaleoliticheskihk cherepov.*

[340] *Sungir': antropologicheskoe issledovanie.* Moscow, 1984.

Finally, the prominent modern Russian anthropologist, A.A. Zubov, made a completely open declaration in the article, *Several Findings of Odontology and the Problem of the Evolution of Man and his Races*,[341] from the anthology, *Problems of the Evolution of Man and his Races*.[342] He wrote: "Dental morphology does not contradict the supposition of the possibility of the independent, parallel development of the races, from various regional groups of paleo-anthropoids."

What is this, if not a renewal of the idea of polygenism, through the efforts of Russian science? In the very latest monograph, *The Paleo-anthropological Geneaology of Man*,[343] Zubov develops and sharpens his concept, which he named **"a model of genetic flow and mixing."** The author emphasizes, that except for hybridization on the paths of settlement, the racial groups had a place of partial or complete isolation, the disintergration of these isolates, and the extinction of separate groups. Here it is worth adding also, that constantly changing demographic situations were caused by war, famine, epidemics, and "demographic explosions", too. In a word, the real picture of the interconnectedness of the links in evolution, and the differentiation of racial groups, appears far more complex than it can be portrayed in any theoretical scheme. The constant struggle of forces in the "gene game" of the evolutionary process begat that variety of biotype forms, among them the openly pathological, and the extreme—both from the point of view of morphological structure and survival strategies. Judas, Mowgli, and Chikatilo are no exceptions, but genetic regularities of their kind, a product of the stratagems of the DNA spiral.

"The reticular, net-like evolution of Upper-Paleolithic Mankind begat a wide spectrum of 'intermediate' anthropological types"—asserts A.A. Zubov.

8   The Problem of the "Boundary" between Man and Animal

The chain of archaeological discoveries at the end of the 19th and early 20th centuries allowed such a prominent German anthropologist-polygenist as Herman Klaatsch to create a principally new model of the rise and evolution of the human races. In his book, *The Condition of Man in Nature*, he indicated: "Just like the ancestors of the man-like apes had already differed from each other when **simianization** began, so too, human groups were not completely identical, since differences were visible among them also, even before the start of **hominization.** Proceeding from there, one can explain the racial differences of Mankind to a significant degree. Thus, the question is one of the divergence of a common group of ancestors into branches, from which each arose as the human races, just like the man-like apes."

At the Congress of Anthropologists in the city of Lindau in 1899, Klaatsch was the first to lay down his views, which follow: Man is an independent branch of the primates, and the man-like apes are a parallel, developing branch of primates—but they are not the ancestors of Man. Not one of the living primate types can, in his opinion, be considered as a close relative of Man. In many respects, Man is closer to the lower forms of ape, than to the man-like apes; many of his traits go back to even the lemurs, and a hypothetical primitive form of primates. Man is a direct descendent of lower primates, while those very primates themselves are primitive members of a genus of mammals.

In 1910, at the Anthropological Congress in Köln, he developed his views on the theory of race genesis. From his essay, under the title, *Polygenesis of the Races and the Common Origin of Man and the Man-like Apes*, it comes that at a very early stage, the primitive groups of the higher primates diverged into western and eastern branches, and each of them in turn, diverged into its races and its species of man-like apes. He openly asserted: "Between two apes and two breeds of people, there is less kinship, than between each of the apes and Man, which developed from one stem. The Neanderthal and the gorilla are members of the western branch, and Aurignacian Man and the orangutan are members of the eastern branch." In the American publication of his book, *The Evolution and Progress of Mankind* (1923), he concluded: "According to all laws of probability, it is very strange that Man occupies an exclusive position in comparison with other

[341] *Nekotorie dannie odontologii k probleme evolyutsii cheloveka i ego ras.*

[342] *Problemi evolyutsii cheloveka i ego ras.* Moscow, 1968.

[343] *Paleoantropologicheskaya rodoslovnaya cheloveka.* Moscow, 2004.

animals. It is erroneous to consider Man, in all respects, to be at a higher stage of evolution. Many of his traits are more primitive, than [those] in apes."

The views of Klaatsch were supported and developed by the distinguished Italian anthropologist-polygenist, Giuseppe Sergi (1841-1936). In the book, *Species and Variations of the Human Genus*, published in 1900, he wrote: "Man: is he one species or several? My observations in Europe and in Africa allow [me] to make a conclusion: there are two species of Man—the Euro-african and the Euro-asian; each of them consists of several races."

In 1892, Sergi made a report at the International Congress of Anthropologists in Moscow. And in 1913, Dmitriy Nikolayevich Anuchin, the classic of Russian anthropologiy, supported his ideas in the books, *The Origin of Man*[344] and *Organic Evolution and the Origin of Man.*[345] In them he worked out in more detail, the concept of a polygenetic origin of "Mankind." According to his version, three independent groups arose from the predecessors of the primates: the **tsercopitheci**, with six independent branches, the **simiadi**, giving four branches, and the

hominidi, also with four separate branches. In turn, these branches are considered by him as four separate genuses: the **eoanthropus, paleonthropus** (Heidelburg Man), **notanthropus** (the Negro branch) and **geoanthropus** (the Mongol branch). All these genuses are not offshoots of a common, earlier, less specialized human type; they are not tied with either the major apes, or the other primates known to us. It is assumed that they already stood apart in time, when groups of hominids separated from their supposed predecessors.

Left: Giuseppe Sergi

Two prominent Russian archaeologists, A.S. Amal'rik and A.L. Mongayt, indicated in a joint work, *In Search of Lost Civilizations,*[346] that according to the results of excavations, one can establish the presence of racial differences, as early as the Late Paleolithic Age. Besides that "together with this, it is noted [that the] exceptional differences in the anthropological composition of these finds—the assumed racial differences—surpass the differences between modern racial types." Thus, it happens that if on the horizon of race genesis, the differences between the basic racial groups was greater than today, then consequently, according to all laws of the logic of evolutionary development, they can only have arisen from different, independent centers; that once more refutes the para-scientific thinking of the monogenists. The prominent Austrian raciologist-polygenist, Ludwig Gumplovich (1838-1909), explained several obvious morphological similarities of the basic human races with a simple conclusion: "Similarities do not always presume a hereditary connection. The same regarding Mankind; the law of evolution and development can be one, when a geneaological tree is not one."

In the 1920s and 1930s, with the stormy beginning of the flowering of biology and the first successes in the area of genetics, the old ideas of the anthropologist-polygenecists received confirmation on a qualitatively new level, and particularly after the rise of the so-called **synthetic theory of evolution**, in the middle of the 20th Century. Even if it is assumed that humanity, according to the opinion of the monogenecists, arose from one pair of ancestors, then the

---

[344] *Proiskhozhdenie cheloveka.* 1913.

[345] *Organicheskaya evolyutsiya i proiskhozhdenie cheloveka.* 1914.

[346] *V poiskakh ischeznuvshikh tsivilizatsiy.* Moscow, 1959.

inequality of the tempos of evolution, including that under the influence of the environment, should have unavoidably separated the races in the process of their isolation, according to hierarchical traits, thus creating "higher" and "lower" [races]. Alongside the differences between the races on a morphological level, they have now added still more indisputable evidence about the bio-chemical differences, and more recently, the genetic differences. There cannot be descendents of a common ancestor, settling in different places all over the earth, everywhere separating from the ancestral group with indentical speed, during the process of development. This is self-evident. Population genetics confirms that the differences in the human races and the populations they comprise [have] an unequal percentage of distribution of atavistic traits, acquired through heredity from animal ancestors. Expressed with the modern language of ecology, **from an evolutionary values viewpoint, a pre-human, a human, and a super-human can exist in one and the same time, within the limits of one ecological niche.** All species of living beings took a different path of development, and in this are their basic differences.

Prominent Soviet anthropologist B.S. Zhukov noted in his monography, *The Origin of Man:*[347]

"Among some lower members of modern Man, traits of some similarity with the Neanderthals are noticed, like the eyebrow ridges of the Australian aborigines, for example. Among them the "wisdom teeth" achieve greater development than the remaining molar teeth; by the structure of their teeth, this brings the Aborigines closer to the higher apes. According to their height, the Neanderthal race is most of all closer to short modern peoples, like the Laplanders, who live in northwestern Russia, and in northern Finland."

In his book, *Man of the Ancient Stone Age,*[348] American anthropologist Henry Fairfield Osborne (1857-1935) expressed it this way: "We cannot assert that in a group belonging to the species *Homo Sapiens*, there was never an admixture of Neanderthal blood. It is interesting to note, that in the moment of first contact with Europeans, the Tasmanians were in the stage of a flint culture, highly similar to that which was widespread among the Neanderthals of the Mousterian Age. The last members of this primitive race died out on the Island of Tasmania in 1877." Another prominent American anthropologist, Alyosh Grdlichka, also fairly asserted: "Traces of Neanderthal blood and the physionomical traits of that race are encountered even among modern Europeans." Still another American anthropologist, Lauren Eisley, remarked: "Homo sapiens walked the Earth together with the Neanderthals."

A genuine revolution in evolution theory was carried out by the Polish anthropologist, Casimir Stoligwo, who discovered a Neanderthal skeleton in a Scythian burial mound near the settlement of Novoselki, at the beginning of the 20th Century. By virtue of this, official science, which thought that all Neanderthals had died out by this historical period, was compelled to introduce a new term: "post-Neanderthaloids." In one of his articles in 1937, he came to the definitive conclusion, that besides pre-Neanderthals and the classical members of the Neanderthal race, "all remaining descending Neanderthaloid forms, known up to the present time, date to periods far later than the Mousterian; [they date] to the Upper Pleistocene, and also to later times—to the prehistoric, the proto-historic, and also to modern times."

Soviet biologist Lev Semenovich Berg (1876-1950) also held to the given concept. In his work, *Homogenesis, or Evolution on the Basis of Regularity,*[349] from the anthology, *Works on the Theory of Evolution,*[350] he wrote: "Neanderthal Man, HOMO NEANDERTHALENSIS, is observed [to have] a number of traits, which in their development, went further than in modern Man, or HOMO SAPIENS. We observe that in accordance with new views, there is no proof that H. NEANDERTHALENSIS was the ancestor of modern Man. The molar teeth in Neanderthal Man went further in their development, than in the chimpanzee or modern Man—who retained in his molars the traits of ancient apes. The brain of the Neanderthal, by its volume, was not second to the brain of the modern European, and even exceeded it; judging by the skull from LA CHAPELLE, the volume of which was 1,625 cubic centimeters. Primitive man received a brain from Nature that was far more complete than was necessary for the mere maintenance of his existence."

---

[347] *Proiskhozhdenie cheloveka.* Moscow, 1928.

[348] *Chelovek Drevnego kamennogo veka.* Leningrad, 1924

[349] *Homogenez, ili evolyutsiya na osnove zakonomernostey.*

[350] *Trudi po teorii evolyutsii. 1922-1930.* Leningrad, 1977.

However, Charles Darwin wrote: "We can be certain, that there are numerous, long-hidden traits in any living being, that are ready to develop in the right conditions." Therefore, the tales of various peoples that tell about vampires and werewolves, are not—from the evolutionary viewpoint of atavism—the inventions of illiterate people. Cattlebreeders and zoologists know well that in the process of the fertilization of several organisms, there is a return to lost traits. William Ripley and Henry F. Osborne were the first to analyze the language of the Basques—the most primitive and undeveloped [language] in Europe. They came to the conclusion that the Basques' ancestors most likely borrowed the language directly from the Cro Magnons. Therefore, in our view, the source of the modern Basque separatist movement should be sought in the characteristics of their anthropology. The same correlation is also obvious with several of the Caucasus peoples of Russia: for example, the Chechens, who only acquired literacy in the 20th Century; their separatism does not yield to any intelligible explanations. But their eternal inclination toward the slave trade, and thirst for blood—which they demonstrate in the torture of prisoners—also enables us to make a conclusion about the very ancient, archaic nature of the given atavistic manifestations, which no doubt have a genetic basis. The system of Chechen *teypov*, of feuds between themselves, also testifies in favor of primeval, communal tribal principles in the organization of these peoples, which is hereditarily strengthened in the traits of its racial structure.

But, as we showed, the picture is essentially complicated by the fact that besides the races having different evolutionary values, each also consists of qualitatively unequal populations, which in turn splinter according to the same principle, right down to separate individuals. Thus, we again speak of the fact that at one and the same time, the pre-human, human, and super-human can exist within the framework of one or another ecological niche. The author formed the given opinion through analysis of a great volume of works done by modern specialists in the area of anthropology.

In her article, *The Problem of the Discovery of Sapiens and Neanderthal Lines in the Early Stages of Evolution,*[351] Russian scientist Yu.D. Benevolenskaya writes: "The hypothesis of the evolutionary transformation of the Neanderthals into neo-anthropes, all the more yields its place to the idea of the crowding out of the former by the modern-type Man; this was also accompanied by cross-breeding between the two."

A.A. Zubov also indicated in an article:[352] "We can talk about the net-like character of the evolution of the genus *homo* in all the stages of its evolution... It is important to note that a "net" can include different evolutionary "stages", interacting between each other and bringing their genetic contribution to the general, united pool of evolutionary diversity in the genus *homo.*"

In translation from bombastic, academic language to common speech, this means that members of the "higher" human stages entered into sexual connections with members of the "lower" Neanderthal stages; the result of this cross-breeding was that "mongrel-mutants" were brought into the world, and then numerically set apart to the level of entire peoples and races, which gave rise to a general, "evolutionary diversity in the genus *homo.*"

And again there is nothing surprising in this fact. From the descriptions of ancient Greek and Roman historians, numerous orgiastic cults, and in general, a very unrestricted sexual life in ancient Europe was known, but little is known about serious sexual diseases, which have only struck white people in the latest times—from the beginning of the Age of Discovery, when Europeans came "into contact"—in the direct sense of the word—with members of the colored races; for their part, they caught the given venereal diseases from animals. Thus, for example, syphilis was brought to Europe from America; the Europeans got it from the Indians, and they in turn got syphilis from local llamas. There is an opinion, that AIDS came from Africa, where Negroes acquired it from monkeys.

---

[351] *Problema viyavleniya sapientnoy i neandertal'skoy liniy na rannikh stadiyakh evolyutsii.* Kur'er Petrovskiy Kunstkameri. Vip. 8-9, Sankt-Peterburg, 1999.

[352] *Problems of Intra-species Systematics of the Genus Homo, in Connection with Modern Ideas of the Biological Differentiation of Humankind,* (Modern Anthropology and Genetics in the Problem of the Races in Man). *Problemi vnutrividovoy sistematiki roda homo v svyazi s sovremennimi predstavleniyami o biologicheskoy differentsiatsii chelovechestva.* Sovremennaya antropologiya i genetika i problema ras u cheloveka. Moscow, 1995.

Left: A Neanderthal in modern clothing. A reconstruction (according to Carlton Stevens Coon).

American biologist Anthony Barnett also argued in the book, *The Human Genus*, that "people of the modern type appeared approximately at the same time, if not earlier than Neanderthal Man, and developed parallelly. The intermediate types between modern peoples and Neanderthals could be the result of either cross-breeding, or of earlier phases in the differentiation of Neanderthals from the line which led to modern Man. " But after all, it is completely obvious that these "intermediate types" did not disappear, but exist today among us today, comprising their own particular social, ethnic, and racial groups **on the basis of an instinctive lust,** as this has its place among the animals, and we out of ignorance trace them to the "modern Man" type, and this is the source of many of our social ills and political disillusionments. Among the "intermediate types" there is also and intermediate morality—in the direct sense of the word, "pre-human." In evaluating them as similar to ourselves, we fall into scientific error, the result of which we ourselves turn out to be the sacrifices.

In the book, *Man, his Origin and Evolutionary Development*, Swedish anthropologist and anatomist, Wilhelm Leche, wrote: "Just as physical traits manifest in separate individuals, assumed by way of heredity from some very distant ancestor, so too, individuals who commit certain anti-social or immoral deeds against those who are near, or against all society, can be considered by way of heredity from an ancestor, as not possessing or weakly possessing social feelings, in which connection these spiritual defects were not suppressed with education or good breeding. Just as natural selection necessarily causes not absolute, but only relative perfection of an organism, so the notion of morality can reach a higher or lower development. That is why in different times and in different peoples, the idea of morality was—and is—so different. That humanity will one day be freed from all of what we call rudimentary organs is unlikely, because, this disharmony is an indissoluble traveling companion of each evolutionary process."

Jeane-Joseph Virey also turned attention to the distinct differences in the physique of the members of the main human races, in connection with rudimentary organs. "Among Negroes, the gray substance of the brain has a darker color. But the main thing is that in Negroes, the peripheral nervous system is far more developed, than in Europeans—while the central nervous system on the other hand, is less developed. It appears that the brain in Negroes goes partially into the nerves; literally, the animal side developed at the expense of the intellectual side. In some animals, there is a third eyelid. In Man it is rudimentary, but in Europeans it is far less expressed than in Negroes, who in this respect are closer with orangutans. The distance between the European and the Negro is not great, in comparison with this gap, which separates Man and the man-like apes. However, the physical forms of Negroes are to some degree, intermediate, between Europeans and the apes."

Also, Joseph Arthur de Gobineau remarked in one of his letters: "Some modern mixed races arose from beings that were intermediate between Man and ape, as a result of mixing them with people." His devotion to the basic ideas of polygenism are clearly expressed in his main essay, *Experiments about the Inequality of the Human Races*:[353] "It appears completely logical to announce, that the groups of which humanity is composed, also differ from each other, like different species of animals in the world of wild nature. The fact that Adam is the ancestor of our White race is not subject to doubt. But despite everything, there is no evidence that the first

---

[353] 1853.

editors of the Adamic geneaology added other peoples to this group, which did not belong to the White race."

In the work, *Embryonic Traits in the Physique of Man*,[354] from the above-cited anthology, L.C. Berg pointed to qualitative differences in the embryonic phase in the members of the different races, and correlated these differences in people with the analogous phases in apes. His conclusions turned out to be shocking. In whole, the logic of Berg was such: one of the most glaring differences of Man from the man-like apes, is the absence in Man of a thick covering of hair. In Man there is only thick hair cover in certain places of the body; besides that hair is sparsely scattered almost over the entire body. In a child in the last months of uterine life, there is a total, continuous, and satisfactorily thick covering of thin and short hair, which disappears, however, before birth. In the embryonic gorilla and chimpanzee, the body is covered by hair, but it is so short, that the skin appears bare. Notably, in the embryonic chimpanzee the hair is short and curly, and sits in tufts, like on Negroes; in the embryonic gorilla, they are straight and grow evenly, as in Europeans. The hairs on the head of the embryonic gorilla form a part that goes to the brow; in the embryonic chimpanzee there is no part in the hair, and it has a bare lobe. L.C. Berg made this conclusion: "Man does not repeat the complete hair covering of his embryonic life, after his birth; on the other hand, the gorilla and chimpanzee in their embryonic life go through a temporary stage, which in Man remains for life. The disappearance of the hair covering in Man is not the result of the influence of external conditions or an adaptation, but occurs under the influence of some internal factors. The ears of an adult human also partly retain traits more primitive, than in the embryonic gorilla. The Mongoloid race possesses the greatest number of embryonic traits."

One of the first theories of the hybrid nature of Mankind was advanced by the Russian racial theorist, Vladimir Aleksandrovich Moshkov, in his magnificent book, *The New Theory of the Origin of Man and his Degeneration*.[355] According to its basic assertion, humankind is a hybrid species, arising from the miscegenation of the White man with Pithecanthropus. As a result, different concentrations of this mixing produced the modern races, but Europeans are not free from the weight of different animalistic atavisms. Moshkov wrote: "If the lower races, in contrast to Europeans, have more Pithecanthropus blood in their veins, then clearly, we should search for the traits in them, which distinguish the latter [the Pithecanthropoids]." He substantiated his point of view [with the fact that] the mass of the brain of Pithecanthropus was larger, than in the members of modern races, and as a result of miscegenation, shrank to the modern level. The famous French anthropologist, Paul Broca, also wrote: "The average size of the capacity of the skull in civilized peoples should decrease somewhat, as a result of [civilization] preserviing a significant number of persons with weak intellect and bodies, who would perish among savages." This was confirmed by Charles Darwin: "The conviction that there is a relationship between the volume of the brain and the degree of intellectual capability in Man, is based on comparison between the skulls of the savage and civilized races, ancient and modern peoples, as well as on analogies of a number of vertebrates."

---

[354] *Embrional'nie cherti v stroenii cheloveka.*
[355] *Novaya teoriya proiskhozhdeniya cheloveka i ego virozhdeniya.* Warsaw, 1907.

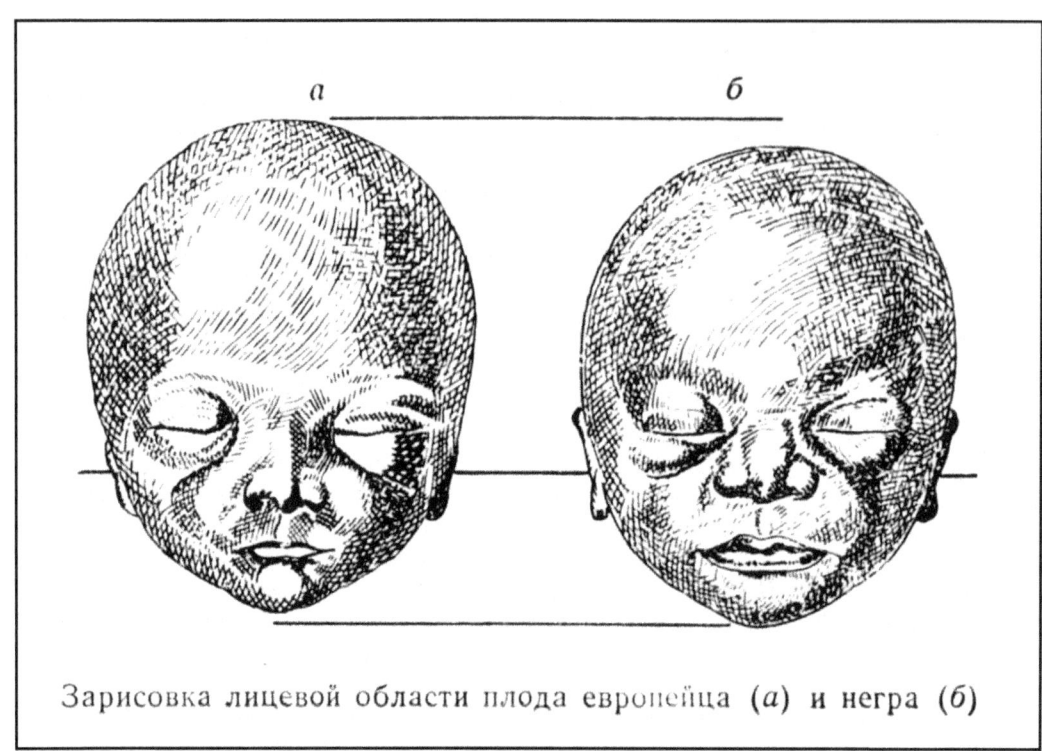

Зарисовка лицевой области плода европейца (*а*) и негра (*б*)

Sketch of the facial sphere of a European embryo (a) and a Negro embryo (b).

In Moshkov's opinion, this mixing of Man with Pithecanthropus had fatal results for all of humanity, both for his social life, and his moral character. Contact between "evolutionary stages" produced utter upheaval of awareness in descendents and caused dissension and doubt in all their spiritual organization. Thrown between angel and demon, in the soul of Man it only produced a mirror image of the revolt of the animal nature in his body. Moshkov wrote: "The female of the Pithecanthropus, which were made the wives of the White man, and the children of the latter, this could not be seen as something different than a breed of his domestic animals, who could, depending on necessity, be adapted to some kind of work or eaten, or traded for something with neighbors. Here then, the foundations of slavery, which today outrages, were laid. In the very beginning, it did not have in itself anything outrageous, and only later became such, when Mankind mixed more strongly, and the differences between slaves and masters decreased. Through several generations, the white race declined, and former slaves, from mixing with noble blood, gradually achieved equality with their masters. In the end, modern Mankind took shape, as a mongrel of ancient species. And there was the cause for the transformation of Man into a worse condition. That is why the brain capacity of the skull of modern Man is lower than the capacity of the original, Neolithic one."

With the discovery of blood groups at the beginning of the 20th Century, this question, which was considered hypothetical, began to be debated with complete seriousness. In 1918, the German biologist, Herman Orleder, discussing the possibility of the miscegenation of Man with some anthropoid, that in the case of successful artificial insemination with a female chimpanzee with human spermatozoa, a hybrid could be born into the world, and without the necessity of turning to a Caesarian section birth.

In the book, *Primatology and Anthropogenesis*,[356] Soviet anthropologist M.F. Nestrukh also emphasized: "The placenta, the morphological and biological traits of the blood, and the spermatozoa of the chimpanzee is very similar to that of humans. Experiments with blood transfusions from Man to chimpanzee and vice versa, were successful. In an experiment, the chimpanzee was **susceptible to syphilis.** This suggests that by means of artificial insemination,

---

[356] *Primatologiya i antropogenez.* Moscow, 1968.

288

one can produce a cross-breed of a chimpanzee and a human, and get hybrids with intermediate traits. Speaking in favor of this opinion are the not-uncommon cases of **successful inter-species and inter-genus cross-breeding** among apes; [there is] a blood kinship and particular similarity between ape and human sexual cells."

As an authentic scientist, Nestrukh turned reader attention to the enormous "interest" toward apes in the ancient world, among the members of the equatorial races. In Ethiopia, Mauritania, Libya, and Egypt, apes were considered an obligatory part of war trophies, on a par with precious items and war prisoners. A bas-relief that portrays them is on the temple of Der el Bakhri, created 3,500 years ago; there is an analogous portrayal of that time, on a burial vault of Vizier Rekhmir in Thebes. In Egypt the Hamadril baboon was a sacred animal, and personified the God of the Moon and the sorcery of Thot. Besides that not only the Pharoahs, but also apes were subjected to mummification—but not servants, wives, or military commanders. Aristotle wrote about the anatomy of apes, and the famous Ancient World anatomist, doctor, and physiologist, Claudius Galen (130 A.D-200A.D.), left a detailed description of the dissection of man-like apes, which he called "mixed copies of people." The first detailed description of the dissection of a human was only carried out by Andrei Vesaliy in 1543. An Egyptian terra cotta relief in the Berlin Antiquarian (inventory No. 31276) portrays a sitting female gorilla with a human face, which is embracing two cubs; one of a simian type, the other human. The Hindu ape-god Ganuman has a human body and and ape head; he is portrayed in the Temple of Suami (in the city of Una, India).

Classic Russian anthropologist Dmitriy Nikolayevich Anuchin was one of the first to turn attention to the myths about the origin of several peoples, from the mixing of Man with animals. A totem ancestor was not considered by him as an elaborate allegory from the land of legends, but namely as an anthropological fact. Its original zoogeneaology also starts from with apes. Anuchin wrote: "In general, one can say that the thought of the possibility of a close kinship or mutual transition between Man and apes uses sufficiently significant dissemination between both half-savage peoples, and between cultured peoples, with only those differences, that in the final instance, an ape origin is attributed usually to more brutish tribes or separate families."

One royal Indian family considered itself the descendents of apes, and its members carried the title: "Tailed Rana", since according to legend, the father of this august family was provided with this appendage. In 1867, the English representative in India gave an order for the slaughter of 500 holy apes; the natives asked for a cancellation of the order, on the grounds that they recognized their ancestors in the apes.

The ancient Greek historian, Diodorus of Sicily (80-29 B.C.), also told of another royal family in Africa, that had a tail, like a natural appendage of the body, that was passed down from family to family, in a succession of many generations. Legends about the original tailedness of separate peoples are encountered in Africa, Asia, and America. The ancient Chinese historians pointed to the Ting-Ling people, who inhabited Enisee, which had green eyes and could have arisen from monkeys, and therefore looked very much like them. Chinese chroniclers geneaologically listed many peoples that came from apes; precisely also did the Indians explain the origin of the Tibetans. In this regard, Moshkov wrote: "Many lower Negro, Malaysian, and American Indian tribes regard apes, in particular the higher apes, as real people who do not speak, only out of fear that they might be put to work." The Kaffirs believe that apes have human spirits. One Tibetan writer reported that Buddhism was spread not only among people, but among several species of ape, also. The Greeks and Romans regarded apes as demonic beings, and the ancient Babylonians regarded them as "servants of fetishes."

The Malaysian name, "orang-utan", has come into general use today as a name for a breed of apes; in point of fact, it translates as "forest man", and is applied by the Malaysians themselves in great part, identically as a designation for apes and for the primitive peoples inhabiting those areas. Inhabitants of the islands of Fiji, Tasmania, and also some tribes of South America, regret to this day that they lost [their] tails, for according to their legends, people became irritable and evil after this. In modern India, the term "ape people" is still applied as a designation for some primitive tribes, which for their part believe that many species of ape arose from people, who committed sins against God. Similar superstition exists among the Arabs in regard to the marmoset. The Kaffirs believe that one of their tribes was turned into baboons. Among Muslims there is also a legend that the inhabitants of one of the Judean cities were turned into apes for breaking the Sabbath.

In the work, *Anthropomorphic Apes and Lower Types of Mankind*,[357] Russian anthropologist D.N. Anuchin wrote: "Often these lower tribes are even mixed in notions about apes to such a degree, that sometimes real apes are accepted as people, and vice versa, real people are described as apes. Often the opposite possibility is assumed in this, so to say, of regressive metamorphisis; that is, the transformation of people into apes. Examples of notions similar to it can be found often enough among the very different peoples."

To this day, many Negroes, Indians, and Malays think that apes can speak, but are only hiding this ability. Others add that among apes a social structure and bureaucracy exists, similar to humans. Finally, among many peoples there is a widespread superstition, according to which apes love to carry away human women for themselves, to live with them and raise children. The ancient Egyptians even believed that the silvery baboon could be trained in writing and music.

It is a fact that today the White Race is accused of all conceivable and inconceivable crimes of racism; this is one of the most unprecedented informational crimes. We insistently emphasize that basic ideas of the inequality of the human races, and the notions of biological closeness of some of them to apes, were not invented by the Europeans, but on the contrary, were drawn on from the aboriginal peoples themselves, in the Age of Discovery. When the first chimpanzees were brought to England in 1700, the English were completely convinced, that they were members of the Pygmy tribes. How could they know, who this actually was, if they had never seen them, nor others? Obviously, the real racist system of values and definitions came to Europe from outside—that is, from aborigines themselves.

Racism is a deep, non-European phenomenon, since it manifested beyond the limits of Europe, among non-European peoples; this is graphically testified about by the chroniclers of the "Great Age of Discovery."

Besides the apes, people with dog-faces occupy a conspicuous place in zoogeneaology; they are called "cynocephalics", "kynokephalics", or "cynamoni". In one early Coptic Christian legend, it tells of how Jesus converted one such being to his faith. In the Russian north, in village churches in the Olonets Region, one can encounter icons with an image of Saint Christopher, who is also portrayed in the form of a man with a dog's head. Ancient Greek and Roman historians talk repeatedly of similar beings. The Ainu of northern Japan are convinced to this day, that they came from dogs. The Aleuts believe that the mother of their tribe was a bitch by the name of Magakh, who conceived hybrids from some old man, and that his laid the beginning of the tribe. Among the Khirgiz there is a legend, that traces their lineage to a "red, male wolfhound and one tsarina, with her forty servants." On Fiji, a legend tells about the god Denge, who once looked into a clear brook, and was struck by the full ugliness of his cross-bred appearance. Numerous other beasts, such as the bear, the fox, the wolf, as well as completely unheard-of magical creations are counted by various peoples among their ancestors, which are reflected for example, in their national symbols.

Because of a lack of apes, many tribes consider beaver, crows, the crane, elephants, fish, and turtles to be their ancestors. Peoples of northeast Africa trace their lineage from crocodiles; inhabitants of the Antilles islands—from ants; some American Indian tribes—from worms. From this it follows that many peoples of the Earth are not the least bit supportive of modern convictions about the wave of "white colonial racism", as they themselves openly spread the information about their low biological origin.

An incredible number of similar legends are dispersed according to their color, but ethnographers, summarizing this material, persistently avoid the biological aspect of this sinful fall, as well as the detailed descriptions in the holy writings of various peoples. Collectors of the "cultural diversity" of humankind are afraid to honestly declare, that many peoples of the Earth are proud of a mixed origin, tracing their geneaology from some act of sodomy. Some do not hide this shameful fact, and brag about it in every way, giving a pompous mythological imagery to it. But "enlightened Mankind" obediently heeds the sorcery of academic ethnographers, summoned to this zoophilia to search for some lofty, esoteric which promotes lucidity and the mystical unity of the family of people.

Once more, we stress our position. We do not at all think that at the modern stage of evolution, one can get a hybrid of a man with some animal, for example, a man-like ape. Let

---

[357] *Antropomorfnie obez'yani i nizshie tipi chelovechestva.* Priroda [Nature], 1,3,4, Sankt-Peterburg, 1874.

geneticists decide this question, if it has in general some practical significance. We are struck by something else altogether, namely: why, in turning to the geneaological tree of some tribes, is it absolutely necessary to search for some anomalous facts, and place an accent not on human nature, but on the rudiments of zoological origin? Why in the search for an animal ancestor— even if it is not a reality, and only in the imagination of this tribe—have animals copulating with people become a trait of good tone in ethnography and cultural mythology? Do we really need such sciences in general, that we are oriented to the legalization and mythologizing of zoophilia? Is it not better to rely on the brilliant intuition of Friedrich Nietzsche, [that] to search for "the human is too human?"

The great Roman philosopher, Titus Lucretius Carus (99-55 B.C.), wrote: "In her stresses, the Earth produced a multitude of monsters, of strange and monstrous forms: they were androgenous, dual-sex, and belonging to neither one sex; some were legless, devoid of a mouth, faceless, and blind beings; there were also monsters that were so bent, that they were not in a condition to walk about at wil." Yes, this is an allegory, but modern mass orgies of homosexuals, transvestites, and other perverts, "belonging to neither one sex" are today a fact of our public life; and the birth of infants with tail appendages, animal ears, excessively hairy faces, and similar biological atavisms, from time to time, in various parts of the world, are a biological fact.

One of the great pagan sages of the Late Classical World (2nd Century A.D.), Sextus Empiricus, so-named for planting a spirit of clarity and practicality in philosophy, advanced the thought that not one trait existed, on the basis of which one could draw a sharp distinction between Man and animal. In this key question of important theological significance, he was even supported by the early Christian writers Arnobius (3rd Century A.D.) and Lactantius (4th Century A.D.). With the start of the Age of Discovery, when Europeans first became acquainted with the native inhabitants of distant lands, the **atheist** movement arose in Europe. [This movement] believed that people located in a stage of primitive communal living got their start from the Earth, arriving in a state of decay from the decomposition of the bodies of apes, pigs, and frogs in the earth; with this [belief] they attempted to explain the similarities that existed, according to their opinion, between the physiques and inclinations of these animals, and members of the colored races in the newly discovered lands. Dutch scientist Hugo Grotius (1583-1645) created an entire theory, in which he substantiated the possibility of the mixing of peoples with animals, the result of which numerous, "questionable" wild peoples appeared in the world in transitional forms.

The presence of animal traits in man as such does not interest raciology, but their collective and quantitative distribution in separate peoples and races does.

A.A. Zubov, a modern classical Russian anthropologist cited by us repeatedly, wrote in his article, *Discussion Questions on the Theory of Anthropogenesis*,[358] that: "Ideas about the formation of hominids and Man, that fully developed toward the middle of our century, are presently subject to serious revision. It is assumed, for example, that bipedal forms did not arise frm quadrupeds, and vice versa: that the man-like apes—the ancestors of the chimpanzee and gorilla—were the descendents of bipedal, erect-walking hominids."

The materialistic Marxist-Leninist theory of anthropogenesis traces the transition of Man from an ape, by means of the development of work skills. This undoubtedly leads into a blind alley, for it cannot explain which part of Man's ancestors evolved to the level of modern HOMO SAPIENS, and for another, along the path of humanization, why one part went back from bipedal to quadrupedal movement—a reverse step in evolutionary terms. Let's recall that an analogous concept and conclusion expressed 100 years ago by German anthropologist Herman Klaatsch was ridiculed. Today, Zubov certifies: "The firmly established fact of the absence of a firm connection between the type of stone inventory and the evolutionary stage of this or that member of the genus HOMO also belongs to the number of new achievements in the science of the origin of Man."

Thus, the old thesis that "work made Man from ape" should be conclusively removed from the daily bulletin as unscientific. But if not work, then what? Zubov continues: "Anthropologists certify the presence of three variants of fossil peoples in the said period of anthropogenesis in Europe: 1) Neanderthals; 2) people of the modern type; and 3) intermediate forms."

---

[358] *Diskussionnie voprosi teorii antropogeneza.* Etnograficheskoe obozrenie [Ethnographic Review], N6, 1994.

These "intermediate types", being the result of ordinary sodomy between modern types and pre-humans, split from the evolutionary process. Hybrids, more inclined to zoophilia by virtue of a high concentration of animal traits in them, returned to an initial animal state by means of sexual selection, and transformed into the modern chimpanzee, the gorilla, and others. The other part, by virtue of the smaller concentration of animal atavisms, rid themselves of exotic sexual predilections, and later evolved to the level of modern Man, but retained the very fact of their origin from animals, in the mythology of many peoples. The first to turn attention to this ethnographic sodomy was Russian anthropologist Dmitriy Nikolayevich Anuchin; and Vladimir Aleksandrovich Moshkov developed his ideas. Thus, not work, but specific sexual behavior is the propeller of evolution, accelerating and directing the process of sexual selection. Besides that no single-axis, forward motion of development can be spoken of, for as we are convinced, devolutionary branches, which lead to degradation, can exist together with evolutionary branches. All these facts can become the object of study of a new science, under the name of **evolutionary zoophilia.** Today, any criminologist is able to certify, that among various peoples a great explosion in the statistical percentage of sexual acts with animals is observed. This fact perfectly supports our hypothesis of anthropogenesis. The "Mowgli Phenomenon" only says that a wolf pack very rarely takes in human cubs, and only accepts them on the basis of animal closeness, by virtue of a certain concentration of non-human traits in them. Cases of wolf attacks, and even of savage domestic dogs against people, are recorded far more often. A.A. Zubov further writes: "The process of Neanderthals mixing with people of the modern physical type, as some anthropologists have already long since proposed, originated in Near Asia." By virtue of the convenience of migratory paths at the dawn of anthropogenesis, this territory served as a ready-made bridgehead, supplying subjects of the sub-human, "transitional type." Sodomy in this region (of which there are even indications of in the Old Testament), is encountered far more often to this day, than among the peoples of Northern Europe. In the ancestral land of the so-called "Black Eve" in Africa, this is also a normal, ordinary phenomenon.

In the book, *Essays on the Evolution of Man*,[359] Russian scientists E.N. Khrisanfova and P.M. Mazhga write: "The genetic similarity of Man and chimpanzee concerns the basic structure, and not the regulatory part of the genome. Molecules of organisms often evolve with varying speed, and this means that the tempo of molecular and phenotypic evolution does not coincide. Also, the evolutionary transformation of the Neanderthal skull to Sapiens skull, could have occurred without any essential violation of the genetic balance. Remaining within the limits of systematics, one cannot isolate a 'ready Man', from the long chain of predecessors. The difficulty of the problem lies in the fact that morphological evolution was characterized by inequality, by 'patchiness.' As M.F. Nestrukh thought, to the very earliest pore, the formation of work activities as 'a morphological boundary' between Man and animal was dynamic, because for earlier reasons the process could stop and even go in the reverse direction." In the brochure, *Stages and Interstadial Differentiation in the Evolution of Man*,[360] V.P. Yakimov also came to the conclusion that: "The process of the rise of Man of the modern kind took place on a rather wide territory, and was conditioned by the mixing of Sapient forms of paleo-anthropoids and transitional forms of ancient peoples."

[359] *Ocherki evolyutsii cheloveka.* Kiev, 1985.
[360] *Stadii i vnutristadial'naya differentsiatsiya v evolyutsii cheloveka.* Moscow, 1967.

Thomas Henry Huxley

It is namely the copulation of human forms with pre-humans that conditioned the course of the evolutionary process in some races back in the reverse direction. The distinguished English biologist-evolutionist, Thomas Henry Huxley, emphasized in his article, *The Place of Man in Nature*,[361] that: "The anatomical differences between Man and the higher apes is less significant, than between the higher apes and lower apes." And Herman Klaatsch pointed in an article[362] from the same anthology, that "Between certain human races and definite man-like apes, there is a tie of kinship not only in general, but even in particular. Man does not at all represent a crowning work, but combines very ancient traits with others that were slowly perfected, and tertiary [traits] that were recently acquired."

---

[361] *Mesto cheloveka v prirode.* From the anthology, *Evolyutsiya cheloveka.* Moscow, 1925.
[362] *TheState of Man in Nature* [*Polozhenie cheloveka v prirode*].

Left: Ernst Gekkel

There is still another eloquent proof that in the process of the formation of races, incidences of sodomy had their place, and caused as a result, different concentrations and distributions between them, of so-called ape or pithecanthropoid traits. Swedish anthropologist Gustav Retzius discovered that the sperm of an orangutan was not as close to human [sperm], as the sperm of lower apes."

It is namely in the context of the given argument that it is worth remembering again that ape traits are, as Darwin pointed out, the result of a reverse recurrence to the original animal ancestor; they do not arise in and of themselves, but are the result of the copulation of higher forms (in terms of evolution) with lower forms. Primitivization of the structure of any organism is the result of bad heredity, when not in compliance with the principles of **evolutionary racial hygiene.** In this regard, Herman Klaatsch distinctly formulated the problem: "Neither between anthropoids and Man, nor between some now-living apes and Man is it necessary to search for a connecting link; the bridge to the "crowning work" should be thrown across from the lowest breed of primates, called the great-ape ancestor."

The genetic weight of the most primitive animal's heredity was scooped up in the process of evolution. Man of the modern type cannot rid himself of the base passions, which literally tear his being to pieces. The chasm of differences separating the races is a very obvious confirmation, which shows all the dissimilar paths of development, arising from the depths of the pockets of race formation. In this regard, Eugen Fischer indicated in his book, *Anthropology*, that: "The processes of humanization and race formation were originally identical." **All races humanized in different ways.**

The absence of a distinct morphological boundary between Man and animal cannot help but have a psychological effect on a psychological level. The distinguished German psychologist, Wilhelm Bundt, emphasized: "Animals are beings, the awareness of which differs from the human, only by degree of attained development. Between Man and animal there is not a larger boundary, than what is encountered within the limits of the very Animal Kingdom."

Modern researchers in the area of neurology and neurochemistry confirmed this position of psychology. E.N. Khrisanfova and T.P. Mazhga pointed out in the book, *Essays on the Evolution of Man*: "The mass of the brain, as a 'crude' structure, in and of itself does not determine human status in the early stages of anthropogenesis. The restructuring of the brain in the hominid direction could appear in the beginning on a cellular, neurochemical, and even a molecular level. This is indirectly confirmed by the findings of comparative studies of the nerve tissues in modern primates. In part, it was discovered that in phylogenetically new territories of the occipital cortex, shifts in the chemistry of the nerve tissues are entirely possible in the absence of differences in the narrow structure of the brain, not speaking already about its cruder changes."

From this it follows that racial psychological differences lead to differences in evolution at the molecular and neurochemical levels of organization in the brains of the members of these races, in the process of humanization.

Soviet paleo-psychologist professor Boris Fedorovich Porshnev's book,[363] *On the Beginning of Human History,*[364] is highly interesting and significant in the scheme of the researched question. In it, he remarks: "The science of anthropogenesis, it seems, should finally become a science about the concrete biological relationships of people, and the preceding form, from which they branched [off]. A scientific absurdity is the view that all specimens of an ancestral species turned into people. It is still more senseless to think that they ceased to be born into the world, since some by way of mutation become people. Paleoanthropoids should have obligatorily tried this or other significant changes, if not in the structure of their bodies, then in these or other essential functional traits and characteristics. Man did not gemmate from Paleoanthropus in the simple sense; that is, he did not appear next to him, and rise from the split of Paleoanthropus, and develop in certain respects with another half of an emerging form."

Porshnev emphasized that under "concrete biological terms", he did not have in mind any evolutionary abstractions from a school course on natural history, but a physical elimination of "pre-humans" by Man, or sexual contact between these biological forms."

This was more than courageous for a Soviet scientist; B.F. Porshnev's thesis is successfully illustrated by findings in the Assyrian-Babylonian *Epic of Gilgamesh*, in the *Veda* and *Avesti*, and in The Old Testament, and in the Hindu *Ramayana*, everywhere observing the structural integrity of information about morphological and behavioral differences between Man and "pre-human"; that allows one to speak not of the fiction of the ancient authors of these texts, but about reliable sketches from nature. "Thus, one may think that the chain of evidence about the pale-anthropoids is unbroken, from the Paleolithic to the first written sources, and stretches even further"—wrote Porshnev.

[In his] understanding of the formation of a racial worldview in the process of anthropogenesis, to his credit, Porshnev also demonstrates it without reference to Marxism-Leninism: "Much in the ancient history of Mankind receives additional illumination, if it is remembered that peoples developed by contrasting themselves to the "anti-people", the "un-people", and the "non-living", that lived on the near or far periphery. This contrast became more conscious. It was the reverse side of the self-awareness of ethnic groups." Initial biological determinism lay at the basis of the formation of the worldviews of races as such. Forming stable societies, peoples thought in terms of the dichotomy: "Us and Them." According to Porshnev, "race formation, at the very least, is the formation of the original great races and their subdivisions—it is a fact that correlates with artificial isolation. By active selection, the Mongoloids, Europoids, and Negroids split, detecting in each other some kind of involvement with anti-people. By way of artificial selection, they eliminated undesirable offspring in this regard, and put a stop to any cross-breeding (along with any contact) with the members forming the "contrasting" race. They particularly energetically segregated themselves from one another, as far as possible. At the basis of race-genesis lies the intense efforts of "peoples" [to guard], by means of isolation, against penetration into their lives by "anti-peoples."

Until the 16th Century, nearly everywhere "people" were considered to be those who composed their own ethnic core, and those further on the periphery, were considered as something less than human; all the stranger hybrids and monsters were recognized in their nature."

Numerous ancient authors, upon whose testimonies modern historical science is based to this day, left us descriptions of man-like beings, created with such unbiased naturalism, and deprived of any touch of the supernatural, that it seems now and then, as if they were all written from one tracing paper. Thus, for example, according to the words of Plutarch, a "satyr" was once brought to the Roman officer Sulla. Desiring to question the mysterious being, he summonded a number of translators, but, as was to be expected, the representative of the lower stage of the evolutionary ladder was not capable of articulate speech. Sulla and his retinue were genuinely frightened by the mooing, bleating, and other howls, so the satyr was expelled. In general, such descriptions of satyrs, fauns, caryatoids, and similar monsters which possess this or that fragment of human appearance, comprise one of the most essential attributes of ancient literature. But to get away from the mythological interpretation of the given subjects, and move

---

[363] Professor Boris Fedorovich Porshnev, 1905-1972.
[364] *O nachale chelovecheskoy istorii.* Moscow, 2006.

the problem in the direction of modern, realistic science, then very quickly it is discovered that all the "enchanting" creatures are for some reason, grouped according to biometric traits within the limits of a distinctive quality, but a nevertheless steady classification. Besides that the given kind of subjective canvas, abundant with man-like personages, is preserved in the European epos until the 15th Century.

Even the religious reform of Zoroaster, in opinion of B.F. Porshnev, was intended to increase the evolutionary distance between people and "non-people", with the help of the social-political levers of the ruling class. Descriptions of relic, man-like beings are preserved in Middle Ages Arabic and Tibetan medical tractates, and also in the tales of the Russian North and the Scandinavian peoples. The first scientific description of a live Paleoanthropoid was left by the Dutch anatomist, N. Tulp, in the 17th Century. The described specimen was caught in the mountains of Ireland, and was a youth; numerous, beast-like traits were observed in his appearance. Not possessing intelligible, articulate speech, he did not possess the most elementary skills of human society. His behavior was akin to the manners of a small, wild animal. As to the distinct, anthropometric traits of the lower stage of evolutionary development of this organism, the Dutch scientist placed the "lobe, constricted and low; the occipital [bone] bulging and cone-shaped", as first among them. Incidentally, according to ideas of modern science, the presence of such a fallen brow ridge and low skull arch characterizes Pithecanthropus and Australopithecus, and among the modern races—Australoids.

At approximately the same time, the first tractates containing the results of the autopsies of such "suspicious subjects" appeared. The authors conscientiously observed deviations in them of the physique of the skull, and in the interior organs of the subjects; again, this is found in the channel of development of classical evolutionary theory, and in the findings of comparative morphology.

All this abundance of information laid the basis of the first racial classification, compiled in 1746 by Karl Linneaus, where along with the customary races divided into independent species, *homo ferus* and *homo monstruosus* were also present as transitional types from ancestral forms, to the modern type of man.

B.F. Porshnev's main conclusion lay in the fact that in the process of the humanization of the modern races, the "un-people" did not disappear someplace, but were subsumed by us. Besides that according to Mendel's Laws, some of them sometimes regain their primeval appearance, and some form durable, persistent communities, and others fall away from commonly accepted norms of behavior. We ourselves add that pirate and bandit republics, rising at times here and there in history, and also such colorful specimens as Mowgli and Chikatilo, absolutely confirm the position of the famous Soviet scientist.

Soviet anthropologist G.A. Vasilyev, appearing with a report at the 7th International Congress of Anthropological and Ethnographic Sciences in 1964, very firmly expressed in the same spirit: "The question of the possibility of examining the phenomenon of mutation, in the process of the transition of our ancestors to meat foods, is very important, because as is known, definite damage arises in the brains of apes, from their preference for meat food, over all other kinds.

In the process, the change in taste that occurs is not isolated, but is only a component of an entire syndrome, where, besides a preference for meat, it leads to a loss among apes of inborn social reactions, and hypersexualism. The fattening ape does not respond to the threatening facial expressions of other apes, and tries to rob food not only from the weak, but from strong apes, and despite blows, continues to conduct himself as before. Hypersexualism manifests in that such an ape strives for sexual intercourse, not only with all apes of any sex and age, but even with other animals that approach it in a cage, and it does not observe seasonal times.

If one compares the behavior of Man and ape, then one cannot help but notice, that Man appears more similar not to the normal ape, but to the ape suffering from Kluver-Bucy Syndrome. In point of fact, Man does not have inborn social, situational reactions, which highly saturate the behavior of ape troops, for example, the Hamadryas baboon; Man gets satisfaction from the use of meat; he does not have seasonal sexual cycles, and is less discriminating than apes in sexual desires, judging by the crimes against morals that existed as early as Biblical Times."

Therefore, it becomes completely obvious, that propaganda of all forms of sexual perversion, from an evolutionary point of view, by consequence brings a distortion or deceleration in the direction of the development of the group of people, amongst whom this propaganda is realized.

For its part, the different degree of admixture of animal blood in the members of different races precisely reinforces the theory of polygenism. If all races rose from the same pocket of formation, then the degree of concentration of animal rudiments would also be distributed equally between them all. However, any morphologist can easily show you, that all modern races are endowed with atavisms of animal origin, in varying degree.

Anthropologist Franz Weidenreich (1873-1948), whose name is connected with the formation of a whole school of science, which received the name **polycentrism**, observed with complete political incorrectness: "The similarity of Man with the ape is by no means concentrated in this or that people, but is distributed in different parts of the body in different peoples, such that each of them is supplied with some hereditary part of this affinity; of course, to one more, to another less, and even we Europeans cannot have claim that we are completely alien to this tie of kinship." In an official textbook of the Soviet era, *History of Primitive Society*,[365] authors A.I. Pershits, A.L. Mongayt, and V.P. Alekseyev wrote: "The archaeologically fixed, uninterrupted transition from the Early Paleolithic to the Late, on all continents of the Old World, and the presence of parallelism in the geographic distribution of the modern races, and of different morphological forms of the Neanderthal type, tips the scales more in favor of the polycentric hypothesis."

The fallacy of the concept of the origin of Man from one pocket of race genesis is also criticized by Soviet archaeologist P.I. Borisovskiy, and another Russian anthropologist—G.F. Debets. And American anthropologist Carlton Stevens Coon dedicated a thorough study of this theme to the great work, *Origin of the Races*,[366] in which he substantiated the "lower" nature of the equatorial races, and divided all "humankind" into five independent evolutionary branches.

In his program article, *Toward Evolution of Theoretical Thinking in Biology: from Monocentrism to Polycentrism*,[367] K.M. Khaylov wrote: "In every concrete instance, monocentric thinking only accounts for part of the real world, when the vital needs of Man are tied with all of its parts. The new systemic approach in theoretical biology lies in the fact that it allows critical revision of the principles of monocentrism and renounces it in favor of the far more fruitful idea of polycentrism, from the point of view of which, the composite elements of living nature—sub-organism structures, organisms, species, communities, and ecosystems—are examined as its equally important elements. This means that in the biochemical scheme, the idea of monocentrism should give up its place to polycentrism."

We emphasize our position, that the evolutionary, systemic approach to large ecosystems automatically spares us from any accusations of racist propaganda. Racism is the variety of fragmented, monocentric thinking that talks about "bad" and "good" races. A polycentric, systemic approach removes the very principle of this artificial division. The multi-centeredness of the origins of the human race leads us away from primitive, dualistic thinking, because the complex dynamics of the evolutionary system are impossible to qualitatively characterize in that paradigm. After all, we do not say that a fox is better than a hare, but rather, that each has its own "ecological niche". The very same applies to human communities, where each race, by virtue of the uniqueness of its origin, submits to its own laws of development, not subject to vulgar unification.

The given moderate point of view is reflected in a UNESCO declaration, *A Proposal According to the Biological Aspects of the Racial Problem*,[368] where it reads in the first article: "The controversial question of how and when different groups of people took form remains."

But if we accept such a highly tolerant formula as the basis, and add to it the stated task of modern evolution theory, then the *UNESCO Declaration on Race and Racial Prejudices*[369] is left completely up in the air, where it states: "All peoples are born free and equal in their dignity and rights"—for it becomes completely incomprehensible, that these are "people", if all are mixed—in various proportions—with apes. It is nowhere indicated in this document, which percentage of "ape" blood is considered permissible for determination of "dignity and rights", or not. The table of

---

[365] *Istoriya pervobitnogo obshchestvo*. Moscow, 1982.

[366] 1963

[367] *K evolyutsii teoreticheskogo mishleniya v biologii: ot monotsentrizma k polytsentrizmu*. From the anthology *Systemic Research [Sistemnie issledovaniya]*. 1973.

[368] Moscow, 12-18 August, 1964.

[369] Paris, 26 September, 1967.

distribution, according to peoples, of this very "animal" admixture, with which anthropologists agree, but which is stubbornly not recognized by privileged internationalists, does not have occasion for clarity. There is presently a distinct gap between theory and practice, fraught with numerous legal mistakes, including [those] of an international character. The legal subjectivity of many authoritative institutes, even on the very highest level, can be subject to doubt and disputed in court. The prospect is revealed for example, in the rejection of the composition of a court, as not corresponding to the minimal-permissible criterion of "humanness", on the basis of racial-biological indicators, and the disagreement of the defendant with the procedural publication of the sentence, handed out to him by the members of the lower—in evolutionary terms—"hominid groups." The character of testimonial evidence also turns out to be unsatisfactory, in light of the evolutionary retardation of the organs of perception in the associative apparatus of the witness. And the selection of the people's sworn representative should be implemented, according to the principle of commonality of phenotypical traits. The subject of examination in a court should not be the deed itself, but the competence of the subject, in relation to his racial-biological status.

John Randall Baker (1900-1984), one of the prominent biologists and physical anthropologists of the 20th Century (and a professor of Oxford University), emphasized in his fundamental monography, *Race*,[370] that: "Each who supports the apparently completely self-evident truth, presented in the American Declaration of Independence, that all people are created equal, should ask himself the question, what does the word "equal" mean? On the basis of what criterion can one make a European and an Eskimo, or a Negro and a European, equal subjects of taxonomy? How to neutralize obvious differences in the physical characteristics of the members of different races? How to carry over the primacy of one race to the primacy of another, in various groups of traits? For example, competition in the Olympic Games demonstrates with all obviousness the biological specialization of the basic racial branches of humankind."

American biologist Anthony Barnett explained that the majority of problems in the humanitarian sphere are connected, since community leaders and sociologists incorrectly interpret biological concepts. "The biological question about inborn differences often gets mixed up with the political question about equality of opportunities. Despite the majesticness of the expressions and the noble intents, the American Declaration of Independence asserts the right of each person to life, liberty, and the pursuit of happiness, but it does not have any reference to the inborn differences between individuals." This same author made the assertion: "The word 'savageness' customarily means early social formations. This term is completely scientific and has no degrading sense. It relates to the form of life of all peoples, before the appearance of *homo sapiens*." But then it will turn out that the numerous tribes of Equatorial Africa and Australia, living in primitive communes, are a wild form of life in the biological scheme, and do not belong to the species *homo sapiens*, being in the direct sense of the word, "pre-human" or "sub-human."

The given traits are also completely scientific and do not carry any sense of degradation, completely reflecting the natural scientific condition of things, and supporting evolutionary theory. In accord with the previous conclusions, we intend to emphasize, that "pre-humans" exist within all races and populations, including among the Europoids; but everywhere they comprise a different concentration and are represented by different forms of savageness in the evolutionary scheme.

The main conclusion of modern evolutionary theory, plunging all "humanists" into a state of shock, lies in the fact that there are no clear and concrete boundaries between Man and animal, but that between races there are [such boundaries].

Many anthropologists, neuromorphologists, and evolutionists think that the very process of humanization is not marked by changes in the area of morphology, but only in the area of behavior. In the monography, *The Origin of Man*,[371] T.O. Bazhutina synthesized a multitude of data and asserted: "The inarguable qualitative difference of the human psyche, lying in emergence beyond the limits of the animal stereotype, from the psyche of higher apes, is based on the insignificant quantitative difference in the morpho-physiological substrata of the brain. It is impossible to find and record the distinct morpho-physiological criterion of 'humanness'." In his

---

[370] 1974

[371] *Proiskhozhdenie cheloveka*. Novosibirsk, 1993.

fundamental work, *Problems of Anthropogenesis*,[372] Y.Y. Roginskiy wrote that: "One should not speak of any sharp morphological criterion between the 'last animal' and the 'first man', because it is impossible to guess the very moment of the origin of human consciousness." The given point of view was shared by Russian anthropologists V.P. Alekseyev, B.F. Roginskiy, M.I. Uryson, G.F. Khrustov, and also the world-known authority in the area of evolutionary theory, American scientist Ernst Mayr. The 7th International Congress of Anthropological and Ethnographic Sciences was held in Moscow in 1964. In the framework of that congress, a symposium was held, which was characteristically titled, *Problems of the Boundary between Animal and Man*. The majority of the participants in the discussion, which unfolded in the course of this scientific undertaking, also came to the conclusion, that in evolution no distinct, fixed boundaries exist between the last anthropoid and the first man. From the above-mentioned logic, it follows that **"more animalistic" and "less animalistic" races exist.**

The founder of Russian racial theory, Stepan Vasil'yevich Eshevskiy (1829-1865), wrote in his fundamental work, *On the Significance of Race in History*,[373] that: "Modern science gives us the opportunity of destroying the deep boundary between Man and animal in general, drawing on the other hand, a still sharper boundary between the Man of a higher race and the Man of lower organization—the being still transitioning from the world of the proper animal, to a world undoubtedly human in its higher meaning. The more a researcher becomes familiar with the different tribes and the more the quantity of ethnological material increases, the more divided the division becomes, and it extends in his conclusions to the proposition of the creation of the human genus, according to tribes."

In the book, *The Physique of Man from a Comparative Anatomical Point of View*, German anthropologist Robert Wiedersheim came to the following conclusion, based on extensive statistical material: "In the attempt to establish the original Man, that is, to find traces of the ancestors of Man, it is necessary to dwell on other points of view, which compel examination of the organization of Man among the animals, as developing partly progressively, and partly regressively." And he was not alone in his summarizations. Thus, it becomes completely obvious, that it is necessary to regard any talk about a single axis of evolution for the entire human genus, as extremely naïve and not conforming to reality. If humankind really had developed in one direction, then the number of vanished peoples and civilizations, with which books on history and archaeology are replete, would not exist.

It is completely obvious, that at the same time some racial groups evolve progressively, others, in conditions of this or that ecological niche, are degradating, even being merged by commonality of social and political life.

Besides that one should absolutely not talk about any universal "unity" of the human body, since, in the process of evolution, one group's organs may develop, while another in this process is located in a state of atrophy. German anthropologist Baron Egon von Eichstedt asserted in his fundamental monography, *Raciology and the Racial History of Mankind*: "The point is that during the formation and growth of the accumulation of racial traits, one or another morphological process of race formation (thanks to the rise of new mutations) goes faster, another slower. This process of development is not in any case spontaneous, and does not occur in empty space, but is located, as with all the quick and the organic, in direct harmonious interaction with the living situation and the zone of habitation. From here there is a definite stratification in the course of evolution, which occurs in all Mankind."

In the opinion of the author, races with accelerated tempos of development, which produce specific **human traits** at the expense of physical and psychological differentiation, are **progressive.** In contrast to them, **primitive** races develop **infantile and animal traits**, because of a lack of growth and specialization.

---

[372] *Problemi Antropogeneza.* Moscow, 1969.
[373] *O znachenii ras v istorii.* 1862.

9. The Fiction of a Single Humankind

Thus, the achievements of modern biology do not contradict the concept of the polygenetic development of "humankind". The fundamental theory of homological series in hereditary changeability, advanced by N.I. Vavilov in the 1920s, has independent scientific merit to this day, and besides that in the opinion of V.P. Alekseyev, it is completely applicable to the theory of anthropogenesis. In a monography titled, *The Law of Homological Series in Hereditary Changeability*,[374] the Soviet biologist formulated two basic rules of his scientific system: "Close genetic species are characterized by parallel and identical series of traits, and as a rule, observe [these] conditions: the closer the genetic species, the sharper and exacter the identity of series morphological traits manifests. Close genetic species consequently have identical series of hereditary changeability. Flowing from the first, the second law in polymorphism says that not only genetic species, but genuses display identities in a series of genotypical changeability."

From there it precisely follows that "humankind" consists of different species—that is, human races; and outer similarity is expressed in an identical number of arms, legs, and so on. This speaks not about a commonality of their origin, but about a closeness of the principles of parallel development. Vavilov makes a legitimate, practical conclusion: "Already in the present time one can figure an expedient definition of polymorphism in species, not by the number of descriptive and possible combinations, but by the number and peppering of **racial traits**, according to which species differ from each other." The given conclusion is found to be in exact accord with the ideas of classical racial theory. Descriptive hereditary racial traits lie at the basis of the similarities and differences between races, which comprise the basis of the theory of polygenism—not abstract intra-species mutations.

Vavilov also developed highly meritorious positions on the **mimicry** of living organisms; that is, about the imitation by one species of the forms of another, which are quite often observed among parasites and weeds, which imitate a basic breed; he also wrote about **convergence**, or similarity of traits. We observe proof of the truthfulness of both of these theses every day, for not only separate individuals, but sometimes entire peoples imitate the outer appearance and behavior of state-building nations, leading a parasitic existence within them.

American biologist-evolutionist Garret Miller emphasized in the article, *The Struggle of Opinions on the Question of the Ancestors of Man*:[375] "Several scientists assert that all living and fossil members of the family of hominids are not at all as one customarily thinks of them, as comparatively recent offshoots from a common trunk. According to the opinion of these scientists, homogeneity of the physique of all the peoples known to this time is to far less degree a result of very close kinship, than of convergence—the protracted action of some or other formative forces on organisms, which are dissimilar by nature."

European anthropologists Hermann Klaatsch and Giuseppe Sergi, specializing precisely in studies in the area of **comparative anatomy and evolutionary morphology** of the human races and fossilized hominids, also supported this concept.

These positions are supported and developed by modern science. In part, biologist F.M. Sheppard indicates in the book, *Natural Selection and Heredity*: "Inasmuch as traits exist, providing adaptation to different conditions in the environment, it is no surprise that many groups of organisms have similar directions of evolution—the enlargement of the body, for example. In definite, specialized conditions, similar adaptations often arise. Horses, for example, arose twice. Mimicry is another example."

Soviet biologist G.F. Gauze developed the basic positions of Vavilov's theory, indicating: "It is known that there are very many examples of parallelism in the outer manifestation of modifications and mutations. According to Schmalhausen, this effect is called **"genocopies."**

Here again do not forget the words of Giordano Bruno, which he spoke 500 years ago, that some people are not people, although they resemble them outwardly. A concrete example is the infamous maniac, Chikatillo. At the time of investigative experiments, he was discovered to have

---

[374] *Zakon gomologicheskikh ryadov v nasledstvennoy izmenchivosti*. Leningrad, 1987.

[375] *Bor'ba mneniy po voprosu o predkakh cheloveka*. Sbornik "Evolyutsiya cheloveka", Moscow, 1925.

a blood type and sperm type that were incompatible, which is not possible for a normal person. He was a "non-human" in the direct sense of the word, and committed inhuman crimes.

In his work, *Chemical Structure of the Biosphere of the Earth and its Environment*,[376] V.I. Vernadskiy focused attention on the unequal tempo of the evolution of species, and also their polygenetic origin: "The notion of a single ancestor, or several ancestors of the modern organic population of the Earth is not only not supported by the direct findings of paleontology, but it contradicts what is known to us about the geological significance of living matter. Heterogeneous living matter is always observed in the biosphere."

Finally, N.V. Timofeyev-Resovskiy, N.N. Vorontsov, and A.V. Yablokov wrote in the joint work, *A Short Essay on the Theory of Evolution*,[377] that: "Formulaic notions of 'species' are not only limited according to content, but are difficult to apply in practice. The content of the concept of species, may, in various situations and various groups of organisms, be very different, reflecting both a type of beginning, and a history and further fate of corresponding species. It is thought, that if the concept of rigid morphology is subsequently applied to all the taxonomies of a higher family, then we may soon come to the conclusion of polyphyletic origin in all large groups, since it is completely natural, that a group of species related at the level of the family, may often not develop from one, but from several species. Clearly, homologous genes and entire gene groups are retained in related species, genuses, and families, and they can homologously mutate.

Undoubtedly, it is necessary to reject the concept of *rigid monophilia* in regard to major taxonomies."

Left: Ernst Krieck

But this refusal is fraught with the destruction of the very notion of "humankind", with all the irreversible consequences that flow with it. All human rights defenders will lose not only the physical, but even the metaphysical basis for action. How to skillfully measure the suffering of beings, in whose lives non-human blood flows, but who imagine themselves to be *homo sapiens*? N.I. Vavilov wrote: "Identical changes in the phenotypical order can be caused by various genes." And V.P. Alekseyev maintained that outwardly, similar peoples have completely different racial origin.

In the 1920s, Soviet biologist U.A. Filipchenko suggested earmarking genotypic parallelism, which is observed in related species, including the primary traits of related species and genuses. Besides that he set apart anatomical parallelism, flowing from identical opportunities of development, lying in the organs. A.A. Zavarzin established commonality of the histological structure of analogous organs, in different classes of animals, independently of genetic relationships. All this also speaks in favor of polygenism in the development of the basic races. Vavilov remarked in this regard: "The number of facts of convergence in living organisms grows with each year. Mutations exist which go in different directions, but are joined under a revealed common law." Namely therefore, peoples that have different racial origins, not uncommonly reveal a similarity in outer appearance. Thus, an even number of arms and legs, and also articulate speech, are not at

[376] *Khimicheskoe stroenie biosferi Zemli i ee okruzheniya.* Moscow, 1965.
[377] *Kratkiy ocherk teorii evolyutsii.* Moscow, 1969.

all criterion for affiliation with a single "human family". Studies in the area of the biochemical differences in the basic human races, also confirm as a whole the theory of polygenism. In the monography, *Historical Anthropology and Ethnogenesis*,[378] Soviet anthropologist V.P. Alekseyev asserted: "Today it is precisely established, that for the extent of all the history of humankind, there were not single-vector changes in blood type factors; and in this regard, the local specific features of the groups of a population were preserved. Thus, in each definite moment of history, Mankind was a commonality—heterogeneous, according to the genes of blood groups."

But if there never was a commonality according to blood group genes, and along with this, there was never their single-vector changes over time, then consequently there cannot be any talk about a single source for the rise of "humankind", and as a result, of a "humankind" as such in general. Namely therefore, the German philosopher and racial theorist, Ernst Krieck[379] said: "Mankind—this is the most doubtful whole of all." Even such a luminary of Russian anthropology as Bunak, recognized in his fundamental monography, *The Genus Homo, his Rise and Subsequent Evolution*:[380] "Hybridological criteria as a single basis of systematics is particularly unacceptable in the taxonomy of primates, since certain viable admixtures are as different in form, as some species of macaque and baboon." Paul Broca also emphasized: "Physiological phenomena of the fertility of hybrids does not serve as a basis for differences in species, nor for the determination of their origin."

Therefore, the "free cross-breeding" of which geneticists speak, concerning the members of different races, is not a basis for assertions of a specific unity of the human genus. Fatal morphological differences between species of peoples, passed down from generation to generation, manifest themselves in a natural way, in their psychologies and behavioral strategies. V.A. Moshkov wrote: "In psychological terms, the differences between the extreme boundaries of humankind are as great, as those between mammal predators, like the lion or tiger, and the ram."

Soviet biologist I.I. Schmalhausen believed that no less than 25% of species in the Animal Kingdom are complete parasites, which by the way, can be observed without effort in day-to-day social practice. It has been established by geneticists, that some mutations in an organism can transform an entire genotype into an anomaly. But this is valid for entire populations, for we also observe entire, insane peoples.

Not in any way wanting to insult this or that people, we again suggest faithfully turning to an expert on the question. E.N. Khrisanfova and P.M. Mazhuga pointed out in the book, *Essays on the Evolution of Man*, that: "There is a great interest in the use of ethological traits in phylogenetic concepts. Classification schemes according to behavioral reaction agree sufficiently enough with the based-upon morphological traits. This means that the specific features of the behavior of a living being are always strengthened anatomically. The hunter and the prey are different biological species; the outward, exterior furnishings of an organism unmistakably determine its biological fate.

One of the important morphological changes of the races in the process of evolution is the so-called phenomenon of **gracilization** (from Latin *gracilis*—slender, gentle), as a result of which a decrease in the general mass of the skeleton occurs, [and a decrease] in the slope of the forehead, the pronouncedness of the eyebrow ridges, the longitudinal diameter of the skull, and the width of the face.

Studies by G.F. Debets show that in Europe, gracile changes began before the Neolithic Period, during the time people transitioned from hunting-gathering to farming.

N.N. Cheboksarov and I.A. Cheboksarova considered it necessary to note in the book, *Peoples, Races, and Cultures*:[381] "However, the process of gracilization did not have a place with all peoples, not even those transitioning to farming; thus, for example, studies by Soviet anthropologist M.G. Abdushelishvili showed that for the extent of their history, in many peoples of the Caucasus an increase in the width of the face occurred, rather than a decrease, and the eyebrow ridges became somewhat larger, and the slope of the forehead decreased."

---

[378] *Istoricheskaya antropologiya i etnogenez.* Moscow, 1989.
[379] 1882-1947
[380] *Rod homo, ego vozniknovenie i posleduyushchaya evolyutsiya.* Moscow, 1980.
[381] *Narodi, Rasi, kul'turi.* Moscow, 1985.

The main conclusion of the authors is that the transformation of traits in various territories, in various racial groups, and in various periods, had differing tempos and directions of changeability. "Gracilization naturally did not have a place with those peoples, which until recent times retained archaic, appropriation forms of economy."

But if you apply Abdushelishvili's logic to the natives of those places, it completely clearly turns out that "many peoples of the Caucasus", particularly the northern Caucasus, practice "appropriation forms of economy", or more simply said, economic parasitism. The economies of many of the republics of the Caucasus are subsidized. Alas, it is certifiably significant that the narco-trade, speculation, the slave trade, and banditry are widespread there. And these criminological phenomena, being typical examples of "appropriation forms of economy", are physiologically consolidated in the appearance of many of the members of the given tribes. **Anti-gracilization**, according to Abdushelishvili, is a result of an evolutionary pre-disposition to parasitism, and as a result, banditry.

The type of appearance and the character of life activity are always inter-related. Studies in the area of genetic aggressiveness speak about this also. It is not necessary to be an astute person, in order to differentiate the inherently creative form, from the inherent parasite. Any zoologist-systematist can easily formulate the evolutionary-biological traits, by which given types of organisms differ.

It is particularly worth mentioning, that our plans do not include pointing and disparaging the "many peoples of the Caucasus", for it is completely obvious, that "appropriation forms of economics" exist not only in this region. In general we do not take upon ourselves the audacity to argue with an émigré from the Caucasus, to whom it is naturally more obvious.

French scientist George Tesse (1900-1972) is one of the founders of the **synthetic theory of evolution.** In 1937 the concept of an **artificial population** was substantiated by him. Naturally, all living things have their inborn system of values, and something artificial will unavoidably have artificial worth. Today we can easily observe all this in the conditions of the modern "melting pot", or the multi-cultural society, with its unhealthy socio-biological climate. And many Hollywood films about replicants, mutations, maniacs, and simple genetic copies, are also created on the basis of the theoretical and practical achievements of modern evolutionary theory. Already put forward and substantiated is the **particular theory of evolution,** which considers the independent development of populations and races, in isolation from all remaining [populations] and according to their particular laws. The most ancient known strategy of this kind is the Jewish Diaspora scattering, which embodies the idea of a separate world of Judaism, living according to its own laws and evolutionary criteria. The given phenomenon is described in detail in Kevin MacDonald's book, *A People That Shall Dwell Alone: Judaism as a Group Evolutionary Strategy with Diaspora Peoples.*

In the conditions of the modern political, financial, techno-genic, and informational diktat, the prospect of a fall of biological rivals down the ladder of evolution is created. Modern pop-culture and the mass-media net, which propagandize the cult of cynicism, toughness, and unscrupulousness, carry within them the single object of awakening the rudimentary animal inheritance in Man, and thereby retarding the evolutionary development of objectionable rivals. Soviet anthropologist M.F. Nestrukh pointed out in the book, *The Origin of Man,*[382] that: "Miscegenation led humankind to a distinctive, biological unification, to a retardation and suspension of species evolution."

Therefore, it is completely obvious that the very idea of "humankind" belongs to those mimics and parasites, who feed off the idea of an imaginary, single people. Common human values—this is only just a gimmick that has the goal of diverting the unsophisticated into an evolutionary bog.

A fundamental discovery of the modern age—the decoding of the human genome—threatens the use of **genomic blackmail and genomic terrorism**, directed against biological rivals. Mankind, (or more accurately, what is meant by the word) is an assembly point of variously characterized evolutionary groups, naturally having different worth and different tasks. Gradation according to the principle of "high-low" extends to entire races, just like it does to separate populations, and then has its effect at the level of separate individuals. If we return to the logical

---

[382] *Proiskhozhdenie cheloveka.* Moscow, 1970.

conclusions of Timofeyev-Resovskiy, Vorontsov, and Yablokov, it becomes obvious that imposed, custom-made **political-monogenism,** when carried to its logical completion, sooner or later becomes its own negation.

For a better understanding, it is sufficient to turn to the latest classical work by E.N. Khrisanfova and I.V. Perevozchikov, titled: *Anthropology.*[383] In order to literally pull the facts in past the ears and into the lap of the favorite, "one-ape" theory of anthropogenesis, the authors of the book were literally compelled to create a miracle of intellectual equilibrium: "Now many new findings appear in favor of a single (monophyletic) origin of the apes of the Old and New Worlds. It is assumed, that on the boundary of the Eocene and Oglicene epochs, there was a chain of islands in the South Atlantic, connecting South America with Africa, and extending from Southern Africa to the Falklands Plateau; there could have been a chain of volcanic islands along the mid-Atlantic ridge, also serving as a path of migration for primates and rodents."

Most of all, we note that to build such a grandiose, synthesized, ideological construction, like the origin of Man, on admissions of a partly archaeological character, without proof from the results of related disciplines, is inadmissible according to all laws of logic, something that anthropologists are unfortunately not taught. Finally, can you picture for yourself, the process of a purposeful transmigration of apes and rodents from island to island, along a path of several thousand kilometers? If only they and others could have navigated by the stars; and if only apes could have learned how to swim; and if rodents grew temporary gills, they could have made their way there by day, and then get rid of them, as if they never had them... Come on, what does this say here? A certain children's book in Russia—*Neznayka on the Moon*—is the peak of realistic art, in comparison with this textbook.

Furthermore, the authors write that "a factual gap exists in the paleontological record, between 8 and 5 million years ago."[384] And then they write: "The most plausible hypothesis now sets the line of human evolution relatively earlier—from 8 to 5 million years ago."[385] So it happens that their most plausible hypothesis proceeds precisely from an absence of supporting paleontological facts.

The Roman Catholic Church Father, Tertullian, once uttered the phrase: "I believe, because it is impossible." Centuries later, the great philosopher Hegel stated: "If my theory does not agree with the facts, then that is too bad for the facts." Speaking for ourselves, we do not have a right to count on a different logic in people, who just twenty years ago were studying the fundamentals of Marxist-Leninist scholastics, mixed in abundance with Hegelianism, in order to gain their professorships and scientific degrees. Paul Broca therefore correctly noted that "logic is unknown to monogenists."

In order to explain the process of the morphological transformation of apes into an intermediate ancestor of Man, they rely on an origin in Eastern and Southern Africa, with radiation leading to the sought for mutations. Of course, it does not explain where this radiation came from, and why some exposed apes stood on their feet, and others did not. Besides that there is not one supporting fact for the manifestation of a new biological species, as a result of the influence of radiation, never mind that the majority of mutations are harmful for an organism, and do not lead to its evolutionary development, but to a lethal outcome.

To this day, in the Roman Catholic ethic, in order to conceal an obviously stupid or unseemly deed, there is still the notion of indulgences—the paid-for remission of sins. In modern biology, the concept of mutations serves the same goals. Its power is limitless, and its direction of action is analogously selective.

We have already shown that Lamarck and Darwin, themselves being evolutionists, ironicized over the notion of "species", on the basis of which they built their theoretical constructions. E.N. Khrisanfova and I.V. Perevozchikov acted in this same spirit, for in the end of their book, they wrote: "In the modern state of affairs in anthropology, the situation is such that facts can be construed to mean anything."

Of course, dwelling over one's own lengthy, epistolary legacy is the personal business of authors, but the trouble is that the given "masterpiece" is recommended by the Ministry of

---

[383] *Antropologiya.* Moscow, 1999.

[384] Page 41.

[385] Page 45.

General and Professional Education of the Russian Federation, as a textbook for students at institutions of higher learning, who are studying in the field of "Biology." One cannot now doubt, that the "field of Biology" is doomed by its new unsinkable helmsmen, among them Lysenko and Prezent.

Equally, by carrying racial logic to the area of human culture, we can observe the same pattern with particular ease. Culture is not taken from some "abstract milieu", as culturologists impress upon us. Only a race creates everything: everything around, and worthy and civilized; and even transforms the landscape. Konrad Lorenz, an Austrian philosopher and specialist in the area of ethology, indicated in the book, *The Other Side of the Mirror*, that: "The unity of human "civilization" as such is a fiction, like the unity of the phyletic tree of life. Each branch, each twig, each species grows at its own risk, in its own direction—and so precisely does a separate culture behave! Thus, human cultures do not arise, as the standardized philosophy of history postulates, in a linear sequence and according to a single law, but independently of each other; precisely also do species of plants and animals arise—a student of evolution would say that they arise polyphyletically. The leveling of all peoples has a destructive effect: if all peoples of all cultures struggle with the same weapon, if they compete with each other with the help of one and the same technology, and attempt to outwit each other on one and the same stock exchange, then **intercultural selection loses its creative effect.**

In his monograph, *The Origin of Man*, Soviet anthropologist B.C. Zhukov transferred the pattern of the evolution of the human races to their cultural achievements. In the same sense, he expressed with complete correctness: "The development of the lateral branches of the human tree could not lead to a gradual perfection of human traits; in order to make [the lateral branches] close to the modern members of the higher human races, they had to either die out, like some fossil human races, or they had to almost remain in transition, like lower modern races, which do not participate in the cultural progress of Mankind."

In light of the above-mentioned facts, a simple conclusion invites itself. Why haven't evolutionists, biologists, and anthropologists across several centuries come to a consensus on the question of the history of Mankind? Why do they re-examine evolutionary concepts all the time, while advancing new ones? For a simple and completely obvious reason: they are attempting to prove the existence of something that does not exist, and has never existed in principle, namely "humankind" itself. **That which by custom we call humankind, is only some evolutionary enclosure, in which organisms that have completely different origins, have competed for millions of years already, and as a result, have differing worth.**

If all peoples had a single origin, they would unavoidably have one tactic for biological behavior, and similar goals in life; and one and the same strategy for reproducing offspring. But we observe nothing similar in the history of "humankind", where some races disperse without a trace, creating cultural worth that is garbage in the eyes of different races, [some] completely unconcerned about the creation of culture, states, or even literacy. Really, one can see something similar in a population of wolves, among which one specimen eats meat from a kill, another indulges in devouring scraps, and a third turns to vegetariansm, eating grass. In the meantime, this phenomenon of extremely narrow specialization in the foods of entire peoples and races, to which we turn no attention at all, and the visiting of national cuisine restaurants, should serve as a first indicator of the diversity of "humankind."

And can you picture for yourself a flock of swans, part of which keeps its marital vows, and raises young, another part indulging in debauchery, trading its own children to be killed for donor organs? Have you ever seen ants, one of which is occupied with building an ant colony, the other destroying it? Where have you seen bees, some of whom are collecting nectar, and another part living for generations at the cost of handing over his percentage to a neighboring beehive? Is it possible in nature, when we say that all these species have an origin from a single source? In the meantime, these and a multitude of other contradictions in the daily behavior of various peoples and races do not compel us to doubt in the authenticity of that which according to misunderstanding, is called "humankind". The notion of "humankind" is a vile trick, an outrageous myth, which should be exposed, once and for all.

If one applies the axiom of the unity of humanity, as postulated by "anthropologist-humanists", then in questions of sexual mating behavior strategies, a single species would not demonstrate such shockingly wide forms of behavior. Just try to picture for yourself badgers, which in one region adhere firmly to monogamous marriage, practice polygamy in another region, and in a third region practice guest marriages. Any zoologist-evolutionist, on the basis of these elementary facts, would not hesitate to consider these animals as different species. But if we start a conversation about Man, for reasons of false political correctness, the principles of systematics suddenly retreat into the background.

Russian psychologist V.P. Osipov correlated the findings of reflexology with vast cultural-historical material, and came to the conclusion that mass homosexuality as a social-biological phenomenon had its origin exclusively in the East. Ancient Europe had no notions about it. Another Russian scientist, ethologist O.V. Mil'chevskiy, certified that the custom of castration originated in the harems of Ethiopia and Libya, as a result of the hatred and jealousy of husbands toward male guards, then spread further into Egypt, Judea, and Assyria. All exotic forms of lewdness, in Mil'chevskiy's opinion, also had their primary, initial origin among the southern equatorial races; for a long time, they were completely unknown among the White cultures of the peoples of Europe. Without exception, the lofty, almost God-like attitude toward virgin girls exists in the mythology of all peoples, which belong to an Indo-European cultural circle, which was formed on the biological foundation of the Nordic race. At the same time, among the Negroes of Equatorial Africa, to this day a custom exists, whereby the high priest of the tribe uses a stone knife to publicly cut away the virginity of girls who have barely achieved sexual puberty. Virginity is an impermissible burden for tribes, which by tradition practice orgiastic cults. Among Mongoloid tribes of northeastern Eurasia, the customs of guest marriage and "renting" of wives is highly widespread. A podium with an essential, vertical pole, designated for whirling around in a dance by unclothed dancing girls, has today become the attribute of any strip-tease club; but for some reason, today they are solidly associated with the European system of values. In point of fact, this traces back to elements of the Asiatic temple prostitution of Ancient Babylonia; German historian Hugo Winkler was the first in European science in modern times to direct attention to this.

The difference in forms and styles is a manifestation of a basic instinct—the instinct to continue the species. This best of all points to an absence of a single evolutionary-biological foundation, as should be present in a single human species.

Practically every book on forensic psychiatry and sexology, obligatorily mentions **heterochromofilia** in an enumeration of sexual perversions; this is when a sexual partner can only be a subject with a different skin color (a variety of fetish).

But then one can logically come to the conclusion, that mixing races is the result of the steady sexual perversion by genetic refuse from the number of original pure races, which have fallen into fetishes in the process of degradation. And if we recall the assertions of modern evolutionists about the net-like development of the basic races, where the "upper floors" are located in direct contact with the "lower floors", then it becomes obvious, that these mixed races are the result not simply of a persistent sexual perversion, but a very grave form of inherited zoophilia, when the object of sexual desire is a subject that occupies an intermediate position between Man and animal. Now it becomes understood, why in the ancient world, apes were called "servants of fetishes", for they and their transitional forms appeared among many peoples as objects of sexual desire; this is graphically reflected in the great number of legends in all parts of the world.

Our conclusions serve as illustrations and as a confirmation of the concept of the hybridization of the human genus, consisting of species of unequal worth in the evolutionary-biological scheme, and carrying within themselves the various manifested forms and different degrees of concentration of human and animal traits. The given was first argued by Russian scholar V.A. Moshkov, in his fundamental work, *The New Theory of the Origin of Man and his Degeneration*.[386] For our part, we think it a great credit to appear in support of this scientific theory, after almost an entire century, and relying on new facts to again prove its reliability. You will immediately experience huge relief, the moment you find yourself free from the weight and judgement of the harmful myth. The world becomes transparent and understandable in its very essence, and the majority of day-to-day problems find their own resolution, when you just barely begin to sort "people" according to a racial-biological scale and according to their worth. There will be fewer disappointments, mistakes, and spiritual wounds. Your faith in yourself will not

---

[386] *Novaya teoriya proiskhozhdeniya cheloveka i ego virozheniya.* Warsaw, 1907.

decrease because of the mosquitoe's bite or the bark of a mongrel watchdog. Precisely so then, it will be necessary to enter into [analogous] relationships with those, whom by mistake you earlier considered among one species of "Man." Understanding now that this specific unity does not exist in principle, you endure the cardinal break-up of moral-values landmarks. But this does not need to be feared, for this is the regular process of purging the biological basis of conscience of slag, which hinders the effective use of natural, racial self-value.

Regarding this, Paul Broca wrote: "Inasmuch as there is not one fact that supports the theory of monogenesis, and not one monogenist hypothesis explains the origin of [different] types, it remains to be assumed, that these types arose by various paths. Monogenists also attempted to talk about the moral unity of the human genus. But all studies in this field have shown the opposite: the intellectual and moral differences of the basic human races are more than anatomical differences. Without a doubt, the human group is a genus. If this genus consisted of one species, this would be a unique exception in nature. Therefore, it is natural to think, that this genus consists, like all others, of several species."

The great Goethe said: "Mankind: this is an abstraction. Since olden times, there were only people, and there will be only people." Houston Stewart Chamberlain formulated it thus: "Then Mankind, about which so many have philosophized, suffers from one heavy ailment: it does not exist." In the book, *The World in the Mirror of the Racial Spirit*,[387] German philosopher Kurt Brenger wrote: "We made race a basis of our worldview, and place it in the center of our spiritual life. By the same, we consciously resisted those peoples, who dream of 'Mankind'. If they do not recognize racial differences, or even want to overcome them, then we firmly know that there is no 'humanity', and never will be. It is not like biological unity, because people are divided into groups, which in science are called *races*. Today, teachings about heredity and raciology distinctly and synonymously prove to us, that the desire to merge all peoples into one whole is the product of humanitarian fantasies. Astute and realistic thinkers always knew that 'humanity' is an empty word, a naked diagram."

A modern French philosopher, Pierre Chassard, also declares that "humanity is a myth that does not correspond to reality." One of the conspicuous theoreticians of National-Judaism, Aaron David Gordon, asserted in his *Letters from Palestine*: "Humanity, of which much is said, is only an abstraction, a term taken from thin air that does not have any correlation with that which really exists on Earth."

Naturally, with the absence of a 'single humanity', there is also an absence of such a thing as 'common human values'. Therefore, turning to the high passion of the great Nitzsche, one may assert that "a reappraisal of all values" is approaching with all inevitability. "Joyful science" will come down to nothing, and a "too human humanity" will give itself to know, and the "new dawn" of the coming "super-man" will approach.

---

[387] 1941

Racial-Ideological Neurology

"One genotype-one instinct"
N.P. Dubinin

"Man more easily notices inequality, than equality"
D.N. Uznadze

The ancient Greek philosopher, Alkmeon of Croton, asserted around 520 B.C., that "the brain is the seat of the soul." However, it is a paradox in the development of European natural science, that a detailed, serious study of the brain—the organ responsible for the function of the mental abilities of Man, only began near the turn of the 19th Century. Still, educators and humanists, as well as the philosophers and naturalists, who composed the cadre of ideological thought for the Great French Revolution, seriously asserted that the brain was only a "formless bundle of tangled hoses," and served as a "protective membrane for the lower cranium." It is not surprising that slogans about universal equality and brotherhood quickly sprouted from such fertile, "nature-philosophical" soil.

1. The Brain as the Seat of the Soul

The first to establish order in this 'heavenly' paradise of rising democracy was the outstanding German doctor and anatomist, Franz Josef Gall (1758-1828), who created studies of the localization of the different mental functions in these or those parts of the cerebrum, and who substantiated a completely revolutionary position (for that time), by which the strength of the mental and intellectual abilities of a person depends on the degree of development of his brain. Besides that he announced that free will, religious instinct, and even the ability for moral judgement were dependent on "hereditary predispositions." Gall wrote: "We come to the conclusion, that in the organization of peoples and races, basic differences exist, which are often of extreme importance. The more we get away from common attributes and delve in the particulars of separate phenomenon, the closer we are to knowledge of nature. The assertion that instincts are blind, is incorrect, for the reasons for action are prescribed, just like with animals...Anyone withdrawing into himself should feel and experience, that will and freedom are one and the same thing. The will is the collective, muscular mechanism of the organism...Never doubt that the human family is furnished with a special organ, by means of which it recognizes the Creator of the universe, and is amazed by Him. God exists, because each organ recognizes and is amazed by Him. The physical condition can halt the development of that part of the brain, by means of which it pleases the Creator to be opened to the human family. If there was a tribe, the organization of which was not completely developed in this regard, then its feelings of religion and knowledge of God would be no more capable than that of animals. For those beings, the organization of which is not developed in the sense of certain, definite abilities, there is no God."

Gall set apart twenty-seven zone organs (as he called them) of localization of the higher mental functions, the degree of development of which conditions the basic mental and cultural differences, between separate individuals, tribes, and entire races. "The brain is the instrument of every feeling, each thought, each desire. The intellect is the result of the simultaneous action of all the spiritual capabilities. By this, one may explain why one person may possess quick and reliable views, relative to some subjects, but be dull relative to others; thus, one may have a fruitful and lively imagination relative to subjects of one genus, but cold and meager for other subjects. One organ may act with more energy than another, and give reason more strength. It is also known that peoples with a large brain rise above peoples with a small brain, that they conquer and oppress them, as they wish. The brain of the Hindu is significantly smaller than the brain of a European, and it is known to all, how several thousand Europeans conquered, and now hold, millions of Hindus in dependency. Precisely also, the brain of the Native American is smaller than the brain of the European, and with America the same thing happened, as with India."

Tirelessly verifying all his brave hypotheses in practice, Hall calculated that the capacity of the skull of the White race was from 75 to 109 cubic inches, while in the Mongoloid race it extends from 69 to 93 cubic inches. Being the son of his own revolutionary times, and a genuine scientist-reformer, Franz Josef Gall was completely unafraid to transfer the findings of his anthropological observations to the social-political sphere, and also to the area of applied rights studies. "All skills and inclinations are inborn; therefore, equality before the law is a great injustice."

Franz Josef Gall   **Johann Mueller**

However, society of the time was subject to poetic views on the nature of Man, and it still categorically thought of the heart as the original cause of all mental activity; and clerical Christian education also resisted the spread of "materialism" with all its strength. Hall was ridiculed, and his ideas were banned from scientific use. But a stormy dawn of natural science in Europe had already begun to do its work.

Johann Mueller (1801-1858), one of the pioneers of physiology, asserted: "A psychologist can be none other than a physiologist." In 1866, English scientist John Ben published the book, *About the Study of Character*, in which he indicated: "The spirit, in all its manifestations, essentially depends on the brain; the more developed it is, then the more developed is the brain, and it weakens with deficiencies and illnesses of the brain." The famous German naturalist, Lorenz Oken (1779-1851), undertook an attempt at correlating the traits, of the psycho-physiological organs of the senses, in the members of the different races, proceeding from the degree of their development. His classification of races, according to the five basic senses, looked like this: 1) the Skin Man—the Black African, corresponding to the stage of the lower animals; 2) the Tongue Man—the brown Australoid and Malaysian, corresponding to the stage of the dog; 3) the Nasal Man—the Red American Indian, corresponding to the stage of the bear; 4) the Ear Man—the yellow Asian, the Mongol—corresponding to the level of the chimpanzee; and 5) the Eye Man—the white European, standing at the stage of Man.

The given psycho-physiological racial classification did not at all need to consider some manifestation of a particularly cynical, racist obscurantism, for even Engels, in his notable work, *Anti-Duehring*, positively valued the contribution of Oken to the development of biology and materialism in general.

And only with the beginning of studies by such scientists as Friedrich Tiedemann (1781-1861), Pierre Grazioli (1815-1865), Karl Vogt (1817-1895), and Dmitriy Nikolayevich Zernov, did a conscious and purposeful study of the specific features and form of the structure of the brains of different human races begin. In part, Karl Vogt was the first to discover that a striking viscosity in the substance of the brain of Negroes, exceeded the viscosity indicator in white Europeans.

Finally, in the second half of the 19th Century, so-called "psycho-motor centers" were discovered, giving rise to the theory of "new phrenology," also called the "geography of the cerebral cortex." Gall was the first to discover the speech center in the left temporal section; this area was later called "Broca's Area," in honor of the distinguished French anthropologist, Paul Broca. This great scientist was the first to start a deliberate campaign for the rehabilitation of Franz Josef Gall's scientific theory. Another luminary of anthropology, the German, Johannes Ranke (1836-1916), also emphasized in his fundamental monography, *Man*: "The first attempt in modern times to return to Hall's path, in the sense of a precise, limited localization of the functions of the cerebral cortex of Man, belongs to Broca. Nevertheless, we should note that with reference to the localization theory, with all its completeness in tracing the "geography of the cerebral cortex," it unavoidably leads not only to the localization of the will and consciousness, but also to their separation—correspondingly—to different centers."

Left: Lorenz Oken

Thus, the basic rational kernel of Gall's whole nature-philosophical system survived all the same, and with time planted seeds, for anthropology, psychology, and physiology, arrived at one and the same conclusion, by various paths: the performance of the basic mental functions of Man are realized separately, and any abstract conversations about some "single spirit" do not have any biological foundations of their own. The prominent German anthropologist-evolutionist, Herman Klaatsch, also considered it necessary to note:[388] "In modern times, Gall's attempt is worth far more, than was done before in a long time. Now they are aware that Gall's mistake is numbered among those that lie on the path to success in learning. And since each scientific conquest, expressed in definite form, has only temporary significance, then on the path to truth, scientific research marches from mistake to mistake. It's not the first time that an idea, thrown out as unnecessary garbage, comes up again after a long time to take an honored place. If the teachings about the skull, which were given by Gall in such a crude form represent an error, then in the foundation of it, a view which now plays a leading role in psychology and psychiatry, was already located in the womb. According to this view, there are definite centers for physiological functions of the parts of an organism in the brain, and in part, for the organs of the senses, such centers are the defined sectors of the so-called gray cortical matter of the large hemispheres."

And thus, Gall's basic idea was supported and developed; that there is a definite, solid connection between the degree of development of the separate zones of the cortex of the hemispheres of the brain, and the strength of the development of the corresponding mental functions.

2. Reasons for the Inequality of Ideas

However, before the start of the age of serious discoveries in the area of neurology, the founders of racial theory postulated that the quality of ideas are completely determined by the quality of the structure of the brain and the sensory organs of their carriers. Josef Arthur de Gobineau wrote in his main book, *Experiments about the Inequality of the Races*, that: "In the ugliness of forms, one can never blame the nature of the abstract thought, because the entire

---

[388] In the book, *Origin and Development of the Human Genus*.

business is in the layout of the eyes, and in the intellects and the hearts, toward which these forms are turned. Black and yellow aborigines can only understand the ugly and hideous—it was created for them, and it became necessary for them." However, to the enlightened public of Europe, for a half a century it was necessary that conclusions of a similar nature be made obvious.

Left: Friedrich Tiedemann

Russian scientist Vladimir Mikhailovich Bekhterov (1857-1927), published a number of works on neuropathology, psychiatry, psychology, morphology, and the physiology of the nervous system, and then occasioned to bring the studies to a qualitatively new level. Rejecting the subjective and highly emotional approach of the science of the time, V.M. Bekhterov was the first to formulate the principle of "objective psychology", according to which it was first of all necessary to "study the biological foundations of mental activity."

In his fundamental essay, *Objective Psychology*,[389] he indicated: "Mental functions stand in connection with the condition of the brain's blood circulation and the composition of the blood, which feeds the nerve cells. It is also known that different functions during common disease processes, and equally, pathological changes of the composition of the blood, change the root form of mental functions. There is not one mental process, which is only subjective or spiritual, in the philosophical sense of the word, and is not accompanied by definite material processes."

Thus, the specific features and structure of the nervous system, in different peoples and races, always manifests itself in the abundance of the forms of mental and material activity. Monuments of culture are never just taken out of nowhere, but are created by people of a concrete racial type, made from the nature of the racial-specific features of the structure of the nervous system.

No abstract spirit has ever created anything, for all creations in our world bear the stamp of the races of their authors.

Bekhterov wrote: "The objective psychology of Man, is not needing of introspection and has in mind only some objective facts and data, which are the result of his nervous-mental activity. Mentally conditioned movements, speech, facial expressions, gestures, deeds, and acts are traced from there, and in a wider sense, this actually constitutes the subject of the objective psychology of peoples, languages, morals, customs, and ways of life of separate tribes, their laws, and social arrangements, their industry and science, their philosophy and religion, their poetry and fine arts—in a word, all that characterizes the outward activity of separate and entire peoples."

Left: Pierre Graziole

Another internationally known classic of Russian science, Ivan Petrovich Pavlov (1849-1936), also emphasized: "dynamic phenomena, playing themselves out in the central nervous system, should be timed with the very fine details of the construction of the apparatus."

Soviet psychiatrist Viktor Petrovich Osipov pointed to the possibility of qualitative and quantitative evaluations of the parameters of the nervous system, and in part of the capacity of the brain among the members of different races and peoples, for creating culture. He wrote in his fundamental monography, *A Course of the General Teachings about Spiritual Diseases*:[390] "The cerebrum and primarily, the cortex of the large hemispheres of the brain, are the substrata of spiritual activity. Proof that spiritual activity is

[389] *Obektivnaya psikhologiya.* Sankt-Peterburg, 1907-1910, T3.
[390] *Kurs obshchego ucheniya o dushevnikh boleznyakh.* Berlin, 1923.

concentrated in the cerebrum, and particularly in the gray cortex of the large hemispheres of the brain, is drawn from different sources: from the area of comparative anatomy, embryology, from the area of physiology, anthropology, pathological anatomy, and histology. Comparative anatomy teaches that the higher the mental organization of the individual, the better the development [that] is observed in his central nervous system, and in part, in his cerebrum. Studies by anthropologists show that to the degree that Man happens to improve natural spiritual abilities, his cerebrum becomes larger and heavier; the brain increases [in size] with the intensification of the struggle for survival and the growth of culture; and uncultured peoples, staying at that degree of development, at which ancient Man was located, possess a smaller brain by volume and weight, in comparison with cultured peoples."

At approximately the same time, Pavlov advanced the position by which "all revealed details of the construction of the brain should sooner or later find their dynamic significance." By way of numerous laboratory studies, it was revealed that the traits of the structure of the brain should correspond to its functional characteristics. Thus, in neurology, the **concept of the unity of structure and function** experienced powerful development. In this regard, Soviet scientist L.A. Orbeli definitively emphasized in *Lessons on the Physiology of the Nervous System*,[391] that the higher the organism—from an evolutionary point of view—the more development the principle of the localization of the functions of the central nervous system receives. Also, L.S. Vygotskiy, pointing to the structure of the centers of the higher functions of mental activity, stated that: "The localization of the higher functions cannot be understood as some genetic timer; that is, the relationships which are characterized for separate parts of the brain in "cultured" and "uncultured" peoples, of "higher" and "lower" races, has in itself evolutionary-biological foundations.

Higher culture develops as a result of a narrower specialization of separate zones in the brains of its creators. A direct and clear functional connection exists between the specific features of the structure of the brain of different races, and the material culture created by them. Race and culture are neurologically interconnected, since each biological type always strives to express itself in recognized, adequate forms, which legitimize its worldview style, in the process of the struggle for survival.

The teachings about the localization of the mental functions were applied in full measure in an explanation of the culture-creating abilities of the brain. Thus, Bekhterov wrote as early as 1907, that the lobe portions of the brain are tied to "psycho-regulatory activity," and also to the correct evaluation of exterior impressions and the expedient, directed choice of movements conforming to the said evaluations."

In general, in Bekhterov's opinion, "the development of the lobe portions went parallel with the development of intellectual abilities, in the ascending order of animals." In 1949 Pavlov also pointed to the special relationship of the lobe portions of the brain with the most difficult forms of conditioned-reflex activity, which lie at the base of "expedient behavior." Academician P.K. Anokhin asserted that "acceptor activity" is tied with the functions of the lobe portions of the brain. English scientist K. Pribram established that difficult processes of self-regulation, which govern acts of behavior, are carried out at the base of the lobe portions. In the book, *The Brain and Activization*,[392] E.D. Khomskaya formulated a thesis, according to which, from a morphological point of view, "command potential" and the "capacity for action" are concentrated within the lobe portions of the brain.

But with all obviousness, it follows that the difference in cultures and even separate ideologies needs to be sought in differences in the structure of the brain, including in its parts. Relying on the results of practical research, V.P. Osipov clearly stated: "Nervous shock is always easier to direct along an already-beaten path of least resistance, which serves as an expression of the general law of energy, that is realized in the area of nervous-mental energy, like a derivative of a single world energy."

Thus, under the influence of this law, racial styles arose in art, science, religion, politics, and economics; for each stable biotype strives to express itself in forms and behaviors characteristic to it only, identifying the reactions of different biotypes according to the principle of "us-them."

---

[391] *Lektsiyakh po fiziologii nervnoy sistemi.* Leningrad, 1935.

[392] *Mozg i activizatsiya.* Moscow, 1972.

Thus, the main function of culture is to signal neighbors about the racial origin of its creators. The type of an idea is always functionally tied with the construction of the brain of its carrier, and mental pathology unavoidably springs up with disharmony.

Joseph Arthur de Gobineau turned attention to the following: "For Asiatic philosophers, true wisdom lies in submission to the strong, in not resisting the inevitable, and being satisfied with what is; Man lives within his thoughts or in his heart; he comes to the earth like a shadow, and walks along it indifferently, and leaves it without regret. The thinkers of the West do not preach such truths to their students. They urge them to savor earthly existence in full measure and for as long as possible. Refusal to accept poverty is the first condition of their law.The chief maxim is to beware the follies of the heart and mind; to take pleasure in [living] is the first and last commandment. The Semitic philosopher makes a desert of the rich Earth, with sands daily advancing on fertile soil; together with the present, they absorb the future. In opposition to it, Aryan doctrine says: furrow the earth with plows, and the sea with ships; then, on one fine day, scorning intellect with its illusory happiness, create a paradise here on Earth, and in the end, go down to her."

Basic stereotypes of the behavior strategies of the different races are recorded on the level of morphology, in the dynamic functioning of the nervous systems of their members; they manifest themselves in all the diversity of living forms, stylistically, flawlessly pointing to the primary source—a concrete race. Therefore, it is completely useless, for example, to explain the philosophy of Schopenhauer in a gypsy encampment, since they are racially-biologically incompatible.

A founder of Russian eugenics—the science of the improvement of the human breed—N.K. Kol'tsov, clearly staked out the idea in his program article, *Genetic Analysis of the Mental Traits of Man,*[393] that: "the characteristic of each separate culture, is the characteristic of that constitutional type of nervous-mental characteristics, which play the main role in the creation of the corresponding culture."

Soviet scientist V.I. Vernadskiy adhered to a similar system of proof; in the article, *On a Scientific World View,*[394] he wrote: "Recognizing great beauty of artistic origin, we clearly understand and unavoidably recognize, that the relationship of human individuals toward it can vary strongly. Whole classes of people can exist, in whom this or that work of art should arouse completely distinctive, unusual impressions. A different example of this is the history of music. Among different peoples, or in different epochs of life, one or another has manifested completely different, basic tone scales in its music. For example, in the history of high development, music that is foreign to us—Chinese or Japanese, for example—lacks two to seven of the basic tones in our music scale. In this regard, the impression made by our music on the European-educated Japanese, is highly instructive. But music close to us, such as the complex compositions of the Hindus, seems foreign to us. In the history of peoples, the most basic ideas have sharply changed, as we see in the history of Greek music, where the basic scale has changed several times. The ancient, discovered hymns seem strange and unmusical to us.

The ideal of beauty in the works of Greek art were created to a significant degree, under the influence of the structure of the body of the Nordic or Mediterranean race. These works cannot summon the same feelings in the foreign physiques of the highly artistically developed peoples of the Mongol race, such as the Japanese, as they do in us.

We can also completely see and constantly observe the same thing, in regard to systems and compositions, ideals, and concepts of religion and philosophy.

A follower of some religious or philosophical teaching cannot demand that they be considered unquestionable and irrefutable; or that it be so recognized by any other person who sincerely treats these questions."

The recognized classic of Russian science also fairly reconciled the logic in the processes of development in the areas of philosophy, religion, art, and anthropological aesthetics. Canons, norms, and stereotypes that form the given roles of being, do not arise out of nowhere in the

---

[393] *Geneticheskiy analiz psikhicheskikh osobennostey cheloveka.* Russian Eugenics Journal, 1924, T1, Issue 3-4. [Russkiy evgenicheskiy zhurnal, 1924. T.1. Vip 3-4].

[394] *O nauchnom mirovozzreniya.* From the anthology: *Works on the History of Science.* [Sbornika: Trudi po istorii nauka]. Moscow, 2002.

human psyche, according to the capricious dictates of social surroundings, but on the contrary, they condition it, drawing their beginning from the group hereditary mass of its creators.

Modern geneticist Yuriy Ivanovich Novozhenov also emphasized in the book, *The Adaptiveness of Beauty*:[395] "Thanks to isolation, even music did not become a universal language of art, understood in equal measure by Europeans, Chinese, and Hindus, or the peoples of Central Asia, within the boundaries of the former [Soviet] Union. To the Russians arriving in our former Central Asian republics - now independent states - the music broadcast by radio seems mournfully monotonous, arousing feelings of monotone wailing and plaintive groaning. At the same time, European music seems cacophonous to the Uzbeks, Tadjiks, and Turkmen, irritating their Central Asian ear for music. Different types of musical language, to the extent of the internationalization of the humanity, leads to a condition of sharp antagonism, or even incompatibility."

In the sensational philosophical-political book, *The Death of the West*, the prominent, modern American politician, Patrick J. Buchanan, relying on data from official crime statistics in the US, and data on the cultural achievement of the races, made the following conclusion: "History shows there are not, and never were, absolutely equal peoples, cultures, and civilizations. Some achieved grandeur regularly; others never came close to it at all. Ways of life, religion, and ideas differ; you will not find equality anywhere."

Russian psychologist and neuromorphologist A.S. Arkin substantiated the thesis that our worldview depends on our spiritual organization, in the book, *The Brain and the Spirit*.[396] He came to this weighty conclusion on the basis of long years of practical experience. Thus, in the article, *On the Racial Traits in the Structure of the Brain Hemispheres of Man*,[397] he wrote: "The middle frontal fissure is the fissure, which to a greater degree than the other fissures of the cerebrum, is subject to changes; and among the members of different races, it has different contours." Besides that proceeding from foreign materials, Arkin spoke for the extent of his article about "brains, rich in convolutions, which as is known, are considered more completely organized."

A.S. Arkin's principle revelation in the article was that the "most characteristic racial differences are noted in the area of the associative centers." These centers develop later, in comparison with other sections of the brain. In them, the outer, morphological differences in the structure of the brain, of members of "higher" and "lower" races, are easily seen. The comprehension of other cultures, and in equal measure, consciousness of one's own culture, is closely entailed with the development of these associative centers. The language of a concrete culture, its style, a certain refinement, or on the other hand, barbaric rudeness, the depth and frequency of experiences characteristic to it, thus have clear physical outlines.

The conclusion in Arkin's work is simple and persuasive: "Racial differences in the structure of the cerebrum have preferred fissures and convolutions, where they manifest more often, and in more relief."

Another Russian scientist, R.L. Weinberg, brought to light the racial differences in the structure of the Roland and Sylvius fissures, in the article, *Toward Teaching about the Form of the Brain of Man*.[398] German anthropologist Karl Vogt also wrote in this regard: "The Sylvius fissure has a more vertical direction in Negroes, and like manner, the Roland fissure."

French anthropologist Paul Topinard emphasized in his fundamental book, *Anthropology*: "The convolutions are thicker, wider, and less complex in lower races. The nerves of Negroes, and primarily the nerves at the base of the brain, are thicker, and the substance of their brains is not so white, as in Europeans." Possessing thicker cranium bones, of which the ancient Greek historian Herodotus wrote, members of the Negroid race therefore regularly have a lower threshold of pain sensitivity. The Association of Boxers pointed to this neurological fact in the

---

[395] *Adaptivnost' krasoti*. Ekaterinburg, 2005.

[396] *Mozg i dusha*. Moscow, 1923.

[397] *O rasovikh osobennostyakh v stroenii mozgovikh polushariy cheloveka. Zhurnal nevropatologii i psikhiatrii im. S.S. Korsakova, 1909, kniga 3-4.* [Journal of neuropathology and psychiatry, named S.S Korsakova, 1909 books 3-4].

[398] *K ucheniyu o forme mozga cheloveka*. Russkiy antropologicheskiy zhurnal, 1902, N4.

second half of the 19th Century, when they rejected the participation of black-skinned sportsmen in [their] matches, on the basis that they were less sensitive to pain, than whites.

In the book, *Man and his Place in Nature*, Karl Vogt pointed out: "The skull of a Negro is as strong as ivory. In a fight, the Negro attempts to strike the chest of his opponent with his head, and both negroes resemble lunging rams in the process."

On the biological foundations of culture, Paul Topinard expressed with still greater clarity: "Impulses, inherent to the brain substance, are so durable, that despite education and civilization, they are preserved even after cross-breeding and miscegenation, and help to identify the latter...Then the question about the differing traits of races, dependent on their brain organization, is significantly simplified, and can really prove that the method of activity in the brain yields differing traits, like the form of the skull, or the characteristics of hair...Suffice it to say, that ideas of morality can constitute physiological differences between races. Comparing fables and allegories, which lie at the base of mythologies, science traces back to the knowledge of mutual contact, in which peoples found themselves, and consequently, segregated acquired traits from their own racial differences...In a broadening of the problem, science staked out the past stages of borrowings in intellectual life, which some races did among others...There are languages that deeply differ from one another, and that demand a particular arrangement of the larynx for conversation in them, and special interpretation for understanding them...It is worth turning attention also to the different methods of perceiving musical scales, in the five parts of the world. That which is harmonious for the hearing apparatus of the brain for some races, is unpleasant to the ears of others. Upbringing is irrelevant here, since the very fact first has an anatomical foundation. The differences in systems of [numerical] calculation are also to be so treated. Peoples known as "Aryans", in general differ in the understanding of calculations, and by aptitude toward mathematics...Aptitudes for drawing also differ. There is a tribe that only knows how to draw circles and sticks; some of its members do not even know how to differentiate between a drawing of a head and a drawing of a tree or a ship...The races differ quite deeply by way of life and understanding of social conditions. There are peoples that seemed destined for a nomadic life, like Gypsies, the Jews, or the Arabs."

In more recent times, science has completely confirmed these deep nature-philosophical summarizations of a culture-biological character. In the article, *The Brain of Man and the Mental Processes*,[399] Soviet psychologist A.R. Luria pointed out that the "synoptic node" is that unit, at the level of which "the soul influences the body." Thus, it becomes completely clear that any phenomenon of a culture has, in its base, the characteristic, morphophysiological traits of the nervous system of its creators. In another of his books, *Essays on the Psycho-Physiology of Writing*,[400] he noted: "The occipital and occipital-parietal area of the cortex of the brain is the central apparatus, which enables realization of the complete visual perception of the individual, transmitting visual perceptions to the complex optical forms, to preserve and differentiate the optical images, and in the end, to realize the most complex and generalized forms of visual and spatial knowledge."

It is namely this which conditions the differences in aesthetics and artistic style among the different races, in the final analysis.

Differences in language and writing are also completely determined by differences in the structure of the brain. German scientist Friedhart Kliechs, wrote in the book, *Arousing Thoughts*:[401] "The Bushmen of Central Africa have language forms and sound formations similar to a parent language, at the transitional stage of development, from animal to Man." Carrying out full-scale tests on the Bushmen, a scientist established that their language is very poor in words that designate numbers; besides that many vocal expressions are substituted with gesticulations. Finally, the language of the Bushmen is characterized by reduplication (a form of repetition), for when speaking about one person, they say "tu"; when speaking about many they say "tu-tu"; and about a large number of people, they say "tu-tu-tu". The man-like apes, by virtue of a specific

---

[399] *Mozg cheloveka i psikhicheskie protsessi.* From the anthology, *Philosophical Questions of the Physiology of Higher Nervous Activity and Psychology.* Moscow, 1963. [*Filosofskie voprosi fiziologii visshey nervnoy deyatel'nosti i psikhologii*].

[400] *Ocherki psikhofiziologii pis'ma.* Moscow, 1950.

[401] *Probuzhdayushcheyesya mishlenie.* Kiev, 1985.

feature of their brain, also are easily trained in a similar system of counting. But then again, de Gobineau prophetically asserted in his main essay, *Experiments in the Inequality of the Human Races*:[402] "Not one people can have a language that stands at a higher stage, than the people itself. The hierarchy of languages is located in strict accordance with the hierarchy of the races." German raciologist and ethnologist Otto Reche (1879-1966) came to this conclusion, in the article, *Race and Language*: "As a result of the combined efforts of different the branches of science about Man, there is a resolution of the problem, 'race and language': the human races arose in a state of isolation, as the product of hereditary dispositions, selection, and endogamous pairings; and that same isolation begat a homogeneous type of language, as a product of the physical and spiritual traits developing in the race. Thus, the original race and type of language always coincide. Language was, so to say, one of the spiritual racial traits. Each race created a surprisingly harmonious instrument, in the type of language characteristic to it; an instrument adaptive in its very fine spiritual movements, an instrument which it cannot throw away unpunished. And if in later times, because of the spreading and mixing of races, the original picture, which was so clear, became clouded, this does not change anything in the basic fact of the spiritual inter-connection of race and language: language is part of the racial spirit."

Each people has that language, of which it is biologically worthy; therefore, the absence of a written language among many peoples points to a hereditary racial-evolutionary backwardness in the corresponding ideo-motor centers in the brains of their members. There is nothing racist or insulting in the given assertion, for this is a fact which is confirmed by the research of a majority of classical ethnographers, culturologists, and religious researchers. English ethnographer Robert Marett wrote: "In large part, ritual begat myth, but myth did not beget ritual. The savage's religion does not so much think out, as dance out." German psychologist and scientist, Wilhelm Bundt spoke in this regard, in the following manner: "The essential springs of mythological thought—this is not an idea, but a fit of passion, accompanying the idea everywhere. Thus, any myth creation originates from a burst of passion, and proceeds from like strong-willed actions."

Consequently, the differences between the basic religious systems need to be sought, not in the different cultural myths that begat them, but in the differences in the structure of the nervous systems of the peoples that created these myths, and also among those that bow to them. It is completely obvious, that in Asia, the traditional religions of the East are practiced by psychologically normal and socially disciplined people of sound mind, while the Europeans that are attracted by any kind of "Eastern exotic" are very often people who are originally from the marginal layers of society, with clearly expressed degenerative asthenic psycho-types. The degree of the adaptivity and assimilability depends on the degree of the identicalness of the type of idea, to the type of construction of the brain of he whom it is intended for. For his part, Pavlov pointed out that the differences in psycho-types, in the end comes down to differences in the types of structure of the nervous system. Namely therefore, some slogans and ideas, easily finding their adherents in one part of the world, do not enjoy popularity in another.

In his fundamental monography, *Anthropogeography*,[403] German scientist Friedrich Ratzel, also a founder of the science of **sociology**, emphasized: "It is worth considering the basic position of anthropogeography, that the range of ethnographic subjects can only progress through Man, with him, in view of him, on him, particularly in him, that is, in his spirit, like an embryo forms and idea. The ethnographic subject moves together with his carrier." His countryman, the ethnologist and religious studies scholar, Leo Frohbenius (1873-1938), formulated this brief, basic principle: "A culture does not have legs, and therefore it compels a person to carry it himself. An individual or a people is the porter of culture." And the world-recognized ethnologist, Bronislaw Malinowskiy (1888-1942), brought the formula of the thesis to completion, proclaiming: "Culture is a biological phenomenon."

---

[402] 1853-1855
[403] 1912

316

3. A New Concept of Intellect

The development of such sciences as molecular biology, psychogenetics, and neurochemistry essentially strengthened the positions of bio-determinism as a system of views in the nature of human communities. The claims of so-called culture in the role of ecumenical arbitrator were reduced to a minimum. Of which culture does a man in general have a right to speak about, if he is not biologically capable? And this generalization can easily be applied to entire ethnoses, populations, and races, for one cannot mix up or confuse high culture and folklore, since otherwise we would unavoidably be compelled to place an opera by Wagner on the same level as the rapping tom-toms of African natives. Such kind of equalizing by cultural anthropology booms out of university pulpits and departments, through the efforts of the "Sorosites", and those like them; you can't say anything about this, except that it is the degeneracy of the minds of the adepts of such disciplines.

A luminary of modern psychology, Hans Juergen Eisench, worked out a new concept of he genetic aspects of intellect in the work, *Intellect: a New Look.*[404] He experimentally measured intellect with the help of an electro-encephalograph (EEG), average attributable potential (AAP), cutaneo-galvanic reaction (CGR), and time of reaction (TR). He wrote: "Tests on the measured time of reaction (TR) have indisputable advantages: they are biologically far more fundamental, and are more independent of the influence of culture, than IQ tests, which are unavoidably distorted by cultural, educational, and socio-economic factors, of this or that kind." By the most natural method, it is said that the hereditary biological status of intellect far better and more accurately characterizes cultural abilities in an individual, than all abstract discussions about culture.

"During the processing of information in the cerebral cortex, mistakes occurred, possibly at the level of the synapses, and the higher the inclination of the individual toward such a type of mistake, then the lower his IQ." Culture, even the highest, can never change the morphological structure of a person's brain, while it strives for, and cannot change the intensity of nervous reactions. **Culture cannot instruct, it can only conform, since "culture is a biological phenomenon."**

G.J. Eisenck observes that what concerns other measured traits of hereditary intellect, is that the "results of studies show, that palmar/volar resistance grows with the increase of the level of intellect." Besides that among people with a low level of intellect, the amount of glutamic acid is lower, and consequently, the creation of a classification of races according to this important indicator essentially enriches understanding by us of the cause-effect ties in world history.

Although the human brain makes up only 2% of the body's mass, it uses about 20% of the energy expended by the organism. The basic supplier of energy to the brain is glucose; therefore, the main indicator of the energy requirement of the brain, is determined by the expenditure of glucose in milligrams, on 100 grams of brain tissue, per minute; this is directly tied with IQ. It is completely obvious that the races possessing different IQ, can easily be measured according to the level of the energy requirement of the brains of their members, in order to finally understand, which, more than others, "works with his brain"—not in the figurative sense, but in the direct sense.

"Glutamic acid, glucose, and other biological agents, are responsible for energy supply to the cerebral cortex, and are tied with the production of neurotransmitters; they play a leading role in genetically inherited intellect, being the source of that very "mental energy", which serves as the biological substratum of factor G (general intellect)"—writes Eisenck.

Finally, the chronic insufficiency of calcium in the brain is tied with mental retardation, and in part, with Alzheimer's Disease, Downs Syndrome, and Parkinson's Disease. But these massive ailments of humanity have clearly expressed racial and ethnic variations, that speak in favor of the discriminatoriness of the given trait.

The general conclusion in H.J. Eisenck's work is this: "Psychometric intellect depends 70% on biological [factors] and 30% on environmental factors. Thus, a departure from A. Bine's concept

---

[404] Questions in Psychology, 1995, #1.

of intellect occurs, and there is a return to the earlier developed views of F. Galton. Any working determination of intellect should be biological in its basis."

And it can never be said, that such a point of view stands alone in modern science, for authors are found in Russia, who hold to a similar system of proofs. Thus, A.M. Mustafin considered it necessary to emphasize at the very beginning of the article, *Biologically Active Points and Intellectual Awareness*,[405] that he completely shares Eisenck's opinion about the 70% hereditary determinism in intellect, which correlates with physiological and biochemical factors, such as hearing differences, EEG, and the speed of glucose utilization. A.M. Mustafin also stated that a correlation was observed between the termperature in the zone of the biologically active points (BAP) of the ear, and the given psychological testing. The given phenomenon is explained by the regulation of blood circulation in the cerebrum and ears of an individual. A remarkable Russian saying comes to mind: "Fools have cold ears."

Besides that a comparison is revealed in the thermometric and morphometric measurements of the ears, with the type of thoughts examined, and a correlation between temperature and progress was also observed. According to morphometric parameters, ears were selected, according to their length, and also their symmetry, since a difference in their length indicates a cognitive style and type of thinking (left or right hemispheres). The main conclusion is that in the development of the zones of the brain, there is a scene of a certain competition of the faculties. Among different races, the faculties developed unequally, in accordance with the hereditary program.

Besides this, intellectual abilities are registered in the basis of measurements of the temperature index of the skin (TIS), at nineteen points on the head. The lower level of intellect and progress in Negro children, in comparison with Whites, is caused, in the opinion of the author, by a more intensive development of motor skills in blacks, to the detriment of intellectual development, something that is inherently strong in the genetic program of their development. "Progress and type of thought are determined primarily by biological factors"—Mustafin summarizes.

For his part, S.B. Malykh remarks in the article, *Studies of the Genetic Determinization of the EEG of Man*:[406] "On the path from genes to behavior, there is an entire number of levels of organization—cellular, intercellular, the level of functional organs, and the level of the integrated individual. As is known, behavior itself is not inherited; DNA is inherited. DNA codes proteins, from which cells are built (neurons); cells in their turn enter into cooperation, determining the structure and function of the brain. The behavior of an individual is determined by the development of the brain, in the process of interaction with the environment."

Hereditary factors influence the bioelectrical activeness of the brain, which is measured by an electro-encephalogram (EEG); but different races, for their part, have different hereditary modes of behavior, which lead to frequent inter-racial conflicts in major, "multi-cultural" cities. In accordance with this, on the basis of EEG measurements, one can objectively substantiate the degree of biological complementariness of the races in socio-intellect, in order to show, who can coexist with whom, and which obviously cannot live together. "As it turns out, the different parameters of the EEG are tied with a wide circle of behavioral and cognitive characteristics, emotionality, and temperature; the EEG is individual specific and sufficiently stable during the course of the life of the pure substance of the characteristic." Besides that on the basis of this characteristic, the possibility opens of the revelation of the parametric incompatibility of different types of so-called "common culture", for with all obviousness, morpho-physiological and functional differences in the structure of the brain of peoples are observed: some peoples prefer the music of Wagner, and others the songs of a gypsy camp. The maximal significance of the coefficient of heredity for EEG is 7-9%, namely, in dependence on the range of frequency. For their part, racial musical canons consist of different harmonics, owing to this. After all, as we remember, Paul Topinard declared: "What is harmonious for the hearing apparatus of the brain of some races, is unpleasant to the hearing of others. Upbringing means nothing here, since the very fact is initial, and has an anatomical basis."

---

[405] *Bilogicheskie aktivnie tochki i intellektual'naya aktivnost'*. Voprosi psikhologii, 1997, #2.
[406] *Issledovaniya geneticheskoy determinatsii EEG cheloveka*. Voprosi psikhologii, 1997, #6.

The least genotypic influence (46%) is observed for the EEG in the left temporal region of the cerebral cortex, which is responsible for the speech centers. Obviously, owing to this, cultural anthropologists deduce a commonality of ethnoses, on a linguistic basis. One race more easily adapts to the language, the cognitive style of thought, the characteristics of temperament, or the stylistics of spiritual tribulations, of another race.

That which in racial psychology is called "the spirit of race" is in point of fact a commonality of macro and micro structures of the brain in the members in a given group. "Different regions of the brain do not work in isolation from one another; the interactions are coordinated by complex neuron chains, which are interconnected. The number and power of the connections find their reflection in the indicators of coherence. High coherence indicates that the EEG of different zones of the brain are synchronized; they are examined as an indicator of the "connectedness" of the regions of the brain, inasmuch as the synchronized oscillations in neuron nets are observed only in that case, when they are connected between each other."

Inheritability of coherence amounts to 85%, as pointed out by S.B. Malykh, and from there, with all obviousness it is clear, why national songs and dances harmoniously activate neuron connections in the hereditary structure of the brain, and have such a mental consolidating effect in the socio-intellect. Genetically conditioned, coherent biorhythms are a coded expression of a racial type.

It is namely in music, with all its intelligibility and purity, that the specific features and uniqueness of the spirit of a race manifest their inimitable style.

In support of our position, it would be appropriate to bring in the opinion of French psychologist and sociologist, Gustav Lebonne, who wrote in his remarkable essay, *the Psychology of Nations and the Masses*:[407] "The spirit of a race possesses a certain amount of common psychological characteristics, which are as durable as anatomical traits. Like the latter, psychological traits are replicated through heredity, with regularity and consistency. Here, stable mental organization is dependent on the structure of the brain."

4. Culture as a Biological Weapon

Logically summarizing all of the above, we unavoidably come to the key part of our research, namely: the psychological and neuropathological influence of the culture of one race type on another, for the values, norms, and principles of world perception—as we are convinced—are biological constants. [That which is] created by the hereditary mass of one race, unavoidably leads to the distortion of the genetic program for existence of another, during its transference, thus undermining its most basic vitality. The founder of the Russian school of psychiatry, Sergei Sergeyevich Korsakov (1854-1900), emphasized in his fundamental monography, *A Course on Psychiatry*:[408] "One must always weigh the influence of racial traits, because many that are considered an anomaly for people of one race, constitute a normal phenomenon for the people of another race."

Modern American scientist E.F.K. Wallis developed the given point of view in a more concrete form, in the work, *Mental Illness, Biology, and Culture*,[409] from the anthology, *Personality, Culture, and Ethnos*.[410] He wrote: "Do cultural differences stimulate development of different forms of mental illness? Yes, different cultures stimulate the development of different forms of mental illness. "Relying on the new principles of psychological anthropology, John Honigman clearly postulated his outlook in the same anthology. He wrote: "Biological survival is one of the goals which culture serves."

But if culture is the result of the functional metabolic processes in the structure of the brain of the concrete members of a definite race, then it would be completely natural to conclude, that the cultural creativity of one race could never promote the exaltation of another race, and its mental health.

---

[407] 1995

[408] *Kurs psikhiatrii.* Moscow, 1901.

[409] *Psikhicheskie zaboleniya, biologiya, i kultura.*

[410] *Lichnost', kultura, etnos.* Sovremennaya psikhologicheskaya antropologiya, Moscow, 2001.

To each his own: this sacral phrase unavoidably sounds like a password for each combination of genes.

In the context of our narrative, it already completely appears as no surprise, that the very term **degeneration** was first suggested for use in 1803, by the German naturalist, Gottfried R. Treueranus (1776-1837), right after the ideals of the Great French Revolution were asserted. And at the end of the 19th Century, when ideas of **decadence** arose in the morally decaying society of Europe, the leading anthropologists and psychologists undertook to create theoretical concepts, and worked out a complex of practical measures for the struggle against the manifestations of degeneracy in society and culture. The Englishman, Sir Francis Galton (1822-1911), created **eugenics**, the science of the improvement of the human breed. In Germany, Wilhelm Schalmeyer (1857-1919) and Alfred Ploetz (1860-1940), created a similar science, calling it **racial hygiene.** The distinguished Italian scientist, Cesare Lombroso (1835-1909), proclaimed the creation of **criminal anthropology**, announcing that almost every criminal is a degenerate and almost every degenerate is a criminal. His countryman, scientist Enrico Ferri (1856-1929), developed the ideas of his teacher, creating a whole fundamental field, which received the name, **criminal sociology.** According to its concept, revolutions and other social earthquakes are entirely the work of the hands of hereditarily burdened people. Jules Dejerine (1849-1917) headed the struggle against the causes of degeneration in France, and in Switzerland, it was neurologist and psychiatrist August Forel (1848-1931). Russia, engulfed by "revolutionary whirlwinds" near the turn of the 20th Century, did not remain indifferent, and showed the world its active position on a healthy society, yielding a glorious constellation of scientists: S.S. Korsakov, A.I. Sikorskiy, V.P. Serbskiy, I.P. Merzheyevskiy, E.Yu. Petri, and P.I. Tarnovskaya. All of them dedicated a significant part of their scientific activity to the question of a diagnostic of degeneration, and the creation of adequate measures for its prevention.

Practically all of the indicated scientists concluded that unhealthy mental manifestations, both social and cultural, [arise] from disruptions in the morphology of the nervous system; that is, in their opinion, anatomical fact always precedes psychological fact. Cesare Lombroso formulated the main thesis of his teachings thus: "There exists a certain percentage of anatomically disfigured peoples, which as a result of their anatomical disfigurement think, feel, sense, rejoice, and are discouraged, differently than the normal type of modern man. Criminal anthropology proves that there are incurable anomalies; therefore, there cannot be any talk about correcting, or more precisely, curing the criminal."

In the book, *New Successes of Science in Criminality* (1892), Lombroso gave a description of these anomalies. In the brain of hereditary degenerates and criminals, a high percentage of anomalies in the convolutions are observed; partial atrophying of the lobe convolutions; and also an increase in the size of the cerebellum, in comparison with the size of the brain. Among the anomalies of the skull are: asymmetry, premature knitting of the [metopic seams], and disproportionate development of the brow ridges, cheek bones, and lower jaw. Pathologies are also often observed in the skeletons of hereditary degenerates and criminals; for example, a disruption in the number of ribs and vertebrae. A lowering of sensitivity to pain is characteristic to these subjects, as well as smaller subtlety of taste, irregularity in handwriting, and gestures; and in their gait the left stride is longer than the right. The metabolic process is impaired among them, and this is easily determined, according to natural functions. Wrinkles are situated vertically in the middle of the cheeks; in criminal anthropology this receives the name **wrinkle disorder**...Among hereditary degenerates and criminals, one can often observe people with anomalies of the ears." The ear auricle occupies first place among the organs, which indicate degeneracy"—wrote Lombroso. Of enormous interest and timeliness is the indication of a connection between mental and physical traits: "Intellectual degeneration is culled in the hereditarily numerous and diverse forms of its transformations. Intellectual retardation is always accompanied by physical degeneracy. There is not one form of madness, which would not pay tribute to crime."

Lombrose made his most essential discovery in this section, for he begins to analyze **political crime as a phenomenon from an anthropological point of view.** Thus he establishes a direct cause-and-effect relationship between ideas that are the result of the anatomically disfigured nervous system of the degenerate, and its political embodiment.

His book, *Political Criminality and Revolution* (1906), is dedicated to the very study of **hereditary political criminals.** In it, the author remarks: "In history we encounter numerous examples of the combination of political criminality with degeneracy. Yes, for us, unfortunately, no statistics are needed in order to be persuaded of the combinability of very progressive ideas with very criminal tendencies. Born criminals usually appear in uprisings and rebellions, and at the start of revolutions. They infect the weak and indecisive with their example, and give birth to a genuine, copy-cat epidemic.

Moreover, in Lombroso's opinion, "moral idiotism" is inherent in natural-born political criminals. According to his observations, almost all the ringleaders of the Great French Revolution were degenerates, just like the later Communards: "In great numbers, the insane enter the ranks of political criminals, because the inclination to crimes of any type, already conditioned by the absence of moral feelings, is strengthened in them by intellectual instability, by a lack of common sense, an exaggerated conceit, and by ideas of grandeur and persecution. It is sufficient to look at the portraits of some political criminals, and not be a specialist, in order to see that they were insane."

Lombroso thus sounded the same bold conclusion: "Among the anthropological factors of political criminality, the influence of race stands in the foreground; it is distinctly illustrated in comparisons of a sharply expressed revolutionary spirit in some nationalities, with absolute apathy manifested in others, who live in the same climatic and social conditions." The modern criminal situation in the majority of megalopolises graphically confirms the assertion of the Italian scientist. In absolute percentage terms, the leaders are members of dark-pigmented racial groups. Lombroso remarked that blondes have the highest percentage of inventors and scientists, but a far lower percentage of hereditary criminals, including political criminals, in comparison with brunettes. In general, cretins and epileptics are almost unheard of among blondes.

For the definition of mentally degenerate people, including political criminals, Lombroso suggested use of the term **mattoide**, which in translation from Italian, literally means "mad". Political criminals often write memoirs; this occupation has become particularly fashionable among them in recent years. Lombroso left us detailed instructions on how to recognize the mattoid by his narrations, while not being familiar with him personally: "Mattoidism is the combination of weak intellect with the mania of grandeur; extremely developed pride and ambition, on a soil of feeble-mindedness. In their writings, one encounters a yearning for the unrealizable, constant contradictions, and verbosity; and above it all reigns boastfulness. Among all mattoids, deficiencies are noticed sooner than a surplus of inspiration. Demoralized by a superfluous development of their own selves, *"I"*, like true geniuses they are capable of easily breaking from tradition and custom, standing out by intolerance. They are capable of playing a certain political role. The majority of regicidal murderers are mattoids, just like the majority of party leaders."

Degenerate ideas always originate in degenerate brains, and their existence is easily detected by a general degenerativeness in the construction of the body, and in mental habits.

Enrico Ferri, a follower of Lombroso, founded the school of **criminal anthropology**. He essentially developed and expanded the views of his teacher, reconciling degeneracy, politics, and art into one. In the book, *Criminal Types in Art and Literature* (1907), he gave this definition of the essence of the new science: "A positive school of criminal law that carries its studies over from the crime to the criminal; that is, from the judicial essence, to he who carries out the very act."

For a definition of hereditary madness, Ferri suggested use of the term, **pazzia ragionante,** which literally means "rationally/intellectually insane." Today, in order to understand what that means, it is sufficient to turn on the television and listen to debates about the legalization of narcotics and the glorification of homosexuality, or to visit a crowd of art students in a fashionable salon. Today, any presentation of major international prizes in the areas of fashion, cinematography, and literature has the clearly expressed tinge of the political shade of mattoidists, and several minutes of an open election campaign can help even the non-specialist to provide a diagnosis of the organizers of the show. Enrico Ferri wrote: "The political criminal can also be a born criminal, who is concealed by the flag of a political ideal, more or less controversial, satisfying his instincts for fraud and violence. Most often of all, political criminals are insane criminals (in the open or rational sense); they manifest in those moments of social agitation, when bright ideals penetrate into the public consciousness and upset the intellectual and moral equilibrium of people, who are already inclined to anomalies of a similar kind. The 'Party' is the madness of all, at the disposal of a few."

Arguments and debates in political corridors are a normal function of the mattoid, and their aspiration to abolish the death penalty throughout the world is a way to realize this pursuit unhindered. In the modern history of Russia, "perestroika" is a biological process, not having any relationship to politics; in view of that the legalization of the multi-party system in Russia, and a campaign for increasing loyalty with regard to the insane was conducted. Being set free, mattoids poured into dozens of just-registered political parties, for they could do nothing else, besides endlessly prattle on "rational insanity". So-called freedom of political conviction helped to legalize a whole spectrum of existing diagnoses. Only a society, not having the faintest idea of the theory of hereditary political criminality, could talk for more than ten years about the causes of the fall of the USSR. Any inquisitor, just starting to grasp the Middle Ages science of the executioner, and seeing the forehead of the creator of perestroika, decorated with the "mark of the devil", would have instantly told a memorized lesson about the traits of demons, which were subject to the autodafe. In the times of L.P. Beria's NKVD, they took such traits into account when handing down verdicts to enemies of the people—that is, hereditary political criminals.

In the monography, *On the Teachings of the Criminal Anthropology School*,[411] Russian scientist Ignatiy Zakrevskiy carried over the problem of innate pathology to the formation of religious outlooks. "Often, inborn criminality merges with revolutionary activity, opening the chance to satisfy anti-social urges, under the guise of concern for the public welfare. There is a certain form of mental disorder, which it follows to call **political epilepsy**; during this, ingenious thoughts alternate with hallucinations. Mohammed, evidently, can serve as an example of this type."

At the very end of the 19th Century, the concept of **degenerology** was conclusively formed by German psychiatrist Emile Krepelin (1856-1926). He pointed out: "The term, 'degeneracy' means a manifestation of such qualities transferred by heredity, which can hamper or make the achievement of basic life goals impossible." Paul Julius Moebius formulated a practical side to these views, through the creation of **pathography**; that is, a systematic method of portraying the morphological traits of mental degeneracy. He also defined degeneration as a "pointless deviation from a type."

Finally, an authority on the problem like August Morel, indicated that "degeneration and unhealthy deviation from the normal human type, are one and the same." Such notions as **rasse krueppel** (cripples of a race), **stigma degenerationis** (burdensome traits), and **abarten** (variation) came into medical use from psychology, psychiatry, and anthropology.

In the book, *Culture and Degeneration*, German neurologist Oswald Bumke formulated a thesis, according to which, not only separate layers of the works of ethnic and social groups, but entire cultures on a world-wide scale, are degenerative; that is, they are the product of sick minds. He emphasized: "In everyday life, the words 'degeneracy' and 'degenerate' are the equivalent of a moral evaluation, a moral judgement." In his program book, *Art, Sick Nerves, and Education*,[412] his Russian colleague, G.I. Rossolimo, put forward an unequivocal appeal: "Science, fully armed, should stand up for those who are threatened by the dangers of pathological tendencies in modern art: it should point out the harm which can done to the system of morals, just as to the condition of the nervous system, from the unbridledness of the imagination, from the cultivation and development of sick mental processes, and from the extreme predominance of works of fantasy over the activity of the intellectual." Besides that classic Russian, and later, Soviet science, called for the "mass hygiene of aesthetic perception." All modern mass-art, television, fashion, and more widely—style of world perception with value tenets, that is the whole type of a civilization - should, according to the tenets of Rossolimo, be recognized as the degenerative resultant activity of a brain, which is inherent "in anatomically disfigured people, who as a result of their anatomical disformity, think, feel, sense, rejoice, and lose heart, differently than the normal type of modern man", as Lombroso observed. And

---

[411] *Ob ucheniyakh ugolovno-antropologicheskoy shkoli.* Kharkov, 1893.

[412] *Iskusstvo, bol'nie nervi i vospitanie.* Moscow, 1901.

consequently, this type of civilization should be destroyed, as "hindering, or making impossible, the achievement of basic life goals", in the opinion of Krepelin.

A modern science, such as **psychological anthropology**, gives more attention to studies in the area of a phenomenon as **alteration state of consciousness (ASC).** Here it is particularly important to underscore the basic emphasis made on the study of the mental health of the individual and the ethnos, within the context of a concrete culture. Scientist E.F. Wallis, cited by us above, asserted in his book, *Mental Illnesses, Biology, and Culture,*[413] from the anthology, *The Individual, Culture, and Ethnos,*[414] that: "Any physical dysfunction of the brain suggests some mental disfunction. Some physical disfunctions cause disorganization of the nervous system, a large part of the components of which remain undamaged." Therefore, one of the basic tasks of psychological anthropology should be the "study of the anatomy and physiology of the central nervous system, where it is examined as a whole." It is namely such an approach that offers the opportunity to classify cultures as more or less pathogenic. "Thus, in the future, there can be an opportunity to examine the frequency, distribution, and forms of mental illnesses of an individual in a society, as an index of his culture. What is particularly important here is the cultural values in direct connection with the reactions of individuals to mental illnesses; that is, each type of culture variously affects the psyche of different ethnic and racial groups. And the arising mental disfunctions in separate parts of the nervous system lead afterward to global disruptions of the entire system as a whole; this already finds its reflection in the disruptions of the physical functioning of the brain. The majority of **ethnic psychoses**, as such, in Wallis' opinion, are tied with the inability of one ethnic group to adapt to the cultural values of another.

Another author in the given anthology, Erica Burgin'on, observes in the work, *Altered States of Consciousness,*[415] that the symbols of culture act not only at the level of thinking, but on the biological system of an individual. The symbols of his culture synchronize and harmonize the actions of the endocrine system in the members of the given race, at the same time that the foreign leads to an imbalance in the process of metabolism, which manifests itself in an increase in the number of psychoses. The famous writer Alduous Huxley observed in this regard: "Not one person, if he was not highly civilized, could listen to the African drummer or the monotone singing of the Hindu for a long time, and remain a whole, critical, and deliberate person. If the influence of the tom-tom or Hindu singing were sufficiently continuous, then in the end, any of our philosophers would begin to hop and wail with savages." Here one can agree with the recognized master of the literary word, for many modern culturologists, who preach the universalness of "common human culture", are under the influence of a foreign mental influence, which is physically deforming their brains; they have lost he ability to think and feel in the categories of their own race. Namely therefore, the percentage of people with psychological and sexual deviations is higher among propagandists of modern avant guarde art, than among conservatives, who give preference to the aesthetic partialities of [their] ancestors.

Foreign racial culture is the main source of mental infections in the environment of the dominant racial type in a society. A man raised on the philosophy of Schopenhauer or Nitzsche, cannot preserve his mental health, if for a long period his mind will be encircled by Zulu tales, or if he inhales the incense of Eastern occultism, with its esoteric debauchery and licentiousness.

The result of the influence of the culture of one race on another, can be compared to mental combat trauma.

The operas of Wagner, Gypsy love songs, and cabaret songs are incompatible, since they were brought into the world by different types of brains, having different **cultural mechanics.** Once more, we will cite the words of the classical Russian neurologist and psychiatrist, S.S. Korsakov: "Much that is considered anomalous for the people of one race, is a normal phenomenon for the people of another race."

We will now examine in brief, the influence of the so-called "cultural norms" of one race on another. Today it is customary to think of the propaganda of sexual freedom, based on all

---

[413] *Psikhicheskie zabolevaniya, biologiya i kul'tura.*

[414] *Lichnost', kul'tura, etnos.* (Sovremennaya psikhologicheskaya antropologiya. Moscow, 2001). [Modern Psychological Anthropology].

[415] *Izmenennie sostoyaniya soznaniya.*

unthinkable and unnatural perversions, which to a significant degree, have an openly pathological character, as part of the European system of values. However, in his fundamental essay, *A Course on the Common Teachings about Spiritual Illnesses*,[416] Russian psychologist V.P. Osipov, showed that based on the findings of comparative psychiatry and psychoreflexology, mass homosexuality and pederastry came from the East, primarily from the geographical ranges, where the great races bordered one another, where the chaos of mixed blood unavoidably ended up being celebrated everywhere, with the most unnatural forms of behavior, including sexual behaviors. **The mixing of races leads to the mixing of values systems, with an unavoidable, subsequent loss of values.**

Osipov emphasized: "According to the testimony of knowledgeable people, pederastry gained so much acceptance in the East that the boy-dancers (bachi) of the Middle-Asian khanates took part in processions, and Saadi and other poets praise their beauty in their verses. The bachi learned dances, songs, and wore semi-feminine clothes; they were coquettish, and knew how to arouse their clients; the wealthy considered it good form to keep bachi, even if they did not use them. Some used bachi for a lack of women, and left the unnatural way at the first opportunity; others remained pederasts forever. The bachi themselves, the majority of whom belonged to the passive type, usually left their craft at the onset of puberty, sometimes marrying and living normal lives. Some could not free themselves of the Uranian penchant, and in turn ran to the bachi for help. In large part, the bachi developed as the result of a particularly forcible sexual upbringing in a definite direction. They are professionals at sexual perversion, who are subject to and who are cultivated and bred for true homosexual lust, which is characteristic of their degenerative constitution."

It is enough to turn on the television or visit a modern art salon, in order to be convinced that the characteristic look of degenerative physiognomy, in sum total with homosexualist grimaces was made into a trait of social grace; and today no discussion about a high art form can manage long without a similar entourage. **Aesthetism became a parade of vestibules of homosexualism.**

Neither in Russian folk tales, nor in the tales of other peoples affiliated with the original Indo-European cultural circle, is there a hint of sexual perversion. And in such classical Aryan religions as Hinduism and Zoroastrianism, homosexuality is considered the most grave crime, punishable by death—not only in this life—but in all subsequent embodiments. There are special, ancient Persian tractates, which describe in detail the technology for the ritual killing of sexual perverts, and bringing them to sacrifice. In the Waffen-SS, sodomists were shot.

The Turkish Sultan Mehmet II—who was besieging Byzantium in 1453 and turning the temples and palaces of Constantinople into a bed of smoking ruins, and carrying the Word of the Prophet Mohammed everywhere—openly kept two harems during all of this: one consisted of women of different ages, and the second was composed of boys. The given fact did not in any way contradict the "cultural" norms of Islam, and did not prevent preaching of the truths of Allah. V.P. Osipov dedicates several chapters in his remarkable book to a description of all types of sexual perversions, which flourished in Biblical times, and from which it becomes conclusively clear, that for an understanding of the "spiritual beauty" of the Old Testament, one must first acquire a textbook on **criminal sexology.**

The distinguished Italian anthropologist and psychologist, Paolo Mantegazza, declared that the genesis of homosexualism arose as a result of an anatomical physiological anomaly, expressed in the extension of the peripheral nerves of the sexual organs onto the rectum, which is included in this erogenous area. The abundance of magazines which commercialize "Martian" mechanical technology for the satisfaction of anomalous, exotic sexual inclinations, makes this this hypothesis completely reliable. What is characteristic here is that the quantity of these magazines in the major cities of Europe and America, increases in parallel with the general democratization of society and the struggle for human rights; this fits in completely with the contours of our concept, according to which, for the perception and spread of each type of ideology, a definite type of nervous system is necessary. Movements for humanism, feminism, and rights for sexual and other minorities, are in need of [a constituency] with a special type of nervous system, which functions through a corresponding massage of the anomalous peripheral

---

[416] *Kurs obshchego ucheniya o dushevnikh boleznyakh.* Berlin, 1923.

nerves in the sexual organs. Democratic values, according to the level and character of agitation, more resemble the prescription of a proctologist.

The work of the Russian psychologist and physiologist, Grigoriy Nikolayevich Brenev is forgotten today, however, his work, *Prehistoric Colored Civilization*,[417] is highly relevant in the context of the modern cultural situation. Through the prism of sexual reflex and language, this Russian scientist (formerly one of the favorite students of Pavlov), essentially exposed the correlation of the basic instinct to continue the species, with one of the main instruments of culture. This represents a revolution in science. He wrote: "The [sexual] reflex is a signaling device of bioprocesses in the past. The modern Russian reflex of the "cradles" of the White man is not broken, but overgrown, absorbing into it the world ideas of the colored contest of all its colored enemies. The Russian language therefore, is the timeless sentry of all civilizations, of all ages, and all the nervous energy of Mankind. The Russian language has seen everything and decisively, reflexively, endured everything and remembered. Language—this is the embodiment of the mental energy of a people."

Comprehension and interpretation of world history, according to Brenev, in general is only possible through a comparative study of the reflexes of different races, for only then will the motives of their behavior, pursued goals, and cultural values become understandable and obvious. Brenev writes: "The modern situation in Russia can only be understood on a precise knowledge of the **reflex of blood** of the white and colored peoples." For the study of the biological essence of the eternal enemies of Russia, Brenev suggested introducing a special scientific subject—**"colored blood studies."** He accurately termed the spiritual condition of the modern Russian individual, overgrown by the unnatural reflexes of colored neighbors, as a **"colonized psyche."** He stated: "In modern experimental Russia, there is a 'switch' reflex everywhere, and from there, an ocean of Russian torments, blood, and tears."

According to his conception, the sound-symbol of a language serves its kind as a "radio apparatus", and the unconditional reflex of blood of the individual serves as the receiving antenna. Language all the while implements the corrective and tuning of the blood reflex, and simultaneously serves as an indicator of bioprocesses. All this is easily observed in daily speech. Thus, the vernacular expression, "God marks a scoundrel", again testifies about the physical marks on a degenerate, like for example, in the case of the founder [Gorbachev] of "Perestroika." The expression "to play a dirty trick"[418] exposes the essence of ritual sodomy, practiced by some non-white peoples, something known from the Old Testament. The phrase "to hold [someone] in a black body" [to work someone to the bone], testifies to the age-old scornful attitude of white people toward blacks.

In this regard, G.N. Brenev wrote: "The white reflex has never recognized the colored races as equal, and never will, since blood-mixing reflexively poisons the psyche." The collision of different blood lowers the moral worth of the White race, and physically leads to degeneracy, and departure to the oblivion of mongrels. It is the old demand of the Colored International world."

Russian scientist V.P. Osipov tirelessly emphasized: "Without exception, all mental acts develop by way of reflex." This means that racially foreign and also openly perverted, degenerate ideas will always render a negative influence on the development of natural reflexes in the bosom of the predominant race. A man, the brain of whom is constructed for the perception of Duerer's paintings, will always experience physiological discomfort at the sight of a painting by Shagal.

5. Fictionalism

In his work, *The Philosophy of 'As If'*, German philosopher-positivist Hans Vaihinger (1852-1933) advanced the idea that the conduct of an enormous number of people is determined by social fictions, such as for example, "all people are equal in their opportunities", or "all men are brothers", and so on. Here the main point is that people are subject to the psychological influence

---

[417] *Doistoricheskaya tsvetnaya tsivilizatsiya.* 1935.

[418] In Russian, *podlozhit' svinyu*; literally "to stick the pig" to someone.

of these fictions; the greater part of their life's energy is spent, in order to realize the illusion. Therefore, the prevailing fiction is a powerful tool of authority. The given theory received the name of **fictionalism.**

However, in light of the given theory, it becomes completely obvious, that if members of one race start to export fictions in the form of cultural norms, or even complete political doctrines, into the living space of another race, then by this they pervert the values and landmarks of its members, and by the same they will undermine the bioresource of this race.

Another prominent German psychologist, Alfred Adler (1870-1937), also thought that a fictionalist behavior lies at the basis of the behavior of the neurotic; through a compensatory fiction, he takes away all strength in the struggle with his own inferiority. An individual who is psychologically conscious of his inferiority, struggles with it according to the principle of compensation. But after all, this rule applies to entire peoples and races, which can be literally robbed of their strength by instilling an inferiority complex in them, and forcing them, by the same, to use up all their bio-resources on a struggle against themselves.

In the book, *Rousing Thoughts*,[419] modern German scientist Friedhart Kliks carried the discussion of the given problem to a qualitatively new scientific level. He writes: "The basic function of nerve cells consists of generating an impulse in response to changes in internal and external conditions. The reaction depends in significant measure on the distribution of the charge in the cell, and in a definite sense can be considered as recognition. The elementary, independent function of the nervous ganglion lies in the choice of alternatives between stimulation or inhibition. The reflex fading/dying down of some insects is an example of a complex realization of the given function. A characteristic function of the nervous impulse of a cell lies in "recognition" of the goals of the choice of the corresponding program of behavior. In natural conditions, this is expressed in an increase or decrease of activity, depending on what the sensory apparatus "recognizes"—predator or prey. A decision of this kind relies on intercellular nervous damping. Here, recognition and control, as functions of the nervous system, are subordinated to satisfaction of needs. This phylogenetic is a basis for the initiation of the motivation of behavior, since hereditary programs and their realization in the traits of behavior, are tuned to the most probable surroundings, to a certain type of living space."

Thus, all the authentic horror in the influence of foreign fiction on the members of a concrete race, consists not only of the perversion of the values landmarks in his consciousness, but namely in the disruption of the function of the nerve centers, and also in the insertion of unnatural reflexes, and the change of the hereditary program of behavior. Mental viruses in the form of fictions are packaged into humanistic slogans of "equality and brotherhood". With the introduction of these into the bosom of the life activities of an active race, race-predators are capable of turning on a damping reflex, and transforming a hunter-race into a prey-race. **The cultural norms of the sheep can never enable the evolutionary perfection of a wolf.** That which uses so-called lower species of animal favorably, primarily on an instinctive basis, are secured in their genotype"—writes Kliks. This means that the cultural norms of a lower organism, imposed on a higher organism by means of the creation of a fiction, do not become more perfected and more productive evolutionarily, because of this. The primitive stereotypes of "lower" races do not change their primitiveness during the transfer to the environmental habitat of "higher" races, and it is namely in this that their noxious influence is constituted.

In his excellent monography, *A Course on the General Teachings about Spiritual Illnesses*,[420] V.P. Osipov explained the essence of the problem: "**Spiritual infection or mental infection** is included in the number of mental factors that cause, in certain conditions, spiritual distress; the influence of spiritual infection is observed in the instilled influence of the example causing imitation; or of the verbal persuasion, that is perceived without proper criticism; conditions favorable to the rise and development of infection, lie in the easy arousability of the object, which is subjected to it. They correlate to such conditions: youthful age, an hysterical constitution, ignorance of an object, and blind faith in the source of suggestion. The influence of a mass model facilitates the development of the spiritual infection, to which less stable elements are vulnerable.

---

[419] *Probuzhdayushcheesya mishlenie.* Kiev, 1985.
[420] Berlin, 1923.

The mass-dissemination of the most capricious and absurd fashions, serves as a good example of the inclination of people toward unhealthy imitation."

Instances of the spread of mass mental epidemics, both in antiquity and in modern times, are well known. But after all, different forms of mass ideology can also be considered as mental infections. Here, a person who appears as the source of the infection is called the **inductor,** and the infected person is called the **induced**; thus, the psychological illness itself is called **induced madness.** The Gnostic heresies of early Christianity, Marxism, totalitarian sects, the spreading of Messianism, and other fashionable intellectual misfortunes are all typical examples of mass, induced insanity, caused by the purposeful activity of **ideological inducers.** The coordinated actions of the members of one race can be the cause of mass mental infection in the bosom of another, which in the final analysis leads to an undermining of the latter's biological viability.

The practical experience of the many-hours agitational appearances of the Bolshevik leaders from the ranks of non-native nationalities, in front of the Russian peoples at the time of the Russian Civil War, is a typical example of the infliction of mental damage by racially foreign inducers on the members of a basic, culture-creating race, with the goal of the biological suppression of the latter. In this regard, V.P. Osipov pointed out: "The predisposed influence of a race is reflected in the quantitative inter-relationships of the clinical forms of spiritual illness, and on the characteristics of their trend; besides that in several countries forms of the manifestation of spiritual distress are observed, which are not observed in other [countries]."

Also, "cultural norms" and fashion trends can be used here as a formal cover for the spread of global mental epidemics. The so-called "cultural medium", as is evident from world experience, most often appears like microflora for the dissemination of idea pathologies. Circles of intellectuals, spiritual seers, people who are "associated with God", or "who personify themselves as the conscience of a nation", and other ideological inducers, have since ancient times used such breeding grounds for dissemination of mass, induced insanity. The modern propagation of common human values, love and brotherhood, in the spirit of the postulates of the Great French Revolution, are a typical example of steady mental infection, spreading like the plague or some other epidemic. The goal is one: the destruction of the biological potential of a competitor, who occupies the same ecological niche. **A foreign thought is almost always a mental infection,** and the aggressive inspiration of its delivery unequivocally leads to the meltdown of the strength nodes of the archtype of a race, against which the given ideological fiction is applied.

German racial philosopher Ernst Krieck prophetically wrote: "The basic law of race says: any education, all types of legal forms and formations of Man, and also all methods of healing, any forms of influence should in general correspond to a racial type and racial values, otherwise they lead to disease and degeneracy. In recent centuries of its history, Europe has experienced a number of methods of influence, in part, methods of healing, which, being connected in their essence with Asians, can only lead to the sickening of the Aryan peoples. Racial-political education is a system of selection; the creation of favorable conditions for all that correspond to their type and their goal, and the suppression of all that are foreign to them.

6. Diagnosis:
Hereditary Aggressiveness

One of the most important reasons for the incompatibility of cultures is the inherently different conditional degree of aggressiveness in their carriers. Experiments on primates make a crystal-clear picture of this question.

In the article, *The Brain of Man and the Mental Processes*,[421] Russian scientist A.R. Luriya indicated that in experiments on animals, a disruption of analytical abilities was observed in them after the destruction of the area of the hippocampus; increased aggressiveness and unrestrained sexual behavior was observed. The area of the hippocampus can be considered as a functional apparatus for comparison of the expected and real effects of this or that behavior strategy.

---

[421] *Mozg cheloveka i psikhicheskie protsessi.* From the anthology, *Philosophical Questions of the Physiology of HigherNervous Activity and Psychology.* [*Filosofskie voprosi fiziologii visshey nervnoy deyatel'nosti i psikhologii*]. Moscow, 1963.

The major cities of Europe and America are today experiencing an influx of African and Asian migrants, whose social conduct, as far as the indigenous white population is concerned, can be considered as the behavior of racial-biological groups, which are suffering from hereditary damage to the area of the hippocampus. Criminal statistical data graphically testifies today, that among dark-pigmented racial groups, the percentage of unprovoked, unlawful acts, sexual violence, and use of narcotics is essentially higher, than among the white population. Thus, according to official summaries of the Ministry of Internal affairs, it was made clear that the percentage of professional narcotics dealers among Gypsies was 60 times higher than among Russians. If we were to compare the cultural and scientific potential of these two peoples, then statistics would be directly inversed. But from the point of view of primatology, the given fact can be easily explained, for according to the research of A.R. Luriya, damage to the lobe portions of the brain leads to such forms of behavior, which are neurophysiologically described as a "decay in purposeful activity", "loss of initiative", "disruption of critical thinking with regard to one's behavior", "disruption of psycho-regulatory activity", and "replacement of initiative with stereotype actions." It is namely therefore the hereditary under-development of the lobe portions of the brain in the members of separate racial groups, which precedes anti-social behavior; in dropping into the bosom of modern civilization, they remove themselves from cultural life, and gravitate toward social parasitism, and sometimes openly set themselves against the society that has taken them in. We will recall in this regard, the words of another Russian neurologist, L.S. Vygotskiy, who asserted that "localization of the higher functions cannot be understood as anything other than chronogenetic; that is, the relationships which are characteristic for separate parts of the brain, take shape in the course of development." Thus it becomes completely obvioius, that for both the realization of flights into space and the distribution of narcotics, one must be born [to the task]. **These or those genes that flow through history, through the history of a creator, find their favorite stream, and sprout in the form of different biotypes, where they find their full flower in the entire cultural, social, and political diversity of a human breed.**

Modern geneticists F. Vogel and A. Motul'ski developed the studies of hereditary aggression on a qualitatively new level, in their fundamental, three-tome essay, *The Genetics of Man.*[422] They write: "On the basis of received data, the conclusion was made, that the antisocial behavior of men with genotype XYY is conditioned by the presence of an additional chromosome. Men with the XYY sexual chromosome arrangement, [are found] to manifest anti-social behavior and enter into conflict with the law, relatively more often than normal XY individuals. Besides this, criminals with the given anomalous arrangement of chromosomes possess a low IQ. Insofar as chromosomal aberrations often lead to a lowering of intellectual functions, and a frequential increase of behavioral deviations, and inasmuch as an electro-encephalogram, in the instance of these disorders, indicates an anomaly in the development and maturing of the brain, we should be able to calculate, that morphological studies [will] provide us [with] data relevant to those mechanisms, by which such aberrations hurt the functions of the brain. The most common forms of pathology turned out to be: underdevelopment of the forward sections of the brain; defects of the calloused body; changes in the orientation of the pyramidal sector up to 180 degrees, and the absence of some brain convolutions, or a disruption in their orientation."

Thus, modern genetics confirms Pavlov's position, that "all revealed details of the construction of the brain, sooner or later, should find their dynamic significance." That is, all inherited deviations in the structure of the brains of individuals, should obligatorily find their manifestation in their inadequate conduct. [This] rule is valid for individuals, and also applies to the great races and ethnic communities, which was first graphically shown by Lombroso.

Today in Russia, active studies on the hereditary basis of aggression are being carried out. Thus, M.V. Alfimova and V.I. Trubnikov make the following conclusion in the article, *The Pyschogenetics of Aggressiveness:*[423] "Nevertheless, considering that genetic differences make a substantial contribution to the formation of the inclination toward aggressive behavior, with time, when the individual genetic diagnosis becomes accessible, the genetic findings can be used for the evaluation of individual reactiveness and provocation to aggression - and this means for the prognosis and prevention of aggression."

---

[422] *Genetika cheloveka.* Moscow, 1990.
[423] *Psikhogenetika agressivnosti.* Voprosi psikhologii, 2000, N6. [Questions in Psychology, 2000, N6].

For our part, we add that today recommendations of medical-genetic consultations are used everywhere in the birth of children. With the collection of statistics about the psychogenetic basis of aggression among different ethnic and racial groups, one can easily establish which of them is capable of a complementary existence within the framework of a single state, and which are obviously not. One will be able to correspondingly answer the question: which cultures can fulfill each other, and which cannot even mutually endure the racial stylistics of a competitor.

Finally, using the findings of these genetic studies, one can choose honest officials, in order not to be witnesses to regular corruption scandals, and also to select political leaders, who maximally reflect the will of the people—all with a high degree of accuracy.

Finally, the opportunity presents itself for the conversion of the costly and completely senseless, one-time procedure of choices, to a simple and extremely exact global system of comparison, on the basis of the given genomes of the electorate with the genome of a potential leader. Racial biological identification far more accurately reflects the political preferences of the voter, than the method of dishonest befuddlement and marked ballots. One can toss phony ballots into the ballot box, but one cannot infuse foreign blood into the veins of a voter. A leader and his flock should be of one blood, in order to fulfill one another.

Returning to the theme of culture and its carriers, it is more appropriate, therefore, to speak not so much about hereditarily burdensome races, but on the contrary, about inherently facilitative [races]. Already, the first bio-chemical studies of blood in different races set everything in its place. Thus, Russian scientist V.G. Shtefko made a significant conclusion in the article, *Biological Reactions and their Significance in the Systematics of Apes and Man*:[424] "Views expressed on the basis of experimental data, lead us to a highly important and interesting conclusion. The cultured races of Mankind, for example the Europeans, have a more complex structure in the protein molecule, than the lower races. Thus, from a biological, or more precisely, a biochemical point of view, they are more organizationally complex, than the latter." Namely on the basis of such kinds of natural science discoveries, German racial psychologist Erich Rudolf Ensch[425] made a more bold summary, prophetically proclaiming: "Race and blood. Blood and race—this lies at the basis of everything. A single thread stretches from the structure of the capillary networks to the worldview. Studies about blood and race prove the primacy of the general human being, including its elementary, particularly inborn determinations, over the world of ideas. Modern psychology proves the very same thing, establishing an indissoluble link between the most elementary and lowest psycho-physiological processes, and the highest forms of progressive living. Ideas in and of themselves are sterile, if they are not linked with a physical being. Only in pure flesh and blood can pure ideas safely develop. An approach against the old, ridiculous and impotent idealism is necessary. Only strong ideas are worthy of respect, ideas which can dominate and conquer."

In recent times, it was customary to explain aggression, including at the level of confrontation of races, peoples, and cultures, with the influence of social-psychological factors. The causes of aggression were sought anywhere: in economics, in politics, in differences in the level of culture and education, even among the sympathies of angels and demons in the heavens—just not in the heredity of different racial groups. Recent research in the given area has concluded this eternal problem of history. Idealistic notions from prophets and humanists about the essence of good and bad have conclusively lost the ground beneath them. Hard-heartedness and vindictiveness, or on the other hand, loving kindness and forgiveness, are henceforth not the essence of an emotional fit, but simply the result of the influence of genes, the concentration of the frequencies of which is different in all the basic racial groups, and creates basic cultural values. In order to create, for example, the giant sculpture of Buddha on the territory of Afghanistan, or to destroy the same, as the bandits of the Taliban Movement did, one must first of all, be born to do it. The most humane world religion—Buddhism—just like the most aggressive—Islam—is not an abstract divergence in precepts, but an eternal struggle of the genotypes of their creators. A superficial study of the biographies of the creators of these religions is enough, in order to realize the fatal differences in the genetic emanations of their begetters. These are two extremes of heredity, two different

---

[424] *Biologicheskie reaktsii i ikh znachenie v sistematike obez'yan i cheloveka.* Russkiy antropologicheskiy zhurnal, Tom 12, kniga 1-2, 1922.
[425] 1883-1940

genetic pools, and all the sparsely reasoned discussions of culturologists, who attempt to reduce the struggle of races to a mutual misunderstanding of cultural norms, is only an extreme form of untreated idealism.

A beaver building a dam never thinks about the nuances of the psychological life of the fish, who are destined to die during its construction. The dam is the "cultural" norm of the beaver, and the eaten fish are its consequence.

To change the surrounding environment by means of hereditary will is a first task of each living organism, conscious every instant of its tribal affiliation. The factor of hereditary health also plays an enormous role.

In the year 2000, in the Moscow State University named [in honor of] M.V. Lomonosova, Russian scientist Larisa Valerianovna Bets made her dissertation in competition for a science doctorate degree in behavioral sciences, on the theme, *Anthropological Aspects of Studies of the Hormonal Status of Man*.[426] In part, the author of the given study suggested a method of determination of the mental and sexual deviations in Man, on the basis of research on the concentrations of hormones. The method received practical approval, and the reliability of its results approached 100%.

Today, in so-called civilized countries, including Russia, a rising activity of sexual deviants (perverts) of all colors is observed, which renders growing pressure on public opinion and cultural norms. The consolidation of deviants occurs especially on a biological basis, and no ideological orientation is a hindrance to the lobbying of the interests of this clan. In all [social and government] structures of Russia, including the Duma, these multi-gender beings with disrupted hormonal status create parties and factions. Their mentally degenerate, mattoid behavior is a genuine cause of political instability in our country. The grimaces of modern authority—these are not the result of the political immaturity of democracy, as journalists try to convince us, but are the result of the biological disintergration of society, which is only cured through biologically radical measures, and not through the innoculations of sociology, in the form of agitations and disputes. Mattoids, as we remember, love to indulge in intellectual madness, but cannot tolerate biological exposure.

Today we have an effective method for the disinfection of our political corridors. Scientists V.V. Yarovenko and A.N. Chistikin point out in the brochure, *Dermatoglyphics in Crime Detection and Forensic Medicine*:[427] "Genetically characteristic conditions; the functional characteristics of the organism of Man; the inclination toward definite types of professions; the conduct of a person in extreme situations; the predisposition to separate types of diseases; the compatibility of spousal pairs, and other factors, are reflected in the capillary nodes of the hands. This is the subject of dermatoglyphics. Determining the inclination of an organism to commit crime [through study of] its genetic plan characteristics, has essential significance for investigative workers. It is an established fact in medicine, that the reliability of dermatoglyphic diagnostics of predispositions toward diseases approaches 97%, because the factors which cause the development of a given pathology, act at the time of the formation of dermatoglyphic traits."

All this says is that if the ancient magi and astrologers predicted the manifestation of bloodthirsty tyrants according to the stars, then today future manifestations of any hereditary political criminal can be discovered at the stage of the inspection of the fetus of the woman giving future birth. Yarovenko and Chistikin did not feel uneasy writing about the "prevention of crimes" on the basis of their methods. Besides that they observed: "There is a definite tie between biorhythms and papillar forms. 'Definite biorhythms' correspond to definite types and kinds of forms." But as much as 100 years ago, Cesare Lombroso wrote: "Among natural-born criminals, there is a distinctive quality of aura, which precedes the commission of a crime and compels a premonition of it." Modern scientific methods enable them to record the biorhythms of an individual at any level, and also to produce color photos of his aura. Fingerprints of any individual can also be taken, without the knowledge of the individual himself. As we remember, mattoids differ in disease specific features of their body functions and secretions; therefore the placement of corresponding chemical sensors on areas of political criminals that interest us, and an

---

[426] *Antropologicheskie aspekti izucheniya gormonal'nogo statusa cheloveka.*

[427] *Dermatoglifika v kriminalistike i sudebnoy meditsine.* Tyumen', 1995.

anonymous system of measurements, allows for a reduction in the margin of reading errors, and for the prevention of retaliatory actions by a likely biological enemy.

The idea of using fingerprints for the establishment of a diagnosis belongs to American anthropologist Harold Cummins (1894-1976), who was the first to bring the term "dermatoglyphics" into use. His scientific initiative dates back to 1936. Later research in the given area completely confirmed the validity of the way the question was posed. Russian specialist G.L. Khit' points out in the article, *The Termination of Line S in Different Racial Groups*:[428] "Dermatoglyphic traits have not had an adaptive character for the entire length of history of the formation of the Mongoloid and Europoid racial trunks." For their part, T.O. Zhilenkova and L.G. Goldfarb emphasized in the article, *Dermatoglyphics in Vilyui Encephalitis*:[429] "The character and degree of prominence of this or that trait of dermatoglyphics is stored in the genotype and remains unchanging in the course of an individual's life. The presence of some hereditary disease (or predisposition to it) in a separate individual, can exceed the dermatoglyphic bounds of changeability in a given (healthy) population."

Thus, the above-cited authors lead us to the thought, that dermatoglyphic factors allow the recording and diagnosis of the basic aspects of **racial pathology**, with an extremely high degree of precision.

In the article, *Dermatoglyphics in Some Chromosomal Anomalies in Man*,[430] I.S. Guseva and V.I. Kazey form a coherent and well-founded concept, with the help of additional data: aggression has an inborn basis, not only at the individual level, but on a racial level also, and yields to evaluation and quantitative measurement.

The given point of view is visible not only in these works, but in numerous others, in which connection research has gained more public popularity.

The authors of a large thematic anthology, *Papillary Forms: Identification and Determination of Personality*,[431] state that "the accumulation of enormous material, allowing one to speak with sufficient precision about the traits of pathological dermatoglyphics, points to a hereditarily, pre-determined disruption in the formation of the human organism, and first of all, in its nervous system." Besides that, in the opinion of these same authors, "skin relief/contours, alongside with common traits of pathology, have in each case, a number of their very own specific traits, which mark a completely different phenotype of behavior. This allows one to talk about the prospects for the use of dermatoglyphic methods in psychology."

Thus, dermatoglyphics demonstrates bright prospects for the explanation and prediction of social, political, and criminal conduct in this or that racial group, on the basis of a systemic illustration of their dermatoglyphic traits.

N.N. Bogdanov, a modern Russian specialist in the given area, summarizes his own experience with studies and foreign publications, in the monography, *Comprehension of Individuality*.[432] He states that "people who possess patterns in the area of the palms, in part 'Th', actively stand out for greater aggressiveness." **Th** is the index of patterns on the first (inter) fingerpad, which correlates to a number of important dermatoglyphic parameters, and to the statistics of its calculations among almost all national and racial groups, which they have been conducting for dozens of years already. An enormous amount of reliable information, derived in the process of lengthy ethnographic studies, has been accumulated.

We turn to the fundamental essay, *The Racial Differentiation of Man*,[433] by the Russian authorities of dermatoglyphics, G.L. Khit' and N.A. Dolinovoy. It illustrates the triumph of the principles of classic racial theory. Judge for yourself. Among the majority of Europoids, the *Th* index fluctuates within the range of 8-9. Nothing like this fact can better explain why the many centuries of war in Europe hardly changed the ethnographic situation on the continent. Among

---

[428] *Okonchanie linii S u razlichnikh rasovikh grupp.* Voprosi antropologii, vip 59. [Questions in Anthropology, Issue 59].

[429] *Dermatoglifiki pri vilyuyskom entsefalite.* Voprosii antropologii, vip. 64.

[430] *Dermatoglifika pre nekotorikh khromosomnikh anomaliyakh u cheloveka.* Voprosi antropologii, 1970, vip 35.

[431] *Papillyarnie uzori: identifikatsiya i opredelenie lichnosti.* Moscow, 2002.

[432] *Postizhenie individual'nosti.* Moscow, 2001.

[433] *Rasovaya differentsiatsiya chelovechestva.* Moscow, 1990.

the Turkish and Mongolian peoples, the value of this index is already essentially higher, and finds historical confirmation in the raids of the Polovtsy, the Khazars, and the Pechenegs against the Rusi, and also in the extent of the Mongol-Tartar yoke and in the aggressive policy of the Turkish Ottoman Empire.

The original victorious procession of Islam in the 7th and 8th centuries should be no cause for surprise, for among the Arabs, the *Th* Index is one of the very highest in the world (20-21). The modern Jews of Israel will not be able to solve the problem of Arab terrorism, by virtue of the fact that their *Th* index is equal to 12.5.

But then among the Ashkenazi Jewish émigrés from Russia, the index is 21.5, and this in turn explains the causes of Bolshevik fanaticism and barbarity in the time of the Civil War, since many commissars were Ashkenazi Jews.

But the most aggressive in the world, according to this parameter of dermatoglyphics, are the women from the Urus-Martanskiy region of Chechnya, who have participated as suicide-terrorists on the territory of Russia, in countless bandit actions. Characteristically, the name "Urus-Martan" literally means, word for word, "dead Russian head." Not far behind the Chechen women, according to this characteristic of aggressiveness, are the Chechen, Dagestani, and Azerbadjianian men; this is confirmed by the summaries of criminal statistics in the Ministry of Interior Affairs.

Abkhazians (*Th* index 5-7) are less aggressive than Georgians (10); and Armenians (9) are less aggressive than Azeris (12-14) and Turks (12-16); these aspects of criminal dermatoglyphics easily find confirmation in the ethno-territorial conflicts in the Caucasus, for the length of history.

The quiet and inoffensive Europoid Aini (4) were already completely assimilated by the Japanese, by the start of the 20th Century. The Chinese (8-12) quite naturally occupied Tibet, the native population of which has a *Th* index of 3-6.

Today, it is possible to explain all of world history on the basis of the interpretation of the parameters of finger and palm prints of peoples, the creators of history.

Such sciences, like conflictology and aggressionology, could receive principally new biological grounds, and sociology and political science of the old classical form should undergo substantial revision. And the given prospect is more tempting, because the traits of dermatoglyphics are completely hereditarily determined, and not dependent on the influence of the environment. There can be talk about a global, systemic reconstruction of world history on a biological basis.

New discoveries in the area of genetics made for a more precise and correct picture of the origin of aggression. Thus, in 1978 the modern Dutch scientist, Hans G. Bruenner, set apart the "aggression gene"—mono amino oxidaze (MAO). His studies confirmed the concept of a tie between chromosomal pathology and low intellect. People with an insufficient retention of mono amino oxidase are more subject to flashes of unprovoked anger and have an IQ lower than average. The hereditary insufficiency of intellect is compensated by their aggressiveness, and this rule is valid not only for separate individuals, but for entire ethnoses and races. Less-cultured peoples, falling into an environment of creatively more gifted people, take out their inferiority complex against the latter, with aggressive behavior, which finds eloquent expression in the data of revealed criminal statistics.

The famous American politician, Patrick J. Buchanan, introduces staggering statistics from the Department of Justice of the United States, in the sensational book, *The Death of the West*. Thus, in 1987, white criminals committed severe crimes against selected black victims in only 3 percent of cases, while the latter committed no less than 50% of the general number of severe crimes against whites.

In the case of rapes, white criminals did not once (out of 83,000 cases) make an attempt on black women, while the number of rapes by blacks against white women accounted for 28% of the total number of rapes.

In the case of robberies, only 2% of thefts against blacks were committed by whites, in contrast to 73% of thefts against whites, being committed by blacks.

In 1994, blacks committed 90% of inter-racial crimes. Inasmuch as blacks constitute 12% of the US population, it follows from this that the probability of interracial crimes being committed by them is 50% higher, than for whites. For blacks, the probability of gang rape or group attacks committed by them is 100-250 times higher, than for whites. Even in the category of "hate crimes", to which less than 1% of interracial crimes are related, the probability is that blacks turn out to be the criminals, and not the victim, at a rate two times higher, than for whites.

Some political prostitute-scientists try to explain the black race's criminal disposition with some exterior social or even ecological causes, not themselves understanding how ridiculous and absurd their argumentation is. American bio-politician Robert Masters attempted to explain the aggressiveness of Negroes in the USA, with a heightened concentration of lead and other metals in their dwellings (paint lead, lead pipes, and so on), and also by the influence of alchohol and fluoride compounds, applied in the decontamination of water.

There is nothing better than these facts to indicate that in the Negroid race, the nervous system is constructed differently, than in Whites; after all, Europoids who inhale the fumes of lead components and drink the same water and alchoholic drinks, are not subjected to the outbursts of unprovoked aggression, which are observed in blacks.

The growth of ethno-separatism and the usual ethnic banditry on Russian territory, and in the former Soviet republics, has the same nature of inherent compatibility of different racial groups.

The blood-thirstiness and barbarity of Chechen fighters, and the torture of prisoners recorded on video camera, cannot have any socio-cultural explanation, for here we are dealing with the biological fact of **aggression for the sake of aggression.**

7. The Biochemistry of an Ideology

In his fundamental monography, *Biopolitics*,[434] Russian scientist V.A. Oleskin emphasizes: "In modern Russia, the significance of biopolitics is objectively growing. In part, the study of social **aggressiveness** is included in its purview, the extensiveness of which complicates all levels of the Russian social consciousness (from scandals in the State Duma to the actions of fighters, along with other "disturbing factors") and any positive development of Russia at the cusp of the millennium. Presidential elections in Russia are a more fertile soil for biopoliticians, than are analogous phenomenon in the West, for in Russia biopolitical law appears in a less concealed form."

In a book of the same name,[435] American researcher Thomas Thorson emphasized that evolutionary biology should serve as the basis of political theory. In the book, *A Theory of Progressive Evolution*,[436] his countryman, Peter Corning wrote: "Politics arose in the course of the Evolution of Man, significantly earlier than the appearance of specialized institutions of government. I am even prepared to assert, that political behavior was an important prerequisite and catalyst in the evolution of language and the flowering of culture. Politics was an integral part of human society; it is not even a uniquely human phenomenon."

Thus, it is a question of the conscious biologization of the causes of politics, and equally, any other socio-cultural behavior. Concrete political ideas are the product of concrete types of brains. Namely therefore, V.A. Oleskin certifies: "The study of the nervous system has biopolitical significance, since it appears as an important somatic factor of political behavior. The influence of genetic factors on behavior cannot be researched without an understanding of the neurophysiological mechanisms of behavior, on which genetic factors act."

Consequently, propaganda of political ideas is always tied with the neurophysiological functioning of the nervous system of those people, for which it is intended. Any ideology brings in its wake changes in the balance of chemical micro-elements in the nervous system.

Neurotransmitters or neuromediators are chemical communicative agents (signal agents), serving for the transmission of information, from neuron to neuron. Discoveries in the area of the functioning of neurotransmitters, made in recent times, literally allow analysis in detail of the neurochemistry of any ideology.

Acetylcholine is important for the initial memorization of new information and subsequent processes of consolidation of memory. A shortage of **dopamine** in the corresponding sectors of the brain leads to loss of initiative, and a more serious deficit leads to the impossibility of

---

[434] *Biopolitika.* Moscow, 2001.
[435] *Biopolitics*, 1970.
[436] 1983

completing active operations. For its part, an excess of dopamine enables behavior connected with the "quest for pleasure" and the development of hedonistic philosophies.

It has been established that the level of **serotonin** influences the working out and determination of the social status of every living being. Among people with a Macchiavellian type of personality, that is, aggressive and purposeful, social rank in society increases in proportion with the increase of the level of serotonin in the blood. Among people belonging to the opposite bio-psyche type, that is, "compliant moralists", social rank in society falls with an increase of the level of serotonin.

Substances called **neuromodulators** are analgesics; included in their number are **endorphins**, which cause a feeling of satisfaction, and are a psychological, compensatory analog of an inner reward for this or any style of behavior.

These and a number of other chemical reagents, worked out by different parts of the brain, enable achievement by the individual of **neurophysiological homestasis**, that is, a feeling of inner confidence, peace, and gaining a sense of life and belonging to a great commonality.

Only a stylistically adjusted ideology, relying on archaic symbols and activating a biotype, can bestow eagerly sought success in the political sphere. Slogans, in essence, are analogs of neurophysiological reagents of the nervous system, maximally consolidating the hereditary properties of a race.

In this regard, V.A. Oleskin again establishes: "In many ways, politics relies on evolutionarily primitive forms of social behavior, which originated earlier than human language and culture." Namely therefore, all forms of ideology best of all lend themselves to activization on a non-verbal level, through gestures, sounds, symbols, and even scents, since a political leader is skilled at working out a balance in the brains the of his followers.

One can consider ideology as a racial-specific elixir, which heightens the viability of a race. On the other hand, the ideology of a foreign race is harmful and dangerous, for it leads to a biochemical imbalance in the nervous system.

Any form of indoctrination is introduced into a society, in order to activate the biochemical analog of one group of ideas and suppress another. The struggle with the "threat of Fascism", or the movement for the rights of sexual minorities does not appeal to the logic of the electorate, but namely to evolutionarily primitive, pre-cultural social stereotypes of behavior. The highly important role in the process of instilling this or that ideology (indoctrination) is played by hormones—informational mediums, manufactured by the endocrine system and carried by the bloodstream to cells in all parts of the body. The ideology of one race, imposed on the members of another, inevitably leads to disruption of the functioning of the endocrine system as a whole. Soviet geneticist N.P. Dubinin emphasized in his book, *What is Man?*:[437] "The brain of Man possesses genetically determined characteristics. The development and vital activity of Man is impossible without the action of his genetic program. Man possesses defined biological traits, the specific features of which manifest at molecular, cellular, organismal, and population levels. Unceasing biological continuity for the extent of the history of Mankind assures the presence of typological traits in the genetic program of each individual. As an object of genetic study, behavior presents itself as a quantitatively measurable reaction of the nervous system, in response to external influences. In such a case, the question about the role of the genotype is decided with application of the calculation of the coefficient of heredity, which is customary for quantitative traits."

However, Francis Galton indicated that "conscience, talent, and other purely human traits are the biologically determined traits of personality, passed on through sexual cells to generations."

Therefore, conforming with the above, one can boldly assert that the **dissemination of this or that type of ideology, in the bosom of each race today, can be quantitatively calculated and reduced to a sum total of neurophysiolocal and biochemical characteristics, for the purpose of revealing the degree of conformity to an ideology, to a basic racial type.**

---

[437] *Shto takoe chelovek?* Moscow, 1983.

8. The Psycho-pathology of Monotheism

However, not only types of ideologies, but even types of worldviews, determined by race, have a strict, natural-scientific explanation.

Man receives more than 90% of his information by means of vision; however, different races see the world differently, not in the figurative sense here, but in the direct sense. The statistics of racial differences in the area of color perception have been accumulating for dozens of years all over the world, and unintentionally lead us to new thoughts and conclusions. For objectivity of analysis, we went to Russian works. K.B. Bulaeva and S.A. Isaychev wrote in the article, *Population-genetic Analysis of Some Parameters of Color Perception:*[438] "In modern psychogenetics, the basic attention of researchers is turned first of all, to discovery of the role of genetic factors, conditioning the formation and development of such intergral characteristics of the psyche and higher nervous activity, like intellect, termperament, perception, memory, EEG, and traits of the nervous system. The search and selection of an object for a genetic analysis of such components of mental or higher nervous activity, which have a distinct genetic determination and influence the formation of more complex traits and parameters, [which are] functionally tied with these components, can be promising."

Thus, it is clear that quantitative characteristics of the nervous system determine mental traits, and that in its turn is tied with the type of worldview.

This means, for its identification and discovery, a statistical calculation of parameters at a racial and ethnic level is necessary. K.A. Bulaeva and S.A. Isaychev thus develop this thought: "It is known that the genotypic architectonics of polygenic and monogenic systems lie at the base of formation of the structures and functions of an organism; it depends more on the genetic structure of a population to great degree, than on the genetic nature of the trait itself. Therefore, a genetic analysis of any quantitative trait with a polygenic system of determination, first of all demands a calculation of the genetic structure of the population as a whole."

Quantitative characteristics, the genetic nature of which is well-studied, are called **markers.** One of the widely known parameters, which satisfies these conditions, in a phenomenal anomaly of color vision: so-called color blindness. In Europe, the frequency with which this anomaly is encountered runs from 2-8%. One of the important indicators of anomalous color vision is "perception of pure green color", measured in nanometers. "The frequency of anomalies of color vision in some isolated populations differs significantly from the average frequency of anomalies in Europe. A high concentration of a recessive gene for green blindness testifies about a specific feature and intensity, of the course of micro-evolutionary processes in the gene pool of small, isolated populations of Man, and their deep differentiation. The discrete character of the variational distribution of the trait of "pure color green perception"is conditioned by the influence of a genetic factor—by two dominant genes, G1 and G2, having one locus in the X-chromosome"— write the authors of the article; and they cite the results of field studies in Dagestan, according to which, the population of the given territory is affected by Daltonism; pure green color perception is at 21%. But Dagestan is a Muslim territory.

And now if we turn to the fundamental one-tome work, *The Gene Pool and Genogeography of Populations (The Gene Pool of the Population of Russia and of Adjacent Countries),*[439] published by the Academy of Sciences under the editorship of the distinguished Russian scientist, Yu. G. Richkov, we will discover that among all the population of the designated vast territories, the Jews and the Arabs especially stand out, according to the number of those who are genetically affected by Daltonism.

In the case of Islam as a religion, originating namely among the Arabs, it happens that the green color of the flags of the Prophet Mohammed were chosen namely as a marker, to allow them to choose "theirs", according to the principle of color blindness.

---

[438] *Populyatsionno-geneticheskiy analiz nekotorikh parametrov tsvetnogo vospriyatiya.* Voprosi psikhologii, 1984 N4. [Questions in Psychology, 1984, N4]

[439] *Genofond i genogeografiya narodonaceleniya.* (Genofond naseleniya Rossii I sopredelnihk stran.) Sankt-Peterburg, 2000, Tome 1.

In general, a preliminary neurological analysis of basic religious doctrines gives food for thought. With the studying of Biblical texts on a scientific basis, an independent science arose at approximately the middle of the 19th Century—**critical Biblical studies**—in which a field of research was sufficiently quickly formed: **political monotheism.** An unbiased analysis of texts opened up the unnaturalness and artificiality of the very idea of a Single God, which originated with the efforts of Egyptian pagan priests, in the 14th Century B.C., in the court of Pharoah Amenhotep IV. In 1907, when American archaeologist Theodore Davis discovered the sealed tomb of the Pharoah in the Valley of the Kings, one could clearly see on the frescoes (according to the tradition of portraying the life path of the King), that from childhood, he was a sickly and feeble boy, with an excessively large head, (in comparison with the body), which had heavy, drowsy eyelids, a sentimental, weak-willed expression on the face, and plump, feminine lips. Professor Elliot Smith performed a medical evaluation of the mummy of the Pharoah, and gave an unequivocal determination: according to the unnatural form of the skull, one could conclude that Amenhotep IV suffered from epilepsy and died an unviolent death at the age of 30. Historians re-constructed the life of the ancient Egyptian pharaoh. The fact that one of his three daughters died in childhood, without any visible physical causes, gives us a conclusive answer—the Pharoah was a hereditary degenerate.

In practically every textbook on psychiatry, it indicates that the basic symptom of epilepsy is delirious monomania. It is not surprising, therefore, that this Egyptian pharaoh's court was first in the flowering of religious chauvinism and intolerance: portrayals of different gods began to be destroyed, and religious books were burned.

Notably, the cult of One God first developed and formed in the Nile River Valley; that is, in a natural trench. We will focus attention on this geophysical fact.

The religious revolution of Amenhotep IV, however, suffered a collapse, simultaneously with his sudden death; but the pernicious religious heresy of monotheism, begat in the court of an epileptic pharaoh, did not disappear, since approximately 100 years later, the fugitive pagan priest Moses revived it—he was banished from the collegium of pagan priests for murder. In northern Arabia, in the oasis of Kadesh, he concluded an agreement with a tribe of fugitive convicts, the Habiru, and re-animated the project, again giving them the idea of a Single God.

In the opinion of many independent researchers of the Old Testament, Moses was also an epileptic; in the Book of Exodus, he says of himself: "I am slow of speech and tongue-tied."

S.N. Plekhanov, a modern Russian writer and religious historian, writes: "Monotheism arose not in the least because of a higher level of comprehension of reality—it rather reflects the sterility of the world that begat it. Let us recall where it originated—in a monotonous and dreary desert. Among the savage nomads led by Moses, the fugitive pagan priest from Memphis, the conviction formed in the course of decades, that some single force ruled this unified nature. In those places where the cultured peoples of antiquity lived, the landscape was richer, with forests, mountains, seas, and rivers. That is why other religious views took shape—the world was seen not as a single-party creation, but as a symphony, an endlessly lasting performance, a huge arena of the struggle of many forces."

The French religious historian, Albert Reville, also stated: "On the bare slopes of Sinai there were no elements at all for a rich mythology." For his part, his countryman, Ernst Renan, wrote about the rise of the idea of a Single God as the "monotheism of the desert."

While all the Aryan religions were born in elevated places, mountains, or plateaus, the religious creation of the Semitic spirit in contrast sprang in depressions: the mouth of the Nile, the Oasis of Kadesh in Arabia, and the shores of the Dead Sea, where Christianity was born. The Aryan spirit always aspired to populate the majestic mountains with a multitude of Gods: Greek Olympus; the plateau of Central Asia, which was the epicenter of origin for Zoroastrianism, and finally, the mountains of Nepal—the homeland of Buddha. The ancient Germans, Slavs, and Celts professed polytheism and also built burial mounds with cult aims. It is not surprising that in the meager saucer of an oasis, surrounded by scorching sands, only one God could fit; others simply could not find a place.

It is worth searching for the traits of jealous capriciousness in the Jewish God Yahweh in this; for all the powerful bravado nevertheless creates the impression that he simply does not have enough space. All actions in the Bible, which have the participation of supernatural forces, remind one of a squabble in the kitchen of a communal apartment—maximum passion in minimum space. The great German poet, Henrich Heine, being a Jew by nationality, mourned the religion of his ancestors in a moment of anger, as a "tenacious Nile Valley infection, an insensible faith of ancient Egyptians." The distinguished ancient Roman historian, Cornelius Tacitus, wrote with indignation that the idea of single, exclusive god was one of the most revolting ideas in the world. English historian Arnold Toynbee stated that Biblical monotheism is the source of the modern ecological crisis. Gods, peoples, and nature comprise a single whole in any paganistic pantheon; in the Bible, God does not reveal himself in the world, and therefore nothing depends on him. The God of the Old Testament is a capricious despot, free to do anything he pleases with his creations and the world. Oswald Spengler considered monotheism a product of a special spirit, which developed a specific, magical concept of a dual universe, another world, the world of God. According to this conception, the sense of events occurring in this world, extend beyond its bounds. The modern pagan, Belgian philosopher Christopher Gerard, thought: "To speak of singularity means to be blind to certain realities, and in this sense, monotheism is a genuine spiritual abnormality, heightened by its authoritarianism in practice. The piety of the Pagan eras did not have anything in common with monotheistic dishonor, which leads to destructive nihilism." Mexican writer Octavio Paz called monotheism "one of the greatest catastrophes of humankind." Finally, in the period of his religious quests, even Lev Nikolayevich Tolstoy expressed the following radical thought: "Some strange god-savage, half-human, half-monster, created the world with a whim, as he pleased, and said it was good…But much bad happened. Man and all his generations fell under a curse."

The most clear and consistent attitude that was expressed toward the concept of a single God was that of Russian scientist N.M. Nikol'skiy: "The problem before us is to clearly and consistently decide on the destruction of the tendentious conception about ancient Jewish monotheism. Not only is Jewish monotheism a legend, but any monotheism is a legend. Judaism was monotheism in theological formulas, but it was not monotheism in its essence. In reality, the dogma of monotheism is one of the most fanastic legends, one of the very greatest deceptions of religion."

Examining the psycho-physiological creators of that deception, Russian scientist V.M. Kaytukov quite accurately called these people "negative zealots."

In the case of a subsequent Semitic religion—Christianity—the situation appears analogous. Italian Biblical scholar Abrogio Donini, frankly described the pathological environment of its origin: "The zone is situated not far from the ancient city of Jericho, to the south of the Jordan River Valley, at 300 meters below the level of the Mediterranean Sea. The soil of this desert is infertile, and divided by poisonous deposits of salt, which remained after the sea waters cover this place at one time. Its surface is furrowed by numerous torrent channels, that are dry in summer and meager of water in the brief rainy season. This is the same place, where according to legend, John the Baptist preached, and Jesus of Nazareth went and isolated himself for a 40-day fast in the desert."

The Dead Sea is a region that is unsuited for life; the geography is not one of stirring colors that healthily arouse the imagination.; The deadly salt sediments, the sharp transition from blazing heat to shadow, and also the [geographical] depression, situated below sea level, are not stimulating to the development of mythological thoughts. Even from a course of school physics and nature studies, it is known that such an unhealthy atmosphere could influence the ecstatic, fanatical minds of the first followers of Christ, which very often consisted of the outcasts of society. Also, when a person's nervous system is protractedly located in a place situated below sea level, it suffers irreversible changes, which lead to the development of numerous pathologies.

The Jews considered Christ deranged and insane, and so did the Roman provincial governor. The creator of the Christian Church, the apostle Paul, was an epileptic, just like Mohammed, the creator of Islam. One doesn't need to be an expert in history, in order to know elementary facts, which testify about the intolerance and hardness of the adepts of Christianity and Islam. But in light of a generalization of natural-scientific and religious studies information, it naturally comes about that **the origination and spread of the very idea of a single God, is first of all tied with pathological factors in the structure of the nervous system, which predetermine the development of delirious monomania, and also with a heightened aggressiveness and Daltonism. Namely a combination of these negative traits allowed the creation of a pyramid of monotheistic philosophy. Unconscious and groundless hardness, together with a genetic inability to differentiate color and tones, automatically prods people into the** bosom of the concept of a single God, truth, devil, or absolute. Before any oneness, it is not necessary to mature, but to sink, even into a sense of ideological poverty. The problem of Aryan polytheism and Semitic monotheism is not a problem of freedom of religious choice, but first of all of the racial-archtype of the construction of the brain.

9. "Right" and "Left" Races

We will examine still another aspect of the inborn differences of the racial archtypes, particularly important by virtue of their everyday, unconscious evaluations in daily life.

"Stand on the right", "keep to the right", "that is right", "you are right"—we hear these customary words often enough, not in the least delving into the metaphysical sense, whereas linguistic philosophy, armed with a theory of the linguistic picture of the world, explains that subconscious choice of the right perspective for all peoples of the Indo-European language group, is conditioned by their archtype. The direction of writing from left to right and right side driving on roads—this is a common heritage, which we receive during birth, as something that goes without saying. The vector of any thinking and moral efforts of an Aryan, projecting onto a metaphysical plan, irresistibly attracts to it the Kingdom of the Right—the holy world of purity, right, and higher thinking. "You are right", "you are justified" [in Russian, the word "justified", *oprav*dani, contains the root for "right", which is *prav*]—these words sound like a conclusive verdict, not needing any clarification. The words "right" and "truthful" [the Russian word for "truthful", *pravdiviy*, contains the root for "right"] fulfill such an all-embracing meaning in our consciousness, that it covers all facts, phenomena, sensory states, metaphysical categories, even entire cultures and civilizations with ease, and their expressive capacity has not weakened in the least from endless use.

Movement to the right—clockwise, that is—is a particular, distinctive trait of the Aryan archetype. And after all, the apparatus that holds eyeglass lenses on our noses, and helps us to see, carries the name "frames" for a reason [the Russian word for eyeglass frames, *oprava*, contains the root, *prava*—"right"]. To hold a wedding [*Spravit' svadbu*] or run a show [*pravit' bal*]—these words are especially important in an emotional and aesthetic plan of phenomena that are also expressed with the help of the terminology of *Right*.

In the book, *Poetic Views of the Slavs on Nature*,[440] Russian researcher A. Afanasyev wrote in this regard: "According to the conviction of the common people, a good angel stands on the right side of a person, and evil on the left; one should not spit to the right, in order not to drive away your guardian angel—spit to the left and you will strike the devil; and while rising from bed, they advise to spit to the left side, and rub the saliva with your feet: by this means you will drive away the unclean, and in this day he already will not make a note behind you of your sins... Rise from bed with the right leg; if you rise on the left leg—all day you will be out of sorts, grouchy, and melancholy; put on and take off shoes, starting with the right foot; he who is walking into a home enters on the right foot, in order to expect a good welcome; in ancient fortune-telling, Slavs observed with which leg the sacred horse stepped over positioned poles—the right or the left, and in the first case they expected success, and in the second, failure... Under the influence of these views, the word **right** signified all that was morally good, fair, and powerful (truth [**prav**da], rule [**prav**ilo], and justice [u**prav**a]); the same correlation of concepts is set down in the German and French languages.

Modern Russian scientist Aleksandr Vasiliyevich Podosinov dedicated the fundamental monograph, *Ex Oriente Lux! Orientation according to the Countries of the World, in the Archaic Cultures of Eurasia*,[441] to the study of the given problem. In it, he emphasized that in India, around 1,500 B.C., that is, in the period of the conquest by the Aryans, a system of racial-archtype orientation was established, since in the *Laws of Manu* it is said that pagan priests were ordered not to eat that "which is in the left hand."

Holy Vedic ceremonies were also performed by a circular procession, from left to right. And the ocean, according to the ideas of the ancient Aryans, "flows from left to right."

---

[440] *Poeticheskie vozzreniya slavyan na prirodu.* Moscow, 1994.

[441] *Ex Oriente Lux! Orientatsiya po stranam sveta v arkhaicheskikh kul'turakh Evrazii.* Moscow, 1999.

Precisely such a picture is observed in the culture of the ancient Aryans of Zoroastrian Iran.

In ancient Greece, during rites of the pagan priest, it was considered auspicious when a bird flew from left to right, but from right to left was considered unlucky. In general, in Greek mythology everything on the left was tied with evil and destruction, and everything on the right with virtue and good. The great Pythagorus dedicated entire tracts to the given question, and for his part, Plato wrote that in devoting high honors to the Olympian Gods—"everything odd, first, and right" was required; and to the Gods of the Underworld went "everything even, secondary, and left." This preference for the right side is even reflected in the wearing of clothes. Thus, in one of the compositions of the famous comediograph, Aristophanes, Poseidon turns to the barbaric god *Triball*, who is behaving himself badly and rudely [and says to him]: "You queer bird, you have thrown your cloak over to the left? Toss it to the right, as our custom dictates."

Among the ancient Etruscans, the left was tied to everything evil and unfavorable, and the term which meant this (*laevys*), later smoothly migrated into the Latin language.

For a strict preference for the right side, Vitruvius recommended constructing temples with odd-numbered steps in Rome, so that the right foot, with which one began to ascend to the platform of the temple, would be placed on the platform first, [when a person got to the top of the stairs].

According to Arnobius, the pagan Romans only sacrificed black victims to the Gods of the left, and white victims to the Gods of the right, because the right side is of the Heavenly Gods, and the left is of the Underworld [Gods]. Virgil described two paths in the Underworld Kingdom of the Dead, one of which, the right [path], leads to blessed Elysia, and the other, to the sinners in torment in Tartar.

And thus, we see it is characteristic of peoples belonging to the common circle of the original range of the Aryans, to relate positively to the right, and together with this, [to relate] negatively to the left. The existence of the given binary system of preferences cannot be explained by the influence of any abstract cultural influences, but only by a specific archtype, having its roots in the original racial construction of the brain of the creators of the given spatial system of values.

However, if we undertake to study the archaic cultures of Eurasia, located beyond the extent of the range of the ancient Aryans, then we easily discover how the picture of preferences changes to the direct opposite.

The great ancient Greek historian, Herodotus, described the "strangeness of different peoples." He mentions: "The Hellenes write their letters and count from left to right, and the Egyptians, from right to left. And all the same, while doing so, they assert that they write to the right, and the Hellenes—to the left." Another famous ancient Greek historian, Plutarch, told of one Egyptian cult song of sorrow, in which it mourned all born within the limits of the left side of the Nile, and died within the bounds of the right side. In the Book of the Dead, one of the goddesses of the Kingdom of the Afterlife, says of herself that she will reunite "in the right side of Heaven." The Nile—the giver of life, according to the mythological ideas of the ancient Egyptians—is also born on the left side, dying on the right.

Historians of the modern era are of the opinion that the notion of the right side as a symbol of death, was characteristic of the Egyptian mentality in the Age of the Pharoahs.

The pagan priests of ancient Babylonia also considered the left side as preferred over the right, and all their astronomical calculations and magical acts were fulfilled from right to left, that is, counter-clockwise. The residents of Sumer in Assyria got dressed in accord with a leftward movement: they enveloped their body with a cloth, starting with the left armpit, and bringing the cloth behind the back to the right shoulder, again extended it to the left arm.

The main book of Daoism—*Dao dey Dzin* (The Canon of the Way of Abundance)—enlightens us on the question of the significance of the left and right sides in ancient China. Its authorship is attributed to Lao-tszi (6th Century B.C.). In it, the following is said, word-for-word: "A perfect person usually considers the left side as more honorable, but in time of war, the right side. In the case of holidays, the honorable place is on the left side; in the case of mourning, it is on the right side."

The Jews, who wrote from the right to the left in the time of Solomon's Temple, also patrolled the Temple in a counter-clockwise manner. The Arabs also write from right to left.

Dispersed all around the world, with time, the Jews rendered more and more influence on the cultural, political, and economic life of Europe. In newer times, this influence achieved its apogee in a series of bourgeois revolutions, first of all in the French Revolution. The Jews received emancipation in the conditions of the new social-political structure. At first tight-lipped from the conditions of the Diaspora, Jewish communities joined European communities on an equal footing. But accepting a European look, they nevertheless brought the traits of their racial mentality to everything around them. The very word "revolution", meaning a radical tearing down of the existing patriarchal system and conservative system of values, began to be steadily associated with the left side of the political spectrum. The words *left, humanistic,* and *progressive,* gradually became [political] synonyms, just as *right, conservative,* and *reactionary* [became political synonyms]. The unnatural Egyptian-Chaldean-Jewish system of spatial orientation blossomed in a magnificent color in the middle of white European peoples, who were brought up under the basic influence of the racial stereotypes of the ancient Aryans. And with the enthronement of the ideals of Karl Marx, the grandson of a rabbi, communism was conclusively identified as a leftist ideology, and proclaimed as the hope of all progressive mankind; and all those who opposed it were written off as right-wing obscurantists and fascists. Thus, the archtype of the White race was desecrated and disfigured.

The given picture of the world was confirmed and worked out in detail, with the development of modern neurology and related sciences. Canadian bio-politician J. Laponce showed that left-handed people differ from right-handed people in a political respect, in his book, *The Left-hander and Politics.*[442] In politics, the right-handed hold to the right relatively often, that is, they have conservative views; but the left-handed gravitate toward leftist viewpoints (reformists, modernists, and revolutionaries). From the works of Laponce it also follows that a number of officially-registered left-handers can be considered as their own kind of yardstick of "democracy": the more democratic the government, the higher the percentage of left-handed persons who comprise its population—according to an official census. Laponce suggested that in undemocratic, for example, totalitarian states, a spirit of conformism reigns, and left-handed people are compelled to be trained to hide their left-handedness; therefore, their numbers are essentially understated in census figures.

Modern biopolitician A.V. Oleskin also emphasizes in his book, *Bio-politic*: "Different cultures in human society gravitate toward preferred development of the 'logical' left brain hemisphere or, on the other hand, the 'graphic' right hemisphere. In political leadership, two styles can also be set apart—the left-hemisphered, oriented to logic and sober calculation, but often 'uninspired', and the right-hemisphered, oriented to a graphic vision of resolution of political problems, often fixed toward the future."

However, we should establish with all obviousness, that the planting of principles for the functioning of one [racial] archtype into the bosom of a different race, unavoidably leads to structural disfunctions, expressed in an increase in the number of nervous-mental deviations.

Cesar Lombroso, the father of criminal anthropology, cited convincing statistics, according to which, among habitual criminals and jailhouse prisoners, left-handers are encountered more often, than among normal people in the real world.

John Skarn, an expert in games of chance, wrote in his book, *Tips for Gamblers,* that left-handers are often encountered among professional cardsharps. The German racial theorist, Hans F.K. Guenther, pointed out that among those who write with the left hand, there is a far higher percentage of bisexuals. And actually, new studies have confirmed the fact, that namely the right hemisphere is responsible for the orientation in a person's own body.

Russian scientist V.V. Ivanov pointed out in his book, *Even and Odd (Asymmetry of the Brain and Character Systems),*[443] that formalism in art is the result of activity in the left hemisphere. Ruben's paintings were created by a right-orientated construction of the brain, but Malevich's *Black Square* [was created] by a left-oriented [one], since the right hemisphere as a whole answers for visual forms, while the left retains their symbolic shells.

---

[442] 1976

[443] *Chet i nechet (asimmetriya mozga i znakovikh system).* Moscow, 1978.

Ivanov's main conclusion in his book was that "only the assertions and commands of the right hemisphere can be true; those of the left are false."

Therefore, it seems completely obvious, that in the first place, the conflict of cultures leads to the development of constructions in the brains of their carriers. The very principle functioning of the brain of Man does not allow a conversion of the values of right-orientated races, into the values of left-orientated races, for they have different neurological features.

New York photographer David Eisendrat once made 50 photographs of theatrical scenes into two mirror-symmetrical variants, and proposed an experiment, whereby he asked random passersby to choose an image they liked most: the correct one, or the inverted one. 75% of those who chose the correct image, wrote in left-to-right in English. Those who wrote in right-to-left Hebrew chose the inverted image.

European comics are positioned from left to right, but Japanese makimono—well-made strips with a story in pictures—are in the opposite direction.

The modern American researcher, M. Gardner, stated in his book, *This is a Right, Left World*,[444] that many daily trifles constantly irritate a left-hemisphered individual in a "right" world, and vice versa.

The reason for the disparity in the "right" to "left" worlds, was explained in the first part of the 19th Century, when French chemist Jean Baptiste Bayeaux discovered that quartz crystal was able to rotate the plane of polarized light. And his fellow scientist, the world-renown Louis Pasteur, found that the properties of chemical substances could be identified, with the exception of one—the ability rotate the plane of polarization of light. "An asymmetrical living organism chooses namely those forms of tartaric acid for food, which answer its demand, and undoubtedly somehow correspond to, his own inner asymmetry; but another form he leaves without change, either completely, or in large part"—Pasteur wrote.

It is namely in this bio-physical principle that one needs to search for the roots of wine prohibition among the Arabs, and not in some distinctive cultural and religious norms, since they are naturally not able to digest this form of tartaric acid, which is so favored by Indo-Europeans.

Further studies established that the majority of organic substances going into the make-up of living organisms are optically active, and the solutions of many chemical substances on the other hand, are optically inactive. "Living agents" is what Pasteur named substances that are optically active. Now it becomes understood why in Indo-European mythology, peoples of the Nordic type are invariably called "people of the sun", or "sons of light". **Saturated by sunlight, the blonde with sky-blue eyes is not a poetic metaphor, but a biophysical fact. Correspondingly with this, the worldview of the members of the Nordic race has a certain, real, and natural basis, unlike the members of other races.**

M. Gardner states: "Asymmetric atoms go into the composition of a molecule, but the molecule itself can nevertheless be symmetric as a whole. It is namely on this simple level that one needs to search for the fundamental contradictions between Aryans and non-Aryans.

Racial differences are differences at the level of elementary particles.

Chemical substances which are dextrarotary, or turn the plane of light rightward, are called **D**, for the first letter of the Latin word, *dexter* (right); and left-turning substances are called **L**, for the first letter in the Latin word, *laevus* (left). Gardner emphasizes: "Almost all matter encountered in living organisms, is the union of carbon, with asymmetry laid in it, thanks to the asymmetry of atoms of carbon." **Different concentrations of left and right unions of carbon are the key to unraveling the secret of racial differences.**

"It is difficult to imagine, how it could be; evolution, creating such diversity, managing without carbon, surpassing all others in the ability to form practically endless numbers of different combinations, each with special individual features. The combinations of known carbons are two times greater, than the combinations of all remaining elements taken together. The tissues of all living beings on Earth, even of those not visible to a microscope, even from the virus to the elephant, consist of matter containing carbon. Some biochemists go so far [as to say] that life itself is defined as some complex property of carbon combinations. Carbon is the great builder, inasmuch as on the exterior shell of each atom of carbon, there is a place for four additional

---

[444] *Etot praviy, leviy mir.* Moscow, 1967.

341

electrons. These atoms therefore can join with each other, forming chains of endless length; in each atom of carbon there still remains two points, to which other atoms or groups of atoms can "attach", like pendants on a necklace"—writes Gardner.

Therefore, in the final analysis, the struggle of races is the expression of the struggle of different carbon combinations in the biosphere.

Two molecules which contain exactly identical numbers of atoms of each sort, but which differ by the method of their combination, are called "isomers". If two molecules have an equal number of atoms of all sorts, which combine between each other identically, then can such molecules all the same turn out different? Yes, in conditions that one becomes a mirror reflection of another. Such isomers are called "stereo isomers." In all cases, when a molecule has asymmetrical structure, it should exist in a different mirror-symmetric form. If, for example, five asymmetric carbon atoms go into the combination, then each of them can be right or left, and the complete number of different possible stereo isomers in this case is very high.

Almost any combination of carbons that goes into the composition of living tissues, is a stereo isomer, turning the plane of polarization of light in a definite direction. It is namely this physical property, which predetermines the existence of right-oriented and left-oriented races, and not any abstract cultural norms.

The most complex and numerous of all carbon combinations are called proteins, and they also set racial differences. In the human organism, it is estimated that there are around 100,000 different proteins. In Man, there can be up to 1,000 different enzymes in the composition of a single cell; the enzymes facilitate the flow of thousands of different chemical reactions, and each enzyme is a protein. Hormones which control the activity of various inner organs, also react to a number of proteins. Not one part of our organism, including bones, blood, muscles, sinews, hair, and nails, can be produced without proteins, which constantly regenerate racial differences.

Gardner observes: "Right or left modifications of any optical combinations possess absolutely identical chemical properties. The differences manifest themselves very specifically. When an asymmetric substance is swallowed or enters the blood by way of infection, it also enters into interaction with the asymmetric combinations, of which the organism consists. Stereo isomers of a definite type are digested by the organism, and its mirror double is set aside, like garbage. In other cases, the organism digests both stereo isomers with varying speed, reacting to them differently. Almost all asymmetrical carbon combinations that go into the composition of living organisms—and such combinations are in the millions—exist only in one or two possible optical modifications.

Namely therefore, the cultural norms of right-oriented races are not compatible with the norms of left-oriented races.

The crystals of one kind are "dead" in one environment, but if this crystal is introduced into a plant or animal of some kind, which is struck by this virus, it immediately begins its deadly activity. When the virus attacks bacteria, its protein membrane fastens to it from the outside and remains there; and the nucleic spiral passes through the wall of the cell, like a drill, and begins to build a new order in the mechanism of reproduction. Soon the cell begins to make not copies of itself, but of the invader—the virus.

The noxious influence of the left-sided virus on the right-sided white race can be observed in everyday life, in the form of pervert sexual propaganda, degenerate modernist art, and other forms of democracy. Even the mode of left-hand writing has this same viral nature.

The sphere of culture is also subjected to the principle of interaction of asymmetric systems. If, for example, left-adrenalin causes 12 times the [amount of] constriction of blood-carrying vessels than its mirror double, and a "rejection" form of Vitamin C renders practically no influence on the organism, then the stereo type behaviors and moral norms of one race can lead to the mental and biological damage of another race.

In Russia, V.I. Vernadskiy can be considered a pioneer researcher in the given area; in an article he wrote in 1938, which was characteristically titled, *Rightism and Leftism*,[445] he indicated: "In space that is occupied by life, that is, [space] within a living organism, it exists under the influence of a cause, tied with the same state, a special state of physical-chemical processes, in which right and left phenomena turn out different. The usual laws of symmetry for such a space,

---

[445] *Pravizna i levizna.*

for example, are disrupted. This manifests in very different traits in a living organism; during metabolism in an organism, distinct crystal combinations form that can yield crystal lattices, in which right and left isomers are observed, sharply differing by the number of their biochemical properties. But it manifests in a countless number of different phenomena—large and small. It manifests in rightism and leftism, so sharply characteristic to the human organism in general. "In his conclusions, Vernadskiy proceeded from the discoveries of Pasteur and Curie. Besides that he did not fear carrying over the positive sciences to the area of metaphysics. Thus, in one of his work notebooks dating back to that time, he prophetically wrote that the metem-psychosis of eastern religions is a distinctive notion about life as a reversible process, proceeding from the principles of dissymmetry; that is, about rightism and leftism as a global and all-embracing phenomenon of organic nature.

Louis Pasteur prophetically wrote that left-right asymmetry is the key to the secret of life. M. Gardner asserts in his book, *This is a Right, Left World*: "The life opened up to us is a product of the asymmetry of the world and its consequences...I even think that all kinds of life, in its initial structure, in its inner forms, is a **result of cosmic asymmetry. The Universe is asymmetrical.**

In 1962, American physicist John Rush expressed the proposition in the book, *The Dawn of Life*, that initially in the universe, in the original plasma, there existed self-replicating molecules of both types of asymmetry. Each of them could feed only on molecules of its type of asymmetry.

One can suggest that organic life itself arose in the universe as a result of the interaction of asymmetric molecules of protoplasm. Consequently, the origination of the very nature of racial differences should be sought at the evolutionary stage that precedes the origin of organic life in the universe. The inequality of the races is thus laid in the very nature of life, on an elementary particle level. In general, life is the result of this inequality.

The cold-blooded realization of this fact unavoidably brings a revision of all philosophy and values systems in its wake, and also leads to a rejection of the very notion of "humanity", for with all obviousness it is made clear, that planet Earth, is after all, only one of the polygons, upon which various assymetrical substances battle, forcing their morals and gods on each other, and also other asymmetric molecular constructions in the form of material culture. And there is nothing fantastic in this conclusion. The struggle of two worlds, having different natures, is laid at the base of the cosmogony and ethics of such ancient Aryan religions as Zoroastrianism. The multi-level competition of the kingdoms of Pravi and Navi comprises the structure of Russian folk tales, and of the absolute majority of myths and legends of the Indo-European peoples. The binary division of the world into "right" and "left" was also laid at the basis of the esoteric system of Pythagorus, in which the numerical ratios, attendant to this division, were considered from the very beginning, to be a God-given providence, lying in the very nature of all things. These basic principles of the ancient Aryan worldview have already found their fruitful embodiment in modern times in science, and stimulated the development of **molecular biology** and **biosymmetrics**, as a result of which, the structure of DNA, RNA, and some proteins were discovered. In modern philosophy, this principle found its embodiment in the review of viewpoints, and is used now in the daily practice of political struggle, and also creates prerequisites for the rise of qualitatively new branches of knowledge, like the **general theory of systems (GTS)**, for example.

The structural division of the universe at an elementary particle level has received the name **dissymmetry** in modern physics. Yu.A. Urmantsev, a prominent Russian scientist specializing in questions of the given problematic, writes in the book, *The Symmetry of Nature and the Nature of Symmetry*:[446] "Such objects are called dissymmetric, which a) change in a mirror reflection in some respects, right down to contrast; which b) do not combine with their mirror relations, as a result of this; and which c) exist in one, two, or more than two modifications." In 1962-1963, it was proven in biology that the demands of a combinative inversion in living nature disrupt, since in a number of cases, in the transition from a D (right) to an L (left) bio-object, some traits of the right change; and furthermore thus, the properties of the D-form cannot remove the properties of the L-variety, through any symmetrical and anti-symmetrical operations. And this again testifies in favor that "right" and "left" races have different structures at the molecular level, from the very beginning.

---

[446] *Simmetriya prirodi i priroda simmetrii.* Moscow, 1974.

In the book, *Physical Causes of Dissymmetry*,[447] another Russian scientist, V.A. Kizel' quite clearly points out: "This is the problem of biophysics: the existence of undoubtable dissymmetry in right and left forms of living systems and their "structural materials"—simple molecules. Dissymmetry is traced in all living systems, from bacteria to higher organisms and Man, and in very diverse living functions and body functions, right down to the sphere of the psyche, and is also reflected in the morphology of an organism."

It is namely different concentrations of left and right forms of molecules in the structure of organisms, which in the final analysis yields that diversity of races, which we have today. Differences of culture which are the result of differences in the inborn mental styles of their creators, also have a molecular foundation.

In chemical reactions, many biochemical antipods have identical physical-chemical properties; it is namely therefore, why so many racial differences have such an indistinct and concealed nature. However, some molecules and enzymes in an organism possess discriminatory functions, attracting only elements of the same sign and repelling opposites. The biochemical system of recognition, "us-them", creates a special code at all levels of hierarchy in a living system, which is passed down to posterity. The differences between rightism and leftism always only increase in the process of evolution. Molecules of one sign have carried on a constant struggle with molecules of another sign, for the entire length of evolution, and in transition from stage to stage, until it has finally become a struggle of organisms for existence, and grown to a war of races. The affiliation of an organism with this or that sign gives it advantages in the struggle for survival.

Therefore, expressed in the educated language of Gardner, one can say that **the "left" world is not recommended to "white" races, and vice versa.**

Some physicists expressed the proposition about the origin of dissymmetric life in various special cosmic pockets. "Left" and "right" forms of life are thus born in various parts of the Universe; that is explained by the varied action of the radiation spectrum of the electromagnetic poles on active, pre-biological forms of matter. In the daytime sunlight, right-rotating, polarized components predominate, and on a lunar night, left-turning polarized components [predominate].

Namely at this original biophysiological level one can explain the presence of sun symbols in Aryan cults, and moon symbols in Semitic cults. In an explanation of this problem, culturologists again end up in a blind alley. Radiation from the sun and moon is nothing but frequent instances of the varied frequency of the electromagnetic spectrum, testifying that in the process of sacred activities, "right" races are attracted to clockwise polarized light, and "left" races are attracted to counter-clockwise polarized light. In this regard, the great Russian philosopher V.V. Rozanov called the Jews "people of the moonlight", in a composition of the same name. In the prominent ancient Aryan religion of Mithraism, which was a former basic competitor of Christianity, the chief diety—Mithra—was always called "sun-born."

The main conclusion in V.A. Kizel's book sounds completely in the spirit of our general discussions: "We see that the question of the origin of dissymmetry of living matter in living systems is far from decided and opens a wide field for studies in the most diverse, and as it would turn out, areas of science most removed from each other. At one time, the school of the Pythagorans taught that the Universe has a right and left side. We see that this question, on another ideological and scientific basis, continues to occupy the human mind to the present day."

The given argument of the author is not a contradiction of new tendencies in modern anthropology, for the prominent Russian scientist, V.P. Alekseyev, points out in the book, *Geographical Pockets of Formation of the Human Races*:[448] "Localization of the very sources of cosmic radiation justifies setting apart the cosmic factors of race formation into a special group."

Thanks to the prophetic genius of Alekseyev, in general the creation of a principally new perspective of science comes to light—**molecular sociology.** "The chemistry of Man, or anthropochemistry, embraces processes originating in the diapason range, from molecular to planetary, and unites them all in a powerful synthesis, which is realized on the boundary of social laws with pure chemical processes of atoms and their combinations."

Just reflect, after all, that the progress of science can give us real opportunities of explaining complicated cultural, political, and social processes in society, not only on a level of the physical

---

[447] *Fizicheskie prichini dissimmetrii.* Moscow, 1985.
[448] *Geograficheskie ochagi formirovaniya chelovecheskikh ras.* Moscow, 1985.

racial differences of peoples in their involvements, but to bring in and mathematically count these patterns at the level of elementary particles. From atoms to their combinations, which comprise the unique specific features of the structure of a race, arises the opportunity to build a direct path to the creation of a similar design of its worldview. Humanitarian disciplines can become precise sciences. Once and for all, one can put an end to speculation in the spiritual sphere. The famous principle, "to each his own", can triumph at all levels of the existence of a race.

10. Prevention of Degeneracy and Resistance to Disintegration

Finally it becomes clear, that racial theory is the key to mastery of the whole system of the universe. But the concept of a worldview founded on it will obviously be incomplete, if in order to please some abstract norms of academic impartiality, we do not consider the question of values. Here it is best of all for us to rely on the time-tested postulates of the classics of science. One of the founders of Soviet psychiatry, V.P. Osipov, emphasizes in his fundamental monography, *A Course of General Teachings about Spiritual Illnesses*:[449] "Preventative measures should be undertaken not only in regard to physical factors, which cause spiritual distress, but in regard also to mental infection, which can also spread epidemically."

For his part, the famous German racial philosopher, Ernst Krieck (1882-1947), wrote in *Happiness and Strength*:[450] "The problem of heredity in general, inasmuch as purposeful human will cannot influence it, is a political problem. Before the doctor stands a question, which demands acceptance of a political decision. Either he will be guided by the fiction that all illnesses are at the very least, curable in principle, or he will think that hopelessly rotted individual lives should be amputated for the health of the whole, like a surgeon does with the diseased members of a body, or a judge does with separate members of society."

Modern geneticists F. Vogel and A. Motulski emphasize in the three-tome book, *The Genetics of Man*:[451] "In contrast to the majority of European states, the societies of the two world powers—the United States of America and the Soviet Union—consist of a majority, belonging to one ethnic group, and several smaller, but stable groups. The existence of national minorities creates tension and conflict situations. The simple method of resolving these conflicts would be absorption of the groups comprising the minority, by the majority."

The political authority of the Soviet Union did not recognize the hidden genetic threat of a multi-national state, refusing for decades to support Russians as a pivotal, nation-forming ethnos. The result of this short-sighted political tolerance was the disintergration of the Soviet Union. It is completely obvious that an uncontrolled wave of democracy will lead to the political non-existence of another colossus—the United States of America.

It is time to conclude our discussion with a quotation, which is perceived by us as manifest. N.K. Kol'tsov, one of the pioneers of Russian genetics, prophetically wrote in the article, *The Influence of Culture on Natural Selection in Mankind*,[452] that: "The process of the degeneration of cultured peoples can be halted, if the measure of the threatening danger can be timely recognized by wide segments of the population, and if due attention is given to this danger, in a common, social-economic government policy. Political figures and political parties should be filled with the conviction, that if they want to build firmly, and not for just one or two generations, then they should take care that succeeding generations of those genotype elements, which are most valuable to them, are fully sufficiently represented. A corresponding evaluation of the groups of a population can only be produced by political parties and government authority.

If a government authority so estimates the hereditary qualities of this or that group of a population, then of course, it can, with definite measures, place valued groups of a population in conditions favorable for increased reproduction. It is only necessary to remember, that one improvement of the material well-being of a given group cannot obtain completely favorable

---

[449] *Kurs obshchego ucheniya o dushevnikh boleznyakh.* Berlin, 1923.
[450] 1943
[451] *Genetika cheloveka.* Moscow, 1990.
[452] *Vliyanie kul'turi na otbor v chelovechestve.* Russkiy evgenicheskiy zhurnal, 1924, Tom I, Vip. 3-4. [Russian Eugenics Journal].

results. It is necessary that the improvement of the well-being [of a group] be tied with the presence of a definite number of children. The character of such measures is already mapped out by the administrators of the Eugenics movement in different countries.

But their success depends to a significant degree on how the selected groups of the population themselves are filled with the awareness of their debt to future generations, and [that they] do not give in to the temptations of Malthusian [thinking]. In the presence of a conscious attitude toward their debt, on the part of the more valued groups of a population, the spread of an easier, Malthusian life among the less valued groups of a population should encounter no obstacles. The segregation and sterilization of sharply defective elements in a population can also have certain significance.

Natural selection, which played a governing role in the evolution of the entire organic world, and in the first Man, is weakened under the influence of [harmful] culture and is even perverted by unnatural selection. The time has come, and it is still not too late, to replace it with a consciously worked out, definitively planned system of **artificial eugenic selection.** For the time in human history, a culture has achieved its flower in the presence of a definite knowledge of the relatively immense significance of selection. Do they really not know how to use this knowledge?!"

The Racial Theory of Time

"A man belongs to his century and his race, even if he declares war on both of them."
Ernst Renan

"Destroy the perception of differences, and you destroy time."
Jean-Marie Guyau

It is possible that the declared problem in the above title of the given work seems too exotic to the respected reader, and even partly fantastic. The modern world, to the limits of universalized time, is thought of by Man as a certain substance, the same for all; it is [thought] to be sufficient to move the hour hand of a clock forward or back, when crossing time zones in the cabin of a comfortable liner, in order to become a participant in another culture.

Without hesitation, the modern, average-statistical man will say "Time is the same for everyone", and will shrug his shoulders in bewilderment, if you suddenly decide to assert that this same time has racial differences.

However, one can only abstract from reality. Two completely obvious facts, which lie at the surface of the public consciousness, place the thesis of the universality of time in obvious doubt.

In the first place, all the peoples of the world have completely different notions about fate, and as a result, have diverse eschatological concepts in their national mythology. In the second place, the basic human races have statistically differing times of sexual maturity, which is reflected with all obviousness in the law-making practices of the peoples, to which these races belong. The time of sexual maturity is one of the most important biological characteristics of any organism, by attainment of which his behavior cardinally changes.

If we analyze the very sense of the word "time" in different languages, then its postulated universalism, [as put out by] the mass media, in general disappears without a trace. *The Etymological Dictionary of the Russian Language,*[453] by A.G. Preobrazhenskiy, shows that the word "time" [*vremya* in Russian] originally sounded like *veremya*, and meant *rotation*, or *circular motion*. The word *bremya*, which means *weight*, is close in sound; from it comes the common saying, "*nesti svoye bremya*" [carry his weight]. It is paradoxical, but a fact: it happens that the original sense of the word, *vremya*, was understood from the set expression, "*nesti tyazhest' svoyego vremeni*"—"he carries the weight of his time."

Such a conclusion does not appear to be the least bit artful, if we turn to the language that is the oldest, and closest to the Indo-European root: Sanskrit. In Sanskrit, *bharma* also means "weight"; in Avestani [sometimes called *Zend*], "I carry" is *bareman*; it conveys the same sense with accuracy.

Besides that the Russian word *vremya* [time], comes close in sound to the Indo-European base *wertmen*, and the Sanskrit *vartman*, which means path, rut, or wheel track.

In reality, however, the word in India that designates time has no tie with *vartman*. In Sanskrit there is a word with a completely different root: *kalah*. However, according to sound, this is extremely close with the ancient Russian word, *kolo*, meaning 'rotation'; from there comes the word, *kolovrat*, serving as the designation for the symbol of the ancient Russian eight-radial swastika. There is also the common Slavic base, *vertmya*; with the disappearance of the letter "t", the original sense of the word, *vremya*, also means 'rotation'. There is also a word in the Breton language—*vreman*—which means "now"; and *pred* means "time" in the Breton language.

As a result of this, the majority of linguists have come to the conclusion, that the concept of time in the Indo-European languages was originally tied with a definite outer manifestation of a certain process. Besides that in the course of evolution, this word changed gender. For example, the ancient Irish *ge* was neuter in the beginning, and later was a feminine-gender word.

The German word, *die zeit*, is tied with the verb, *ziehen*—"to pull". In Russian, we say *vremya tyanetsya*—"time drags"; the Germans put this same sense into the phrase, *die zeit zieht sich*.

---

[453] *Etimologicheskiy slovar' russkogo yazika.* Moscow, 1959.

The English word "time" comes from the Old English word, *tima*, which comes from the Old German word, *timon*, which also means "to stretch out", or "to extend".

The Latin equivalent of this word is *tempus*, which is sometimes directly tied with either the verb, *tendo* (to drag, to pull), or with the verb, *teneo* (to remain, to last). The closeness of the Latin words *tempus* and *templum* (temple) is significant, for originally the latter referred to a sacred space, around which the Etruscan pagan priests drew a line at noon.

Numerous facts point to the sacredness of the category of time. The Chinese hieroglyph, *shi*, which means "time", is represented by a variant of the hieroglyph—*si*—which also means "temple." The name of the Greek god *Chronos* speaks for itself in this sense, just as the name of the Hindu goddess, *Kali*, is derived from the above-mentioned Sanskrit word, *kalah*. The ancient Roman god *Janus* also deified time, just like the ancient Egyptian *Tot*. In the esoteric part of Zoroastrianism—zervanism—or "time", *Zervan* deified the high beginning. The ancient Germanic goddesses of fate, the *Norni*, also served as the embodiment of time. Their names, *Urd*, *Verdandi*, and *Skuld* mean, respectively, "past", "present", and "future." They do not simply "pull" some thread, they spin the yarn of fate. It is characteristic that the Greek *chronos* is close in sound to the word, *khreon*, which means "fate, necessity."

We can observe a strikingly similar picture in the mythology of the ancient Mayans. All the obelisks and altars in their cults were erected with the aim of perpetuating periods of time. And the same intervals of time were portrayed in the form of a burden, carried on the backs of the hierarchical god-porters; thus it achieved personification of days, months, and years. Calculating namely which god would be marching on a given day, the pagan priests of the Maya could determine their joint influence, and by the same, predict the fate of humankind.

We do not at all wish to explain all nuances and encroach into the precise area of linguistics and culturology with this short excursion. In light of the declared theme, it is enough for us to make the following conclusions: notions about time trace back to the most ancient and remote layers of the human psyche, in connection with which, its sacralization is no accident. Each group of the Indo-European languages has its designation for time, but not a single term. Beyond the limits of this language group the picture is more varied still. Time is thought of, not as universal, but namely as a subjective category, sometimes as a process, which allows the intrusion of Man into it. Time can drag, stretch, and be carried on the shoulders, like fate. Time can decide what is most important. In ancient philosophy, the study of time began namely under such a perspective. The founder of the Milesian school [of natural philosophy], Thales, said that "time is wisest of all, for it reveals everything." His student, Anaximander, was the first to attempt to formulate a substantial concept of time, and Parmenides and Heraclitus separated it into past, present, and future, in their views.

Despite the obvious revolutionariness (for that time) of Plato's philosophy, the metaphysical essence of his assertions are archtypical, for he spoke of time as a negation of eternity. In his [play], *Timaeus*, he wrote: "for the beginning should be demarcated into two things: what is eternal, not having an origination of being, and what is eternally arising, but never existing. That which is comprehended with the help of reflection and explanation, is obvious and is eternally identical existence; and that subject to opinion and irrational sensation, arises and dies, but never exists, in point of fact."

Further, with a clarity characteristic to Plato, he reproduces the architecture of his worldview: "Thus, time arose together with Heaven, so that simultaneously born, they would simultaneously fall, if a downfall for them comes; for eternal nature served as the prototype for time, so that it came to resemble her, as much as possible." In accordance with his views, God created "a certain movable likeness of eternity"—this is the quintessence of Platonic philosophy's ideas on time.

Aristotle also made an appreciable contribution to the study of the problem of time. In his famous *Physicae*, he expressed very deep thoughts on the nature of things: "...time is a measure of the movement and being of the body in this state...for movement 'to be in time' means to measure time itself and its existence."

But time can flow, and not identically for all; it can differ for bodies placed within it. The materialists of different ages did not completely turn attention to Aristotle's grandiose thought, choosing from his composition only those universalist ideas most comfortable to them. One need not think, that located in time, it is also necessary to move, like all found in movement: after all, time is not movement, but a number of movements; and rest can be included in movement. Namely, anything not motionless rests, and being by nature able to move, is without it..."

Finally, the elite materialists diligently avoid the basic adage in Aristotle's book: "Without spirit, time cannot exist."

And thus, the subjective-psychological approach to comprehension of the problem of time is on hand.

The greatest optimist of all time and peoples, Epicurus, also thought that in the measure of duration, "one should control the immediate impression, in accordance with which we speak about a long or short time, and research this impression, attaching it to time, like we attach it to other objects."

The great philosopher of antiquity, Titus Lucretius Carus, for some reason also enrolled in the materialists, in his composition, *On the Nature of Things*, he generally called time "some special kind of random property."

Here, the subjective approach at the basis of differentiating the perception of time, sounds like a triumphant accord. This is already racial theory in its purest form:

"Also time is not in and of itself, but an object
Itself leading to the sensation of that which happened in the centuries
What happens now and what will follow later."

In this passage it is sufficient to change the word "subjects" to "races", and it can be boldly quoted in any textbook on neurophysiology, which studies in part the problem of the distinctive nature of the perception of time among various peoples, the speed of reaction, the threshold of sensitivity, and many others, relating to the area of competence of classical raciology.

Stoics magnificently developed the idea of cyclical time. Endless in and of itself, in their conception it appeared as a receptacle of the periodically changing world, which again and again arises, passing through definite stages of development, and dies as a result of a regular, universal fire. However, principly new worlds do not arise; everything returns to its own circle, with insignificant changes.

On their path, racial theoreticians of the early 20th Century already confirmed this thought of the ancient stoics, showing numerous historical examples of how peoples are born, blossom, wither, and disappear into the depths of time; but races live eternally, finding their historical, cultural, and political embodiment in all things new, and in new forms with unchanging biological content.

Epicurus, that modern esoteric, called time the "attribute of attributes", and Aenesidemus, a member of the school of ancient skeptics, thought of time as a function of the activity of consciousness. After 2,000+ years, it is necessary for Mankind to experimentally prove these simple truths. The neo-Platonist, Plotinus, again returned to the insights of the "Divine Plato", stating that "time is the life of the soul, living in a transitional motion, from one living manifestation to another."

But the ancient cycle of development predicted by the sages of philosophy came to an end. Sextus Empiricus, distinctly satiated by the abundance of keen concepts about the nature of time, in general denied it the right to existence: "Time is nothing."

"Blessed" Augustin, who with his work celebrated the transition from the philosophy of Antiquity to Middle Ages scholastics, a work literally achieved in the last moment of flight of the traditional arch-typical representation of time. The chain of invigorating transformations of being is severed, and the Crucifixion and Resurrection of Christ is announced as the initial and absolute point in history; henceforth time is described as linear-progressive. To weave fate, to manage time, to carry its weight on one's shoulders, Man was denied all this henceforth. The great mystery of the struggle of peoples and races was announced to be at an end, and all should disappear in the crucible of a single atonement. Therefore, time is now pronounced as single and universal for all. The chamber of weights and measures of the Christian upper hierarchy, leads to a common standardization and unification of time: from the Creation of the world to the First Coming of Christ and the Second Coming. The fires of the Inquisition wait for dissidents who make attempts on the new standards.

More and more in our time, representatives of the modern precision sciences, and also some lovers of esoteric exotics, turn to the works of Middle Ages mystics and alchemists, in search of "higher" revelations.

For our part, we should emphasize with all clearness, that for racial theory, and equally for the problem of the study of time, this was a completely empty and useless interval of European history. And the reason here is completely objectively concluded, for church theology disfigured the inert consciousness of the man of the Middle Ages with dogmatics, to such a degree that the notion of "historical process" only manifested with great tardiness in European philosophy, at the boundary of the 17th and 18th centuries. It was namely Isaac Newton, who was the first to grasp the Law of Universal Gravity and say that time is an intergral part of space and the "boundless sensorium of God."

Gerder was the first to introduce the notion of causality; Fichte developed the highly important notions of determinism, and Hegel persuasively presented history as namely a process.

In the introduction to his brilliant composition, *Anthropology*, Immanuel Kant already considered it necessary to emphasize: "The notion of a 'worldview' is particularly applicable to knowledge of Man, in his tribal attributes."

Thus, Kant was in essence, the first in the modern era to place a racial basis under the historical process of the development of knowledge. Time differs for all, and he points to this thought in his passage from *Criticisms of Pure Reason*: "Time is not a certain thing, that has existed in and of itself, nor is it inherent to things like an objective attribute...time is worth considering as real—not as an object—but as a way to perceive." Finally, Kant always spoke of the "subjective reality of time," and emphasized that "time itself does not change, but [rather] something found in time [changes]." But what is located in time, besides the different human races, each "subjective", according to its inherent "way of perceiving?"

150 years before the first discoveries in the areas of neuro-chemistry, neuro-physiology, biophysics, and the theory of information, he intuitively understood that in each biosystem, among which the human races belong, time flows in its own way. The great Kant heroically rescued the ancient, in part Platonic, ideal notion of time from the complete oblivion of the Middle Ages.

Another genius of philosophy, Arni Bergson, developed success in the given field with all his strength at the close of the 19th Century, for he also connected time with living nature only, and rejected its existence in the inorganic. In his noteworthy book, *Time and Freedom of Will*, he states: "Criticism of Pure Reason rests on the postulate that our common sense is only capable of Platonization; that is, the injection of any possible experience into pre-existing forms." It is easy to understand, that in the plane of experience, a pre-existing form is a racial archtype or a "racial trait", as Kant called it. Namely therefore it is understandable, why Plato emphasized that knowing is half of remembering. Remembering is relying on the experience of one's race.

In another of his books, *Duration and Simultaneity*, Bergson made the following characteristic conclusion, in the same spirit: "Between time and space there exists only one difference: our consciousness moves along time."

But different races possess different consciousness, and this is a biochemical fact, which no philanthropist can ignore.

In the context of developing the declared theme, we will now examine the views of Jean-Marie Guyau, the French philosopher and sociologist at the cusp of the 19th and 20th centuries, who in our view was unjustly consigned to oblivion. His innovative book, *The Origin of the Idea of Time*, was written more than 100 years ago, but is still relavent today, in all the clarity and weightiness of a narration. Nevertheless, it produces the impression of a revolutionary, non-conformist composition.

Most of all, Guyau was a high classic historian of religion, who to this day gives depth to all his social and philosophical summaries, and leaves an impression of the infallibility of the racial intuition of the author.

He starts his composition with an analysis of the definition of categories of time in the Indo-European languages, which differentiate past, present, and future in verbs; this language itself involves us in the comprehension of the structure of time and cause-effect ties. Strengthening this or that object, phenomenon, or form in our consciousness, language records them in a strictly defined part of time, which of itself points to the evolution of notions of time, for undeveloped languages are able to indicate movement, without its participation.

"Desiring and acting in the direction of our desires, we simultaneously create space and time; we live, and the world, or what name we give this, is created before our eyes. The force of will in particular produces the stability of memory."

Further, Guyau gives a truly staggering image, as never more accuate, which reflects the origin of the archtype of racial symbols: "Abstraction is the riverbed of time, which is created from its current." He also emphasizes, that from a psychological viewpoint, any memory means consciousness of something that Man is already not able to change, although here it is his integral part. Thus, it stresses the stability of the racial psyche, which, according to the opinion of Guyau, begets a specific evaluation of the length of time, as a "result of an inner eye." From a philological analysis of the evolution of the idea of time, the French philosopher naturally crossed over to an analysis of biological evolution. Time is the result of the struggle for existence among all other living beings, for even in our memory, sensation and recollection incessantly struggle with each other, and they survive either more crude and sturdy, or more evolutionarily adaptable. In nature, one species always strives to subdue others that occupy one and the same ecological niche. Precisely also, in the organizational structure of the memory, one feature of a fact memorized by us unavoidably strives to displace and erase another; in the process, time smooths out and idealizes everything, pushing ahead in the harsh competitive struggle, only the most hardy and deep traits of the memory. Therefore, time is undemocratic in its principle essence, for it begets in us a spirit of competition and contest.

"The perception of differences and likenesses, the first condition of the idea of time, has by its result the notion of duality, and with the help of duality, it creates the *days*. The idea of *day* is originally nothing other than the perception of differences in similarities."

In our opinion, it was namely in the course of the natural struggle of the races for existence, that the first numerical magic arose, which reflected their [respective] genetic [codes]. The astronomy of Stonehenge, the Pythagoran numerical concept, the mathematical mysticism of the Maya, the Kabbala, and the numbers of stages and degrees of the Masons—all this is a precise and reliable reflection of the genetic codes of the races, which created the given teachings.

The algebra of a race is time. It is included in race at the level of a genetic program. "Differences, similarity, number and degree or intensiveness; it is namely these factors which lie at the basis of the idea of time."

On the basis of time, like the basis of any struggle, lies the conscious choice of an active position; therefore, Guyau asserts: "The future is that toward which we move, and not that which moves toward us."

Time admits work with one's self, when, for example, a person feels pain and reacts to it, in order to remove its source; then he already begins to divide time into parts. It is no accident that in world statistics, information for the evaluation of the well-being of peoples is begun with years that precede wars, epidemics, revolutions, and global catastrophes.

Time is able to beget diseases of consciousness, in which connection even whole peoples [are afflicted]; one disease is the "false memory". "Holy people", "chosen by God", just like "'Divine abandonment" and other sclerotic spasms of recollection about the "fall into sin", graphically confirm this. Incidentally, modern discoveries in the area of neurochemistry and physiological psychology prove these propositions. "End of the World Syndrome" is inherent in many zealous forecasters, mystics, and politicians; this only testifies about damage to the parietal/sincipital parts of the brain, and has no relationship with real aeschatology.

In the conclusion of his remarkable book, Jean-Marie Guyau makes a conclusion, which was confirmed by the findings of experimental science only decades later:

"All the preceding leads us to the conclusion, that time is not a condition, but a simple product of consciousness; time is not a part of the nature of consciousness, but is a result of it. Time is not outside of desires and recollections. In our opinion, time is only one of the forms of evolution; instead of being the cause of the latter, it comes from [evolution] itself. Actually, time is the result of the transition from the homogeneous to the heterogeneous; this is differentiation entering into everything. Life and consciousness suggest diversity, and diversity begets the extent of time. Destroy the perception of differences, and you destroy time."

Speaking in the spirit of Guyau, and taking into account the achievements of modern precision sciences, we can make such a conclusion, which was not possible in Guyau's time, for its shocking simplicity.

Time is the natural result of the racial differentiation of Man.

Even the Bible, which depicts the racial homogeneity of Mankind (Adam and Eve), testifies, that in the Heavenly Garden of Eden, there was no time, apparently because there was no division of people into races.

The following phase in the development of notions about time also prefaced the development of entire scientific disciplines, such as biology, the theory of information, and genetics, which confirmed the intuitive insights of the romantics of naturalism.

It is worth mentioning a completely forgotten book by Russian thinker Valerian Muravyev, titled, *Mastery of Time*.[454] This is again a revolutionary composition, many passages of which can be easily observed today in daily public-political life. The maxims of the author, first proclaimed more than 80 years ago, possibly appear to make many mouths sore today, but the clarity of the production of the task, still at the dawn of systematic studies and inter-disciplinary research, is able to summon respect today for this regular, forgotten Russian of exceptional natural talent. From the very beginning, Muravyev literally took "the bull by the horns", asserting: "In general, all attempts prepare the real technical mastery of time: they reveal its subordination, its secondariness, and by the same token they place human will and reason in the position of enemies of a sort, if not the rulers of time."

Like a nomad thinks up verses about the steppes, and a pirate thinks the same of the endless expanses of the sea—he considers them as natural genetic accomplices in a successful enterprise. Muravyev saw a comfortable element in time, an area on which one can carry on a struggle. And as a conscientious colonel of Napoleonic temperament, he set about to gather his army.

"The key to the overcoming of time lies in our ability to manage a plurality of things, for plurality logically precedes time. The theory of the mastery of time, first of all leads to a theory of the formation of collectives of living beings: first of all, collectives of humans; and then of theories of the interaction of these collectives, on to the so-called inanimate multitude." But it is namely genetic consolidation, which lies at the basis of the racial idea, which is capable of this "collective association." The author excellently understands that even an unbridled revolutionary should know to rely on the experience of preceding generations; therefore, his following conclusion is similar to an appropriate choice of flag, understood to all. "In each culture, philosophy and science serve as weapons for the surmounting of time."

More than 50 years before the start of serious work in the area of genetic engineering, control of the human consciousness, and psychological weapons, the author of the book created a hierarchy of systems for the mastery of time:

1) genetics
2) politics
3) sphere of material production

Today, the basic mass of people is occupied in the sphere of material production, and constantly feels its dependence on the caste of politicians, but they in their turn, feel their own powerlessness in the face of the new pagan priesthood, which possesses the secrets of human genetics. It is completely obvious that one mediocre developer of genetic weapons is worth a dozen excellent politicians, who think nothing about genetics, as striking any one of the latter would neutralize the will and intellect of a thousand "political activist" creators of material values.

By its strength, clarity, and super-human insight, Muravyev's next passage literally sets off tremors, for he produces an impression of the discussions of a mathematician, who decides to practice magic in his spare time; this is not a Platonic idea of the denial of eternity, this is a scheme of its destruction:

"The overcoming of time demands a plurality or collectivity of figures. Indeed, the strength of action is conditioned exclusively by the unification of active elements, comprising the whole and

---

[454] *Ovladenie vremenem.* Moscow, 1924.

jointly producing this action. Therefore, time-forming action absolutely demands, besides the plurality of the object of action, the plurality of its subject or figure. In this act, the inner plurality of the object merges with the plurality of the subject. By such a path, a new, widening subject of action is created. But besides the plurality of the figure and the object of action, the condition of the mastery of time is also a full coordination of the activities of all elements of the active system. Time can be conquered, but only by one path—the path of establishing equal effectiveness of all manifestations of the elements. Unification of their efforts should go expanding from element to element, from small multitudes to large ones. Action according to the mastery of time is measured by the attributes of the acting multitude—its power and degree of inner commonality; and victory over time is strictly proportional to the degree of this commonality."

Without difficulty, we can guess that in the ancient Gnostic evangels, at the dawn of the rise of Christianity, this algorithm of actions portrayed bombastic forms and esoteric symbols, for rallying the "brotherhoods." At present, they prefer natural scientific terminology, which does not in the least hinder the rallying of their ranks, the observation of whose purity is also followed according to the corresponding method; so also to drain competitors according to the defined system:

"Inasmuch as some members of a system are conscious subjects and create their time, a part of the time of the system is subject to them, in the form of sequential phenomena dependent on them. The remaining part of the time of the system remains coercive for the members of the system, for it intrudes on them from outside. The role of consciousness in the formation of time most clearly comes to light, if this process is studied in the different degrees of consciousness of the members of the system. General time becomes subject to the union of the people. The equal effecting of the time-forming activities is the equal-effecting of the time subject to them. On the other hand, the discord of members of the society destroys this general authority. The time of a system is forced on them, only in the case when their strife lacks coordination."

An explanation of the structural organization of the *phenomenon of Kashchey the Immortal*—a famous character in Russian folk tales—becomes completely possible in light of the given conceptual prescriptions. Kashchey the Immortal bought the value of the struggle and discord of other mythological characters, artificially brought together, and collected their time-forming actions, which in the middle of "passive members of the system", were destroyed without a trace, passing into the possession of the active member of the system—Kashchey, who thus mastered time. Simply by using the organization of the topology and metric of space, he literally robbed others, then fed his immortality. Time in the biological environment of Kashchey flowed more slowly, than in the biological environment of his competitors, although they all belonged to one exclusive, energy-information system.

The modern struggle of the races occurs exactly along the same principle. Victory as a result passes to those in the single energy-information system of Earth, who know how to structurally quarrel with competitors, robbing from them of the same primary resource of living matter—time. Muravyev also scientifically proved the resignation of the knowledgeable; even the discernment of lone individuals is unable to prevent the defeat of biological systems that have a more narrow level of organization.

"The change in the relationships of things, in the sense of their isolation or unification, is always the changing of time. The assertion of self and the strengthening of one's duration, like the existence of a united center, is created by the conscious overcoming of exterior time. In place of the begotten relationships of things, new dictates are created by an objective reasoning process. By means of intellect, we constantly influence time, and not its modification. In this, the sense is of intellectual actions. The so-called irresponsible elements or things, which forcibly participate in time, become uncontrollably engrossed with its blind current. Alert beings, although they are subject to time to a certain measure, nevertheless possess the ability to make [time]. Any act of reality is a struggle of coercive time and subjected time. It is necessary to stop relying on a prepared eternity and to begin **to make time.**"

However, we also remember from the tales, that *subjected time* ended for Kashchey the Immortal: Little Ivan the Fool broke an egg, in which a needle was enclosed, then caused the agony and self-liquidation of Kashchey. All the above-presented can be explained in the following way: our great fairytale countryman was the first in world history to employ genetic weapons, striking the heredity of the enemy, symbolized by the egg, and by the same destroyed his time - his arrow of time, symbolized by the needle. In the given case, it is necessary to understand in the defeat of Kashchey's genome, not only the vulgar virus or some harmless lichen, but the variety of radiation which causes a mutative effect and includes on the cellular level, a mechanism for the self-destruction of the biological system. Incidentally, the very finale of the tales is for some reason, described very detailedly, eloquently, and accurately; an illiterate yarn-spinner did not create it, as they assure us in school, but rather at a minimum, a Nobel Prize winner in genetics.

However, as they say, a fairytale is a lie, with a grain of [truth].

But, if we return from the bosom of fairytale fiction to modern science, then priority in the given area is given to Italian scientist Vito Volterra—the founder of the science named **mathematical ecology**, who worked out its corresponding scientific apparatus. We urgently recommend that all philanthropists and humanists read his life's main work, before sleeping: *The Mathematical Theory of the Struggle for Existence.*[455]

The subjugation and destruction of an enemy as a biological species is considered here with the help of integrals and differential equations—mathematical functions which also have temporal measurement. Of course, the offensive names of several peoples, of whom the adepts of world conspiracy theories so love to train themselves, are not here. As prescribed, a struggle not to life, but to the death occurs between the 10th and 11th centuries. But substitute them for suitable names—a matter of taste and free choice for each one familiar with Gauss' equations, all the more if given the distinct instrumentation of mathematical models: "prey-people", "predator-people", "donor-people", and "parasite people". **As with a child's game, it is only necessary to choose which side you are on in the beginning.**

In the beginning, Vito Volterra defined the idea of the phenomenon, "struggle for existence", by which he understood a competition of biological communities, living in a single, common environment, where in conditions of limited space and material resources, one biological species must necessarily destroy another. Further, detailed calculations of the confrontation follow, considered namely as a **struggle in time.** "When two biological species, conditionally named *prey* and *predator* coexist in a limited environment, the first will develop more slowly than the larger individuals that exist of the second species; and the second [will develop] more quickly than the multitude of the first species. If two species destroy equally and proportionally to the number of their individuals, then the average number of prey increases, and the predators decrease."

Further, the founder of Mathematical Ecology, namely from the position of mathematics, gives an important explanation of the general theory of evolution: "It is impossible that species could have coexisted with limited changes."

First of all, this means that if the *prey* desires to change its ecological status, is should itself subject all its instincts for survival and consciousness to mutation. Biological victory from a mathematical point of view is a cardinal changing of the hierarchical principles of organization, and also the criteria of the assembly of the whole system. In mysticism, this genetic ceremony of transmutation is called *initiation*, by which, from a medical point of view, it follows to understand the change in the biochemical parameters of the blood; in turn, this brings in its wake a global restructuring of consciousness. Therefore, the lazy *predator*, *losing his sense of smell*, can turn into the *prey* in an instant, and not himself know it.

Vito Volterra leads a logical chain of discussions to a substantiation of the two basic rules of mathematical ecology. From the first it is apparent, that the more complex the society, then the more stable it is. The second states that in a biological community, the competitive struggle of species most strongly manifests near the condition of equilibrium of the structure of the society, and this condition of equilibrium is determined, not by the characteristics inherent to the species, but by the characteristics of the inter-species relationships. This means that in conditions of a single ecological niche, the greatest chances for victory go not to the species that is strongest in biological terms, but to the species more capable of successfully realizing its advantages, in relation to its competitors. Consequently, the racial-biological education of one's species allows skillful use of the strongest of his side, and the weakest sides of the enemy as a biological species; that is the key to survival. For our part we add, that only from the point of view of mathematical ecology, does one of the ancient plots of world mythology finally become

---

[455] *Matematicheskaya teoriya borbi za sushchestvovanie.* Moscow, 1976.

understood: when the *predator* takes pleasure in seeing the death throes of the *prey*, prolonging its torment. Everything is very simple. In postponing the inevitable death of the enemy, the victor becomes satiated by the energies of the agony of the organized structure of the enemy. These seconds are already short, but exclusively of the mental confrontation of the *prey* and *predator*, sheds light on the entire depth of racial relationships. From history and myths, we clearly remember that in duels, members of the white race easily and naturally destroy members of the dark race, and do not experience any pleasure in their death throes. This is obviously because, in the process of a struggle, white people recognize the energy-informational superiority of their structure of organization, over the members of the lower dark race, the sufferings of whom do not represent for them any evolutionary-biological value.

On the other hand, when a member of a dark race ends up in the role of victor, then it is shown as a gigantic historical event; he can never refuse himself the satisfaction of taking pleasure in the torture of a White Man, who is already doomed. The man of the dark race literally likes to drink the White Man's sufferings by the shot glass, in order to somehow acquire the principles of biological superiority, which are unattainable for him.

"Predator-agony-prey"—this universal logical order decodes the essence of the worldwide confrontation of the races. The struggle and its tragic finale—here is a universal biological marker, by which the hierarchical value of any living structure is defined. **Taking pleasure in another's agony is the fate of lower races.**

Finally, time, as we are convinced, is that element onto which the finale of any struggle is projected. Victory means to have the future in a trans-personal sense; that is, as a member of one's species, one's race.

At approximately the same time, in the 1920s, Russian scientist V.I. Vernadskiy created the concept of living (biological time), emphasizing by the same, the unequal significance of its course in different biological systems. He namely called living beings the source, cause, and carrier of any "internal time." Namely because of this, the specific features of the origin of time are to be searched for in the hereditary traits of an organism. Time is not an independent substance, but a trait or an attribute, and an accompanying indicator of the state of any living being, having an affiliation with a concrete species. Life activity or metabolism, is unthinkable without space-time relationships. In the first place, in Vernadskiy's opinion, heredity here means the organization of biological codes in time.

Grasping the grandiose ideas of Plato, Kant, and Bergson, the Russian scientific genius gave them an already qualitatively different substantiation. In his works, metaphysics gradually began to turn into physics, colored by the tones of enduring values.

Parallel with the development of anthropology, psychology, and biology around the turn of the 20th Century, this time was marked by progress in the area of the theory of information and the development of means of communication. The introduction of the Morse Code revolutionized the scientific notion about the idea and essence of information as such. In essence, the first information in world history was recognized as a powerful tool, a self-sufficient value, and a factor of progress. And again, the philosopher's sudden flash of insight preceded the discoveries of practical science.

In 1909 German biologist Jakob Johann von Uexkuell introduced the notion of "umwelt" into everyday life; word-for-word, it means "the surrounding world", or "surrounding environment", and defines an energy-informational niche in space, which a concrete biological species occupies. Thus, only species which belong to one umwelt can appear in the struggle for survival. Birds and amphibians, for example, or insects and mammal predators have different umwelts, and therefore the spheres of their vital interests do not intersect. The same main idea in the concept lies in the fact that what is important energy-information for one species, is of no value to another. The mole easily manages without keen eyes, while the eagle cannot exist without them. This same principle extends to human populations, where entire peoples can exist without the development of higher mathematics or transcendental philosophy, and get by with begging, robbery, light trade, and other forms of social-biological parasitism. A shallow cardsharp can easily cheat a prominent physicist in a game of cards, not at all because his intellect is superior, but only because the avid cardplayer is in his umwelt, his element, and the physician is not. Changing the energy-information system together with its social protection, we easily secure the triumph of the system of intellect over the shallow tricks of the card-playing wire-puller.

By the middle of the 20th Century, the outstanding Soviet biologist, I.I. Schmalhausen, brought the given problem to a qualitatively new level, explaining the transfer of hereditary information in the structure of DNA, with the help of the Morse alphabet. Thus, genetics received the precise, methodological base of the theory of information. Schmalhausen's work, *Cybernetic Bases of Biology*,[456] is a fundamental composition, for with its help the members of different scientific fields received a single conceptual apparatus, and natural sciences were enriched by the mathematical precision of calculations. Schmalhausen wrote: "Hereditary information is transferred in a coded form, and furthermore, all at once in a spatial connection, and not in a temporal succession. In the deciphering of this code in the processes of individual development, the information also unfolds in space, although its transformation is undoubtedly completed in temporal sequence."

Thus, biology, armed with mathematics and the theory of information, unequivocally attested to the fact that the genetically inherited traits—including racial traits—of each living organism are realized in time and space. In any biological system, including the system of human races, time is derived from the function of the realization of hereditary parameters.

In support of this important thesis, it is also worth citing the works of another Russian scholar: A.G. Gurvich. In his book, *Theory of the Biological Zone*,[457] he developed and substantiated a theory, according to which each living organism generates a **biological field** around itself, in the process of vital activities; this [field] supports the specific, and more precisely, the racial specificity of the entire organism as a whole. Each biological field is continuous and successive; this means that the conditions of living systems do not exist without the field, which realizes and supports the specific and distinctive nature of the organism, from generation to generation. The biological field puts the unifying molecular structure of the organism in order, in the process of vital activities and reproduction; it dynamically possesses the vector of orientation in time and space.

The biological field accompanies the entire evolutionary process of development of a species, supporting its racially distinct nature. Besides that in the process of division of an organism's cells, it transfers the encoded hereditary history of the species to them, and also imparts a genetic program of further, optimal development.

In light of the given studies, it becomes obvious that the synthesis of natural and precise sciences in the 20th Century confirmed the ancient metaphysical generalizations of Plato, Kant, and Bergson, that time is a function of the realization of the hereditary racial traits of an organism, and does not exist in and of itself, like an abstraction. Time is genetically determined by the biological system of an organism implementing its vital activities. **Time is a function of race, not the reverse.** I.I. Schmalhausen summarized: "Life is a struggle. The struggle is against equilibrium. Life is metabolism, the exchange of matter, the exchange of energy, and the exchange of information with the surrounding environment."

For their part, these theories successfully fit into recent discoveries in the areas of physiological psychology, neurochemistry, biochemistry, and neurobiology.

In the article, *The Chemical Continuum of the Brain as a Mechanism of Reflection of Reality*,[458] academician P.K. Anokhin wrote: "All the details of organization of the diverse organs of sense are 'adjusted' by precision methods to the energetic characteristics of the outside world." This means that new studies in the area of higher nervous activitiy again completely agree with the postulates of mathematical ecology, and also confirm the fact that each biological species has its umwelt, that is, its fundamental niche.

"On the basis of neuro-chemical molecular processes, modern neurophysiology shows that the vital significance of separate events is represented in the brain, even in the specific chemical processes of the brain, as if they plot the "steps" of these vitally important events. Thus, for example, we have different chemistries for suffering, melancholy, fear, joy, and other essential emotional experiences and events in the life of a person."

Today, the opinion is firmly established in science, that the differences between races are conditioned, in the first place, by differences in biochemical processes in the exchange of matter, called **metabolism.**

---

[456] *Kiberneticheskie osnovi biologii.* Moscow, 1968.

[457] *Teoriya biologicheskogo polya.* Moscow, 1994.

[458] *Khimicheskiy continuum mozga kak mechanism otrazheniya deystvitel'nosti.* Voprosi filosofii (Questions in Philosophy), N6 1970.

Anokhin therefore made the following conclusion: "An organism seeks precisely the missing components of matter that are programmed by his metabolism. Consequently, in the protoplasm of nerve cells, the brain has an authentic chemical continuum, which reflects the continuity of events of the exterior world, that is, the space-time continuum."

From the above one can make the following conclusion in light of the theme declared by us:

Hereditary information about specific racial traits is passed on in the form of encoded biochemical reactions, from generation to generation; in the process of the transfer of this information, the biological field of the living system promotes the maintenance of the racial uniqueness of the entire system in time.

Besides that it is namely the biochemical reactions of the organism, which condition the specific features of the energy-information exchange, which correspond to the umwelt of his race.

With the help of new methods in neurobiology, it was established that the speed and complexity of the thought processes in the brain of an individual depend on the quality of the neurons, the level of complexity of their organization, in fact, [they depend] on the constructive particularities of the neurons, and also on the intensity of the chemical exchange processes. All these parameters differ in the members of different races. If an individual is not able to grasp higher mathematics or transcendental philosophy, then it is necessary to blame not the methodology in textbooks or the mediocrity of teachers, but the slow flow of the chemical exchange processes, and the unsatisfactory spatial organization of the neurons in his brain. **Like only comprehends like.** The philosophy of Nitzsche can only be grasped by a person with a similar genetic-biochemical constitution of the brain, and not some unmeasurable abstract of pan-humanist values. To each his own.

Academician P.K. Anokhin states that ideas, among them political ideas, have a biochemical character and are reflected in the protoplasm of the nerve cells. For their part, ideas have an inborn, inherent genetic structure. Plato, DesCartes, and Leibnitz developed the concept of "inborn ideas."

In the final analysis, any outer political program in a society serves as an adaptation of the inner biochemical processes in the organism of separate individuals, which have a genetic predisposition to these or those "inborn ideas." "Right" and "left" ideologies are as old as the world, just like democracy and tyranny. Demand for "civil rights" or a "strong hand" are not ideological, but a biochemical need of an organism, which seeks balance between the genetic, inborn structure, and the biochemical reactions in nerve cells, caused by outer political events. Enthusiasm for anarchy or fascism, or for democratic or Christian values, is, after all, only a method of providing maximum comfort to one's own brain, in correlation with its genetic code. Slogans do not teach anyone—they only fulfill the role of catalysts of biochemical processes, in this or that direction. Any political program without a remainder decomposes into microelements, and is removed bit by bit, to this or that part of the electorate.

In the final analysis, according to the outward appearance of the political leader, one can easily establish how much iron, potassium, iodine, and other elements are retained in the blood of his potential voters, and on the other hand, proceeding from the biochemical structure, they choose a leader with the corresponding set of slogan-reagents. Anokhin summarizes: "The continuum already immediately begins to influence the construction of the whole, or purposeful behavior, leading to an active search for such components in the continuum, which somehow determine the satisfaction of the needs of the organism."

Thus, new achievements in the research of higher nervous activities essentially enriched racial theory and politology.

At the end of the 19th Century, German philosopher F.A. Lange spoke about the *a priori-ness* of "psycho-physical traits", which serve the perception of time, in his book, *The History of Materialism and Criticism of its Significance in the Present Time*. But we now know, that this *apriornost'* is conditioned by nothing other than the genetic differences of races. Lange also wrote: "Psycho-physical traits, by virtue of which we are compelled to contemplate things in space and time, exist in any case, before any experience." But before [there is] any experience, a genetic program, which is determined by race, is laid within a person. In the 1920s, Russian scientist P.M. Nikiforovskiy created the concept of the "physiology of time," and in the 1960s Soviet researcher Ya. F. Askin observed: "The specific features of the stream of time in different conditions, is tied with the notion of an inherent time for each system of reference (that is, of time changing in a given concrete system of reference). In the transition from one physical system of reference to another, they demand alteration of not only the spatial coordinates, but the temporal coordinates."

Vernadskiy spoke earlier about "biological time," and Askin more precisely stated this category, introducing limitation "of a physical system of reference" in living structures, which are races.

Askin also asserted: "According to its origin, 'physical time' is the result of psychological time, which is considered as the sum total of intuitive information." But after all, intuitive information is taken, not from the vacuum of space, but is the result of the experience of peoples of a concrete race.

Soviet scientist N.E. Vvedenskiy analogously pointed out that time is directly connected with the speed of physico-chemical reactions of stimulation, originating in the brain of a person. But after all, these physico-chemical reactions are also genetically encoded, and each race carries its own code. Researcher B.M. Teplov stated in his book, *Problems of Individual Differences*,[459] that "Time which a psychologist observes in his 'immediate present', is not the same as the time which a physicist measures with his fluctuating atoms."

Thus, modern sciences, which are oriented to the study of the nature of Man, clearly show us, that no universal time exists in nature, which is the same for "all individuals." Abstract physical time—this is the time of non-living material. Russian philosopher N.Ya. Grot consciously upheld this concept in the second half of the 19th Century; to the present day it has been experimentally confirmed many times. The modern, mass philosophical-psychological school of **behaviorism** also stands on the positions of biological determinism, in the interpretation of the nature of time.

In his book, *The Perception of Time*,[460] Soviet scientist D.G. Elkin explained the materialistic view of the given problem, in accordance with which the perception of time is a reflection in the brain of an objective length, speed, and sequence of occurrences in duration.

Racial theory has nothing against the way [such] a question is put, only stating more precisely, by virtue of this, that in different races [there is] a different structure of the brain, and consequently, the perception of time among them should be different.

In connection with this, Soviet neurophysiologist E.K. Sepp stated: "The ability to record the sequence of processes is a basic ability of the cerebral cortex. On the basis of this ability lies the dynamic polarization of neurons." In whole, the orientation of time in a person is realized with the help of the cortical sections of the brain. However, it is worth emphasizing, that any professional anthropologist can easily and lucidly explain that the skulls of different races differ constructively. Besides that as early as 200 years ago, the solid rule was established that differentials in the process of the growth of the brain of any living organism, itself forms the outer construction of the cranium, and never the opposite. Consequently, between the form of the skull and the specific features of the perception of time, one can establish a certain correlation, as described in the notions of both anthropology and psychology.

D.G. El'kin gives to understand: "The perception of time is a complex mental process, which only in its basis reveals a physiological and biochemical basis."

And thus, anthropology, psychology, biology, and neurophysiology in turn came to certify one and the same fact: the time of a living system, like a human race is, in part depends namely on the hereditary specific nature of this system, and is measured in correspondence with its principles of organization. In confirmation of the given thesis, the author writes: "An increase in temperature decreases the accuracy of the perception of temporal intervals and imparts to it a distinctly expressed tendency toward underestimation. With the lowering of temperature, [the likelihood] of a mistake in the perception of time becomes the opposite of that which is observed in the conditions of increased temperature."

In connection with the given scientific fact, the particular inborn punctuality, which is characteristic of members of the Nordic race, comes to mind, as does the extreme carefree manner of southern races, in relation to the priceless resource [of time]. In classic raciology, the Nordic race is defined by such qualities as activeness, dynamism, quality, and initiative.

---

[459] *Problemi individual'nikh razlichiy.* Moscow, 1961.
[460] *Vospriyatie vremeni.* Moscow, 1962.

In this plane, Andri Bergson did much to prepare the grounds of the racial theory of time, for he placed the duration in dependence of the action completed by a person. Time is the product of action and the product of effort, which is subjectively experienced by a person.

Another modern Russian researcher, A.P. Levich, wrote in his summarizing theoretical article, *The Scientific Comprehension of Time*:[461] "The true growth of a system can be measured not on an astronomical scale, but only on the scale of the inherent time of the system. But for this, an 'inherent scale' should be substantiated and organized."

The development of such modern sciences as genetics and molecular biology, today permit the design of individual scales for the measurement of time in different biological systems, with an imperial degree of accuracy. The unit of time, equal to the interval between similar phases of cell division, is called a **detlaph.** Besides that, time in populations is measured by the quantity of replaced generations. From there it becomes completely understood, why the time of sexual maturity in different races is not identical: because the rate of metabolism and the intensity of cell division in the organisms of peoples differ, and as a result, [so does] their biochemical activity and psychological perception of time.

The differences in the structure and specific nature of the perception of time by natural means, begets a spirit of competition and contest in the members of different biological species for its mastery, as the most vitally important resource.

Nobel Prize Laureate Ilya Prigozhin, who developed the position of the modern science of **synergetics**, repeatedly turned attention to the works of alchemists, who gave themselves the task of manipulating time; that is, the subjugation of the biological time of an organism to one's will. In the book, *Time as a Physical Phenomenon*,[462] N.A. Kozyrev also developed the concept of active traits of time. In the meantime, psychologists P.V. Simonov and P.M. Ershev, by means of modern science, successfully formulated this classical position on ancient alchemy: "The need of economy of force, which motivates the individual to seek the shortest, easiest, and simplest path to his goals, belongs to the number of biological needs. The need for economy of strength is close to the need for weapons."

Thus, what was formulated centuries ago as the insights of philosophers, became clear later with the help of mathematical models, and is finally clearly revealed by means of modern experimental science, as it applies to living beings: **time is a weapon.**

This is not simply a vital resource, not simply a factor in the process of leadership—this is a powerful element, in which very real wars for evolutionary prospects unfold. Time, like all traits of a race, is a manifestation of the never-ending struggle for existence.

In light of our discussions, we again make a small excursion into the area of linguistics, for the modern world distorts the significance of some archtypical categories to the point of distortion, and modern so-called "academic" science profaned their sacral meaning.

The ancient term, *potential* (potential, possibility), was understood by the ancient sages, not in the modern abstract-theoretical sense, like a certain foundation of energetical processes, but namely as a **cause of expansion**; that is, a condition of the biological proliferation of one race in the *oikoumene* of another. For its part, the term **expansio** (extend, expand) precisely signifies the age-old Aryan idea of time.

In connection with this, ancient mathematicians derived the **vectoral equation of expansion:**

$$\bar{R} = \bar{S} + \bar{U}$$

where

$\bar{S}$ is the variable enzyme-forming value, and $\bar{U}$ is the vector of symmetry of a biological object, or of a system as a whole. This means that the expansion of biological systems, including races, will be maximal only in such a case, if the biological exchange processes within a race, as a sum total of an identical heredity, will flow in an optimal regime. Besides this, the biosymmetry of a race should coincide with its outer forming, that is, its social, ideological, political, cultural, and religious institutions.

---

[461] *Nauchnoe postizhenie vremeni.* Voprosi filosofii, N4, 1993 (Questions in Philosophy, N4, 1993).

[462] *Vremya kak fizicheskoe yavlenie.* Leningrad, 1991.

Only then, when the natural biological traits of a race are found to be in harmony with the conditions of existence, do we have a right to expect maximal expansion from it. Time is a manifestation of expansion in its exclusively racial-biological sense.

As a result of this, the transition of a competing race, by method of occult-political or genetic influence, to a regime of non-optimal functioning, decreases the vector resultant of its expansion. It is namely such a struggle, which the pagan priestly castes of the conflicting races are engaged in, on the basis of ancient mathematical equations. The vectors of the collective efforts of a race move in time, as we have repeatedly shown; therefore, a deviation of these collective efforts from the axis of the biosymmetry of race, reduces its expansion. Aristotle set forth the sense and conceptual basis of this equation; like many scientists of antiquity, he was a mage, who was initiated in the highest secrets of the pagan priestly caste.

Chronobiology is such a science that engages in the study of the principles of biosymmetry, that is, of the optimal functioning of all vital manifestations of an organism, and groups of organisms (races).

On the basis of experimental data, the genetic regulation of the biorhythms of an organism was revealed. Modern author Yu. A. Romanov writes: "The temporal organization of a biological system is formed by the sum total of all its rhythmic processes, interacting and coordinating in time between themselves, and with the changing conditions of the environment." That is, we again return to the vector equation of expansion, expressed by other means, now in the form of a biological time system. N.I. Moiseyeva and V.M. Sisuev observed: "A biorhythm reflects the current of time in a living system, which 'refracts' through the regularity of its temporal organization."

Again, from Aristotle's equation, but now through means of modern sciences, we come to the rationale of the racial theory of time. **Each race has its (own) time, measured by its biological hours.**

In 1930, J.N. Louis, a specialist in the area of physical chemistry, made a daring and radical attempt to abandon objective temporal orientation. For that he created his concept of the "**arrow of time**", according to which the sense of time is completely conditioned by phenomena of physics and chemistry, in each separately taken concrete consciousness. Thus again, the very idea of an "arrow of time" confirmed Aristotle's vector equation. G. Hogland also wrote about the "chemical basis of our sense of time," and in 1936 Lecomte du Noiiy developed the concept of **biological time**, tied namely with the intensity of cellular reproduction. "Biological time is the basic phenomenon in the construction of living matter"—he wrote. W. Goody established that on a neurophysiological level, definite parts of the brain cortex fulfill the role of a calculator of time.

We already spoke about the mathematical abilities of the ancient American Indians. Curious is the fact that the pagan priests of the Maya used the juice of the peyote cactus for religious purposes, as it contains the substance **mescaline.** The participants of religious ceremonies, who drank this juice fell into a trance, which was accompanied by hallucinations, the absence of sensations of pain, but most of all, by loss of any ability in the individual to reckon time.

The world-famous mathematician, Norbert Viner, established the diapason/range of the frequencies of the brain, which fulfill the function of "living clocks."

In living cells, metabolism is a general process of replacing molecules in the composition of cells. The general process for multi-cellular organisms is growth, during which new cells manifest, and existing ones are replaced, or vanish. The dynamic of size, summing up the birth and death of a specimen, comprises a general process for a population. The time for metabolic processes, measured in "detlaphs", is species-specific; that is, racially conditioned, and very specific features of the metabolic processes within each race form up its genetic archtype. Therefore, Viner emphasized that in real time, the main problems of biology are tied with systems and their organizations, in time and space." In 1923, Soviet scientist V.G. Bogoraz emphasized: "In essence, each person, each living individual has its own time. People with an excitable termperment have one time, and phlegmatics another; and with melancholics [there is] a third [time]." But after all, the same is valid for whole races, which have their [own] types of character. Modern Russian science researcher, V.A. Kanke, also thinks that "the problem of the revelation of the qualitative heterogeneousness of time" has moved to the forefront, "and that it is necessary to abandon the notion [that] time is monotonous, an indifferent essence. If matter develops, then with the same necessity, it should develop its attributes, among them time."

N.N. Bragina and T.A. Dobrokhotova assert that individual time is inalienable from the brain of a given person, but, after all, the brain of a person has inborn racial parameters. In the collective anthology of works, *The Problem of Unity in Modern Biology*,[463] G.A. Yugay brings attention to the fact that in each organism there exists a special physiological time, tied with the specific features of inner-cellular organization. Modern biology, in the opinion of this author, directly poses the question of the localization of the mechanism of the biological clocks in this or that structure of the cell. Besides that Yugay declares: "Biological time reflects the relationship between living systems." But under living systems, we in part understand human races. **In connection with this, our original exotic thesis, that time is the result of the interaction of races, is confirmed by the same unequivocal method.**

Biological clocks are present in each living cell, launching the whole complex aggregate of vital processes, in accordance with a set genetic program. Time-forming activities of separate cells are added to the synchronized time of separate human organs, and those in turn are joined at the level of the biorhythms of the organism as a whole. In connection with this, biological time has a multi-level base and hierarchy.

Peoples and races also have their living "biological clocks", supplying their maximal viability, and as a result, potency—expansion, that is. The temporal organization of any biological species leads back to the genetic program to a maximal increase of its biomass. Therefore, any malfunction of the "biological clocks" unavoidably brings a reduction of the vitality of a race in its wake. Namely this simple rule lies at the base of all the occult wars, which humanity has conducted for the length of world history. To impose one's system of numbering years on the enemy, one's sacred geography, one's archtypical symbols, one's fashion, one's "method of dividing time"—according to the keen expression of Valerian Muravyev—all this leads to the malfunction of the biological clocks of a competitor.

Modern, patriotic-minded sociologists and politologists, who complain that the biomass of the Russian people is shrinking, suggest numerous absurd prescriptions to us, for the "healing" of the nation; but they do not understand the very essence of the mechanism of degeneration. As long as the Russian people live according to the Christian calendar, and make prayers for Jerusalem, and also visit Lenin's Tomb, and follow foreign fashion and lean to foreign symbols, there can be no talk of any racial recovery. Until that time, when the quality of standards in our lives is of foreign manufacture, there can be no increase in the biomass, and as a result, potency or expansion, is impossible.

One of the visible theoreticians of Zionism, Harold Fisch, wrote in his book, *The Jewish Revolution*,[464] that in the creation of the State of Israel, problems arose with the calendar, and the question was decided, not in favor of the European calendar, but the traditional Jewish calendar. "Figuratively speaking, the non-Jewish calendar's poisonous stinger was withdrawn." By "poisonous stinger", the Zionist theoretician precisely understood the "arrow of time" or the "vector equation of expansion", which has a clearly expressed racial orientation. "Never live according to foreign time"—this key rule is known to ideologists of all world religions, occultists, and racial leaders.

English scientist J.Whitroy wrote in the book, *The Natural Philosophy of Time*: "The history of natural philosophy is characterized by the interaction of two competing philosophies of time: one of them sets its goals of 'exclusion', while the other is based on faith in its primevalness and irreducibility."

And modern American scientist Jeremy Rifkin characteristically titled his book: *Wars of Time. The Main Conflict of the History of Man*.[465] In it, the author expresses completely analogous ideas, asserting that there are no two cultures which think alike, therefore there are no identical concepts of time. Each culture in the plane of organization of time is unique. Rifkin writes: "Each thought, event, and epoch of an independent culture is defined in terms of the original definition, sequence, structure, plan, and duration, by its norm of synchronization, appeal, and temporal perspective. Each culture possesses its standard for evaluating phenomena. Everywhere, time possesses a different value."

---

[463] *Problema tselostnosti v sovremennoy biologii.* Moscow, 1968.

[464] This book might be titled *The Zionist Revolution* in American book works.

[465] New York, 1986

But after all, cultures do not rise in an empty place, they are created by people of a concrete race and type, in connection with which Rifkin validly talks about **"anthropological zones of time."** This means that each pocket of race genesis has its unique "biological clock", which is conditioned namely by the hereditary traits of the given race.

Symbols, architecture, religion, the furnishings of a dwelling, musical stylistics, national costumes and dances—in a word, everything that so moves culturologists, is no more than the outer packaging of the idea of time of a given race, subjective to its experience. Strength, duration,

"Race War"—An ancient Bushmen Cliff Painting

and intensity of these or those psychological experiences, common to the whole race, are encoded in all its cultural manifestations in the form of style, and serve as a protective shell from foreign or other racial influences. Culture is a derivation of the idea of time, and as a result, the expansion of a concrete race. Therefore, namely in culture can all enemies and vices, and together with them, the strong aspects of a given race be sensed. Clearly, its past comes to the surface, and the future is easily read, being represented by the potency of the present.

Imposing its time together with its symbols, each culture completes biological expansion in relationship to its competitors. The Greenwich Meridian and the clock on Spasskiy's Tower on the Kremlin—these are cannon barrels aimed at the element of time.

In connection with this, modern scientist Alfred Kortsibski believes that Man is the only being capable of building time, and Daniel J. Burstin in general thinks that time was first discovered by Man, before [he discovered] fire and [before he invented] the stone axe.

Manfred Eigen, Nobel Prize Laureate for contributions to the study of the fundamental problems of the origin and development of life on Earth, develops the same ideas in co-authorship with Ruthild Winkler, in the book, *The Game of Life*:[466] "Only by constantly using an influx of free energy, can a system continuously renew itself, and by this stop its decline to a state of thermodynamic equilibrium, which Erwin Schredinger keenly called a state of death. Characteristic for the processes of life, a dynamic order can be supported only at the expense of constant compensation of the production of entropy."

This rule is applicable, in his opinion, namely to the functioning of race, for which, as for all living systems, the logic of biological processing of information is the main thing. "Here orientation in time, characteristic for all evolutionary processes of self-organization, is the basis of our inner sense of time." As early as the 20th Century, German scientist Karl Friedrich von Weizsacker formulated two basic theses of **bio-informatics as a science:**

1. Only that which is understood, is information.

---

[466] *Igra Zhizni.* Moscow, 1979.

2.   Only that which produces information, is information.

The methods and techniques of the occult wars are precisely oriented to the destruction of bio-informational processes in the racial structure of the enemy. The main effort here is delivered against the racial archtype, being, as we showed above, an expression of the metabolic exchange processes inherent in a given race, of its family, the key which opens its biological essence.

In connection with this, Eigen and Winkler observe: "The scale of reproductive orderliness, that is, the capacity of storage of information, is limited by the recognizability of symbols."

The symbols of a race namely enable an increase in the bio-informational capacity of its archtypical consciousness. The more recognizable a symbol is, the more archtypical it is, and the more it consequently reflects in greater degree, the essence of exchange processes. Stars of every stripe, crosses, and swastikas, are reflections of this eternal process of the struggle of the archtypes. Besides this, the speed of recognition of these symbols is directly tied with the speed of the evolution of the given race. The archtypes of a race are set down in the outward symbols of day-to-day attributes, through the subconscious influence the speed and intensity of its metabolic exchange processes. Namely therefore, the pagan priestly castes have conducted an incessant war with the symbols of competitors, for the entire length of history. The mastery of a symbol inevitably brings in its wake the mastery of time.

"With the growth of accuracy, the transfer of information and the informational capacity grows. But the greater this capacity, the more varied the possibilities of specialization become. The evolution of the entire biosphere is a majestic process of accumulation of information and the formation of memory. Its orientation in time is the inalienable trait of progressive evolution. This quality is tightly connected with the temporal direction of the growth of entropy, in irreversible processes"—the German authors conclude above.

Alongside occult methods, which are tied with the disorganization and distortion in time and space of the bio-informational processes within the racial structure of an enemy, there have been developments in recent decades in the area of genetic and psychotronic weapons, also realizing their destructive power in the attachment to temporal processes, which originate in biological structures. The same element of time is militarized more and more each day. More and more, time becomes and aggressive and decisive factor in politics, business, culture, religion, and absolutely, in day-to-day life.

"An Aryan Sentenced to be Sacrificed Frees Himself"-
painting on ancient Greek vase, 4th Century B.C.

An international conference was conducted in the USA in 1966, upon the initiative of scientist Julius Thomas Fraser, known as *Interdisciplinary Perspectives of Time*. Afterwards, the *International Society for the Study of Time* arose, of which D.T. Fraser was selected as the founder-secretary. The material of the conferences, which were regularly conducted by the society in various countries, are reflected in the combined periodical publication, *The Study of Time*. The American author's book titles speak for themselves: *Of Time, Passion, and Knowledge* (1975), *Time as Conflict: a Scientific and Humanistic Study* (1978), and *The Genesis and Evolution of Time: a Critique of Interpretations in Physics* (1982).

Alexander of Macedon battles a race of Half-Humans, French miniature, 13th Century A.D.

It is hardly necessary to explain, that the given official initiative, according to the interdisciplinary study of the problem of time, is only the tip of the iceberg, where what is not seen is large—the studies are in the interests of military and political departments.

Modern authors V.D. Tsygankov and V.N. Lopatin clearly write in the fundamental work, *Psychotronic Weapons and Russian Security*:[467] "In psy-problems there is still another aspect—**power.** G.I. Shipova's theory of the physical vacuum combines in itself Einstein's General Theory of Relativity (GTR) and Heisenberg-Schredinger-Dirak's theory of quantum mechanics (QM), and places the key to the creation of a powerful means of influencing the topology and structure of time and space, in the hands of researchers and developers. The possibility manifests of the creation of an unusually powerful and effective means of coherent irradiation and defeat of targets, and also of highly maneuverable and economic flying apparatuses (a kind of flying saucer).

Further, on the basis of U.A. Baurov's string theory of a physical vacuum, the authors of the book explain the mathematical concept of exceeding the speed of light, and changing the metrics of space: "Time overcomes information, it stops the 'stream', since length—the interval—stops existing, and space overcomes instantly."

The very same concludes in the fact that forceful use of the psy-factor influences the genetic code of a living organism, from which it follows that **any global conflict with application of principally new types of armament will now unavoidably have racial consequences.**

Powerful radiation can influence the genetic matrix of a race and its energy-informational envelope, including or excluding, according to the will of a foreign operator, this or that metabolic exchange process, which can bring irreversible mutations, both of a positive and negative character, in its wake. The prospect of creating a race of supermen, with a parallel conversion of racial competitors into a "sub-human" state, acquires visible physical traits, reinforced by mathematical formulas. **The theorem of a bloodless war** once and for all alters notions of the

---

[467] *Psikhotronnoe oruzhie i bezopasnost' Rossii.* Moscow, 1999.

very essence of coming conflicts, and hoary generals with gray hair will again lose a war, for which they are not prepared.

As we recall, the biological field of any object, including a race, carries within it the past history of the species, and the genetic program of development for the future. Thus, by influencing the energy-informational matrix of a race, one can simply "collapse" or "cave-in" a competitor within time, like Kashchey the Immortal did in the Russian folk tales. Besides that by influencing the topology of space, one can reduce all the actions of an enemy to zero, compelling him in the direct sense of the word, "to mark time" or walk in place.

Tsygankov and Lopatin made this orthodox conclusion in their book:

"And that state, that government, which is first to recognize the importance of psy-problems, and knows how to integrate the various science school cadres into a single project of decision, will become the monopolist in control of the community of nations and states."

As German scientist Eugen Fischer wrote in his book, *Anthropology*: "The great battle of peoples, which is still not concluded, has its anthropological side."

In accordance with the above-laid facts, we see that racially foreign, 'Immortal Kashcheys' have already long-influenced the biological body of the Russian people; it is decreasing their ability to expand, and defeating their evolutionary lifespan. The scenario is advanced. It only remains for us to call the bygone Little Ivan the Fool to life. He is not burdened by the "nonsense" of pan-humanistic ideals, and is therefore free to set Kashchey's genetic program to self-destruct, so that in an instant of agony, these parasites will be put out of our time, once and for all.

GPM
**Anthropologische
Instrumente**
für die
Somatologie und Osteologie

*GPM
Anthropological
Instruments*

*for
Somatology and Osteology*

DKSH
Market Intelligence

SiberHegner & Co. Ltd.
P.O. Box 888 8034 Zürich, Switzerland
Phone +41 1 386 7272, Fax +41 1 386 7282

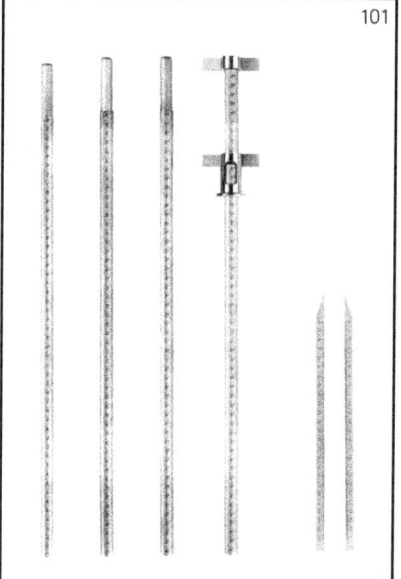

101

Anthropometer. Scale: 0-2100mm (0-960mm) for determination of all vertical lengths on the body.
Weight: 1.450 kg.

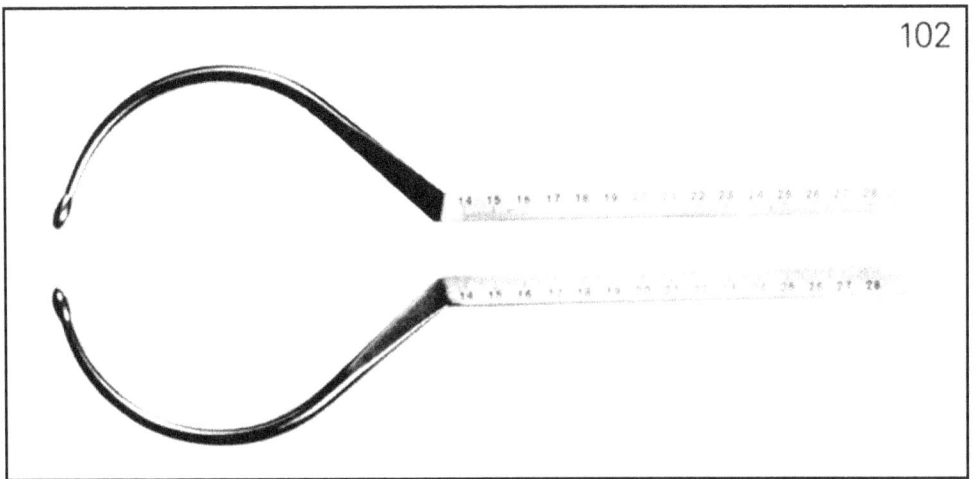

Curved gaging arcs for anthropometics; for example, for measurement of the sagittal diameter the chest. Weight: 0.170 kg.

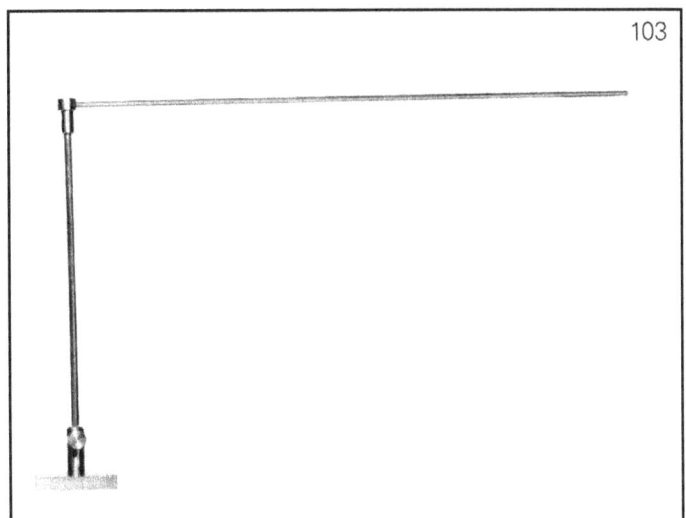

Stylus for measuring the height of the ears.

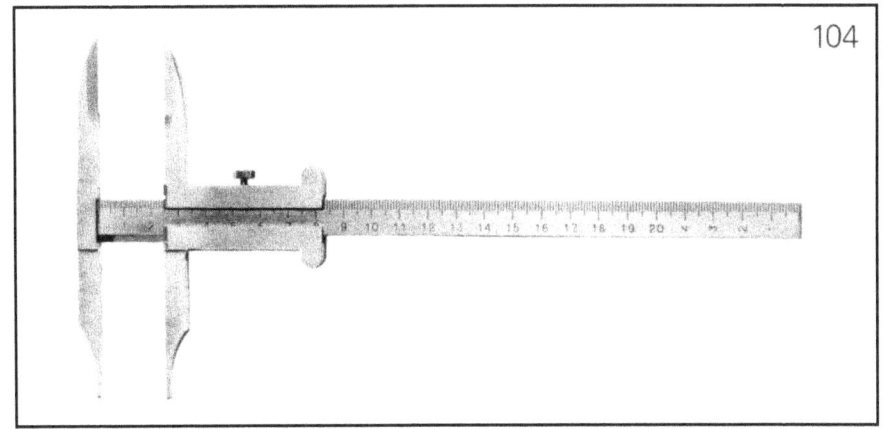

Calipers (according to Martin). Length of the piece: 0-200mm. Weight: 0.170 kg.

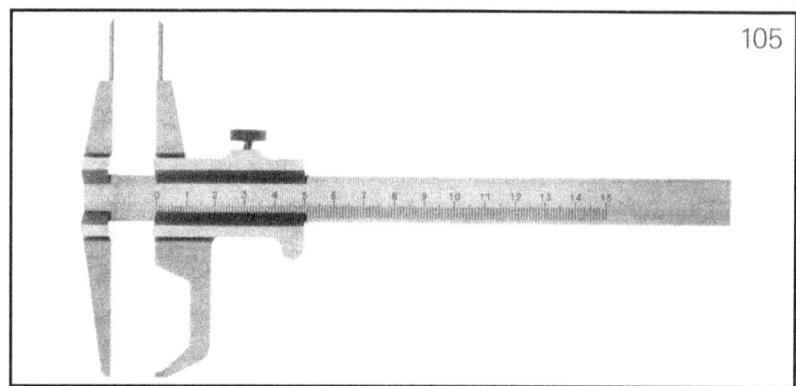

Calipers with vernier (1/10 mm) with needles. Special instrument for shallow measurements. Length of the piece: 0-150 mm. Weight: 0.240 kg.

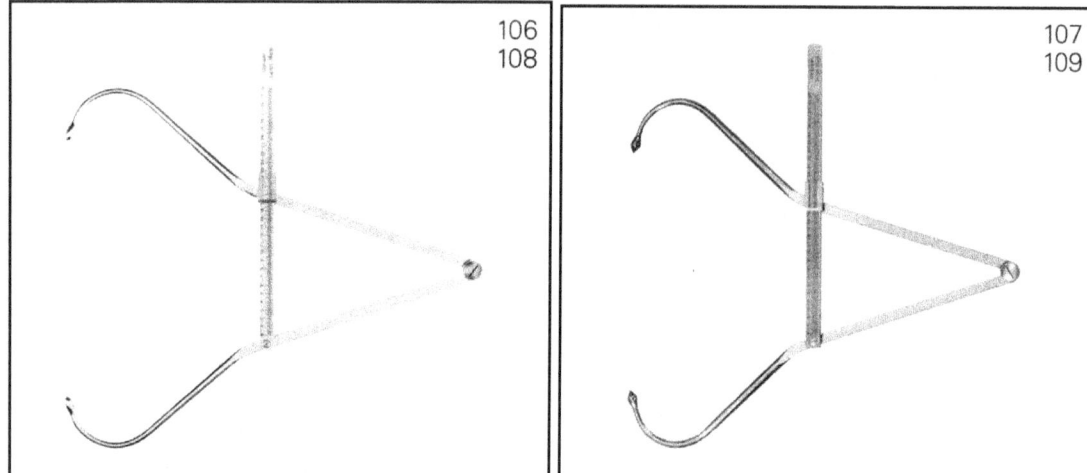

Left: Calipers for probing rounded extremities. Numerical scale: 0-600 mm. Weight: 0.450 kg.
Right: Calipers for probing rounded extremities. Numerical scale: 0-300 mm. Weight: 0.220 kg.

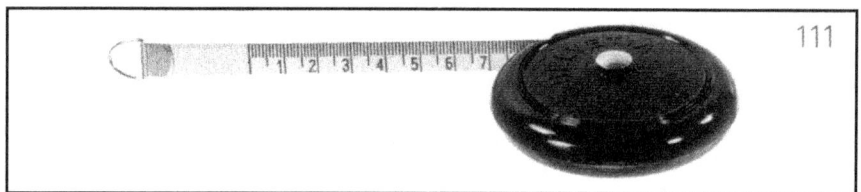

Plastic tape measure. Length: 0-2000mm. Weight: 0.025 kg.

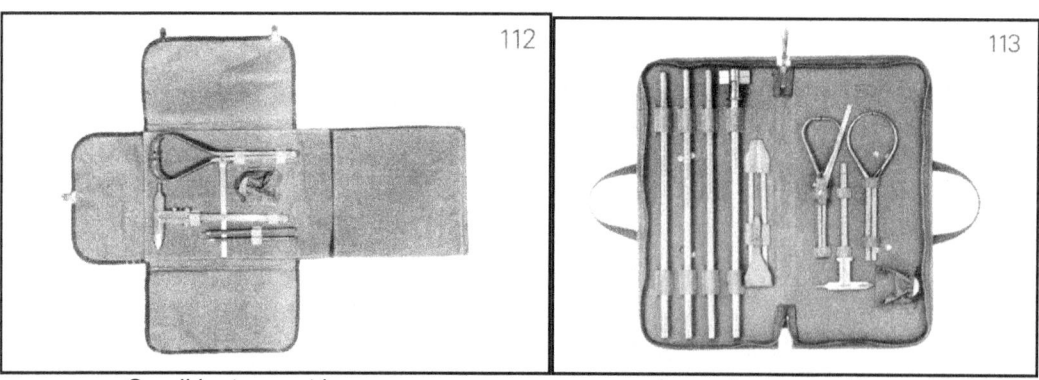

Small instrument bag   Large instrument bag.

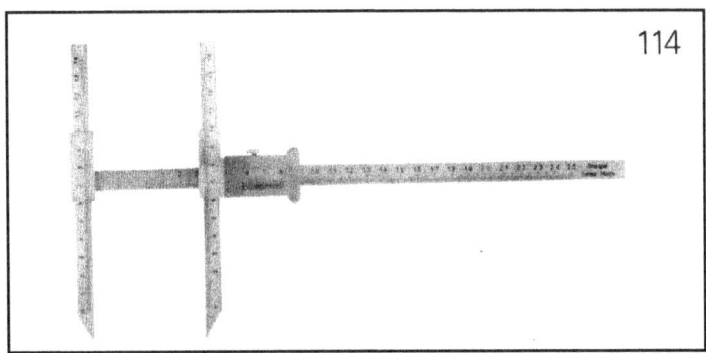

Calipers for measuring the absolute and projected dimensions of the face.
Length: 0-250 mm/0-140 mm. Weight: 0.300 kg.

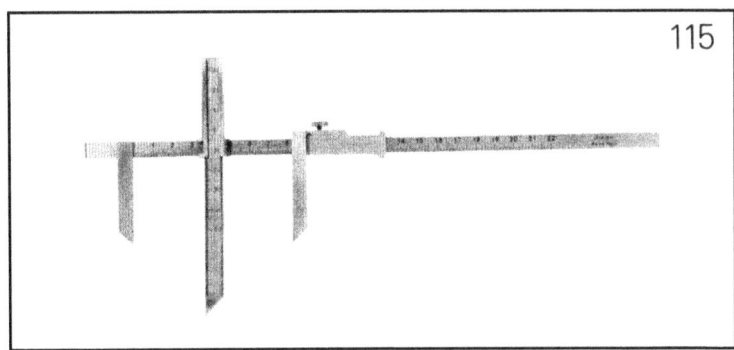

Calipers with coordinates. Scale: 20-220 mm. Weight: 0.160 kg.

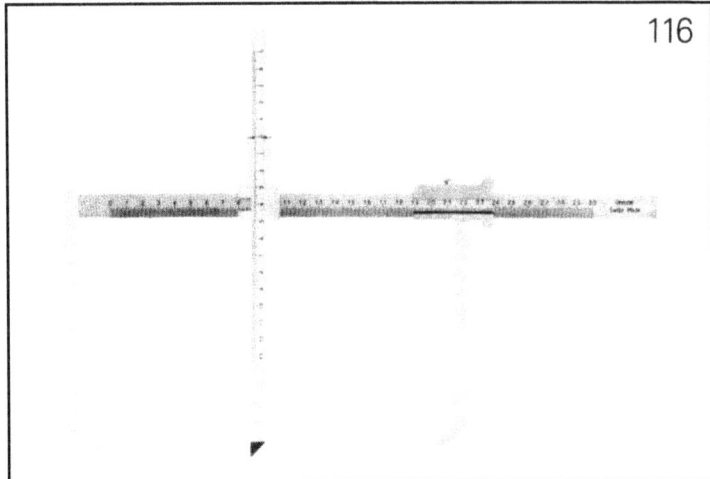

Calipers with coordinates (Eichel). Scale: 20-300 mm. Weight: 0.500 kg.

Goniometer. An instrument for measuring angles. (Mollison).

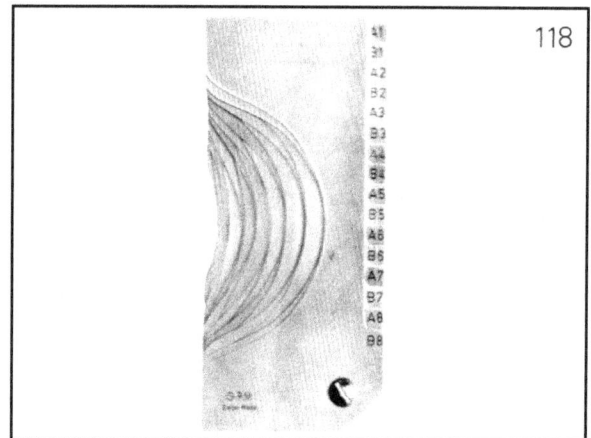

A template/gauge for the breast (Lipz). Weight: 0.700 kg.

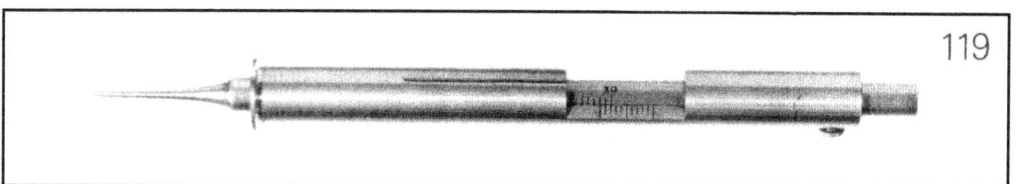

A gauge for the thickness of the skin. Scale: 0-30 mm. Weight: 0.065 kg.

"Lange"—an instrument for wrinkles of the skin (US manufacture). For clarification of the degree of fatiness. Scale: 0-60 mm.

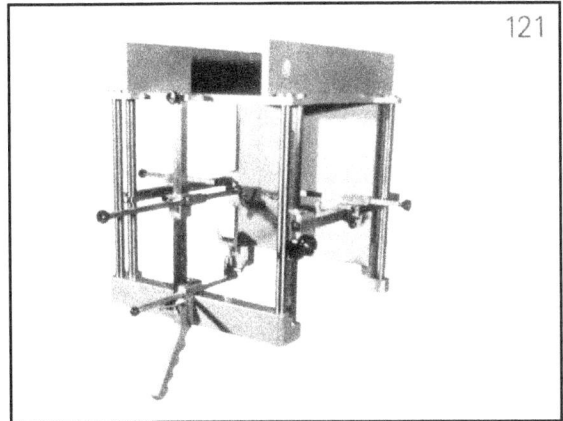

Anthropostereometer for measurement of the head and skull. Weight: 11.08 kg.

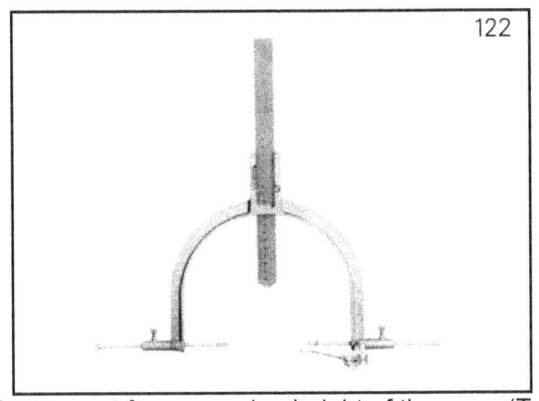

Instrument for measuring height of the ears. (Todd).

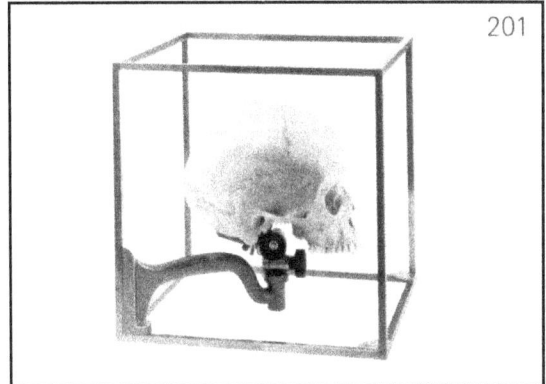

Cubic craniophor. Weight: 2.30 kg.

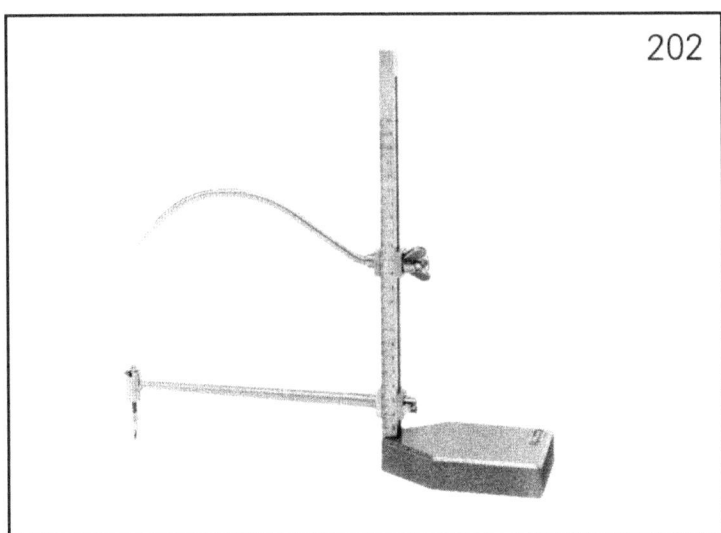

Diagraph (Martin). Weight: 1.700 kg.

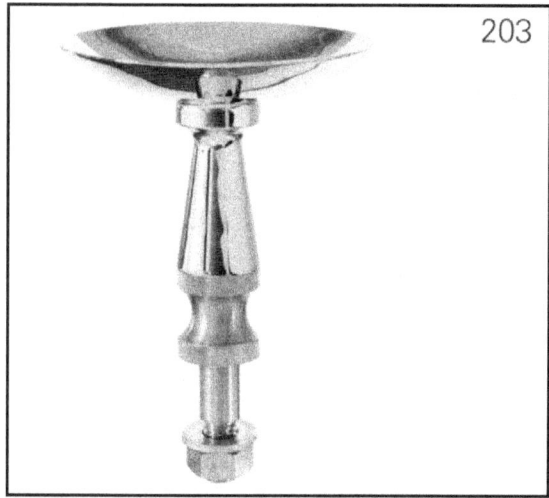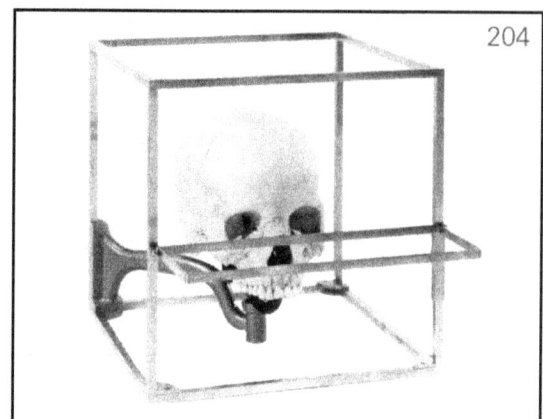

Left: Prop for the skull for a cubic craniophor. Weight: 0.300 kg.
Right: Sighting device or cubic craniophor. (Schlaginhaufen)

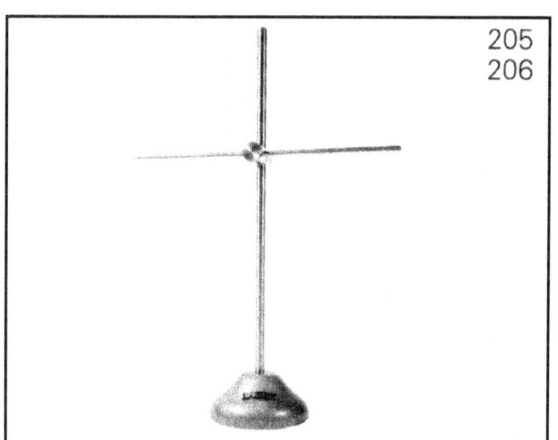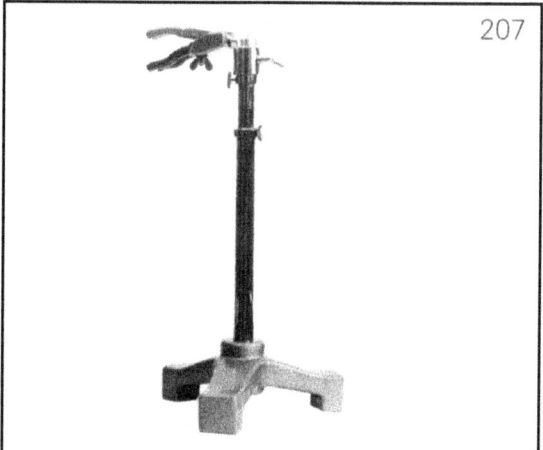

Left: Stylus for guiding horizontal lines. Height: 450-mm. Weight: 1.300 kg.
Right: Tubular craniophor (Martin) with a device for measuring the height of the ears (Black).
Weight: 2.100 kg.

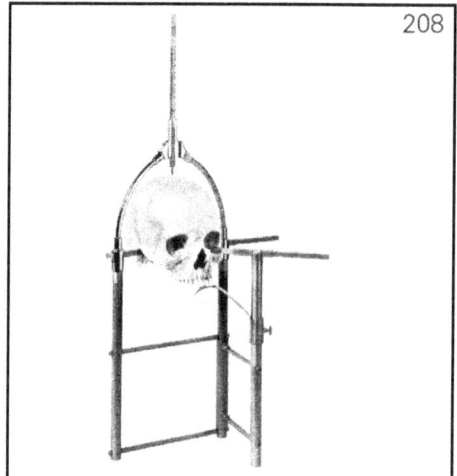

Craniophor (Mollinson) specifically for mounting the skull on the plane of the ears-eyes. Weight:
1.870 kg.

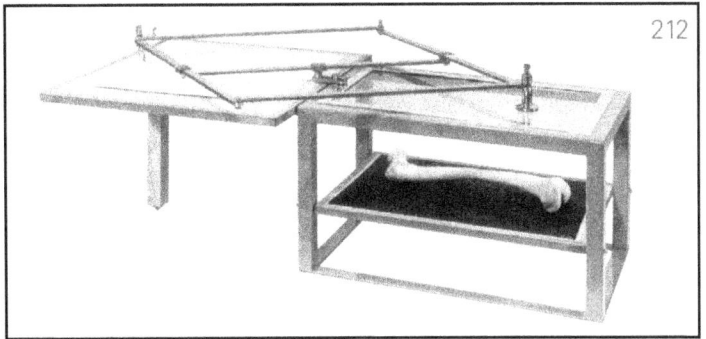

Rectangular dioptograph. Weight: 15.500 kg.

Quadratic dioptograph (Martin). Weight: 10.00 kg.

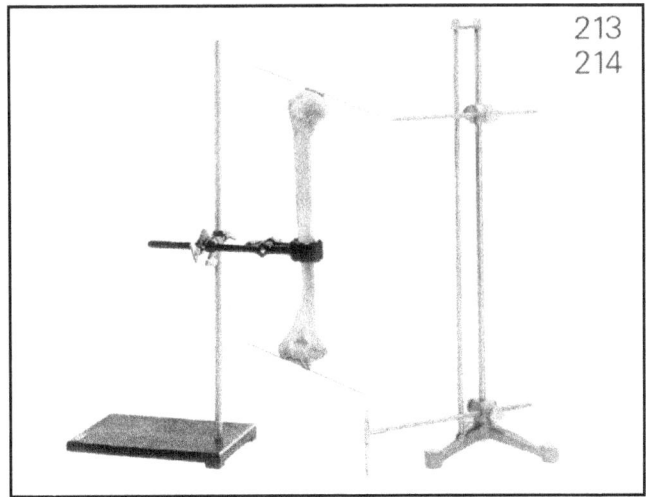

Parallelograph (Martin) for measurement of the angle of the axes of the joints.
Weight: 1.600 kg. Weight of bone holder: 2.400 kg.

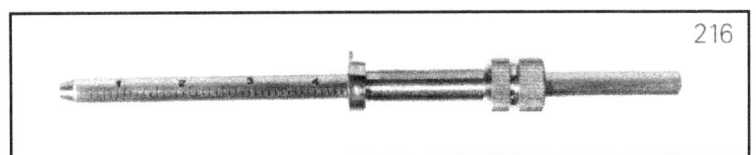

Orbitometer for measuring the depth of the orbits. Weight: 0.015 kg.

From the Translator

I began translating Vladimir Avdeyev's book, **Raciology**, in May 2009. I knew that his work touched on physical anthropology and the major races of the world; having always been interested in history, I looked forward to the project.

Even though I have an interest in anthropology, I had not read a book on the topic since I was 20 years old—when I read a copy of **Origins**, by Richard Leakey and Roger Lewin. This book concluded with the rather unrealistic notion that killing and warfare are not part of Man's nature, but simply a result of the instinct for cooperation being badly manipulated. It seemed to me that a personal or worldview was being imposed on the science of anthropology. Since then, more anthropological theories have been put forward, which seem more to advance a worldview, than science. Of late, the most absurd of these is that *races do not exist, they are merely a social construct*—in other words, we are told that those differences, which our senses perceive every day, really aren't there: we only *imagine* they are there. One does not need to be an anthropologist to recognize that this is science committing fraud.

In a free society, common sense would be enough to demolish such a totalitarian absurdity, but we live in a society whose freedoms are under siege from within. So the cudgel of political correctness extends the life of such nonsense. Even in America, careers can be placed at risk for the offense of arguing that the major races are an objective reality, as our eyes have told us all along. And so the teaching podium is slowly, but steadily, reserved for only those who will not contradict the socialist mantra, that "race does not exist." Alexis de Tocqueville wrote in his book, **Democracy in America** (1835), that "those who hope to bring about revolutions by means of the press, are desirous of confining it to a few powerful organs"; he could have easily extended this observation to public education.

For me, translating **Raciology** was not only an opportunity to apply my language skills, it was also an opportunity to advance those 1st Amendment freedoms which Americans hold so dear, and which are under steady assault by the totalitarian democracy attempting to take shape in these United States. I consider bringing **Raciology** to the English-speaking world as a continuation of that Oath to support and defend the Constitution of the United States; it is an irony that free speech in America is advanced by publishing the ideas of a free-thinking Russian man – a man who saw through the "contradictions" of Soviet science and the Soviet State.

Once upon a time in the Western world, to talk religion was to talk politics; there was no separating the two. In like manner today, to talk anthropology is to talk politics. And just as modern anthropological science today is made to serve government ends, so too, was the science of the Classical world made to suit the politics of its time: the Sun-Centered Theory of the solar system, though correct, was forced to move over for the Earth-Centered Theory of the solar system, because the latter was favored by the religious authorities of the day. And the Earth-Centered Theory held sway for 1,500 years, until Renaissance astronomers disproved it and restored the Sun-Centered Theory to its rightful place.

The "Out of Africa" Theory of Human Evolution has become the sacred cow of Western science and Western governments. Its false precepts provide an ideological support for the tearing down of all that is European, or of European origin, by way of Stalinist population transfers. Can science and the West afford to suffer 1,500 years of the "Out of Africa" Theory?

Patrick Cloutier,
Translator